28th Rare Earth Research Conference 2017

Ames, Iowa, USA
18 - 22 June 2017

ISBN: 978-1-5108-7781-8

Printed from e-media with permission by:

Curran Associates, Inc.
57 Morehouse Lane
Red Hook, NY 12571

Some format issues inherent in the e-media version may also appear in this print version.

Printed by Curran Associates, Inc. (2019)

Iowa State University – Conference Planning/Management
1601 Golden Aspen Drive, Suite 110
Ames, IA 50010
USA

Phone: 515 294-1775
Fax: 515 232-6716

www.cpm.iastate.edu

Additional copies of this publication are available from:

Curran Associates, Inc.
57 Morehouse Lane
Red Hook, NY 12571 USA
Phone: 845-758-0400
Fax: 845-758-2633
Email: curran@proceedings.com
Web: www.proceedings.com

TABLE OF CONTENTS

June 18-22, 2017
28th RERC
Ames, IA

THE 28TH RARE EARTH RESEARCH CONFERENCE APPRECIATES THE SUPPORT OF ALL ITS SPONSORS

Lutetium (more than $5,000)

IOWA STATE UNIVERSITY
College of Engineering

Europium ($3,000 or less)

IOWA STATE UNIVERSITY
Department of Chemistry

IOWA STATE UNIVERSITY
Department of Materials Science and Engineering

COLORADO SCHOOL OF MINES
EARTH • ENERGY • ENVIRONMENT

Rare Earth Research Conference

June 18-22, 2017
Scheman Building, Ames, IA

WELCOME TO THE
28TH RARE EARTH RESEARCH CONFERENCE

Welcome Rare Earthers!

As the local chairs of the 28th Rare Earth Research Conference, it is our pleasure and privilege to welcome you to Ames, Iowa.

The City of Ames was founded in 1864 as a station stop on the Cedar Rapids and Missouri Railroad and was named after 19th century U.S. Congressman Oakes Ames of Massachusetts, who was influential in the building of the transcontinental railroad. Ames was founded by local resident Cynthia Olive Duff and railroad magnate John Insley Blair, near a location that was deemed favorable for a railroad crossing of the Skunk River.

Today the small town of Ames, Iowa is the ninth "Best Place to Live in the United States", has a population of over 65,000, and offers cultural, recreational, educational, business and entertainment amenities more common in larger metros.

In 1856, the Iowa General Assembly enacted legislation to establish the State Agricultural College and Model Farm. Story County was chosen as the location on June 21, 1859. When Iowa accepted the provisions of the Morrill Act of 1862, Iowa State became the first institution in nation designated as a land-grant college. The institution was coeducational from the first preparatory class admitted in 1868. The formal admission of students began the following year, and the first graduating class of 1872 consisted of 24 men and 2 women. Today, Iowa State University of Science and Technology (ISU) hosts more than 36,000 students.

ISU also operates the Ames Laboratory (432 employees), a U.S. Department of Energy Laboratory founded 70 years ago by its first director, Frank H. Spedding, following his participation in the Manhattan Project. Since then, the Ames Laboratory has been dedicated, although not exclusively, to the chemistry and physics of the "Fraternal Fifteen" so named by the late Karl A. Gschneidner in 1966, the "Rare Earths"

RARE EARTHERS — A new term, "rare earthers," was coined for us in 1965 when the Holiday Inn, Ames, Ia. welcomed conferees to the Fifth Rare Earth Research Conference in the manner shown above.

The first Rare Earth Research Conference (RERC) was organized by Eugene V. Kleber at Lake Arrowhead, California in 1960. Ames hosted the 5th RERC in 1965. After 52 years, another conference was overdue, so now alternates on a three-year cycle with the International Conference on f-Elements (Europe) and the International Conference on Rare Earths. The 28th Rare Earth Research Conference in Ames follows the successful 26th RERC in Santa Fe, New Mexico (2011), and the 27th RERC in Squaw Valley, Nevada (2014).

We are grateful to the Program Committee, headed by Eric Schelter, the Spedding Award Committee, headed by Ana de Bettencourt-Dias and to Aubrey Robertson and Jennifer Vit (ISU Conference Planning and Management). We would also like to extend sincere thanks and appreciation to our sponsors.

Thank you all for coming and participating actively in the 28th RERC in Ames, Iowa.

Vitalij Pecharsky, Anja-Verena Mudring, and Gerd Meyer

Dr. Lynda Soderholm, Argonne National Laboratory, USA
Ce^{IV}, Th^{IV}, and Pu^{IV}: Comparative Tetravalent f-ion Chemistry

After completing her Ph.D. at McMaster University (Canada) and a NATO post-doctoral fellowship at the Centre de Research Nucléaire in Strasbourg (France), Lynda Soderholm accepted a staff position in the Chemistry Division at Argonne National Laboratory. Working in the Heavy Element Chemistry group, her background in lanthanide materials was expanded into studies involving the actinides as well, notably the transuranic elements. Today she is the leader of the Heavy Element and Separation Sciences group, which focuses their fundamental research efforts on metal-ion solution speciation and its role in the synthesis and characterization of new materials.

Prof. Paul Canfield, The Ames Laboratory and Iowa State Univ., USA
Magnetic Properties of the $SmNi_{4-x}Fe_xB$ Solid Solution

Paul C. Canfield, Ph.D., graduated summa cum laude from the University of Virginia (Charlottesville) with a BS in Physics in 1983. He received his MS from the University of California, Los Angeles, where he received his Ph.D. in 1990, having researched experimental condensed matter physics. From 1990 to 1993, Dr. Canfield was a postdoctoral researcher at the Los Alamos National Laboratory in New Mexico. In 1993, Dr. Canfield joined the Ames Laboratory at Iowa State University. Dr. Canfield's research is centered on the design, discovery, growth and characterization of novel electronic and magnetic materials. He has made key contributions to the fields of superconductivity, heavy fermions, quantum criticality, quasicrystals, spin glasses, local-moment metamagnetism, and metal-to-insulator transitions. He was awarded the 2011 Department of Energy Lawrence Award for Condensed Matter Physics. In 2014, Dr. Canfield was awarded the APS David Adler Lectureship Award in the Field of Materials Physics, and was named a Gordon and Betty Moore Materials Synthesis Investigator. In 2015, he received the Humboldt Research Award.

Prof. Mark Jensen, Colorado School of Mines, USA
Trending Now Rare Earth Separations

Mark Jensen is a Professor of Chemistry and the Jerry and Tina Grandey University Chair in Nuclear Science and Engineering at the Colorado School of Mines, where he directs the Nuclear Science and Engineering Program. He earned his PhD in Inorganic and Nuclear Chemistry from Florida State University in 1994 for work that included rare earth separations and then spent 20 years as a scientist in the Heavy Element Chemistry and Separation Science group at Argonne National Lab. His research examines the separations, environmental, and biological chemistry of the f-elements. He is an associate editor of the separations journal Solvent Extraction and Ion Exchange.

Prof. Feng Wang, City Univ. Hong Kong, China
Controlling Rare Earth Luminescence in Core-Shell Nanoparticles

Dr. Feng Wang received his BE (2001) and PhD (2006) degrees in Materials Science and Engineering from Zhejiang University. His PhD thesis was focused on the synthesis and characterization of lanthanide-doped fluoride nanomaterials under the supervision of Prof. Minquan Wang and Prof. Xianping Fan. He then carried out postdoctoral research work with Prof. Xiaogang Liu at the National University of Singapore and the Institute of Materials Research and Engineering in Singapore. He joined the faculty of City University of Hong Kong in 2012 as an Assistant Professor, and was promoted to Associate Professor in 2017. His research is focused on nano- and micro-structural materials comprising luminescent lanthanides for applications in biology and photonics.

Prof. Suzanne Bart, Purdue Univ., USA
Manipulation of Redox Properties of F-Block Elements Using Redox-Active Ligands

Suzanne Bart graduated from the University of Delaware with a BS in chemistry (2001), and earned her PhD from Cornell University with Prof. Paul J. Chirik (2006). Following this, she was an Alexander von Humboldt Postdoctoral Fellow at the Friedrich-Alexander University Erlangen-Nuremberg under the direction of Prof. Dr. Karsten Meyer. In 2008, she became an assistant professor at Purdue University, and was promoted to associate professor in 2014. Her research interests include organometallic transformations mediated by organoactinide species, coordination chemistry of f-block elements, and elucidation of electronic structures of metal complexes bearing redox-active ligands. Suzanne has been the recipient of an NSF CAREER award, and has been named a 2012 Cottrell Scholar, a 2014 Organometallics Young Investigator Fellow, and a 2015 Rising Star from the Women's Chemist Committee of the American Chemical Society.

Prof. Jun Li, Tsinghua Univ., China
On the Periodicity of Oxidation States in d- and f-Element Compounds

Prof. Dr. Jun Li received a Ph.D. in Physical Chemistry from Fujian Institute of Research on the Structure of Matter, Chinese Academy of Sciences (1992). He was a postdoctoral researcher at the Department of Chemistry, University of Siegen (Germany) and at the Department of Chemistry, The Ohio State University (USA) from 1994 to 1997. He worked as a Research Scientist at The Ohio State University from 1997 to 2001 and then became a staff member at the Pacific Northwest National Laboratory (PNNL, USA) in 2001, where he was appointed as a Senior Research Scientist and later promoted to a Chief Scientist. He was also an affiliated full professor at University of Idaho and Washington State University. He is now a Cheung Kong Professor at the Department of Chemistry, Tsinghua University. His research involves computational chemistry, theoretical inorganic chemistry, heavy-element chemistry, and computational catalysis science. He has published five book chapters and more than 200 peer-reviewed papers.

AMES MAP

SCHEMAN BUILDING

Scheman, First Floor

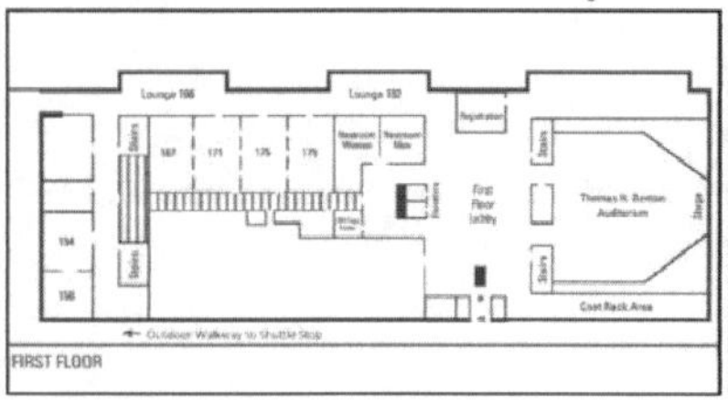

SCHEMAN BUILDING

Scheman, Ground Floor

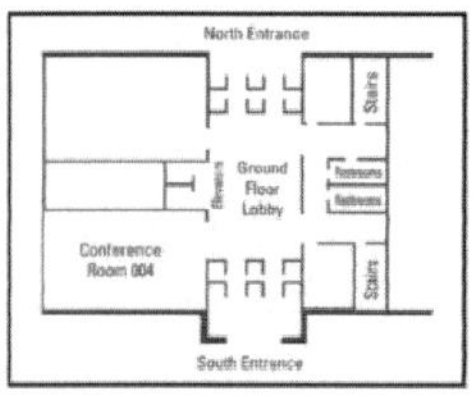

COMMITTEE MEMBERS

Program Committee

Eric Schelter, University of Pennsylvania, USA
Rebecca Abergel, Lawrence Berkeley National Laboratory, USA
Matt Allen, Wayne State University, USA
Ana de Bettencourt Dias, University of Nevada Reno, USA
Jenifer C. Braley, Colorado School of Mines, USA
Julia Chan, University of Texas at Dallas, USA
Yaofeng Chen, Shanghai Institute of Organic Chemistry, China
Trevor Hayton, University of California Santa Barbara, USA
Ross Ellis, Oak Ridge National Laboratory, USA
Richard Layfield, University of Manchester, U.K.
Justin Walensky, University of Missouri, USA

Local Organizing Committee

(All Iowa State University & Ames Laboratory)

Gerd Meyer, Co-Chair
Anja Mudring, Co-Chair
Vitalij Pecharsky, Co-Chair
Yaroslav Mudryk
Durga Paudyal
Denis Prodius
Voldymyr Smetana
Jennifer Vit, Administrative point of contact
Aubrey Robertson, Administrative point of contact

International Advisory Board

Polly Arnold, University of Edinburgh, United Kingdom
John Brennan, Rutgers University, USA
Jean-Claude Bünzli, ETH Lausanne, Switzerland
William J. Evans, University of California at Irvine, USA
Yasuchika Hasegawa, Hokkaido University, Japan
Peter Junk, James Cook University, Australia
Jaqueline Kiplinger, Los Alamos National Laboratory, USA
Marinella Mazzanti, ETH Lausanne, Switzerland
David Parker, Durham University, United Kingdom
John Sarrao, Los Alamos National Laboratory, USA
Thomas Schleid, University of Stuttgart, Germany
Satoshi Shinoda, Osaka University, Japan
Fernando Sigoli, University of Campinas, Brazil
Chun-Hua Yan, Beijing University, China

CONFERENCE AGENDA DAY 1

Sunday, June 18, 2017

5 pm	**Registration & Welcome Reception**	Brunnier Art Museum, 2nd Floor, Scheman
7 pm	**Frank H. Spedding Award Lecture**	Benton Auditorium, 1st Floor, Scheman

The History of the Spedding Award

The Frank H. Spedding Award for Outstanding Contributions to Science and Technology of the Rare Earths has been awarded previously fourteen times. The previous awardees include: W. E. Wallace, Georg Busch, S. Legvold and W. Koehler, A. Mackintosch and H. Bjerrum Moeller, B. R. Judd, Karl Gschneidner, Jr., LeRoy Eyring, Gregory R. Choppin, Brian Maple, Lynn Boatner, John Corbett, William J. Evans, Gerd Meyer, and most recently, Joe D. Thompson from Los Alamos National Laboratory.

Spedding received his Ph. D. from Berkeley in 1929 and was quickly recognized for his work in spectroscopy and rare-earth chemistry, being awarded the Langmuir Award from the American Chemical Society, and a Guggenheim Fellowship to Europe for the 1934-5 academic year. In 1935, he received a two-year Baker Fellowship at Cornell, working with Hans Bethe, and joined the faculty at Iowa State College (University) in 1937.

As the Rare Earth community well appreciates, the "Fraternal Fifteen" elements - so named by Karl Gschneidner in 1966 - comprise a host of interesting properties that make them noteworthy - Spedding noted in his "Spedding Papers" (ISU Archives, 1951) that pronounced magnetic properties arise from the filling of the 4fshell, leading to magnetic properties of great technological importance. And yet, these fraternal fifteen were considered "contaminants" by many.

The demands of the war years required the development of purification techniques on a grand scale, and the rare earth elements that naturally associated with uranium ores were considered contaminants that needed to be separated. Moreover, rare earth elements are found among the fission products of nuclear reactions, and separation is needed to purify unspent fuels and plutonium. Spedding's young Institute for Atomic Research was at the heart of some of the most critical separation and purification processes for the Manhattan Project effort, and it would be the separations developed therein that would kindle the processes that would later benefit all of rare earth chemistry and physics.

And so, the Rare Earth Research community recognizes the accomplishments of one of our own by celebrating under the watchful eyes of Frank Spedding, a pioneer among many who recognized the scientific and technological implications of the "Fraternal Fifteen".

Ken Raymond, 2017 Spedding Awardee

Dr. Kenneth N. Raymond

Kenneth N. Raymond obtained a B.A. in Chemistry from Reed College in 1964 and Ph.D. in Inorganic Chemistry from Northwestern University in 1968. He began his faculty appointment at the University of California at Berkeley on July 1, 1967. There he remained, appointed Chancellor's Professor in 2006. He served as Vice Chair of the Berkeley Chemistry Department (1982-1984 and 1999-2000) and Chair (1993-1996).

Dr. Raymond did early work on siderophores, low molecular weight complexing agents used by bacteria to obtain the iron they need to grow. This led to a general interest in metal ion selective chelating agents for a range of applications. These include lanthanide complexes for use in imaging and assay applications, and radioisotope complexing agents for biomedical use. More recently he collaborated with Berkeley colleagues to use metal-ligand supramolecular clusters from his laboratory as host molecules and enzyme mimics in catalysis.

Dr. Raymond has received many awards and honors, including election to the National Academy of Sciences and the American Academy of Arts and Sciences. In addition to his academic appointment on the Berkeley campus, in 2001 he co-founded Lumiphore, Inc. He is the author of 24 US Patents, 11 International Patents and 569 research publications. In March 2017, he received a Department of Energy grant to continue his research on energy transfer in lanthanide luminescent complexes.

CONFERENCE AGENDA DAY 2
Monday, June 19, 2017

CONFERENCE AGENDA DAY 2
Monday, June 19, 2017

7:30 - 8:15 am	**Registration and Refreshments**	First Floor Lobby
8:15 - 8:45	**Welcome Remarks**	Benton Auditorium
8:45 - 9:45	**Ce^{IV}, Th^{IV}, and Pu^{IV}: Comparative Tetravalent f-ion Chemistry,** *Lynda Soderholm* *Eric Schelter, Chair*	Benton Auditorium
9:45 - 10:00	**Break**	

Welcome Remarks delivered by Sarah Rajala, Dean, College of Engineering; Beate Schmittman, Dean, College of Liberal Arts & Sciences; Adam J. Schwartz, Director, Ames Laboratory

Concurrent Sessions: 10:00 am - 12:00 pm

Time	Theory, Computation and Spectroscopy 1, *Valérie Vallet* — Benton Auditorium	Spedding Award Symposium 1, *Eric Werner* — Room 004-Ground	Gschneidner Memorial Symposium 1, *Gerd Meyer*, Room 150-154
10-10:30 am	**Ab Initio Approaches for the Accurate Prediction of Heavy Element Thermochemistry,** *Angela Wilson*	**Lanthanide Texaphyrin-based Theranostics,** *Jonathan Sessler*	**A Life in Science: Remembering Karl A. Gschneidner, Jr.** *Vitalij Pecharsky*
10:30-10:50	**Covalency of Actinides with Chelating Agents Across the Actinide Series,** *Ping Yang*	**Selective f-Element Extraction Agents based on Tripodal Scaffolds,** *Shannon Biros*	**Rare Earth Chemistry in Ionic Liquids and Molten Salts,** *Anja Mudring*
10:50-11:00	BREAK	BREAK	BREAK
11:00-11:20	**Theoretical Studies of Actinide-Ligand Bonding Interactions,** *Enrique Batista*	**Towards a Better Understanding of the Complex Relationship Between Ln-based MRI Probes and Water,** *Mauro Botta*	**Chemical Bonding in Gallides of the Rare-Earth Metals: from Zintl Concept to Positional-Space Approach,** *Yuri Grin*
11:20-11:40	**Competitive Solvation Effects Within Multicomponent Solutions of Trivalent Ions,** *Aurora Clark*	**Development of Aromatic Macrocyclic Bi-and Tri-Functional Chelators for Lab and Clinic,** *Darren Magda*	**Extraordinary Magnetic and Structural Transformations in HoCo2 via Competing 4f and Itinerant Interactions,** *Durga Paudyal*
11:40-12 pm	**Soft X-ray Spectromicroscopy Studies of f-electron Materials,** *David Shuh*	**Production, Chelation, and Biodistribution of Actinium Isotopes: A Handle on the Future of Targeted Radiotherapy,** *Rebecca Abergel*	**Formation of 1:13 Phase in La(Fe,Si) -Based Alloys,** *Yi Long*
12:00-1:30	LUNCH* — Room 167-179	LUNCH* — Room 167-179	

*Attention graduate students and post-docs: Plan to bring your lunch and a copy of your abstract to the Research Slam, hosted by Colorado School of Mines. Details available at the registration desk.

1:30-2:30	**Magnetic Properties of the $SmNi_{4-x}Fe_xB$ Solid Solution,** *Paul Canfield* *Susan Latturner, Chair*	Benton Auditorium
2:30 - 3:00	**Break**	

Concurrent Sessions: 3:00 - 5:30 pm

Time	Organometallics and Coordination Chemistry 1, *Skye Fortier* — Benton Auditorium	Spedding Award Symposium 2, *Jonathan Sessler*, Room 004-Ground	Solid State Chemistry and Physics, *Tori Forbes* — Room 150-154
3:00-3:30 pm	**Recent Advances in the +2 Oxidation State Chemistry of the Rare Earth Metals,** *William Evans*	**Application of Gadolinium Complexes to Environmental Remediation of Phosphate,** *Valerie Pierre*	**Complex Intermetallics Grown from Rare Earth / Transition Metal Eutectic Fluxes,** *Susan Latturner*
3:30-3:50	**Synthesis of U(VI)bis(imido) Complexes from UCl4: Investigations of Oxidation Processes in Uranium Chemistry,** *Jim Boncella*	**Development of CMPO-Based Ligands for Selective Lanthanide Extraction,** *Eric Werner*	**Orbital Mixing and Multiconfigurational Ground States in Lanthanide Organometallics,** *Corwin Booth*
3:50-4:10	**Non-pincer-type Mononuclear Rare-earth Metal Alkylidene Complexes: Synthesis, Structure and Reactivity,** *Yaofeng Chen*	**Coordination complexes of d0 vs. d10 metals: what can we learn about bonding?,** *David Tatum*	**High Magnetic Field Studies of the CemMnIn3m+2n (M=Co, Rh, Ir) Heavy Fermion Superconductors,** *Eric Bauer*
4:10-4:20	BREAK	BREAK	BREAK
4:20-4:40	**Rare Mixed-valent Rare Earth Metal Complexes,** *Peter Junk*	**Further Investigations of Curium Photoluminescence Efficiency with Hydroxypyridinone Ligands,** *Christopher Andolina*	**Zintl Phases with Rare-Earth Elements,** *Svilen Bobev*
4:40-5:00	**Small Molecule Activation Induced by Lanthanides,** *Peter Roesky*	**Sensitised Ln (III) Emission and Excited State Dynamics of Cofacial 'Pacman' Porphyrin Terpyridine Complexes,** *Evan Moore*	**Interplay between the Sb/Bi Bonding and Charge Transport Properties in Rare-Earth Antimonide and Bismuthide Oxides,** *Yurij Mozharivskyj*
5:00-5:30	**Recent Advances in Triamido-Supported Actinide-Ligand Multiple Bonding,** *Steven Liddle*	**Lanthanide Helicates: From Self-Assembly To Functionality,** *Jean Claude Bünzli*	**Fundamentals of URu2Si2 Probed through Electron Doping and High Magnetic Fields,** *Ryan Baumbach*

5:30-7:30 **DINNER ON YOUR OWN** *Refer to page 21 for recommended dining options. Refer to page 22-24 for shuttles*

7:00-9:00 pm **POSTER SESSION I,** *Co-Chairs: Anja Mudring and Eric Schelter*

Light refreshments will be served. *Refer to page 19 for poster presentation details*

7:30 - 8:30 am	**Morning Refreshments**	First Floor Lobby
8:30 - 9:30	**Trending Now – Rare Earth Separations,** *Mark Jensen* *Mark Antonio, Chair*	Benton Auditorium
9:45 - 10:00	**Break**	

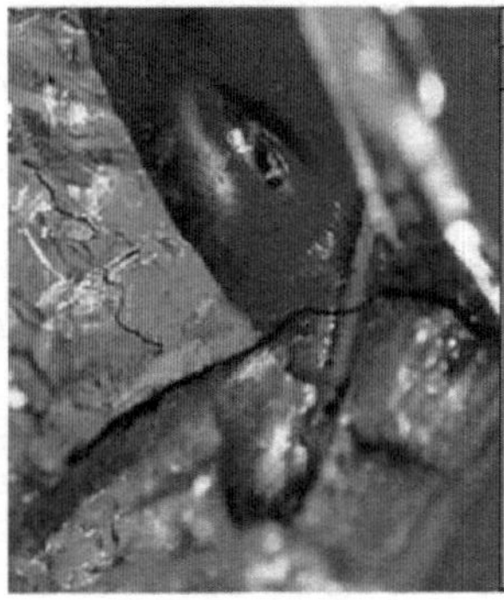

Concurrent Sessions: 10:00 am - 12:00 pm

Organometallics and Coordination Chemistry 2, *Yaofeng Chen*	Benton Auditorium
10-10:30 am	**Follow the Thread: Unravelling the Mechanism of Substrate Reduction by Sm(II)-Water Complexes,** *Robert Flowers*
10:30-10:50	**Homoleptic Organolanthanide Compounds Supported by β-SiH-containing Ligands,** *Aaron Sodow*
10:50-11:00	BREAK
11:00-11:20	**An N-Tethered Uranium-Arene Complex for the Synthesis of Uranium-Metal Bonds,** *Skye Fortier*
11:20-11:40	**Organothorium Pnictogen Chemistry,** *Justin Walensky*
11:40-12 pm	**The Influence of Multi-Configurational Electronic States on the Properties and Reactivity of Organlanthanides,** *Greg Nocton*
12:00-1:30	LUNCH — Room 167-179

Concurrent Sessions: 10:00 am - 12:00 pm

Nanocrystalline Materials 1, *Rebecca Abergel* — Room 004-Ground	Separations Chemistry 1, *Jenifer Braley* — Room 150-154
Enhancing Lanthanide Upconversion with Molecular Concentrators, Micro Pillars and Energy-Looping Nanoparticle-Based Lasers, *Jim Schuck*	**Cerium Redox Chemistry Pumps Nitrate Ion Transfer across Liquid-Liquid Interfaces in Solvent Extraction Systems,** *Mark Antonio*
The Nanoscale Effect in the Upconversion Dynamics of -NaYF4:Yb,Er, *Mary Berry*	**Modifications of the f-element / Aminopolycarboxylate Complex for Efficient Differentiation of Trivalent Actinides from Trivalent Actinides,** *Peter Zalupski*
BREAK	BREAK
Growth Mechanism and Rational Synthesis of Monodisperse Upconversion Nanocrystals based on Sodium Rare-Earth Fluorides, *Markus Haase*	**Outer-sphere Coordination of Counter-ions in Lanthanide(III) Complex with Diglycolamide Ligands,** *Derek Brigham*
Chemically and Structurally Flexible Hosts for Yb-Er Sensitizer-Activator Pairs, *Federico Rabuffetti*	**Re-establishing an American Rare Earth Industry: The Challenges and Opportunities,** *Brock O'Kelley*
On the Magnetic Properties of Ln3+ Based Nanoparticles, *Frank Van Veggel*	**Alternative Ligands for the Separation of Rare Earth Elements Using Solvent Extraction,** *Kevin Lyon*
LUNCH	Room 167-179

1:30-2:30	**Controlling Lanthanide Luminescence in Core-Shell Nanoparticles,** *Feng Wang* *Rebecca Abergel, Chair*	Benton Auditorium
2:30 - 3:00	**Break**	

Concurrent Sessions: 3:00 - 5:30 pm

Actinide Chemistry and Physics 1, *Trevor Hayton*	Benton Auditorium
3:00-3:30 pm	**Small Molecule Activation by Multimetallic Uranium Nitrides,** *Marinella Mazzanti*
3:30-3:50	**Fluorescence Spectroscopy and Microscopy as Tools for Monitoring the Redox Chemistry of Uranium,** *Louise Natrajan*
3:50-4:10	**f-Element Silylamide Complexes with Remarkable Bonding Motifs,** *David Mills*
4:10-4:20	BREAK
4:20-4:40	**Periodicity and the Chemistry of Protactinium,** *Richard Wilson*
4:40-5:00	**The Roles of 4f and 5f Orbitals in Bonding: a Magnetochemical, Crystal Field, Density Functional Theory, and Multireference Wavefunction Study,** *Wayne Lukens*
5:00-5:30	**New Reactivity in Actinide Chemistry Facilitated by Supporting Ligand Design,** *John Arnold*

Concurrent Sessions: 3:00 - 5:30 pm

Biological & Medicinal Applications 1, *Matthew Allen* — Room 004-Ground	Gschneidner Memorial Symposium 2, *Vitalij Pecharsky* — Room 150-154
Photophysical Characterization of a Library of Luminescent Lanthanide Complexes, *Eszter Borbas*	**The Legacy of Karl Gschneidner in Rare Earths and Critical Materials,** *Alex King*
Lanthanide-based Nanoparticles: a versatile platform for optical probes, *Fernando Sigoli*	**Rare Earths-Silicide/Silicon Composite Electrodes with Excellent Anode Performance for Lithium-Ion Battery,** *Hiroki Sakaguchi*
PARASHIFT probes: the importance of coordination environment on the NMR shifts of lanthanide complexes, *Kevin Mason*	**Inspiration from Karl A. Gschneidner, Jr.: Rare Earth containing Zintl Phases and their Thermoelectric and Magnetic Properties,** *Susan Kauzlarich*
BREAK	BREAK
Expanded Macrocyclic Ligands for Actinium Chelation, *Justin Wilson*	**The Crystal Structure and Magnetism of $R_{10}Pd_{21}$ Compounds (R = Pr, Nd, Sm, Gd-Lu, Y),** *Pietro Manfrinetti*
A Structural Analysis of Hydration Across the Lanthanide Series, *Mark Woods*	**A Forty-Year Journey in Rare Earth Science and Technology,** *Jean Claude Bünzli*
Divalent Europium in Magnetic Resonance Imaging, *Matthew Allen*	**Non-existent Intermetallics,** *Gerd Meyer*

5:30-7:30	**DINNER ON YOUR OWN** *Refer to page 21 for recommended dining options. Refer to page 22-24 for shuttles*	
7:00-9:00 pm	**POSTER SESSION II,** *Co-Chairs: Aaron Sodow and Susan Kauzlarich*	*Light refreshments will be served. Refer to page 19 for poster presentation details*

CONFERENCE AGENDA DAY 4

Wednesday, June 21, 2017

CONFERENCE AGENDA DAY 4
Wednesday, June 21, 2017

7:30 - 8:30 am	**Morning Refreshments**	First Floor Lobby
8:30 - 9:30	**Manipulation of Redox Properties of F-Block Elements Using Redox-Active Ligands,** *Suzanne Bart* *Marinella Mazzanti, Chair*	Benton Auditorium
9:30 - 10:00	**Break**	

Concurrent Sessions: 10:00 am - 12:00 pm	
	Actinide Chemistry and Physics 2, *Louise Natrajan* — **Benton Auditorium**
10-10:30 am	**Taking Advantage of the Chemistry of Heterocycles in Actinide Selective Coordination of U(VI) (UO22+) and Th(IV): Crystal packing, Reactivity, and π- π Interactions,** *Anne Gorden*
10:30-10:50	**Solution thermodynamics of hydropyridinonate 4f and 5f complexes,** *Gauthier DeBlonde*
10:50-11:00	**BREAK**
11:00-11:20	**Electronic Structure in Lanthanide and Actinide Dialuminides from Al K-edge XAS,** *Stefan Minasian*
11:20-11:40	**Coordination of Macrocyclic Ligands to the Uranyl Ion,** *Trevor Hayton*
11:40-12 pm	**Uranyl Reduction Facilitated by a Redox-Active, Donor-Expanded Dipyrrin,** *Nicola Bell*

Concurrent Sessions: 10:00 am - 12:00 pm	
Molecular Magnetism 1, *David Mills* — Room 004-Ground	**Critical Materials 1,** *Gisele Azimi* — Room 150-154
Designer Materials: Molecular Nano-Magnets Based on Lanthanides, *Muralee Murugesu*	**Technospheric Mining of Rare Earth Elements From Landfilled Stocks of Industrial Process Residues:,** *Gisele Azimi*
Synthesis and Understanding on Lanthanide Single-Molecule Magnets, *Bingwu Wang*	**Determining Computational Methods to Aid in Rare Earth Element Extractant Design,** *Marilu Dick-Perez*
BREAK	
Spectroscopic and Electrochemical Investigations of Heterometallic Lanthanide-Transition Metal Complexes, *Michael Nippe*	**Chemical Speciation of Rare Earths in Wyoming Powder River Basin Coal Fly Ashes,** *Jinke Tang*
Rational Conception of Dysprosium-based Complex featuring Different Single-Molecule Magnets: Magnetic Interaction and Isotopic Effects, *Fabrice Pointillart*	**Evidence of REE Mobility and Fossil Ion Adsorbed REE Deposits in Pennsylvania,** *Tracy Bank*
Catalytic Stibine Dehydrocoupling versus Antimony-ligated Dysprosium Single-Molecule Magnets, *Richard Layfield*	**Rare Earth Element Distributions and Extraction from Coal Ash,** *Elliot Roth*

12:00 pm	**Boxed Lunch**	**First Floor Lobby**
12:00-6:00 pm	**Free time on your own, refer to pages 15-16 for recommendations**	
6:00 pm	**Reception & Bar**	**Reiman Gardens**
7:00-9:00	**Rare Earth Research Conference 2017 Banquet** *Refer to page 24 for Banquet shuttle information*	**Reiman Gardens**

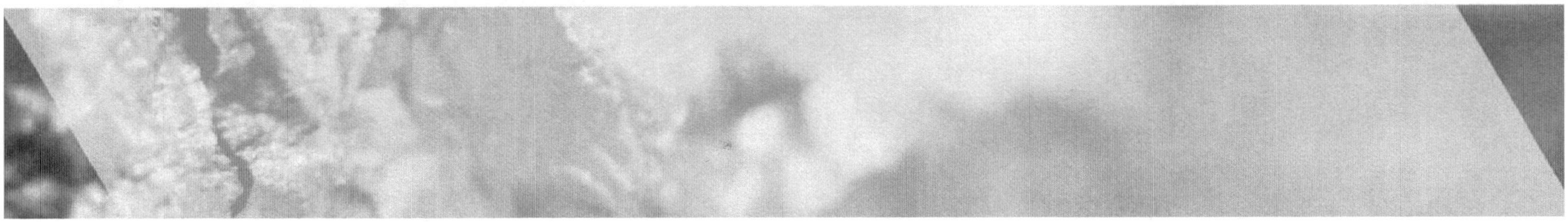

CONFERENCE AGENDA DAY 5
Thursday, June 22, 2017

7:30 - 8:30 am	**Morning Refreshments**	First Floor Lobby
8:30 - 9:30	**On the Periodicity of Oxidation States in d- and f-Element Compounds,** *Jun Li* *Nik Kaltsoyannis, Chair*	Benton Auditorium
9:30 - 10:00	**Break**	

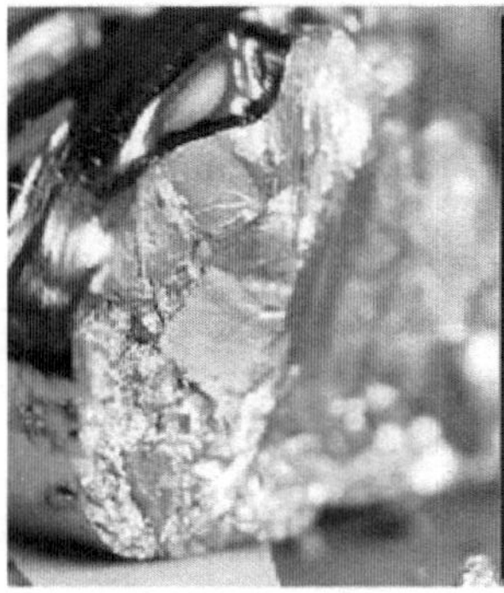

Concurrent Sessions: 10:00 am - 12:00 pm

Time	Organometallics and Coordination Chemistry 3, *Greg Nocton* — Benton Auditorium	Solid State Chemistry and Physics 2, *Karah Knope,* Room 004-Ground	Theory, Computation and Spectroscopy 2, *Andy Kerridge* — Room 150-154
10-10:30 am	**Lanthanide Complexes for Luminescence and Singlet Oxygen Generation,** *Ana de Bettencort Dias*	**Interplay Between Localized and Itinerant Magnetism in Eu-Containing Pnictides of the ThCr2Si2 Structure Type,** *Michael Shatruk*	**Early Actinide Superlatives,** *Nik Kaltsoyannis*
10:30-10:50	**Phosphinoboranate Reactivity and Coordination Chemistry with Uranium and Lanthanides,** *Scott Daly*	**Interaction of Neptunium with Crown Ether and Cucurbituril Macrocycles,** *Tori Forbes*	**Formation and Characterization of Praseodymium Oxides and Nitride-Oxides with Pr(V) Oxidation State,** *Mingfei Zhou*
10:50-11:00	**BREAK**	**BREAK**	
11:00-11:20	**Chemistry at the Edge: Actinide Interactions with (poly)aminopolycarboxylates,** *Jenifer Braley*	**Crystal Chemistry of New Rare Earth Nickel Silicides,** *Marcella Pani*	**Probing the Bonding in Dimers of Tetravalent Cerium and Actinide Homologues,** *Valérie Vallet*
11:20-11:40	**Reactivity Comparison of Complexes of 4fn5d1 and 4fn+1 Ln2+ Ions with Cyclooctatetraene,** *Megan Fieser*	**Visible and NIR-Emitting Lanthanide Doped Bismuth-Organic Materials,** *Karah Knope*	**Covalent Stabilisation of Bonding in M(BTP)3 and M(BTPhen)2 (M = Ce-Lu, Cm, Am; BTP = bis(triazinyl)pyridine; BTPhen = bis(triazinyl)phenanthroline),** *Andy Kerridge*
11:40-12 pm	**Applied Coordination Chemistry for the Separation of Rare Earth Elements Using Thermodynamic and Kinetic Controls,** *Eric Schelter*	**Solution Structure Trumps Energy – Determining Factors for the Luminescent Properties of Multimetallic Kinetically Inert Lanthanide(III) Complexes,** *Thomas Just Sorensen*	**Applicability of Cerium Surrogates to Describe Plutonium Oxalate Precipitation,** *Lindsay Ray*

12:00-12:15 pm	**Closing Session**	**Benton Auditorium**
12:15	**Adjorn**	

CONFERENCE AGENDA DAY 5
Thursday, June 22, 2017

ACTIVITIES IN & AROUND AMES

WINERIES & BREWERIES

DOWNTOWN AMES

»**Olde Main Brewing Company**, *316 Main St, Ames*
Locally owned restaurant and brewery featuring handcrafted entrees, fresh brewed beer and root beer.

»**Torrent Brewing Company**, *504 Burnett Ave, Ames*
Torrent Brewing Company was founded by home brewer Andy McCormick, longtime resident of Ames, Iowa. Opens at 4:30pm.

»**Della Viti**, *323 Main St, Ames*
Iowa's first self-serve wine bar. Beers, cocktails, retail sales, wireless Internet and appetizers available. Open daily at 4pm.

NORTH AMES

»**Prairie Moon Winery and Vineyards**,
3801 W 190th St, Ames
Located 3 miles north of Ames. We are a family owned winery that makes all wines on premise and have 12 acres of our own vineyards. Stop in to try our statewide known Honeymoon Red.

»**Alluvial Brewing Company**, *3715 West 190th St, Ames*
Located 3 miles north of Ames. Our taproom features 10 unique varieties of craft beer dedicated to quality and taste, focused on celebrating our community and our unique piece of Iowa.

SHOPPING

DOWNTOWN AMES

»**Worldly Goods**, *223 Main St, Ames*
Worldly Goods is a retail store assisted by volunteers that showcases fair-trade items from around the world.

»**Ciccotti Art Glass**, *2306 258th St, Ames*
Unique Hand-blown glass work. Demonstrations and tours, gallery and gift shop.

CAMPUSTOWN

»**Iowa State University Book Store**,
2229 Lincoln Way, Ames
Shop the Iowa State University Book Store for your Iowa State clothing and gift needs.

»**Barefoot Campus Outfitter**, *2306 Lincoln Way, Ames*
Barefoot Campus Outfitter - Your one stop shop for all your ISU fan gear!

SHOPPING CONT.

CAMPUSTOWN CONT.

»**Campustown Spirit**, *2320 Lincoln Way, Ames*
Shop Campustown Spirit for licensed Cyclone apparel and spirit merchandise.

»**Dogtown University**, *217 Welch Ave, Ames*
Iowa State apparel. Unique designs. Custom screen printing and embroidery.

NORTH AMES

»**North Grand Mall**, *2801 Grand Ave, Ames*
JCPenney, Younkers, Kohl's, T.J. Maxx and a lineup of your favorite national and local specialty stores.

»**Coe's Floral and Gifts**, *2619 Northridge Pkwy, Ames*
A full-service florist and home décor store.

OUTDOOR ACTIVITIES

»**Coldwater Golf Links**, *1400 S Grand Ave, Ames*
Unique in Iowa with its links-inspired design. Coldwater Golf Links is Ames' premier facility.

»**Veenker Memorial Golf Course**, *2916 Veenker Dr., Ames*
A premier public golf facility owned and operated by ISU. Veenker is a beautiful, demanding and challenging course.

»**Homewood Golf Course**,
401 East 20th St and Duff Ave, Ames
Homewood is a beautiful, well-groomed 9-hole, par 34 golf course in north Ames for players of all ages and abilities.

»**Reiman Gardens**,
1407 S. University Blvd, Ames
17-acre public botanical garden including tropical plant conservatory, butterfly wing with hundreds of live butterflies, and gift shop.

»**Ada Hayden Heritage Park**, *5205 Grand Ave, Ames*
Available rentals include canoes and kayaks for $10/half hour or $13/one hour and stand-up paddleboards for $15/hour.

»**Iowa Arboretum**, *1875 Peach Ave, Madrid*
Enjoy thousands of trees, shrubs and flowers in this 378 acre public garden. Hike woodland and prairie trails. Grounds are open year-round, sunrise to sunset. Admission: $5 per car. Free to members.

»**Ledges State Park**, *1515 P Ave, Madrid*
The Ledges State Park contains a sandstone gorge carved by Pea's Creek, a tributary of the Des Moines River.

ACTIVITIES IN & AROUND AMES

ARTS & CULTURE

DOWNTOWN AMES

» **Ames Historical Society**, *416 Douglas Ave, Ames*
Discover fascinating local stories, explore changing exhibits, hear engaging programs, delve into research materials, select Ames-related gift shop items.

» **Octagon Center for the Arts**, *427 Douglas Ave, Ames*
The Octagon Center for the Arts is a community art center providing gallery exhibits, art classes for children and adults and an art shop featuring Iowa artists.

»**Design on Main**, *203 Main St, Ames*
A full-service florist and home décor store.

ON CAMPUS

»**Brunnier Art Museum**, *Scheman Building, Ames - Rare Earth Research Conference central location*
Brunnier Art Museum is the only accredited museum emphasizing a decorative arts collection in the state of Iowa and one of the nations' few museums located within a performing arts and conference complex.

»**Farm House Museum**,
601 Farm House Ln, Ames
Nestled in the center of Iowa State University, the Farm House Museum sits as a monument to early Iowa State history and culture.

»**Byron Anderson Sculpture Garden**,
603 Morrill Road, Ames
The Sculpture Garden wraps around historic Morrill Hall. Shaded by flowering greenery, winding footpaths lined with perennials and shrubs take visitors through works of primarily American twentieth and twenty-first century sculpture.

»**Christian Petersen Art Museum**,
0003 and 1017 Morrill Hall, Ames
The museum is named after the nation's first permanent campus artist-in-residence, Christian Petersen, who sculpted and taught at Iowa State University from 1934 through 1955, and is considered the founder of the Art on Campus Collection.

OUTSIDE AMES

»**Boone & Scenic Valley Railroad & James H. Andrew Museum**, *225 10th St, Boone*
A historical, educational railroad museum providing excursion train rides across the beautiful Des Moines River Valley. Wednesday Excursion departs at 1:30pm. Tickets available online at www.bsvrr.com

DES MOINES

All activities in Des Moines are
at least a 45 minute drive from Ames

»**Blank Park Zoo**, *7401 SW 9th St, Des Moines*
Blank Park Zoo is a 25-acre zoological park on the south side of Des Moines, Iowa near historic Fort Des Moines.

»**Botanical Gardens**, *909 Robert D. Ray Dr, Des Moines*
Tropical and temperate gardens make us an oasis in the heart of downtown Des Moines. Whether you visit on your own or explore with a group, you will find the Botanical Garden a place where you can learn about fascinating plants, hone your gardening skills or immerse yourself in beautiful gardens.

»**Living History Farms**, *11121 Hickman Road, Des Moines*
Living History Farms is an interactive outdoor museum which educates, entertains and connects people of all ages to Midwestern rural life experiences.

»**Prairie Meadows**, *1 Prairie Meadows Dr, Des Moines*
With more than 85,000 square feet of 24/7 casino action including all your favorite table games, poker tournaments, exciting promotions, and more than 1,900 slots - Prairie Meadows has every game you'll want to play this side of Vegas.

»**State Historical Building**, *600 E Locust St, Des Moines*
Discover Iowa's rich heritage through a variety of exhibitions, tours, collections and unique programs and events throughout the year.

KEY

Easily accessible from Scheman via CyRide

Walking distance from Scheman

Most hotels in Ames loan out bicycles to their registered guests. Ask the front desk staff at your hotel for details.

Drive 35 minutes or more outside Ames

Check hours of operation on the day of your visit.

Actinide Chemistry and Physics

1. **Anastasia Blake,** *Uranium Complexes with Neutral Soft-Donor Chalcogen and Borohydride Ligands*
2. **Korey Carter,** *Engaging the terminal: promoting supramolecular assembly with the uranyl oxo atoms*
3. **Bradley Cowie,** *An Effective Route to Highly Symmetric Mixed Actinide-Lanthanide Complexes via Reduction of* $[UO_2]^{2+}$
4. **Shanna Estes,** *Molecular-Scale Insight into the Structure, Stability, and Electrochemistry of Ceria Nanoparticles*
5. **Emily Hardy,** *Pyridine Containing Schiff Base Ligands for Quick Detection of Actinide Contamination*
6. **Geng Bang Jin,** *Influence of Monovalent Counter Cations on Neptunium(V/VI) Chloride Chemistry in Aqueous Solution*
7. **Christopher Klug,** *Lanthanide Extraction by TBP into bis(trifluoromethylsulfonyl)imide Ionic Liquids*
8. **Artaches Migdissov,** *An experimental study of the stability of uranyl and U(IV) chloride aqueous complexes at elevated temperatures:implications to the mobility of U in natural systems*
9. **Nicolas Reynier,** *Green processing of Canadian REE ores by combining attrition and lixiviation*
10. **Jennifer Wacker,** *U(IV) Chloride Complexes Isolated from Acidic Aqueous Media:* UCl_6^{2-} *and an Unprecedented* $U(H_2O)_4Cl_4$ *Structural Unit*

Bioanalysis and Medical Applications

11. **Lawrence Miller,** *CyTTHA: a versatile scaffold for designing lanthanide imaging probes.*
12. **Sylvie Pailloux,** *In vitro and in vivo detection of bacteria with lanthanide complexes*
13. **Josiane Sobrinho,** *Ratiometric nanothermometers based on* Eu^{III} *and* Tb^{III} *complexes grafted to pNIPAM*

Coordination Chemistry

14. **Amanda Bohanon,** *Syntheses and Crystal Structures of Lanthanide Dithiooxamide Complexes*
15. **Conrad Goodwin,** *f-Element Silylamide Complexes with Remarkable Bonding Motifs*
16. **Yasuchika Hasegawa,** *Effective Photo- and Triboluminescent Eu(III) Coordination Polymer with Triangular Spacers*
17. **Yuichi Hirai,** *Large spectral difference in triboluminescence and photoluminescence of Tb(III)/Eu(III) coordination polymers*
18. **Anna Kaczmarek,** *Low Percentage Doping in Multinuclear Lanthanide Polyoxometalates (POMs) and Their Use as Ratiometric Cryogenic Thermometers*
19. **Naoyuki Koiso,** *Structural and Photophysical Properties of Eu(III) Coordination Polymer with an Axial Chiral Ligand*
20. **Lada Puntus,** *Can the π-Ligands Serve As an Efficient Antenna for the Luminescence Sensitization of Lanthanide Ions?*
21. **Florent Réal,** *Unravelling the excited state dynamics and luminescnce of the aqueous cerium(III) ion by combined experimental and theoretical approaches*
22. **Colin Reedy,** *Syntheses and Crystal Structures of Lanthanide Periodates*
23. **Mona Struckmann,** *New Insights in Rare Earth Triflates*

Coordination Chemistry

24. **Ilya Taydakov,** *Novel highly symmetrical class of lanthanide tetrakis-complexes* $[Na(H_2O)_4][LnL_4]$ *(HL -1,3-bis (1,3-dimethyl-1Hpyrazol-4-yl)-1,3-propandione)–synthesis, crystal structure and unique photophysical properties.*
25. **Eric Villa,** *Facile Linking of Lanthanide Cations with Soft Metals via Thiosulfate*
26. **Satoshi Wada,** *Chiroptical and magneto-optical properties of chiral nona-nuclear Tb(III) clusters*
27. **Anne Willert,** *Low-temperature Electrochemical Deposition of Neodymium and Dysprosium Using Ionic Liquids*
28. **Masanori Yamamoto,** *Syntheses and Photophysical Properties of Novel Coordination Polymers Composed of Luminescent Europium and Metal Joint Blocks*

Critical Materials

29. **Rajiv Chouhan,** *Theoretical Investigation of Crystal Symmetry and Site Dependent Permanent Magnetic Properties of* Sm_2Co_{17}
30. **Kinjal Gandha,** *Enhancement in Nd-Fe-B Coercivity: Dy Diffusion or Heat Treatment?*
31. **Helena Khazdozian,** *Investigating Sm-Co Permanent Magnets as an Alternative to Dy-containing Nd-Fe-B in Direct-Drive Wind Turbines*
32. **Tej Lamichhane,** *Solution Growth of New Rare-earth Poor Ce/Co/Mg Permanent Magnets*
33. **Gregory Nichols,** *Developing Best Practices for the United States Department of Defense Rare Earth Mining and Recovery Efforts*
34. **Meng Pan,** *Crystal structure and thermoelectric properties of layered compound NdOZnSb*
35. **Denis Prodius,** *Breaking the Paradigm: Record Quindecim [15+] Charged Magnetic Ionic Liquids*
36. **Marc Schmid,** *Mitigating Rare Earths Supply Risks: Insights from Japan's Supply Structure*

Magnetism

37. **Francois Guillou,** *Systematic investigation of the crystalline, magnetic and calorimetric properties of* $Gd_{1.02}Sc_{0.98}Ge_{1-x}Sb_x$ *compounds*
38. **Timothy Hackett,** *Electronic Structure, Magnetic Properties, and Magnetocrystalline Anisotropy of Lanthanide Dialuminides*
39. **Shane Harstad,** *Enhanced cooling power of ball-milled* Gd_5Si_4
40. **Manish K Kashyap,** *Magnetocrystalline Anisotropy Driven Magnetostructural Transformations in RMn2 (R = Tb, Ho)*
41. **Xubo Liu,** *Magnetic states and exchange interactions in GdNiAl*
42. **Sathish Malik,** *Normal-inverse-Normal Magnetocaloric effect across the first order transition in the electron doped manganite* $Ho_{0.15}Ca_{0.85}MnO_3$
43. **Adam Neuhouser,** *Understanding of Magnetic Properties and Magnetocrystalline Anisotropy of Heavy Lanthanides*
44. **Henrique Neves Bez,** *Evaluating Bean-Rodbell Assumption on the Linear Relation between the Exchange Constant and the Volume Change in* $La(Fe,Mn,Si)_{13}$
45. **Andriy Palasyuk,** *Development of "GAP MAGNET" in Ce – TM – TM' systems*
46. **Michael Onyszczak,** *Magnetic Properties of the* $SmNi_{4-x}Fe_xB$ *Solid Solution*
47. **Arjun Pathak,** *Magnetic and Structural Phase Transitions in* $Pr_{1-x}Er_xAl_2$
48. **Alessia Provino,** *Crystallographic ordering and magnetic structure in the next-to-ternary* $RCo_{1-x}Ni_x$ *compounds (R = Gd-Lu)*
49. **Nirmala Rathinam,** *Multiple magnetic transitions and magnetic entropy change in pseudo-binary* $Er_{0.75}Dy_{0.25}$ *polycrystal*
50. **Kirk Rudolph,** *Magnetism and Magnetocaloric Effect in* $(Gd_{5-x}Sc_x)Si_{1.8}Ge_{2.2}$
51. -. **Michael C Tringides,** *Metal growth on and under graphene: morphology, growth mode and intercalation*

Catalysis

1. **Matt Bailey,** *Photocatalytic Carbon-Carbon Bond Formation Using a Luminescent Eu(II)-Containing Complex and Visible Light*
2. **Xiaomei Liu,** *Electrochemical Performance of $Ni_{1-x}Fe_x$-SDC Composite Anodes for Intermediate Temperature Solid Oxide Fuel Cell*
3. **Igor Slowing,** *Non-Innocent Behavior of Cerium Oxide Supports in Catalytic Conversions of Phenolics*

General

4. **Shinji Hirai,** *Innovative Use of Light Rare-earth Fibers as Antibacterial Agents*
5. **Jignasu Mehta,** *Thermo-acoustical properties of Eu(III) and Eu(III)-Lysozyme(HEW) in ternary system at atmospheric pressure*
6. **Giday WoldeGabriel,** *Chemical and Mineralogical Contents in New Mexico Coals and Coal By-products - Assessing the REE Resources*

Organometallics

7. **Nicholas Andreychuk,** *Reactivity of a Uranium(IV) Dialkyl Complex Supported by a Rigid NON-Donor Ligand: Alkyl Exchange and C-H Activation of Pyridines*
8. **Kasuni Boteju,** *Synthesis and Characterization of Homoleptic Bis(dimethylsilyl)benzyl Lanthanide Compounds*
9. **David Emslie,** *Cationic Uranium Alkyl Complexes of a Rigid NON-Donor Pincer Ligand: The Impact of Arene-Coordination on Ethylene Polymerization Activity*
10. **Chad Palumbo,** *A New Class Of Ln^{2+} Complexes, [K(2.2.2-cryptand)][Ln[(Oar^{Ad,Me})$_3$mes]]: Synthesis, Structure, Spectroscopy, and Theoretical Characterization*
11. **Smita Patnaik,** *Reactivity of homoleptic trisalkyl lanthanum grafted on mesoporous silica*
12. **Bradley Schmidt,** *Activation and Polymerization Studies of $Nd[C(SiHMe_2)_3]_3$ as a Precatalyst*

Phosphors and Lighting

13. **Tarek Alammar,** *Luminescent Properties of Mechanochemically Synthesized Rare-Earth Containing MIL-78 MOF*
14. **David Böhnisch,** *Luminescence and Energy Transfer in Eu^{3+} Activated $Li_3Ba_2Tb_3(MoO_4)_8$*
15. **Stefan Fischer,** *Investigation of Luminescence Properties of Ce3+-Doped Tb3(Al,Ga)5O12 Phosphor*
16. **Md Yeathad Hossan,** *Inert-Shell Effects on the Upconversion Dynamics of -$NaYF_4$:Yb,Er Nanomaterials*
17. **Jonas Joos,** *Lanthanide-activated ternary sulfides: design of materials and light-emitting devices*
18. **Jing Liu,** *Concentration and Temperature Dependent Upconversion Luminescence of $CaWO_4$:Er^{3+},Yb^{3+} 3D Microstructure Materials*
19. **Kazuyoshi Ogasawara,** *Construction of Energy Diagrams for Ce^{3+} Coordinated with Six O^{2-} Ions Based on First-Principles Calculations*
20. **Guo-Jun Wang,** *Photoluminescence properties and energy level of RE (RE = Pr, Sm and Tb) in $Y_4Si_2O_7N_2$*
21. **Magdalena Wilk-Kozubek,** *Luminescent Properties of Lanthanide-Based Metal-Organic Frameworks with Rigid Tritopic Carboxylate Ligands*
22. **Mark Williams-Wynn,** *Distribution of Yttrium and Europium In Organic-Aqueous Phases Containing Sulphuric Acid and HDEHP*
23. **Guang-Xing Xiang,** *Effect of Mn^{2+} in β-$NaYF_4$: Eu^{3+} on the morphologies and enhanced luminescence of Eu^{3+} by inhibition the luminescence from Eu^{2+}*
24. **Tao Xu,** *Study on the Crystal Face Design and Luminescence Properties of Monodispersed YAG Microcrystal*

POSTER SESSION II
Tuesday, June 20, 2017

Physics

25. **He Feng,** *Crystallization behavior and component optimization study on cerium doped $(Gd_{1-x}Y_x)_2Si_2O_7$ single crystal*
26. **Yong Han,** *Nucleation and Growth Kinetics for Near-Surface Intercalation of Rare Earth Islands in Graphite*
27. **Soham Manni,** *Magnetism in CePtPb and GdPtPb with distorted Kagomé lattice*
28. **Yuttana Mona,** *High-pressure synthesis and thermoelectric properties of partially filled skutterudite compounds $Ce_xCo_4Sb_{12}$*

Solid State Chemistry

29. **Viktor Balema,** *Mechanochemical Synthesis: Novel Materials for Magnetic Cooling Applications*
30. **Horst Borrmann,** *The α-manganese structure in RE_5Mg_{24} type compounds*
31. **Alexander Brown,** *Synthesis and Structure of New Structure Type Ln52Fe2Sn49 (Ln = Gd, Tb, Dy)*
32. **Chris Celania,** *R_3Au_9Pn (R = Y, Gd, Tb, Dy, Ho; Pn = Sb, Bi): A Link between $Cu_{10}Sn_3$ and $Gd_{14}Ag_{51}$*
33. **Katharina V. Dorn,** *Under Pressure: Synthesis and Crystal Structure of the Rare-Earth Metal(III) Hydroxide Oxidotungstates(VI) RE(OH)[WO4] (RE = Dy and Ho)*
34. **Christian Funk,** *Mixed-Valent Europium in Two ortho-Oxoborates Derivatized with Soft Anions: $Eu_5X[BO_3]_4$ (X = S and Br)*
35. **Roman Gumeniuk,** *Are Remeika Phases Simply Cubic?*
36. **Ingo Hartenbach,** *Under Pressure: New Representatives of the Lanthanoid(III) Chloride Oxidotungstates(VI) $Ln_3Cl_3[WO_6]$ with the Smaller Ln^{3+} Cations (Ln = Dy - Er)*
37. **Joshua Konne,** *Mechanistic Study of Crystal Growth Pathways of Pr123 Using Biopolymers*
38. **Hazel Lim,** *The citric acid leaching of rare earths from eudialyte*
39. **Qisheng Lin,** *Synthesis, Structure, Bonding and Properties of $Pr_5Co_2Ge_3$ and $Pr_7Co_2Ge_4$*
40. **Wanyu lv,** *Cd substitution in $Eu_5In_2Sb_6$ enhancing the thermoelectric performance*
41. **Arthur Mar,** *Quaternary Rare-Earth Chalcogenides*
42. **Yaroslav Mudryk,** *Enhancing Magnetic Functionality with Scandium in Gadolinium-based Rare-Earth Alloys*
43. **Jacob Rabe,** *One-step synthesis of nanostructured RE-Nitrides from RE-fluorides via gas-solid mechanochemistry*
44. **Melissa Rhodehouse,** *In Search of New Rare-Earth Rich Intermetallics*
45. **James Ridenour,** *A Systematic Exploration of Rare-earth-p-halobenzoic acid-terpyridine Materials: Structural and Supramolecular Trends in Three Series of Molecular Complexes synthesized via Hydrothermal Methods*
46. **Thomas Schleid,** *Make Europium Great Again*
47. **Volodymyr Smetana,** *Structural Variations and Magnetism in the Transition-Metal-Rich Intermetallics $Eu(T1,T2)_5In$ (T = Cu, Ag, Au)*
48. **Asa Toombs,** *Structural and Electronic Effects of Ga Doping into $GdCo_2$: An Electronic Structure Study of $GdCo_{2-x}Ga_x$*
49. **Bastian Trepka,** *Synthesis of anisotropic hybrid nanoparticles containing f-block elements.*
50. **Elizabeth L. Kunz Wille,** *Rare Earth Solid Solution of $Yb_{14-x}RE_xZnSb_{11}$: Structure and Properties*
51. **Zhan Xu,** *Fabrication of Yttrium-based Fluorides Through Topotatic Transformation of LREHs*

Technological Applications

52. **Youhei Kubota,** *Reversible Phase Transformation of γ-Ce_2S_3 by Introducing Oxygen And Use of the Product as a Wave Absorber*
53. **Ann Lii-Rosales,** *Reactions of Dysprosium on the Graphite (0001) Surface*

A ^{161}Dy Mössbauer spectroscopy investigation of $DyCrO_4$

G.A. Stewart[1], J.M. Cadogan[1], W.D. Hutchison[1] and D.H. Ryan[2]

[1]School of Physical, Environmental and Mathematical Sciences, UNSW Canberra, Australian Defence Force Academy, Canberra BC 2610, ACT, Australia

[2]Centre for the Physics of Materials and Department of Physics, McGill University, Montreal, Quebec, H3A 2T8, Canada, dhryan@physics.mcgill.ca

INTRODUCTION

The rare earth (R) chromates $RCrO_4$ form with the tetragonal zircon type structure ($I4_1/amd$). They are of interest because of competing ferromagnetic and antiferromagnetic super-exchange interactions between the 3d (Cr^{5+}) and 4f (R^{3+}) sites[1], believed to be responsible for the giant magnetocaloric effect observed recently for R = Gd, Dy and Ho.[2,3] The ^{161}Dy-Mössbauer spectroscopy measurements on $DyCrO_4$ reported here were prompted by earlier ^{169}Tm- and ^{155}Gd-Mössbauer spectroscopy results for $TmCrO_4$ [4] and $GdCrO_4$,[5] respectively. In both instances, it was necessary to interpret the Mössbauer spectra in terms of a superposition of two sub-spectra (approx. 80:20 intensity ratio) despite there being only a single crystallographic R(4a) site. In addition, the magnetic transitions exhibited first-order character, which is contrary to bulk magnetic measurements. $DyCrO_4$ is reported to undergo a small crystal distortion to an orthorhombic (Imma) structure between 27 and 40K and to order ferromagnetically at T_C = 22.4K.[6]

EXPERIMENTAL METHODS

The $DyCrO_4$ sample was prepared from the chromium (99.99 %) and dysprosium (99.9 %) nitrates following the procedure described by Sáez-Puche et al.[1] About 2g of a stoichiometric mixture of the nitrates was ground together in an agate mortar and loaded into a porcelain boat. This was then heated in a tube furnace under flowing O_2. The sample was held at 120°C for ten minutes then raised to 150°C over a period of half an hour to drive off the water slowly. The reaction was completed by heating to 580°C and holding overnight. The process yielded ~ 0.6g of the dark green product according to the reaction:

$Dy(NO_3)_3.6H_2O+Cr(NO_3)_3.9H_2O \rightarrow DyCrO_4+6NO_2+15 H_2O$

Sample purity was confirmed using Cu-K_α x-ray diffraction and the pattern was fitted with the tetragonal $I4_1/amd$ (#141) space group using the GSAS/EXPGUI package. Fitted lattice parameters were: a = 7.1355(7) Å and c = 6.2667(7) Å. The 29 mCi, 6.9d ^{161}Tb source for the ^{161}Dy Mössbauer experiments was prepared by thermal neutron (nγ) activation (3 d @ 10^{13} n $cm^{-2}s^{-1}$) of 200 mg of $^{160}GdF_3$ at the OPAL reactor in Sydney, Australia. The Mössbauer source was mounted at room temperature on a sinusoidal drive outside a cold-tail cryostat. Dy metal was used to calibrate the drive. All Mössbauer spectra were fitted using a full-hamiltonian method. The nuclear spin of both the ground and excited states involved in the ^{161}Dy Mössbauer transition is 5/2 and the γ-ray energy is 25.6513(3) keV.

RESULTS AND DISCUSSION

Our ^{161}Dy Mössbauer results show a simple magnetically-split spectrum at 5K (figure 1). Compared to the reference Dy metal spectrum there is a small increase in the line width. However, contrary to the earlier ^{169}Tm- and ^{155}Gd Mössbauer work,[4,5] a second sub-spectral component is not immediately evident. The magnetic hyperfine field at the ^{161}Dy nucleus at 5 K is 555(3) T which converts to a Dy^{3+} magnetic moment of 9.9(1) μ_B, close to the 'free-ion' value of 10 μ_B. The spectra are paramagnetic above T_C with the quadrupole splitting and Wegener relaxation broadening diminishing as the temperature increases to room temperature.

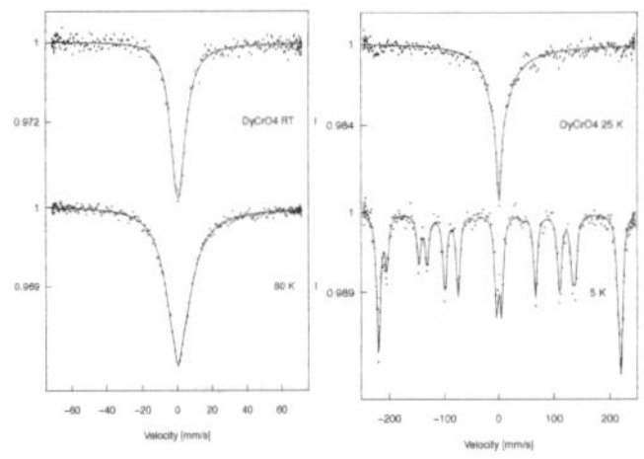

Figure 1: ^{161}Dy Mössbauer spectra of $DyCrO_4$ obtained at RT, 80K, 25K and 5K.

CONCLUSIONS

We have used ^{161}Dy Mössbauer spectroscopy to study the magnetic behaviour of $DyCrO_4$. The Dy^{3+} magnetic moment is 9.9(1) μ_B, essentially the 'free-ion' moment. Unlike in the cases of the isostructural $TmCrO_4$ and $GdCrO_4$ we found no evidence for first-order transitions or mixed-phase behaviour.

REFERENCES

1. Sáez-Puche R. et al., J. Sol. St. Chem. 171: 161, 2003
2. Midya A. et al., Appl. Phys. Lett. 103: 092402, 2013
3. Dong Q.Y. et al., Mater. Lett. 161: 669, 2015
4. Jiménez E. et al., JMMM. 272–276: 568, 2004
5. Jiménez-Melero E. et al., J Phys. Cond. Mat. 18: 7893, 2006
6. Long Y.W. et al., JMMM. 322: 1912, 2010.

ACKNOWLEDGMENTS

The source irradiation was funded by the Australian Institute of Nuclear Science & Engineering under the award of grant ALNGRA-14547 to GAS and JMC.

A Forty-Year Journey in Rare Earth Science and Technology

Jean-Claude G. Bünzli

Swiss Federal Institute of Technology, Lausanne (EPFL), Switzerland
Hong Kong Baptist University, Hong Kong, S.A.R., P.R. China
jean-claude.bunzli@epfl.ch

INTRODUCTION

Both the scientific status and geopolitics of rare earths (RE) have seen dramatic changes during the past 40 years. New RE-based technologies have become indispensable in most of the modern appliances and devices. Moreover, the RE market has seen unexpected and spectacular variations in prices, as well as a shift in technical innovation from the Western World to Asia.

DISCUSSION

At the end of the 1970s, rare earths were still considered as minor elements by scientists working with transition metals although several industrial applications were already well established. They mainly included cracking catalysts, mischmetal for flint stones and metallurgy, optical glasses, and phosphors for fluorescent lamps and television screens.[1]

Yet, tremendous development were to be seen in the 1980's including three-way automotive catalysts, samarium-cobalt and iron-boron-neodymium magnets, erbium-doped fiber amplifiers for optical communications, nickel-lanthanide hydride rechargeable batteries, and time-resolved luminescent immunoassays.

In parallel to technological improvements of materials and products, the discovery of new effects and properties continued in the 1990s onward and several important landmarks were reached. They include the development of super-conducting magnets thanks to new cryogenic techniques, Gd-based bubble memory for computers, the advent of lanthanide-based white-light emitting diodes, the discovery of the giant magnetocaloric effect and associated magnetic refrigeration. Emerging applications are presently in the fields of energy conversion, luminescent sensors and tags, security inks, bioimaging, quantum encryption, optical refrigeration…

The traditional applications based on chemical properties (catalysis), magnetism, and optics have all seen unprecedented advances. In particular, the contribution of lanthanides to lighting has led to an increase in electricity-to-light conversion efficiency by an order of magnitude. Moreover, lanthanide photonics has been fueled by striking improvements in upconverting nanoparticles.[2]

During the same span of time, mining of rare earths, initially mainly achieved in the USA, has shifted to China which now dominates the market with an ≈85% share. This has been accompanied by a shift in scientific and technological development from the Western World to Asia.

Consumption is increasing slightly but steadily and it is forecast that it will reach 200 ktons equivalent rare earth oxides by 2020 (presently 126 ktons). As for prices, there are now back to their 2008 level creating stress in the production chain not only because of the imbalance problem, but, also, because mine operators produce at prices higher than market prices.

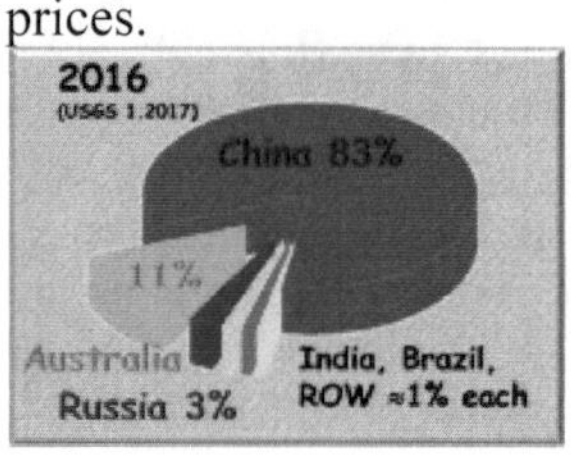

CONCLUSIONS

Rare earths have played a critical role in technological advances during the past 40 years. They will remain indispensable in the future since substitutes are difficult to find. It may also be anticipated that due to rising demand, the market might regain some stability.

REFERENCES

(1) Bünzli, J.-C. G. Lanthanides. In *Kirk-Othmer Encyclopedia of Chemical Technology*; Wiley Online Library: 2013; Chapter 2013, pp 1-43.

(2) Bünzli, J.-C. G. Lanthanide Luminescence: From a Mystery to Rationalization, Understanding, and Applications. In *Handbook on the Physics and Chemistry of Rare Earths*; Bünzli, J.-C. G. and Pecharsky, V. K., Eds. Elsevier Science, B.V.: Amsterdam; 2016; Vol. 50, Ch. 287, 141-176.

A neutron diffraction demonstration of the long-range magnetic order in the quasicrystal approximant $DyCd_6$

D.H. Ryan[1], J.M. Cadogan[2], Tai Kong[3], P.C. Canfield[3], A.I. Goldman[3] and A. Kreyssig[3]

[1]Centre for the Physics of Materials and Department of Physics, McGill University, Montreal, Quebec, H3A 2T8, Canada, dhryan@physics.mcgill.ca
[2]School of Physical, Environmental and Mathematical Sciences, UNSW Canberra at the Australian Defence Force Academy, Canberra BC 2610, ACT, Australia
[3]Ames Laboratory US DOE and Department of Physics and Astronomy, Iowa State University, Ames, IA 50011, USA

INTRODUCTION

One of the interesting questions concerning quasicrystals is whether long-range magnetic order is compatible with the quasiperiodicity. The binary R–Cd system (where R is a rare earth) is remarkable in that it hosts two equilibrium quasicrystalline (or icosahedral) phases, i-$YbCd_{5.7}$ [1] and a new, moment-bearing set of binary phases i-$RCd_{7.75}$ for R=Gd–Tm, Y[2]. These icosahedral phases occur either side of the RCd_6 cubic 'approximant' phase that is constructed from a BCC stacking of icosahedral clusters, the building blocks of the quasicrystalline phases. It is now well established by magnetometry and XRMS measurements that the RCd_6 system exhibits long-range antiferromagnetic order that at times is surprisingly complex, such as in the case of $GdCd_6$ where no fewer than four magnetic transitions have been observed[3].

The 'technique of choice' for studying magnetic structures is neutron diffraction but this avenue has not been pursued, due to the rather high neutron absorption by R elements such as Gd and Dy, and the Cd. In this paper, we present neutron diffraction patterns of $DyCd_6$ obtained at 25 K (above the ordering temperature of 17.8 K[3]) and 3.5 K, using a flat-plate sample mounting method[4]. These patterns clearly demonstrate long-range antiferromagnetic order in $DyCd_6$.

EXPERIMENTAL METHODS

The $DyCd_6$ sample was prepared at Ames Lab. using the method outlined by Das et al.[5] Neutron diffraction patterns were obtained at the NRU reactor, Chalk River, Canada, using a neutron wavelength of 2.37 Å and all diffraction patterns were fitted using the Le Bail method and the FullProf/WinPlotr program. Given the rather large neutron absorption by both Dy and Cd, we used the large-area single-crystal flat-plate mounts that we have used for work with numerous Sm-, Eu- and Gd-based materials[4].

RESULTS AND DISCUSSION

In figure 1 we show the neutron diffraction patterns of $DyCd_6$ obtained at 25 K and 3.5 K, together with the difference plot, emphasizing the strong magnetic contribution. The crystal structure of RCd_6 is cubic $Im\bar{3}$ (#204) at RT and a weak monoclinic distortion occurs at low temperatures[6]. Given the weakness of this distortion, we fitted our diffraction patterns in terms of the cubic Im-3 structure. At 25 K, the cubic lattice parameter is 15.380(1) Å. At 3.5 K, we observed strong magnetic contributions at $2\theta = 6.25°$ and $8.85°$, indexed as the purely magnetic (½ 0 ½) and (100), respectively (figure 2). In terms of the $Im\bar{3}$ group, we require two propagation vectors to describe the antiferromagnetic order, $\mathbf{k}_1$=[100] and $\mathbf{k}_2$=[½ 0 ½], yielding "anti-I" order. We can also describe the magnetic order in terms of the low-temperature monoclinic group $C2/c$ [6] with a single propagation vector $\mathbf{k}$=[100] in this case. The temperature dependence of the magnetic scattering intensity from the two most prominent purely magnetic peaks confirms the magnetic ordering temperature as 18.0(2) K.

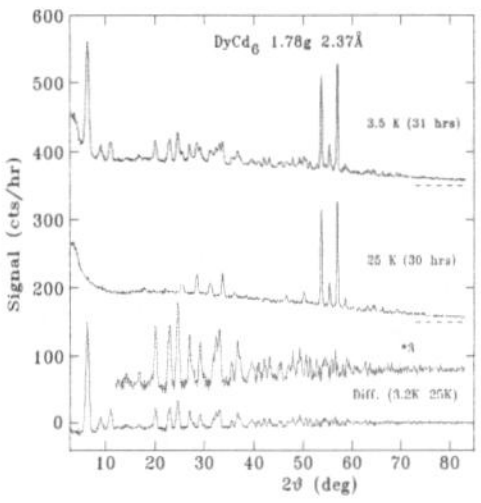

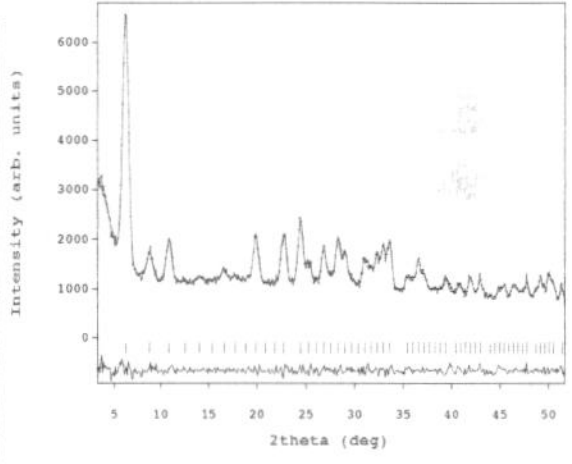

Left: Fig. 1: Neutron diffraction patterns of $DyCd_6$, at 25 K (blue) and 3.5 K (green), together with their difference plotted in red.

Right: Fig. 2: 'Le Bail' fit to the neutron diffraction pattern of $DyCd_6$ obtained at 3.5 K.

CONCLUSIONS

We have used neutron powder diffraction to show that the 'quasicrystal approximant' compound $DyCd_6$ shows long-range "anti-I" antiferromagnetic order below 18 K.

REFERENCES

1. Tsai A.P. et al., Nature 408: 537, 2000
2. Goldman A.I. et al., Nature Mater. 12: 714, 2013
3. Mori A. et al., J. Phys. Soc. Jpn, 81: 024720, 2012
4. Ryan D.H. and Cranswick L.M.D., J. Appl. Cryst. 41: 198, 2008
5. Pinaki Das et al., Phys. Rev. B 95: 054408, 2017
6. Tamura R., Isr. J. Chem., 51: 1263, 2011

ACKNOWLEDGMENTS

PCC and TK (growth and basic characterization) were supported by the U.S. Department of Energy, Office of Basic Energy Science, Division of Materials Sciences and Engineering. The research was performed at the Ames Laboratory. Ames Laboratory is operated for the U.S. Department of Energy by Iowa State University under Contract No. DE-AC02-07CH11358.

A New Class Of Ln^{2+} Complexes, [K(2.2.2-cryptand)]{Ln[$(OAr^{Ad,Me})_3$mes]}: Synthesis, Structure, Spectroscopy, and Theoretical Characterization

Chad T. Palumbo,[1] Megan E. Fieser,[1] Dominik P. Halter,[2] Vamsee Voora,[1] Joseph W. Ziller,[1] Filipp Furche,[1] Karsten Meyer,[2] William J. Evans[1]

[1]*Department of Chemistry, University of California, Irvine, California 92697-2025, United States*
[2]*Department of Chemistry and Pharmacy, Inorganic Chemistry, University Erlangen-Nuremberg, D-91058 Erlangen, Germany,* ctpalumb@uci.edu

INTRODUCTION

The recent isolation of [K(2.2.2-cryptand)][Cp'_3Ln] [Ln = Y and all the lanthanides except Pm due to its radioactivity; Cp′ = $(C_5H_4SiMe_3)^{1-}$] demonstrated for the first time that the entire lanthanide series can form molecular complexes with metals in the +2 oxidation state and that electronic ground state configurations of Ln^{2+} ions are affected by their coordination environments.[1-6] The $(Cp'_3)^{3-}$ coordination environment is unique in that, to date, it is the only environment to allow isolation of Ln^{2+} complexes across the series.[6,7] This work describes recent attempts to move beyond $(Cp'_3)^{3-}$ with the goal of isolating more stable examples of Ln^{2+} using a tris(aryloxide)mesitylene ligand, $[(^{Ad,Me}OAr)_3mes]^{3-}$, in order to study its effects on the electronic structure of Ln^{2+} ions in complexes. Structural comparisons with the U^{2+} analog, [K(2.2.2-cryptand)]-{U[$(^{Ad,Me}OAr)_3$mes},[8] will be made.

EXPERIMENTAL METHODS

New complexes with $\{Ln[(^{Ad,Me}OAr)_3mes]\}^{1-}$ anions (Ln = La, Ce, Pr, Nd, Sm, Gd, Dy, Er, Yb) were synthesized and structurally characterized. Structural changes between Ln[$(^{Ad,Me}OAr)_3$mes] and $\{Ln[(^{Ad,Me}OAr)_3mes]\}^{1-}$ complexes were correlated with UV-visible and electron paramagnetic resonance spectroscopies to create a three class model that assisted in assigning electronic ground state configurations of $\{Ln[(^{Ad,Me}OAr)_3mes]\}^{1-}$ complexes. Density functional theory calculations were also performed to support assignments of ground states.

RESULTS AND DISCUSSION

The new complexes were prepared by the reactions illustrated in Scheme 1 below.

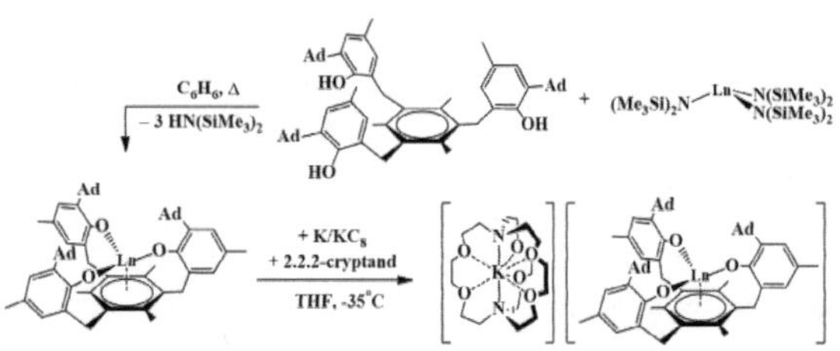

Scheme 1. Synthesis of [K(2.2.2-cryptand)] {Ln[$(^{Ad,Me}OAr)_3$mes]} complexes.

Structural and spectroscopic analyses were crucial in identifying three types of Ln^{2+} ions in the complexes [K(2.2.2-cryptand)][Cp'_3Ln]: (1) traditional $4f^{n+1}$ Ln^{2+} ions (Sm, Eu, Tm, Yb), (2) new $4f^n5d^1$ Ln^{2+} ions (La, Ce, Pr, Gd, Tb, Ho, Er, Lu), and (3) "configurational crossover" ions (Nd, Dy), whose configurations depend on the coordination environment. In the tris(aryloxide)mesitylene coordination environment, we have identified three classes of reduction products: (I) those of metal-based reductions, (II) those of ligand-based reductions, and (III) metal-based reduction products that are supported by π-acid bonding of the ligand. In contrast to $4f^n5d^1$ Nd^{2+} and Dy^{2+} in $(Cp'_3Ln)^{1-}$, it is concluded that these ions are $4f^{n+1}$ in $\{Ln[(^{Ad,Me}OAr)_3mes]\}^{1-}$. The fact that Er^{2+} falls between Classes I and III suggests Er^{2+} may also be a configurational crossover ion.

CONCLUSIONS

A structural, spectroscopic, and theoretical analysis of a new class of low valent rare earth metal compounds has aided in the realization that the $[(^{Ad,Me}OAr)_3mes]^{3-}$ ligand can be used to affect electronic ground state configurations of Ln^{2+} in compounds.

REFERENCES

1. Hitchcock, P. B. *et al.*, Angew. Chem., Int. Ed. 47:1488-1491, 2008
2. MacDonald, M. R. *et al.*, J. Am. Chem. Soc. 135:9857-9868, 2013
3. Fieser, M. E. *et al.*, J. Am. Chem. Soc. 137:369-382, 2015
4. Evans, W. J., Organometallics 35:3088-3100, 2016
5. Kotyk, C. M. *et al.*, Chem. Sci. 6:7267-7273, 2015
6. Corbey, J. F. *et al.*, Organometallics 34:3909-3921, 2015
7. Coles, M. P. *et al.*, Organometallics 31:2682-2690, 2012
8. La Pierre, H. S. *et al.*, Angew. Chem. Int. Ed. 53:7158-7162

ACKNOWLEDGMENTS

We thank the U.S. National Science Foundation for support of the theoretical studies (CHE-1464828 to F.F.) and the experimental studies (CHE-1565776 to W.J.E.). We also thank Jason Jones and Mikey Wojnar for assistance with X-ray crystallography and Prof. A. S. Borovik for spectroscopic assistance. We also thank NSF for providing graduate fellowship support for Chad T. Palumbo (DGE-1321846).

A Structural Analysis of Hydration Across the Lanthanide Series

Mark Woods[1,2*], Katherine M. Payne[1,] Benjamin E. Kucera[3], Victor G. Young[3] and Edward J. Valente[4]

[1*]Department of Chemistry, Portland State University, United States, mark.woods@pdx.edu
[2*]Advanced Imaging Research Center, Oregon Health and Science University, United States, woodsmar@ohsu.edu
[3]Department of Chemistry, University of Minnesota, United States,
[3]Department of Chemistry, University of Portland, United States,

INTRODUCTION
The nature of chelate hydration in lanthanide chelates during dissociative exchange remains a subject of controversy. It has been demonstrated in solution that "partial *q* values" reflect the aggregate of the number and position of water molecules both on and proximate to the metal centre.[1] Despite the fact that the number and position of water molecules must, by necessity, change throughout the course of the dissociative exchange process there is still disagreement about how dissociative exchange might affect the overall hydration state. One point of view takes its cue from crystallography in which (for the most part) water is either bound or not bound to the metal ion. The crystal structures of the octadentate ligand DOTMA and its lanthanide chelates challenge this binary view of hydration. We will present structures of the chelates of DOTMA with ions from Ce^{3+} to Yb^{3+} and discuss the structural and hydration changes that occur across this series. We will conclude by considering what this series tells us about the hydration and exchange in solution.

EXPERIMENTAL METHODS (if relevant)
Relevant experimental details have already been published.[2]

RESULTS AND DISCUSSION
One of the more interesting features of LnDOTMA in the crystal is that in all cases (irrespective of Ln^{3+} ion) twisted square anitprismatic (TSAP) structure is adopted. This feature is unusual for tetracarboxylate cyclen derivatives, a change from TSAP (early Ln^{3+}) to SAP (later Ln^{3+}) is usually observed.[3] DOTMA affords a unique opportunity to examine the structural variations for a single coordination geometry as the ionic radius of the Ln^{3+} ion decreases.
The structures of LnDOTMA chelates fall into two basic groups: those before the gadolinium break; and those after. Before the gadolinium break the structure of the chelates is broadly similar: each chelate has a single water molecule located on the metal ion at a distance consistent with a bonding interaction. After the gadolinium break the structures become more interesting. Multiple chelates are found in the unit cell of each structure. As the ionic radius decreases across the series, the hydration number and Ln-O distance in each of the chelate structure varies in a predictable trend. The hydration state of the chelate within the crystal decreases steadily from Tb^{3+} to Tm^{3+}.

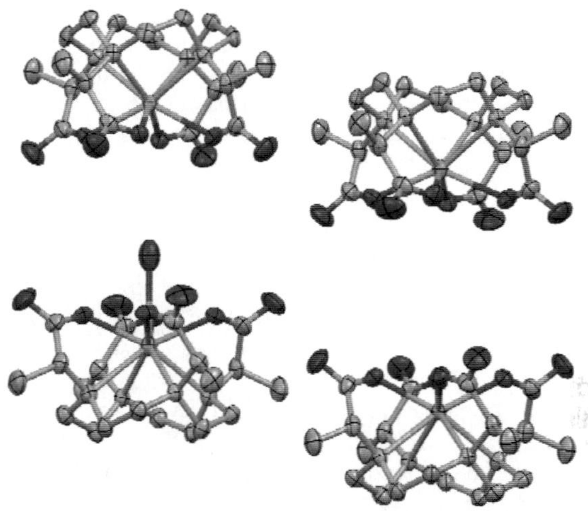

Figure 1. Showing the 4 Er^{3+} chelates in the unit cell of the ErDOTMA crystal. Note that three chelates have no water molecule coordinated to the metal ion but the fourth is hydrated. Hydrogen atoms, non-coordinated water molecules and counter ions are omitted for clarity.

These crystal structures paint a clear picture of the conformational changes that occur as the water molecule is pushed off the metal ion. In this case the driving force for the departure is the reduction in ionic radius but it undoubtedly parallels the changes that occur in the chelate as the water exchanges in solution.

CONCLUSIONS
The effect of lanthanide contraction on the hydration state of LnDOTMA chelates is revealing. There is no well-defined cut-off between hydrated and dehydrated chelates, but rather the transition between these states is gradual across the series affording insights into the nature of hydration during the process of dissociative water exchange in solution.

REFERENCES
(1) Beeby, A.; Clarkson, I. M.; Dickins, R. S.; Faulkner, S.; Parker, D.; Royle, L.; de Sousa, A. S.; Williams, J. A. G.; Woods, M. *J. Chem. Soc., Perkin Trans. 2* **1999**, 493.
(2) Aime, S.; Botta, M.; Garda, Z.; Kucera, B. E.; Tircso, G.; Young, V. G.; Woods, M. Inorg. Chem. **2011**, *50*, 7955.
(3) Benetollo, F.; Bombieri, G.; Calabi, L.; Aime, S.; Botta, M. *Inorg. Chem.* **2003**, *42*, 148.

A Systematic Exploration of Rare-earth-*p*-halobenzoic acid–terpyridine Materials: Structural and Supramolecular Trends in Three Series of Molecular Complexes synthesized *via* Hydrothermal Methods

J. August Ridenour[1], Korey P. Carter[1], Ray J. Butcher[2] and Christopher L. Cahill[1*]

[1]Department of Chemistry, The George Washington University, United States
[1]Department of Chemistry, Howard University, United States
jaridenour@gwmail.gwu.edu

INTRODUCTION

Rare-earth hybrid materials have desirable properties, such as luminescence and magnetism for use in luminescent probes[1] and single molecule magnets[2], which provide ample incentive for solid-state materials research. Previous efforts within our group to explore rare-earth containing metal-organic hybrid materials have been successful, aided by the use terpyridine as a 'capping' ligand to control oligomerization and as a luminescent antenna for sensitized emission.[3-5] The work presented herein is the culmination of three distinct series of rare-earth *p*-halobenzoic acid-terpyridine materials where solid-state samples of sixty complexes were structurally characterized with single crystal and powder X-ray diffraction techniques. Each series of materials presented were synthesized with different *p*-halobenzoic acid ligands (*p*-chloro, *p*-bromo, *p*-iodo) and structural and supramolecular trends between the three series of materials were analysed.[3-5]

EXPERIMENTAL METHODS

All sixty complexes were synthesized by placing the reactants in a Teflon-lined steel autoclave and heated statically at varying temperatures (120 °C, 150 °C, or 180 °C) at autogenous pressure. Small amounts of NaOH were added to some of the reaction vessels to aid in crystallization of diffraction quality crystals. Powder X-ray diffraction was done to analyse the bulk purity of products, while samples for single crystal X-ray diffraction were isolated from those bulk products.

RESULTS AND DISCUSSION

There are six structure types observed within the sixty rare-earth materials, all of which are binuclear tectons that utilize the 'cap and link' strategy; metal centers 'capped' with terpyridine ligands to create discrete binuclear 'tectons', and further coordinated by *p*-halobenzoic acid ligands which provide peripheral halogen atoms for 'linking' *via* halogen-based supramolecular interactions. The three series of materials, which encompass the entire suite of commonly use rare-earth metal centers, provide a suitable platform of materials for analysing structural and supramolecular trends by modulating the size and identity of the rare-earth metal as well as the size and polarizability of the peripheral halogen atom. The observation of a particular structure type, and therefore a specific halogen-based supramolecular motif, is suggestive of a preference for that supramolecular interaction over another difference interaction motif.

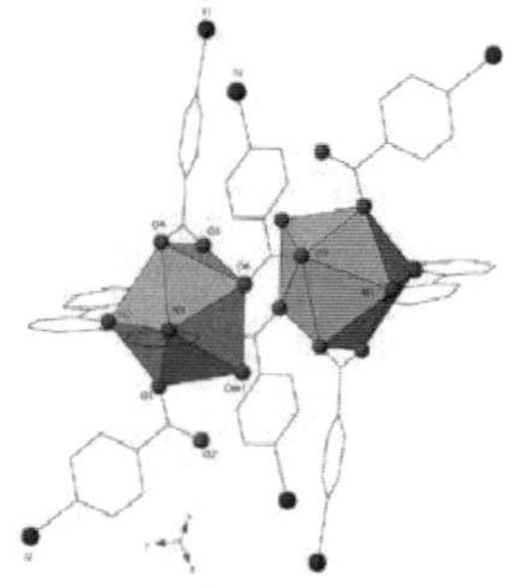

Figure 1 shows an example of a RE^{3+} binuclear tecton.

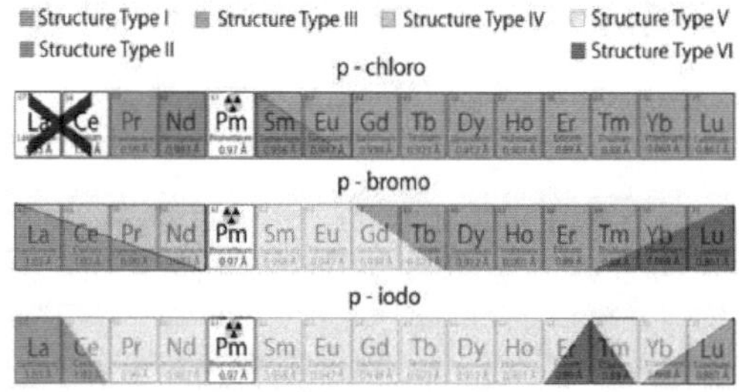

Chart 1 shows the summary of structure types and with what RE^{3+} metal center in all three structural studies.

CONCLUSIONS

Changes in the local coordination environment around the rare-earth metal center have subsequent effects on the offset π-stacking and observed halogen-based supramolecular interactions across a series of *p*-halobenzoic acid materials and across the entire comprehensive study. The propensity for and relative strength of halogen-π interactions is consistent with the greater polarizability found with larger halogen atoms.

REFERENCES

1. S. V. Eliseeva and J.-C. G. Bünzli, *Chem. Soc. Rev.*, 2010, **39**, 189-227.
2. N. Ishikawa, M. Sugita, T. Ishikawa, S.-Y. Koshihara and Y. Kaizu, *J. Am. Chem. Soc.* 2003, **125**, 8694-8695.
3. K. P. Carter, S. J. A. Pope and C. L. Cahill, *CrystEngComm*, 2014, **16**, 1873-1884.
4. J. A. Ridenour, K. P. Carter, R. J. Butcher, and C. L. Cahill, *CrystEngComm*, 2017, **19**, 1172-1189
5. J. A. Ridenour, K. P. Carter, and C. L. Cahill, *CrystEngComm*, 2017, **19**, 1190-1203

ACKNOWLEDGMENTS

This material is based on work supported as part of the Materials Science of Actinides, an Energy Frontier Research Center funded by the U.S. Department of Energy, Office of Science, Office of Basic Energy Science, under Award Number DE-SC0001089.

Activation and Polymerization Studies of Nd{C(SiHMe$_2$)$_3$}$_3$ as a Precatalyst

Bradley M. Schmidt, Aradhana Pindwal, and Aaron D. Sadow*

Department of Chemistry, Iowa State University, United States of America
bschmidt@iastate.edu

INTRODUCTION

Rare-Earth catalysts, especially those of neodymium, have been widely used in the production of polybutadiene (PBD) on an industrial scale.[1] These systems have gained popularity in producing polymers of high *cis*-1,4 content, which is produced on a scale of million tons per year for the production of tires alone.[2] Developing single site binary, ternary, or quaternary catalysts systems for the polymerization of butadiene and isoprene remains a topic of high interest for this reason. Most reported systems make use of neodymium carboxylate species.[3] Here we study the activation of our recently reported tris-alkyl-lanthanides to mono- or dicationic active species and their subsequent reactivity in polymerization reactions. Through a series of NMR experiments and polymerization reactions we attempt to understand the activation to a dicationic species and its role in the polymerization of diene monomers.

RESULTS AND DISCUSSION

Traditional neodymium-carboxylate based diene polymerization catalyst systems often lack an understanding of the active site specifics. We sought to understand the environment of the neodymium-center upon activation parameters leading to polymerization of diene monomers. The synthesis of our precatalyst is shown in reaction (1). As shown the Nd-center is stabilized by 6 β–agostic Si–H interactions. This allows for polymerization to occur without the presence of donor solvents, such as THF, or acids as typical for carboxylate systems, which can hinder reaction performance.

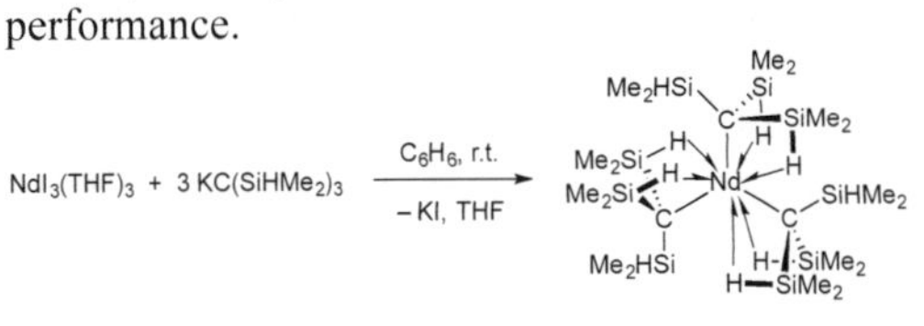

(1)

In the presence of olefin reagents the neutral species remains unchanged. As organoaluminum is a common co-catalyst in these systems we were interested in how the neutral species would react toward these reagents. In NMR studies with the analogues lanthanum species we see rapid alkyl exchange. When studied with trimethyl aluminium we observed a reaction ratio of 1:3 needed to consume the LaR$_3$ completely. EXSY experiments showed dynamic exchange between excess AlMe$_3$ and the unidentified lanthanum species. Activation by Si–H abstraction can lead to mono- or dicationic species as shown in equation (2).

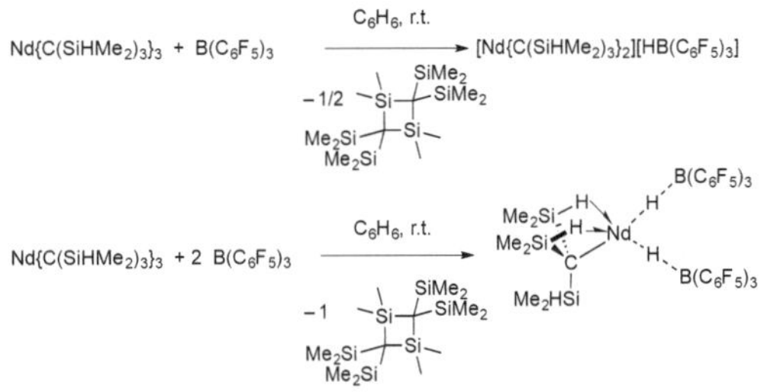

(2)

The dicationic species shows slow insertion of isoprene at 60 °C in C_6D_6. In the presence of tri-isobutyl aluminium (TIBA) the conversion of 10 equivalents of isoprene is completed within 10 minutes at room temperature, with similar results seen for butadiene. When NdR$_3$ activated by Ph$_3$CCl a new activation pathway is observed. 1 equivalent of Ph$_3$CH is observed along with 1 equivalent of an alkyl abstracted organic species shown in reaction (3).

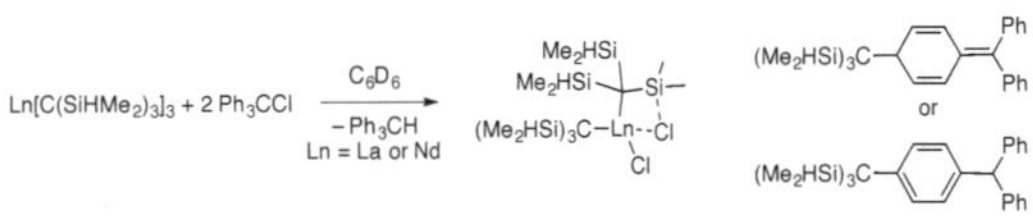

(3)

Crystallization of this species hasn't yet proven fruitful. When subjected to butadiene in the presence of TIBA a highly active catalyst is generated which yields > 90% cis-1,4 selective PBD.

CONCLUSIONS

Activation of Nd{C(SiHMe$_2$)$_3$}$_3$ to generate a dicationic species along with an organoaluminium co-catalyst prove to be highly active toward diene polymerization. The isolable species of [NdC(SiHMe$_2$)$_3$][HB(C$_6$F$_5$)$_3$]$_2$ yields a PBD polymer of nearly 1:1 cis:trans 1,4 additions when paired with TIBA. When activated by a chlorine source such as Ph$_3$CCl an alternative activation pathway is seen. However, its activity toward diene polymerization when paired with TIBA is nearly identical while the microstructure shifts to greater than 90% cis-1,4 selectivity.

REFERENCES

1. Zhang, X. *et al.*, *Polymer* 54:2973-2978, 2013
2. Modi, S. *et al. Chem. Eng. Sci.* 107:256-265, 2014
3. Derek, K. J. *et al. Eur. Poly. J.* 37:2105-2110, 2001

ACKNOWLEDGMENTS

I would like to thank Dr. Pindwal for her role in the initiation of this project. Along with Dr. Arkady for his contributions in crystallography and structure verification.

Alternative Ligands for the Separation of Rare Earth Elements Using Solvent Extraction

Kevin L. Lyon[1*], Mitchell R. Greenhalgh[1], R. Scott Herbst[1], Amy K. Welty[1], Melissa M. Warner[1], Derek M. Brigham[2] and Ross J. Ellis[2]

[1*]Aqueous Separations and Radiochemistry Department, Idaho National Laboratory, United States Kevin.Lyon@inl.gov
[2]Chemical Sciences Division, Oak Ridge National Laboratory, United States

INTRODUCTION

Solvent extraction has, and will continue to be, an important industrial process for the separation and purification of rare earth elements (REE). Current market conditions and economic contstraints related to REE production have driven the need to investigate alternative approaches to improve the separation of REE. Traditional phosphonic acid (i.e. PC88A) solvent extraction processes consume significant quantities of acid and base to obtain high recovery and purity, and the low separation factors among adjacent lanthanides require a large number of stages to facilitate a given separation. Alternative ligands such as Cyanex® 572 and N,N,N',N'-Tetraoctyl Diglycolamide (TODGA) are being examined as alternative solvent extraction systems to reduce capital and operating costs of a REE separations plant. Improvements to current separation technologies may help enable domestic supply chains for REE.

EXPERIMENTAL METHODS

Flowsheet development for PC88A and Cyanex® 572 were performed in collaboration with the Cytec Solvay Group using solvent extraction modelling software described in literature.[1] A simulated mid/heavy REE separation was chosen using Gd, Ho, and Y to compare the separation performance of PC88A and Cyanex® 572. Flowsheet tests were conducted in a 30 stage counter-current mixer-settler cascade.[2] Collection of laboratory equilibrium data and subsequent flowsheet modelling for the TODGA system were performed using techniques previously discussed in literature.[3] TODGA studies primarily focused on light REE to separate Pr and Nd from La and Ce.

RESULTS AND DISCUSSION

A 29 stage mid/heavy REE separation cascade operated until steady state was achieved for both PC88A and Cyanex® 572 solvent systems. The number of stages and the feed solutions were identical for both flowsheets with the exception of the corresponding acid concentrations in extraction, scrub, and strip feeds. PC88A required 3.4 M HCl to fully strip the organic phase in 3 strip stages; Cyanex® 572 required 2.3 M HCl to strip an organic phase similar in REE composition in the same number of strip stages.

TODGA solvent extraction tests show an average adjacent REE separation factor of 2.6 for La-Nd,[4] compared to an average separation factor of 1.19 published for PC88A.[5] Despite this improvement, high recovery and purity may only be achieved using REE exchange reactions in scrub similar to techniques used with phosphonic acids. A preliminary counter-cuurent mixer-settler study revealed TODGA is indeed capable of this scrubbing mechanism, effectively scrubbing Pr from an organic phase loaded with a mixture of Pr and Nd using aqueous Nd scrub solutions. Flowsheet development is currently in progress to demonstrate industry-relevant REE separations using TODGA.

CONCLUSIONS

Laboratory testing suggests that Cyanex® 572 achieves comparable separation performance as PC88A while reducing the strip acid requirements by approximately 30%. Cyanex® 572 is expected to reduce reagent consumption and subsequent operating costs for traditional phosphonic acid REE separation technology, especially for heavy REE separations. Laboratory studies suggest that TODGA provides higher separation factors among the light rare earth elements as compared to phosphonic acid extractants. The neutral solvating TODGA ligand has been demonstrated in the laboratory to undergo metal-metal exchange similar to phosphonic acids; REE separations may be feasible in a reduced number of stages due to the improved separation factors. Furthermore, neutral ligands such as TODGA do not require saponification, which may have a significant impact reducing reagent consumption in REE separation processes. The chemical similarities of adjacent REE still prove challenging for flowsheet development; technoecnomic evaluations of these alternative flowsheets may offer significant reductions in capital and operating costs for REE separations.

REFERENCES

1. Soderstrom, M.D.. *et al.*, COM Proceedings, 2014.
2. Lyon, K.L. *et al.*, 28th IMPC Proceedings, 2016.
3. Lyon, K.L. *et al.*, Ind. Eng. Chem. Res., 56 (4): 1048-1056.
4. Ellis, R.J. *et al.*, Inorg. Chem., 56 (3): 1152-1160.
5. Gupta, C.K. and Krishnamurthy, N., CRC 2005.

ACKNOWLEDGMENTS

The authors would like to acknowledge Cytec Solvay Group for their contributions to Cyanex® 572 flowsheet modelling and development. The authors acknowledge the Center for Advanced Energy Studies for conducting the ICP-OES analysis for this research. This research is supported by the Critical Materials Institute, an Energy Innovation Hub funded by the U.S. Department of Energy, Office of Energy Efficiency and Renewable Energy, Advanced Manufacturing Office.

An Effective Route to Highly Symmetric Mixed Actinide-Lanthanide Complexes via Reduction of $[UO_2]^{2+}$

Bradley E. Cowie,[1] Markéta Suvova,[1] Polly L. Arnold*[1] and Jason B. Love*[1]

[1]EaStCHEM School of Chemistry, The University of Edinburgh, UK, B.E.Cowie@ed.ac.uk

INTRODUCTION

The uranyl ion, $[U^{VI}O_2]^{2+}$, is the most abundant form of uranium in the environment with high solubility and mobility in aqueous media,[1] and strongly covalent and chemically robust *trans*-oxo ligands.[2-4] While one-electron chemical reduction of $[U^{VI}O_2]^{2+}$ provides $[U^{V}O_2]^{+}$, which is unstable towards disproportionation under aerobic conditions, deployment of an appropriate ligand framework under anaerobic conditions enables access to stable and isolable $[U^{V}O_2]^{+}$ uranyl complexes, most often through reductive functionalisation of the uranyl oxo-ligands.[5-9] Since discovering the feasibility of U^{VI}/U^{V} reduction via f-element–oxo group coordination in our U^{VI} uranyl-pacman complexes,[10,11] we have become increasingly interested in understanding oxo-group–f-element interactions. As a result, we sought a versatile/general reaction protocol that would provide facile access to oxo-group bridged mixed actinide (An)/lanthanide (Ln) compounds in the absence of intricate ligand systems.

RESULTS AND DISCUSSION

We have found that reduction of $[U^{VI}O_2]^{2+}$ is achieved through reaction with simple Ln^{II} or U^{III} compounds, forming highly symmetrical complexes with strong, linear oxo-group bridges (Scheme 1). Uranyl chloride ($[UO_2Cl_2(THF)_2]$) reacts with Ln^{II} to provide the trimetallic U^{V}-Ln^{III}-U^{V} complex, $[\{UO_2(py)_5\}_2(LnI_4)]I$. In addition, uranyl chloride reacts with U^{III} to give the trimetallic U^{IV}-U^{IV}-U^{IV} complex, $[\{UClI(py)_4\}_2(UO_2I_4)]$ (**3**), via two-electron uranyl reduction. Alternatively, the uranyl bis-silylamide, $[UO_2(N'')_2(THF)_2]$ (N'' = $N(SiMe_3)_2$), reacts with Ln^{II} to form $[\{LnI(py)_5\}_2(UO_2I_4)]$ via two electron uranyl reduction; all attempts to utilise Ln^{II} salts to achieve reduction to U^{IV} starting from uranyl chloride have been unsuccessful. Finally, by altering the solvent, $[UO_2Cl_2(THF)_2]$ reacts with Ln^{II} to provide the one-dimensional U^{V}-Ln^{III} coordination polymer $\{[Ln(MeCN)_6][UO_2I_4]\}_\infty$. The complexes presented herein form a new class of axially symmetric uranyl clustering, and are robust enough to be characterised by IR and UV/VIS/NIR spectroscopy, single crystal X-ray crystallography, cyclic voltammetry and elemental analysis, thus offering a new way to build strongly coupled homo and heterometallic poly(f-block) chains.[12]

REFERENCES

1. Amme M. *et al.*, J. Nuc. Mat. 341:209-223, 2005. 2. Denning R. G., J. Phys Chem. A. 111:4125-4143, 2007. 3. Szabó Z. *et al.*, Coord. Chem. Rev. 25:784-815, 2006. 4. King R. B., J. Coord. Chem. 58:47-53, 2005. 5. Jones G. M. *et al.*, Angew. Chem. Int. Ed. 51:12584-12587, 2012. 6. Arnold P. L. *et al.*, Nature 451:315-318, 2008. 7. Arnold P. L. *et al.*, Nature Chem. 4:668-674, 2012. 8. Brown J. L. *et al.*, J. Am. Chem. Soc. 132:7248-7249, 2010. 9. Kiernicki, J. J. *et al.*, Angew. Chem. Int. Ed. 56:1097-1100, 2017. 10. Arnold P. L. *et al.*, J. Am. Chem. Soc. 135:3841-3854, 2013. 11. Arnold P. L. *et al.*, Angew. Chem. Int. Ed. 55:12797-12801, 2016. 12. Arnold *et al.*, Manuscript in preparation.

ACKNOWLEDGMENTS

The authors thank the EPSRC-UK, the University of Edinburgh and the EP/M010554/1 and EP/H004823/1 programs for support. BEC thanks the Government of Canada for an NSERC Post-Doctoral Fellowship. The authors also thank Nicola Magnani, Eric Colineau, Jean-Christophe Griveau and Roberto Caciuffo for the magnetometry results presented herein.

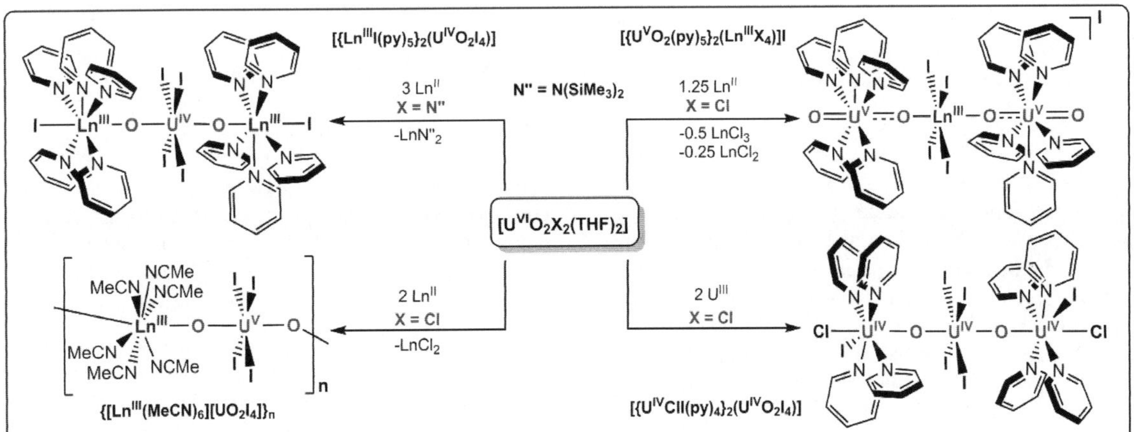

Scheme 1 - Synthesis of symmetric An/Ln complexes from $[UO_2Cl_2(THF)_2]$ and Ln^{II} or U^{III} salts.

An experimental study of the stability of uranyl and U(IV) chloride aqueous complexes at elevated temperatures: implications to the mobility of U in natural systems

Artaches A. Migdisov[1*], Alexander Timofeev[2], Hakim Boukhalfa[1], Robert Roback[1], and Anthony Williams-Jones[2]
[1*]Los Alamos National Laboratory, Los Alamos, US, artas@lanl.gov
[2]Department of Earth and Planetary Sciences, McGill University, Montreal, Canada

INTRODUCTION

Chloride is one of the most abundant ligands in natural aqueous solutions and chloride-bearing hydrothermal solutions are often involved in processes that potentially can cause mobilization and transport of U. As examples, many U ores are known to have formed through hydrothermal deposition from solutions containing up to 52 wt% of NaCl[1]; the Fukushima nuclear accident involved interaction of nuclear fuel with sea water (~0.6 m NaCl)[2]; and, finally, repositories of nuclear waste in geological formations include those in salt domes (WIPP).

Conventional wisdom holds that chloride speciation has little impact on the mobility of U in such systems due to low affinity of U to chloride and, thus, low stability of corresponding complexes. This assumption however has mostly been based on ambient temperature stability data. This assumption has been recently challenged by Dargent et al.[3], who demonstrated high stability of uranyl chloride complexes at elevated temperatures up to 350 °C. The goal of our study is therefore to provide an independent proof for the data of Dargent et al., and to specifically investigate the speciation of U(IV).

EXPERIMENTAL METHODS

Experiments involved determination of the stability of uranyl-chloride complexes using in-situ UV-vis spectroscopy of homogeneous uranyl-chloride solutions at temperatures up to 250 °C and 100 bar coupled with autoclave experiments to determine the solubility of UO_2 and U_3O_8 under controlled redox conditions at in solutions containing variable concentrations of NaCl. The latter experiments were performed in batch series in Ti autoclaves at temperatures 200 to 350°C and saturated pressure of water vapour. Redox conditions were controlled using Co/Co, Ni/NiO, and MoO_2/MoO_3 solid state redox buffers. Solubility experiments provide information on the stability of uranyl-chloride (MoO_2/MoO_3 buffer) and U(IV)-chloride (Co/Co, Ni/NiO) complexes at elevated temperatures.

RESULTS AND DISCUSSION

UV-vis spectra indicate UO_2^{2+}, UO_2Cl^+, $UO_2Cl_2^\circ$ and $UO_2Cl_3^-$ are the dominant species present and no evidence was found for species of higher chlorination ($UO_2Cl_4^{2-}$ and $UO_2Cl_5^{3-}$) as suggested by Dargent et al. Our data indicate that the main uranyl-chloride complex present in aqueous solutions at concentrations of NaCl relevant to natural conditions and T>150 °C is $UO_2Cl_2^\circ$ (Figure 1). Our data from solubility experiments are in good agreement with the spectroscopic data. The data demonstrate that in chloride-bearing hydrothermal solutions at low oxygen fugacity, uranium is present mainly as UCl_4°. Moreover, concentrations of UCl_4° at T>250 °C under reducing conditions are comparable to, or even higher than those in oxidizing systems (Figure 2).

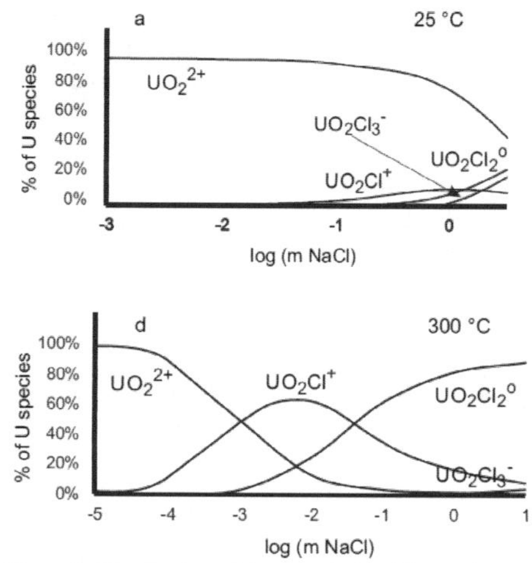

Figure 1. Distribution of the uranyl-chloride species as a function of the logarithm of total molality of NaCl in solutions.

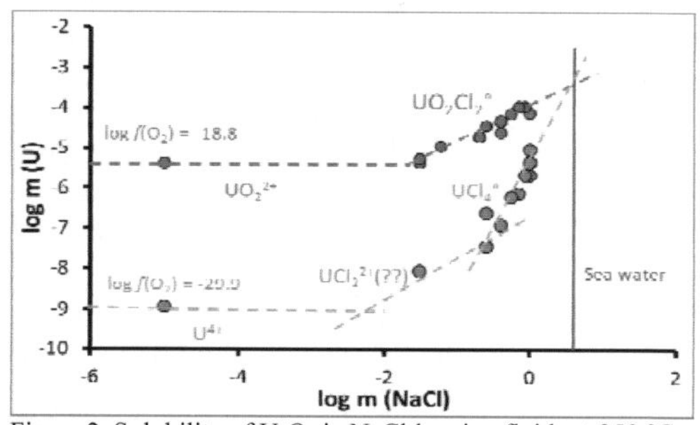

Figure 2. Solubility of U_3O_8 in NaCl-bearing fluids at 350 °C and pH(T) = 2

CONCLUSIONS

Our study indicates that the role of chloride as a mobilization and transport agent for uranium changes drastically with increasing temperature. Whereas at ambient conditions chloride has very low affinity to uranyl this affinity quickly increase with increasing temperature. At 150-200 °C and under reducing conditions, uranyl-chloride complexes are quite stable and important ligands capable of transporting U in natural systems.

REFERENCES

1. Richard A., *et al.,* Nat. Geosci. 5: 142–146, 2011
2. Burns P. C., *et al.*, Science, 335: 1184–1188, 2012
3. Dargent M., *et al.*, Eur. J. Mineral., 25:765–775, 201

An N-Tethered Uranium-Arene Complex for the Synthesis of Uranium-Metal Bonds

Skye Fortier[1*], Katie Harriman,[2] Bess Vlaisavljevich,[3] and Muralee Murugesu[2]

[1*]Department of Chemistry, University of Texas at El Paso, USA, asfortier@utep.edu
[2]Department of Chemistry, University of Ottawa, Canada
[3]Department of Chemistry, Northwestern University, USA

INTRODUCTION

Metalloarene complexes, e.g. $Cr(\eta^6\text{-}C_6H_6)_2$, are an important class of molecules that provide a unique window into the bonding capabilities of a metal, insight that is valuable for better understanding chemical properties and predicting reactivity. Metalloarene interactions can form through coulombic contacts or through a number of covalent bonding modes (η^1-η^6) to give complexes which can feature metal-arene σ, π, and δ bonds. In addition, coordinating arenes can modulate the chemistry of a given metal center, leading to unique reactivity patterns.[1,2]

In our laboratory, we have been exploring molecules that contain uranium-arene bonds in an effort to better understand the role of the *6d*/*5f*–orbitals in bonding and to access new actinide chemistry. Specifically, we have designed a dianionic, terphenyl bisanilide ligand $[\text{bis(2,6-diisopropylanilide)terphenyl}]^{2-}$ (L^{Ardipp}) that enforces a uranium-arene η^6-interaction through N-atom tethering. The synthesis and characterization of the U(III) complex $UI(DME)(\eta^6\text{-}L^{Ardipp})$ and its reactivity with anionic, metal nucleophiles is presented.

EXPERIMENTAL METHODS

The ligand, H_2L^{Ardipp}, can be synthesized in two steps using standard Suzuki and Hartwig-Buchwald protocols. Treatment of H_2L^{Ardipp} with excess KH gives the K_2L^{Ardipp} salt in good yield as a dark, orange red solid. Addition of $UI_3(diox)_{1.5}$ to K_2L^{Ardipp} in DME affords $UI(DME)(\eta^6\text{-}L^{Ardipp})$ in modest yields as a dark brown product (eq 1).

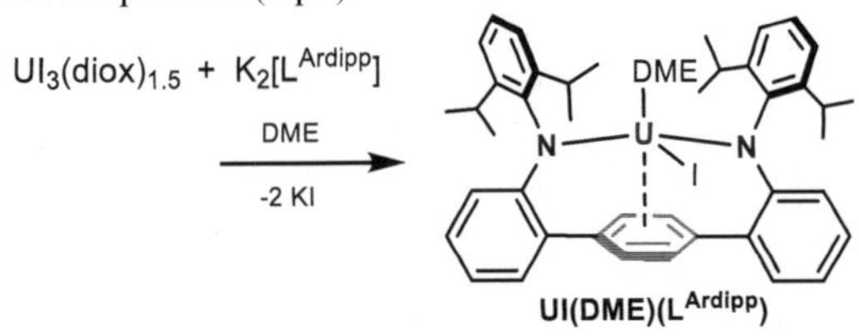

(1)

UI(DME)(L^{Ardipp})

RESULTS AND DISCUSSION

$UI(DME)(\eta^6\text{-}L^{Ardipp})$ is readily crystallized by the diffusion of hexanes into a saturated DME solution stored at -30 °C for several days. The solid-state structure of $UI(DME)(\eta^6\text{-}L^{Ardipp})$ reveals a chelated uranium atom, bound by both nitrogen atoms of the L^{Ardipp} ligand, with the metal center positioned directly above the central arene ring of the terphenyl scaffold. Inspection of the U-C_{arene} distances (2.879(5) to 2.957(5) Å) reveal contacts indicative of uranium-arene bonds consistent with hapto-six binding.[1,2]

The U-arene interaction and the sterics of the flanking 2,6-diisropropylphenyl groups in $UI(DME)(\eta^6\text{-}L^{Ardipp})$ present a unique coordination environment around the uranium center. In a preliminary examination of the L^{Ardipp} ligand's ability to support new reactivity or uncommon uranium bonds, $UI(DME)(\eta^6\text{-}L^{Ardipp})$ was treated with the nucleophilic metal anion [Na(tmeda)][Fp] (Fp = $CpFe(CO)_2$). From this, the new paramagnetic complex $U(Fp)(\eta^6\text{-}L^{Ardipp})$, featuring a rare instance of an unsupported Fe-U bond, is cleanly formed in good yield.

Studies of the electronic structure and magnetic properties of $UI(DME)(\eta^6\text{-}L^{Ardipp})$ and $U(Fp)(\eta^6\text{-}L^{Ardipp})$ are underway.

CONCLUSIONS

We have developed a new ligand system, L^{Ardipp}, that enforces U-arene bonding through N-atom tethering to give the U(III) complex $UI(DME)(\eta^6\text{-}L^{Ardipp})$. The metal-arene interaction is anticipated to play an integral role in modulating reactivity at the metal. Examination of the chemistry of $UI(DME)(\eta^6\text{-}L^{Ardipp})$ and study of its electronic structure will shed valuable light on the chemical properties of the early actinides.

REFERENCES

1. Meyer, K. *et al.*, Inorg. Chem. 48:9419-9426, 2009
2. Diaconescu, P. *et al.*, Dalton Trans. 44:15360-15371, 2015

ACKNOWLEDGMENTS

The authors would like to thank the University of Texas at El Paso, Northwestern University, and the University of Ottawa for support of this project.

Applicability of Cerium Surrogates to Describe Plutonium Oxalate Precipitation

Lindsay E. Roy*, Tom Shehee, and Chris South

Savannah River National Laboratory, Savannah River Site, Aiken, SC 29808, lindsay.roy@srnl.doe.gov

INTRODUCTION

Cerium is often used as a nonradioactive surrogate for plutonium in development work, particularly to simulate several plutonium precipitation processes. There are similarities between cerium and plutonium oxalate precipitation methods but variables related to crystal nucleation and growth need to be quantified when considering a surrogate to describe the production process.

Our interest in metal oxalate precipitation chemistry is to understand the link between process parameters and properties to optimize precipitator performance in the HB-Line Plutonium Finishing Facility. We are currently developing a simulation system that will provide a description of crystallization and agglomeration phenomena for the plutonium oxalate precipitation process. However, a fundamental description of supersaturation and precipitation kinetics to verify the models is important for project success since batch sizes and tank geometries play an important role in dictating the properties of the final product.

In that context, this presentation will be an overview of the modelling efforts at SRNL to integrate multiple time/length scales for a description of the crystallization dynamics of PuO_2 from the production facility. After analysing the Pu experimental results, calculations have started to evaluate the applicability of using cerium analogues to describe the crystal nucleation and growth phenomena. Finally, an experimental research plan for evaluating cerium oxalate precipitation will be presented.

RESULTS AND DISCUSSION

The simulation system used to model the oxalate precipitation process consists of three main development areas: Data generation, data storage, and data analysis. The data generation component involves development of a workflow software tool that will automate and manage the computational steps necessary to determine a specific material property. The data storage component will enable the creation of databases from high-throughput calculations to store and query associated datasets. The data analysis component will involve the development of web applications to allow users to interactively analyse specific datasets. The current phase of the project is data generation.

Classical nucleation theory (CNT) models involve supersaturated solutions at supercooled temperatures where the molecules in solution begin to agglomerate and form microcrystalline structures. Once the clusters reach a certain critical size, molecules in solution will more favourably associate with the crystal surface than the solvent, making crystallization a spontaneous process. The net flux of the nucleation is then related to the free energy barrier of nucleation which is used to determine the crystal growth rate in solution. In nonclassical crystallization, nanometer-sized particles are considered to be the primary units in the construction of larger crystals. The nonclassical growth mechanism helps to explain the nanoaggregates of polycrystalline material during production scale operations.

To determine the free energy of nucleation, calculations focus on plutonium nitrate behaviour in solution prior to complexation of oxalic acid. The solution is modelled as a continuum solvent with the necessary hydronium and nitrate ions to result in the appropriate nitric acid concentration. The next step is to determine if the free molecules in solution will agglomerate on the desired timescale. However, we recognize that comparison to lab scale plutonium experiments may not be appropriate because smaller particles and different morphologies are formed due to lower solubility and higher nucleation rates. Therefore, calculations have begun examining cerium as potential surrogate for experimental validation of the results. If the calculated results are comparable, a single-stage, plant-scale precipitator with appropriate agitator design will be designed to mimic the mass transfer effects in and out of specific zones within the tank using the cerium surrogate.

CONCLUSIONS

SRNL is currently developing a simulation system to integrate multiple time/length scales for a description of the crystallization dynamics of PuO_2 from the HB-Line Plutonium Finishing Facility to optimize precipitator performance. Initial calculations of the free energy of nucleation have started for both plutonium and cerium oxalate precipitation to quantify if a surrogate is appropriate for validation experiments. If comparable, a research plan has been developed to correlate the relationship between supersaturation and precipitation kinetics for the production process.

ACKNOWLEDGMENTS

The authors would like to thank the Department of Homeland Security for providing financial support to this project.

Application of Gadolinium Complexes to Environmental Remediation of Phosphate

Valerie C. Pierre,[1*] Sarah M. Harris,[1] and Sylvie L. Pailloux[1]

[1]Department of Chemistry, University of Minnesota, USA
* pierre@umn.edu

INTRODUCTION

As a key ingredient in fertilizers, significant concentration of phosphates can accumulate in agricultural runoffs. The resulting surplus of nutrients causes eutrophication of surface and coastal waters, boosts algal growth, and creates dead zones that have significant economic and health impacts. On an industrial scale, purification of water can also be performed with porous membranes that can separate out a desired species. A number of such materials have recently been reported for the sequestration of phosphate from aqueous systems, but they suffer from poor selectivity and poor efficacy at neutral pH. Part of the problem hindering the development of efficient and recyclable material for phosphate sequestration arises from the general paucity of receptors that bind inorganic phosphates with sufficient affinity directly in complex aqueous media, with high selectivity over competing anions, and in a reversible manner such that both the device and the phosphates can be recycled. Herein, we describe the rationale behind the design of lanthanide-based supramolecular receptors that can catch inorganic phosphate with high affinity and directly from aqueous mixtures at neutral pH and release it under acidic conditions for efficient recycling of phosphate from polluted waterways in rural area.

RESULTS AND DISCUSSION

Gadolinium complexes are currently used extensively and primarily as contrast agents for Magnetic Resonance Imaging (MRI). Although their medical applications are different than the environmental one of focus in this paper, such complexes present many benefits for phosphate recycling. They are stable in water at neutral pH and, depending on the ligand, also under acidic conditions. They have one to two open coordination sites filled by water molecules that can be readily displaced by certain hard anions, most notably HPO_4^{2-}. Several gadolinium-based MRI contrast agents are already known to bind phosphate in water at neutral conditions. Significantly, and unlike in the zinc and copper-based fluorescent probes, the lanthanide ion does not leach out of these MRI contrast agents upon binding the oxyanion. Instead, phosphate binding results in the formation of a ternary $GdL(HPO_4)^-$ complex which remains water soluble. Moreover, gadolinium is a labile metal and as such phosphate binding and release occurs rapidly within minutes, a necessity to making our recycling scheme efficient. Lastly, these complexes are readily synthesized on a ton scale, as needed for any future translational applications.

Although complexes of gadolinium and other lanthanide ions with open coordination sites are well known to bind phosphates, they do not all do so with the same affinity and selectivity. Even minor differences in ligand can lead to substantial changes in anion recognition. Herein, we discuss how the structure of the ligand L, the geometry of the gadolinium complex GdL, its charge, the presence of a hydrogen-bonding network, and the geometry of such a network all affect anion recognition, binding, and selectivity.

CONCLUSIONS

Our results demonstrate that lanthanide-based supramolecular receptors can be designed for recyclable pH-dependent catch-and-release of phosphate from aqueous solutions. In particular, gadolinium complexes making use of tripodal aromatic ligands can be stable in water both at neutral and acidic conditions. Such complexes readily binds phosphate in water with high affinity at neutral pH and importantly, with high selectivity over other competing anions, most notably carbonate and nitrate. The formation of the ternary $GdL(H_nPO_4)^{3-n}$ complex is highly pH-dependent, with complete release of phosphate observed at pH 2. The high binding affinity of the receptor for phosphate, its high stability both at neutral and under acidic pH, and its ability to withstand multiple catch-and-release pH cycles highlight the potential of this class of compound in the development of material for sequestration of phosphate.

ACKNOWLEDGMENTS

The authors acknowledge the support of the National Science Foundation provided by INFEWS N/P/H2O:SusChEM CHE-1610832.

Applied Coordination Chemistry for the Separation of Rare Earth Elements Using Thermodynamic and Kinetic Controls

Eric J. Schelter[1]

[1]P. Roy and Diana T. Vagelos Laboratories, Department of Chemistry, University of Pennsylvania, 231 South 34th Street, Philadelphia, Pennsylvania 19104, United States, schelter@sas.upenn.edu

Separation of rare earth elements at scale is accomplished by counter current solvent extraction methods, which, despite their low efficiency for separation in a single pass, are highly engineered and optimized to achieve high purities in thousands of extraction and back-extraction steps. Despite the mature stage of solvent extraction technology, this method presents high initial and ongoing costs. We and others have considered targeted separations for pairs of rare earths, especially those used in permanent magnets, catalysts and phosphors, in the context of recycling rare earths from consumer materials. We have developed organic ligands whose metal complex structures enable separations through changes in physical properties, e.g. solubility properties. The structure property relationships of these compounds and their performance in separations under thermodynamic and kinetic control will be described in this talk.

Are Remeika Phases Simply Cubic?

Roman Gumeniuk[1,2*], Walter Schnelle[2], Lev Akselrud[2], Qiang Zheng[2], Ulrich Schwarz[2], Andreas Leithe-Jasper[2]

[1*]Institut für Experimentelle Physik, TU Bergakademie Freiberg, Leipziger Str. 23, 09596 Freiberg, Germany, roman.gumeniuk@physik.tu-freiberg.de

[2*]Max-Planck-Institut für Chemische Physik Fester Stoffe, Nöthnitzer Str. 40, 01187 Dresden, Germany

INTRODUCTION

The primitive cubic (space group *Pm*-3*n*; *a* = 9.676 Å) $Yb_3Rh_4Sn_{13}$ type of structure was discovered in the early 1980s by Remeika and co-workers [1]. Its structure belongs to the group of the so-called cage-compounds, where positively charged filler-atoms are situated inside large voids formed by a covalently bonded atomic framework. Such a combination of different types of chemical bonding together with vibrational dynamics of the filler results in numerous exciting physical phenomena (*e.g.* superconductivity, heavy fermion and Kondo behaviours, charge-density wave, thermoelectric properties *etc.*). All this render Remeika phases an object of intensive investigations. In the past few decades more than 100 ternary intermetallics containing fillers like alkaline-earth, Y, rare-earth (*RE*), some actinides and frameworks consisting of *d*-elements of subgroups 8-10 as well as *p*-elements (*X* = In, Si, Ge, Sn and Pb) have been shown to crystallize with this type of structure [2]. In this contribution we report on the observed structural distortions and physical properties of Ge- and B-containing Remeika-phases.

EXPERIMENTAL METHODS

$M_3Pt_4Ge_{13}$ (*M* = Y, U, Th, *RE*) and $AlPd_{15}B_7$ samples were prepared by arc melting of the elements with the following annealing at 800-1000 °C for 10 days. $RE_3Pt_4Ge_{13}$ Remeika phases were obtained by high pressure syntheses (8 GPa/2 h). High-resolution powder x-ray diffraction (HR XRD) measurements were performed at ESRF in Grenoble.

RESULTS AND DISCUSSION

HR XRD studies revealed numerous distorted structural variants of primitive cubic $Yb_3Rh_4Sn_{13}$ Remeika-type: cubic $Ca_3Pt_4Ge_{13}$ with doubled unit cell parameter, [3] rhombohedral high temperature (HT) $Y_3Pt_4Ge_{13}$, [4] tetragonal $Yb_3Pt_4Ge_{13}$ [3], monoclinic low temperature (LT) $Y_3Pt_4Ge_{13}$ [5] and even cubic anti-type $AlPd_{15}B_7$ (where larger Al atoms occupy the positions of smaller Rh in prototype) [6]. $RE_3Pt_4Ge_{13}$ are shown to crystallize with LT-$Y_3Pt_4Ge_{13}$ structure type, while $U_3Ir_4Ge_{13}$ [7] resembles HT-$Y_3Pt_4Ge_{13}$ structural arrangement. All structures mentioned above are non-centrosymmetric and are structurally related to Remeika-type within group-subgroup schemes. Such a close relationship implies the same structural motif for all studied compounds, which are built up of icosahedra condensed with four trigonal prisms (Fig. 1). Interestingly, the distortions leading to the symmetry reduction and to the non-centrosymmetric symmetry in the structures occur exclusively within the covalently bonded Ge-component framework. Therefore, the main changes are always observed within the icosahedral- and trigonal prismatic units.

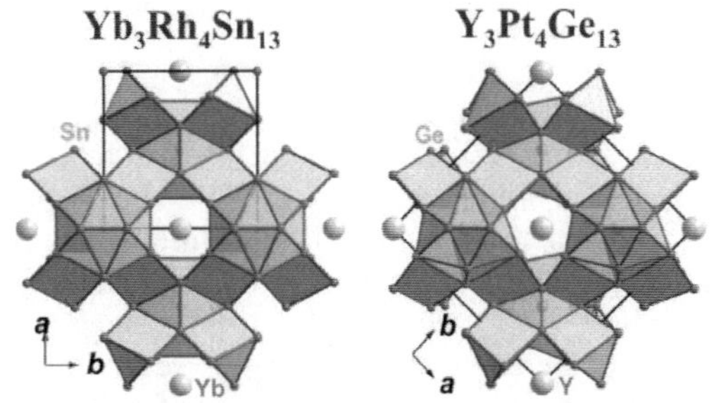

Fig. 1: Ideal and distorted units: icosahedra $Sn1Sn_{12}$ and $Ge1Ge_{12}$ (yellow) as well as trigonal prisms $RhSn_6$ and $PtGe_6$ (light gray) in the studied crystal structures.

DFT calculations reveal the appearance of a dip in the DOS at the Fermi level for complex structures, which indicates them to be more stable in comparison with ideal prototype-like arrangement. LT-$Y_3Pt_4Ge_{13}$ and $AlPd_{15}B_7$ are strongly coupled BCS superconductors with T_C = 4.5 and 2.9 K, respectively. $Sm_3Pt_4Ge_{13}$ is a valence fluctuating system, while $U_3Ir_4Ge_{13}$ is a ferromagnet with itinerant 5*f* states.

CONCLUSIONS

Most of the intermetallics, reported to crystallize with Remeika type of structure [2], are characterized by non-bonding *X*-*X* contacts within the icosahedra and unphysically large displacement parameters for *X*-atom centring the icosahedron. This observation together with the diversity of possible structural arrangements reported here support a considerably more complex structural chemistry for Remeika phases as initially anticipated.

REFERENCES

1. Remeika J. *et al.*, Solid St. Commun. 34:923-926, 1980
2. P. Villars, Pearson's Handbook, Crystallographic Data for Intermetallic Phases, Materials Park, OH 44073, 1997
3. Gumeniuk R. *et al.*, Dalton Trans. 41:6299-6309, 2012
4. Gumeniuk R. *et al.*, Dalton Trans. 44:5638-5651, 2015
5. Gumeniuk R. *et al.*, Phys. Rev. B 87: 224502, 2013
6. Zheng Q. *et al.*, Dalton Trans. 45:3943–3948, 2016
7. Gumeniuk R. *et al.*, Phys. Rev. B 91:094110, 2015

Breaking the Paradigm: Record Quindecim {15+} Charged Magnetic Ionic Liquids

Denis Prodius,[1] Volodymyr Smetana,[1] Simon Steinberg,[1] Magdalena Wilk-Kozubek,[1,2] Yaroslav Mudryk,[1] Vitalij K. Pecharsky,[1,2] and Anja-Verena Mudring[1,2]

[1]Ames Laboratory, US Department of Energy and Critical Materials Institute, Ames, Iowa 50011-3020, USA
denisprodius@ameslab.gov
[2]Department of Materials Sciences and Engineering, Iowa State University, Ames, Iowa 50011-1096, USA

INTRODUCTION

Ionic liquids (ILs) are salts that are liquid at low temperature (<100 °C) which represent a class of new 'green' materials with nonmolecular character and wide range of practical applications. A virtually limitless variety of cation/anion combinations ($\sim10^{19}$) gives us a choice to create an IL for a desired process, device or product. However, one of the design restraining criteria and paradigms in IL research is that a combination of highly charged ions will lead to a material being solid at room temperature (or below 100 °C) because of the strong Coulombic attraction. In our recent work,[1] we demonstrated that the paradigm of using only low charged ions for creating new ionic liquids should be revisited. Complex metal ions can be used as building blocks to create low melting salts. In addition, integrating metal ions into ionic liquids offers the potential to enrich an ionic liquid with additional chemical and physical properties. From this point of view, rare earth cations are well-suited to provide a range of ionic liquids with supplementary features, as rare earth compounds are well known for their outstanding luminescence, magnetism and catalytic activity. Here we report on a new family of rare earth-containing liquids, $\{RE_5\}$ (where RE = Er, Ho and Tm), featuring the novel pentanuclear rare earth cation with the highest charged ion (+15) for an ionic liquid achieved so far.

EXPERIMENTAL METHODS

A family of bis(trifluoromethanesulfonyl)amide-based ionic liquids of composition $[RE_5(C_2H_5\text{-}C_3H_3N_2\text{-}CH_2COO)_{16}(H_2O)_8](Tf_2N)_{15}$ (RE = Er (**1**), Ho (**2**), Tm (**3**); $C_3H_3N_2$ ≡ imidazolium moiety) featuring the cationic, record quindecim {15+} charged pentanuclear rare earth (RE)-containing cation $[RE_5(C_2H_5\text{-}C_3H_3N_2\text{-}CH_2COO)_{16}(H_2O)_8]^{15+}$ has been synthesized and fully characterized (SXRD, TG/DSC, IR, ESI-MS, magnetic and NIR photoluminescence properties, catalytic activity). The ILs overall show a low tendency to crystallize. Crystals for Er (**1**) and Ho (**2**) compounds formed after a few months and whilst crystallization took for the Tm (**3**) compound one year.

RESULTS AND DISCUSSION

To identify the structural units in these complex ionic liquids, single crystal X-ray structure analysis was applied. The Er- and Ho- compounds are isotypic and crystallize in the non-centrosymmetric orthorhombic space group $Pca2_1$ and feature the quindecim charged $[Er_5(C_2H_5\text{-}C_3H_3N_2\text{-}CH_2COO)_{16}(H_2O)_8]^{15+}$ unit which is composed of five crystallographically independent Er atoms. The magnetic properties of **1-3** were measured in the temperature range from 2 to 300 K and in an applied magnetic field of 1000 Oe. The room-temperature χT values of complexes **1**, **2**, and **3** are 56.2, 67.9, and 32.9 $cm^3 \cdot K \cdot mol^{-1}$, respectively. These results are in good agreement with the theoretical values expected for five noninteracting rare earth ions (**1**, 57.4 $cm^3 \cdot K \cdot mol^{-1}$; **2**, 70.35 $cm^3 \cdot K \cdot mol^{-1}$; **3**, 35.75 $cm^3 \cdot K \cdot mol^{-1}$) according to the single ion values at 300K. Using room temperature χT values, we calculated effective paramagnetic moments in the liquid state for **1-3**, which are equal to 21.2 (**1**), 23.3 (**2**), and 16.2 (**3**) μB. Thus these materials are not only featuring the highest charge ever observed for an ionic liquid, but also are magnetic ionic liquids with the highest magnetic moment observed ever. Other achievements of this work are the realization of rather challenging NIR luminescence in the liquid state of materials and highly efficient catalytic abilities.

CONCLUSIONS

In conclusion, we have demonstrated the syntheses and structural characterizations of the record quindecim charged pentanuclear rare earth based ionic liquids which do not only respond to a magnetic field, but also show NIR luminescence and extraordinary catalytic activity in the three-component synthesis of ethyl 2-methyl-4-(2-oxo-2,3-dihydro-1*H*-3-indolyl)-5-phenyl-1*H*-3-pyrrolecarboxylate (loadings level ~30 mmol%).

REFERENCES

1. Prodius D. *et al.*, Mater. Horiz. 2017, DOI: 10.1039/C6MH00468G.

ACKNOWLEDGMENTS

Financial support from the NSF (CHE-1465071) for the materials development and synthesis is gratefully acknowledged. Crystallographic studies were supported by the Critical Materials Institute, an Energy Innovation Hub funded by the U.S. Department of Energy, Office of Energy Efficiency and Renewable Energy, Advanced Manufacturing Office. Magnetic property measurements and crystallographic studies were supported by the Division of Materials Science and Engineering, Basic Energy Sciences Programs, Office of Science of the US Department of Energy under contract No. DE-AC02-07CH11358 with Iowa State University.

Can the π-Ligands Serve As an Efficient Antenna for the Luminescence Sensitization of Lanthanide Ions?

Lada Puntus[1,2*], Dmitrii Roitershtein[3], Evgeniya Varaksina[2] and Konstantin Lyssenko[2]

[1]Kotel'nikov Institute of Radio Engineering & Electronics of RAS, Moscow, Russia, ladapuntus@gmail.com
[2]A.N. Nesmeyanov Institute of Organoelement Compounds of RAS, Moscow, Russia
[3]A.V. Topchiev Institute of Petrochemical Synthesis of RAS, Moscow, Russia

INTRODUCTION

The catalytic olefin polymerization systems based on lanthanide complexes with cyclopentadienyl derivatives allow to obtain linear polyolefins with high degree of crystallinity. At the same time the cyclopentadienyl ligands have not been almost used to sensitize the luminescence of lanthanide ions. Therefore the investigation of the optical properties of such complexes allows for the first time not only to evaluate the possibility to design bright luminescent materials from the complexes but also to carry out in-depth analysis of the Stark structure of the luminescence spectra in order to study "in situ" the structure of the lanthanide complexes in catalytic polymerization reactions.

EXPERIMENTAL METHODS

Lanthanide complexes with polyaryl-substituted cyclopentadienyl ligands of the following types: $Cp'LnCl_2$, Cp'_2LnCl, Cp'_3Ln, where Ln = Tb, Dy, Gd, Lu, Nd and Cp' = $C_5Ph_2H_3$, $C_5Ph_3H_2$, C_5Ph_4H, C_5Ph_5 were synthesized by the original approach (Fig. 1). Photopysical properties were studied by optical spectroscopy including quantum yield and lifetime measurements. Combination of the experimental (single crystal X-ray diffraction study including topological analysis of the electron density distribution function) and theoretical (TD-DFT) methods was adopted in this research.

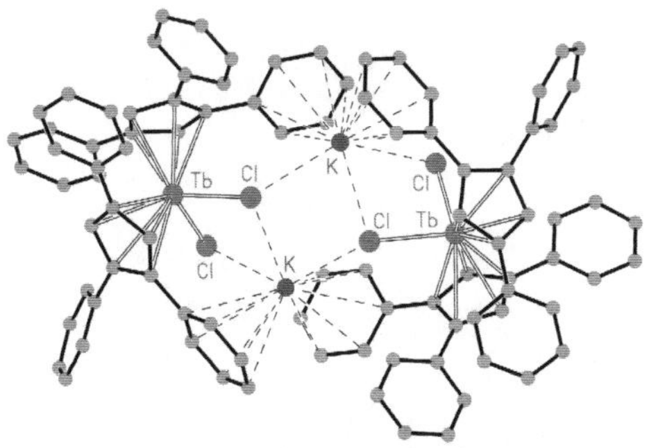

Figure 1. General view of $[Cp'_2TbCl_2]^+$ complex, where Cp' = $C_5Ph_3H_2$

RESULTS AND DISCUSSION

The introduction of phenyl into cyclopentadienyl ligand leads to expansion the excitation channel up to 450 nm. Namely one strong excitation band is observed at 270 nm for unsubstituted Cp while the maximum of the longwave band is 430 nm for $C_5Ph_3H_2$ ligand (Fig. 2). The lifetime of the Tb^{3+} ion in the later complex is 0.95 ms and overall quantum yield is ~5%. In other studied complexes the value of quantum yield of the Tb^{3+} ion reach up to 50%. To understand these data the contributions of the different charge transfer states including π-stacking induced charge transfer (SICT) state[1] or/and H-bond induced charge transfer (HBICT) state[2] as well as back energy transfer process in the luminescence sensitization of the Ln^{3+} ion were analyzed. The scheme of the energy transfer process in the systems considered was unraveled by experimental and theoretical bases.

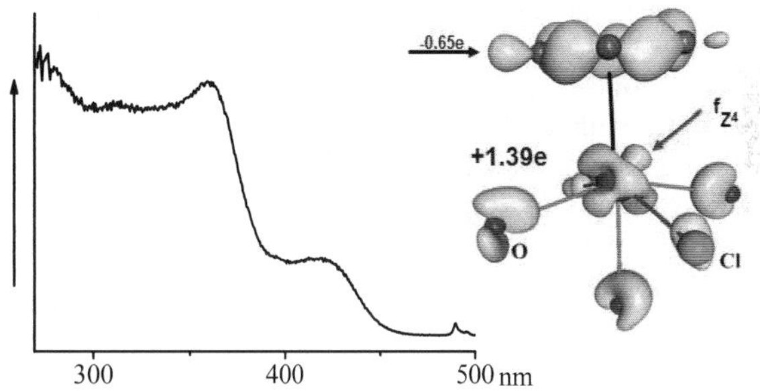

Figure 2. Excitation luminescence spectra of $[Cp'_2TbCl_2]^+$ complex (Cp' = $C_5Ph_3H_2$) (left) and experimental DED map in $CpLuCl_2(THF)_3$ (right)

CONCLUSIONS

In the present work we have investigated the interrelation between structure and optical properties for Ln complexes containing polyaryl-substituted cyclopentadienyl ligands. The total number of aryl substituents introduced into the cyclopentadienyl ring determines the self-organization of lanthanide cyclopentadienyl complexes leading to the formation of bi- or tetranuclear pseudoclusters. A broad excitation channel (up to 450 nm) makes these complexes a perspective candidates for different photonic applications.

REFERENCES

1. Puntus L.N. *et al.*, J. Phys. Chem. B. 113, 9265, 2009
2. Nelyubina Y.V. *et al.*, Chemistry - A Chemistry European Journal, 20, 2860, 2014
3. Puntus L.N. *et al.*, Inorg. Chem. 47, 11095, 2008

ACKNOWLEDGMENTS

The authors would like to thank the Russian Foundation for Basic Research (Grant no: 16-03-00832) and Russian Science Foundation (Grant no: 17-13-01357) for providing financial support to this project.

Catalytic Stibine Dehydrocoupling *versus* Antimony-ligated Dysprosium Single-Molecule Magnets

Fu-Sheng Guo, Benjamin M. Day, Thomas Pugh, Richard A. Layfield

School of Chemistry, The University of Manchester, Oxford Road, Manchester, M13 9PL, U.K.
Richard.Layfield@manchester.ac.uk

Single-molecule magnets (SMMs) are coordination compounds that display magnetic bistability and an energy barrier (U_{eff}) to reversal of the magnetization.[1] Many of the most impressive SMMs are based on lanthanides, particularly dysprosium, with several systems being used in spintronic devices.[2] The vast majority of SMMs contain classical ligands based on *N*- and *O*-donors; our group reported the first organometallic SMMs, and we have recently extended our work to develop the first SMMs with heavier pnictogens.

The SMM properties of a series of dysprosium ring systems with phosphorus-, arsenic- and antimony-donor ligands will be described, as will a magneto-structural correlation that allows the SMM properties to be enhanced in a rational way.[3-7] Whereas the chemistry with *P*- and *As*-donor ligands proceeds largely as anticipated, the analogous antimony chemistry produces two different outcomes, one in which the stibine pro-ligands are dehydrocoupled to give *cyclo*-poly(stibinidenes) and another in which antimony-ligated SMMs form (see Scheme).

We have found that the antimony-ligated SMM $[(Cp'_2M)_3(Sb_4Mes_3)]$ (M = Dy, or Dy/Y) is a by-product in the catalytic stibine dehydrocoupling reaction, and that the SMM itself also catalyses the dehydrocoupling

REFERENCES

1. D. N. Woodruff, R. E. P. Winpenny, R. A. Layfield *Chem. Rev.* **2013**, *113*, 5110.
2. S. Thiele, F. Balestro, R. Ballou, S. Klyatskaya, M. Ruben. W. Wernsdorfer, *Science* **2014**, *344*, 1135.
3. T. Pugh, F. Tuna, L. Ungur, D. Collison, E. J. L. McInnes, L. F. Chibotaru, R. A. Layfield, *Nat. Commun.* **2015**, *6*, 7492.
4. T. Pugh, V. Vieru, L. F. Chibotaru R. A. Layfield, *Chem. Sci.* **2016**, *7*, 2128. (b) T. Pugh, A. Kerridge, R. A. Layfield, *Angew. Chem. Int. Ed.* **2015**, *54*, 4255.
5. T. Pugh, N. F. Chilton, R. A. Layfield, *Chem. Sci.* **2017**, *8*, 2073.
6. T. Pugh, N. F. Chilton, R. A. Layfield, *Angew. Chem. Int. Ed.* **2016**, *55*, 11082.
7. F.-S. Guo, R. A. Layfield, *Chem. Commun.* **2017**, *53*, 3130.

ACKNOWLEDGEMENTS

RAL thanks the European Research Council for the Consolidator Grant 'RadMag' (646740), and the EPSRC for an Early Career Fellowship (EP/M022064/1). FSG thanks the European Commission for a Marie Skłowdowska-Curie International Fellowship (653784).

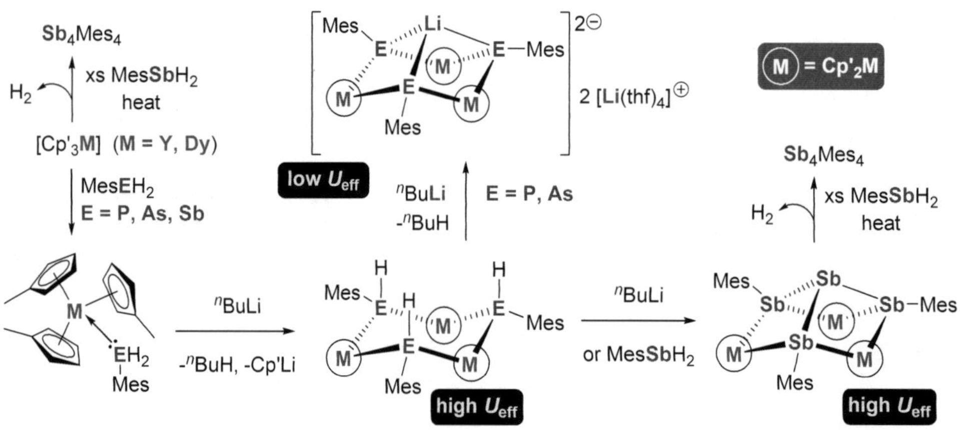

Cationic Uranium Alkyl Complexes of a Rigid NON-Donor Pincer Ligand: The Impact of Arene-Coordination on Ethylene Polymerization Activity

David J. H. Emslie,* N. R. Andreychuk, and B. Vidjayacoumar

Department of Chemistry, McMaster University, Canada, emslied@mcmaster.ca

INTRODUCTION

Previously, the Emslie group reported neutral, base-free thorium(IV) dialkyl complexes supported by the xanthene-based tridentate pincer ligand XA_2,[1,2] and subsequent reactions with $B(C_6F_5)_3$ and $[Ph_3C][B(C_6F_5)_4]$ afforded the first non-cyclopentadienyl thorium alkyl cations.[2] Each of these complexes features a facially-bound arene in the form of a neutral solvent molecule in $[(XA_2)Th(CH_2SiMe_3)(\eta^6\text{-arene})]$ (arene = benzene or toluene), or a benzylborate anion in the contact ion pair, $[(XA_2)Th(CH_2Ph)]$-$[(PhCH_2)B(C_6F_5)_3]$. This propensity to engage in arene coordination highlights the electrophilicity and steric accessibility of the actinide centre. However, benzene and toluene solutions of the aforementioned cations were inactive for ethylene polymerization (1 atm), suggesting that ethylene is unable to access the metal centre in the presence of these arenes.

RESULTS AND DISCUSSION

In attempt to circumvent arene coordination, the XA_2 ligand was installed on uranium(IV),[3] which has an ionic radius ~ 0.05 Å smaller than its thorium(IV) congener; we envisioned that shorter uranium–element bonds in a monoalkyl XA_2 uranium cation would afford a more hindered coordination environment, perhaps disfavoring undesirable arene coordination.

Reaction of $[(XA_2)U(CH_2SiMe_3)_2]$ (**1**) with 1 equiv of $[Ph_3C][B(C_6F_5)_4]$ in arene solvents afforded the arene-coordinated uranium alkyl cations, $[(XA_2)U$-$(CH_2SiMe_3)(\text{arene})][B(C_6F_5)_4]$ {arene = benzene (**2**), toluene (**3**), bromobenzene (**4**), fluorobenzene (**5**) and *o*-difluorobenzene (**6**)}. Compounds **2**, **3** and **5** were structurally characterized, and in all cases the arene is π-coordinated. Compounds **2-4** generated in C_6H_5R (R = H, Me or Ph) showed no polymerization activity under 1 atm of ethylene. By contrast, **5** in fluorobenzene (20 and 70 °C), **6** in *o*-difluorobenzene (20 °C), and **5-Th** (the thorium analogue of **5**) in fluorobenzene (20 and 70 °C) achieved ethylene polymerization activities between 11,200 and 57,600 $g \cdot mol^{-1} \cdot h^{-1} \cdot atm^{-1}$, highlighting the extent to which arene solvent coordination can influence ethylene polymerization activity in sterically open, highly electrophilic metal complexes.

REFERENCES

1. Cruz, C. A.; Emslie, D. J. H.; Harrington, L. E.; Britten, J. F.; Robertson, C. M. *Organometallics* **2007**, *26*, 692.

2. (a) Cruz, C. A.; Emslie, D. J. H.; Harrington, L. E.; Britten, J. F. *Organometallics* **2008**, *27*, 15. (b) Cruz, C. A.; Emslie, D. J. H.; Robertson, C. M.; Harrington, L. E.; Jenkins, H. A.; Britten, J. F. *Organometallics* **2009**, *28*, 1891.

3. (a) B. Vidjayacoumar; S. Ilango; M. J. Ray; T. Chu; K. B. Kolpin; N. R. Andreychuk; C. A. Cruz; D. J. H. Emslie; H. A. Jenkins; Britten, J. F. *Dalton Trans.* **2012**, *41*, 8175. (b) Andreychuk, N. R.; Ilango, S.; Vidjayacoumar, B.; Emslie, D. J. H.; Jenkins, H. A. *Organometallics* **2013**, *32*, 1466.

ACKNOWLEDGMENTS

D.J.H.E. thanks NSERC of Canada for a Discovery Grant. N.R.A. thanks the Government of Ontario for an Ontario Graduate Scholarship (OGS) and McMaster University for a Richard Fuller Memorial Scholarship.

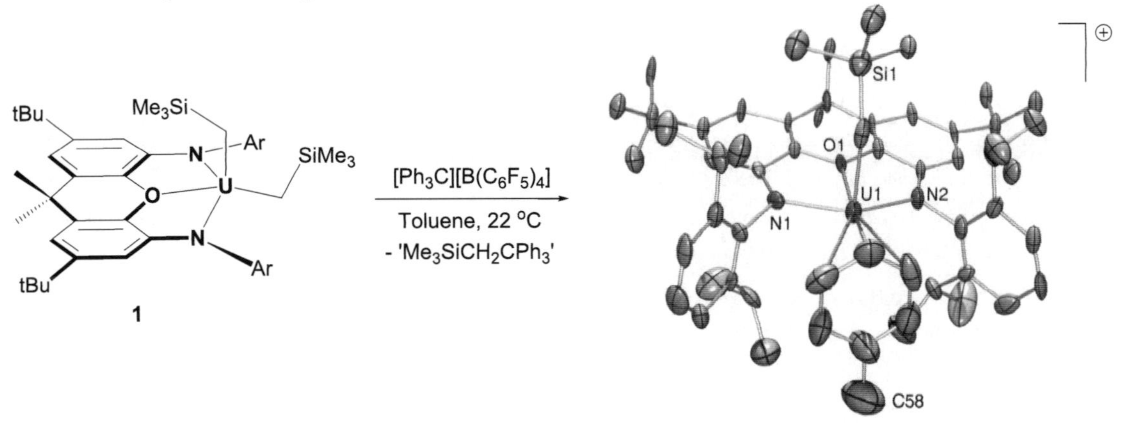

Cd substitution in $Eu_5In_2Sb_6$ enhancing the thermoelectric performance

Wanyu Lv [a], Kai Guo [a], Xinxin Yang [a], Jun Luo [a], Jing-Tai Zhao [a, b]

[a] *School of Materials Science and Engineering, Shanghai University, Shanghai 200444, China*

[b] *State Key Laboratory of Advanced Special Steel, Shanghai University, Shanghai 200444, China*

kai.guo@shu.edu.cn

Low thermal conductivity and tunable electrical properties of Zintl phases make them as promising candidates for thermoelectric application, and among them rare Earth based chalcognides constitute a important part [1-4]. $Eu_5In_2Sb_6$ reported [5,6] adopts $Ca_5Ga_2As_6$-type structure (space group Pbam), which satisfies the classical Zintl concept. Both experimental data and theoretical calculations evidence that this compound is a narrow-band-gap semiconductor. Cd-doped samples with compositions of $Eu_5In_{2-x}Cd_xSb_6$ (x = 0.02~0.1) were synthesized via a direct element combination reaction. With increasing Cd substitution, the $Eu_5In_{2-x}Cd_xSb_6$ system exhibits increased p-type carrier concentration and a resulting transition from non-degenerate to degenerate semiconducting behavior. That improvement of the electronic transport properties by Cd doping was embodied in the increased electrical conductivity with one order of magnitude. The thermoelectric figure of merit ZT of $Eu_5In_{2-x}Cd_xSb_6$ can be enhanced to 0.36 at 620 K via optimization of carrier concentration in comparison to pristine $Eu_5In_2Sb_6$ (0.05 at 440 K) .

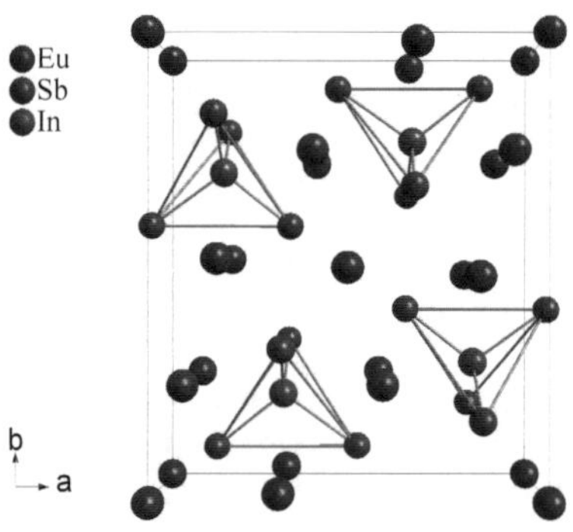

Fig.1. The crystal structure of $Eu_5In_2Sb_6$ with In (blue) and Sb (green) form covalently bonded $InSb_4$ tetrahedra

References

[1] Kauzlarich S M, Brown S R, Snyder G J. Zintl phases for thermoelectric devices[J]. *Dalton Transactions*, **2007** *(21)*: 2099-2107.

[2] Brown S R, Kauzlarich S M, Gascoin F, *et al.* $Yb_{14}MnSb_{11}$: New high efficiency thermoelectric material for power generation[J]. *Chemistry of Materials*, **2006**, 18(7): 1873-1877.

[3] Gascoin F, Ottensmann S, Stark D, *et al.* Zintl phases as thermoelectric materials: tuned transport properties of the compounds $Ca_xYb_{1-x}Zn_2Sb_2$[J]. *Advanced Functional Materials*, **2005**, *15(11)*: 1860-1864.

[4] Wang X J, Tang M B, Chen H H, *et al.* Synthesis and high thermoelectric efficiency of Zintl phase $YbCd_{2-x}Zn_xSb_2$[J]. *Applied Physics Letters*, **2009**, *94(9)*: 092106.

[5] Park S M, Eun Sang Choi, Kang W, *et al.* $Eu_5In_2Sb_6$, $Eu_5In_{2-x}Zn_xSb_6$: rare earth zintl phases with narrow band gaps[J]. *Journal of Materials Chemistry*, **2002**, *12(12)*:1839-1843.

[6] Chanakian S, Aydemir U, Zevalkink A, *et al.* High temperature thermoelectric properties of Zn-doped $Eu_5In_2Sb_6$[J]. *J.mater.chem.c*, **2015**, *3(40)*:10518-10524.

Ce^{IV}, Th^{IV}, and Pu^{IV}: Comparative tetravalent *f*-ion chemistry

L. Soderholm

Chemical Sciences & Engineering Division, Argonne National Laboratory, United States
ls@anl.gov

INTRODUCTION

From a simplistic perspective, the tetravalent f-ions (M^{IV}) are expected to exhibit similar chemistries. Within this guise Ce^{IV} and sometimes Th^{IV} are used as surrogates for the tetravalent transuranic ions. In addition to yielding misleading conclusions, this mindset has served to mask a chemical diversity that reflects not only subtle changes in ionic size but more complex electronic effects that impact structures, bonding, and physical properties. Some of these differences are highlighted by comparing and contrasting monomeric and small oligomeric nitrate compounds.

EXPERIMENTAL METHODS

Speciation of the tetravalent *f*-ions, isolated in single-crystals as simple molecular nitrates, are compared with difference pair-distribution functions (dPDF)[1] of the aqueous solutions from which the precipitates formed.

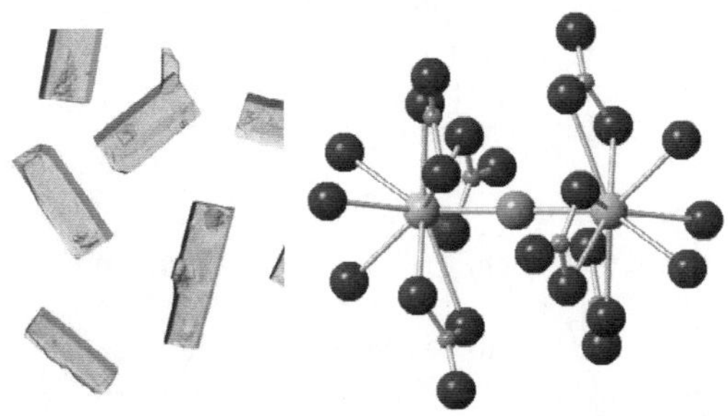

Fig. 1. Crystals of $Ce_2O(NO_3)_6(H_2O)_6$ were isolated from aqueous nitrate solutions. The structure revealed three capping nitrates on each of the oxo-bridged Ce^{IV} in the dimeric complex.[2]

RESULTS AND DISCUSSION

The homoleptic $M^{IV}(H_2O)_n$ complex has been isolated only for the Th analog.[3] The same is true for the simplest of the binary nitrates, $M^{IV}(NO_3)_4(H_2O)_n\cdot m(H_2O)$. Isostructural hydroxo-bridged dimers have been isolated for both Th and Pu.[4,5] In contrast, Ce^{IV} forms an unusual oxo-bridged dimer, which is also present in solution.[6] In all isolated structures the metal-ions are capped with bidentate nitrate ligands. The impact of bidentate-nitrate ligation is to limit the chemical, structural and electronic characteristics of the metal-ion complexes. This behaviour is different from that seen for M^{IV}-sulphates, where the bridging anion allows for the formation of some very novel structures.

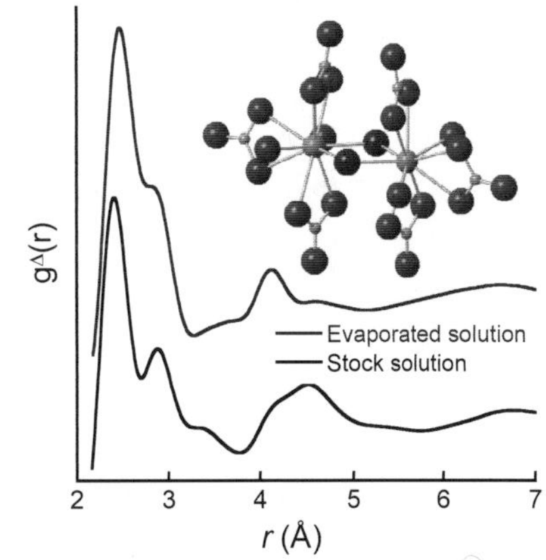

Fig. 2. Atomic correlations in a solid-state Pu hydroxo-bridged nitrate dimer are used to assign the metal-centric correlations seen in dPDFs of the 93mM Pu starting solution and in the solution after significant evaporation (304 mM). No evidence of the Pu dimer are observable in solution, even at high concentration.[5]

CONCLUSIONS

Elucidated comparisons between M^{IV}-nitrate complexes are producing structures inconsistent with expectations based on size effects alone. Whereas the nitrate ions are serving to isolate small clusters, their compositions and structures do not follow expected trends. These results are consistent with an influence of the M^{IV} electronic structure on compounds that form.

REFERENCES

(1)Skanthakumar, S.; Soderholm, L. *Mater. Res. Soc. Symp. Proc.* **2006**, *893*, 411.
(2) Estes, S. **unpublished.**
(3)Wilson, R. E.; Skanthakumar, S.; Burns, P. C.; Soderholm, L. *Angew. Chem.* **2007**, *46*, 8043.
(4)Wilson, R. E.; Skanthakumar, S.; Sigmon, G.; Burns, P. C.; Soderholm, L. *Inorg. Chem.* **2007**, *46*, 2368.
(5) Knope, K. E.; Soderholm, L. *Inorg. Chem.* **2013**, *52*, 6770.
(6) Demars, T. J.; Bera, M. K.; Seifert, S.; Antonio, M. R.; Ellis, R. J. *Angew. Chem.* **2015**, *54*, 7534.

ACKNOWLEDGMENTS

This work is supported by the U. S. Department of Energy, Office of Science, Office of Basic Energy Sciences, Division of Chemical Sciences, Biosciences and Geosciences, under contract No DE-AC02-06CH11357.

Cerium Redox Chemistry Pumps Nitrate Ion Transfer across Liquid-Liquid Interfaces in Solvent Extraction Systems

Mark R. Antonio

Chemical Sciences & Engineering Division, Argonne National Laboratory, United States
mantonio@anl.gov

INTRODUCTION

In the recovery and purification of rare-earth elements, solvent extraction (SX) is an effective and established process. There is more than 60 years of industrial practice that was begun by Oak Ridge National Laboratory researchers, who in 1953 reported the first kilogram-scale separation of gadolinium.[1] Yet, despite decades of science and engineering, there is still a fundamental need to understand ion transfer across water-oil phase boundaries to achieve high-performance extraction. Herein we present a general method based upon electroanalytical chemistry for exploring ion transfer processes in biphasic SX systems with cerium.

EXPERIMENTAL METHODS

As depicted in Fig. 1l, our approach couples a Faradaic response—the Ce(III/IV) redox couple—in an organic phase with the non-Faradaic transfer of a nitrate anion across an aqueous-organic interface and into a bulk phase.[2] The experimental system is shown in Fig. 1r.[2-4]

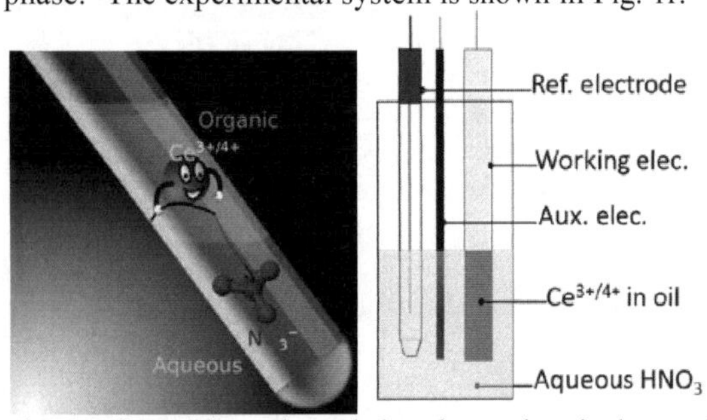

Fig. 1. (l) We investigated the electrochemical transfer of nitrate anions between aqueous and organic phases driven by the reduction and oxidation of Ce(IV/III) extractant complexes in organic phases. (r) A film of the Ce-bearing organic phase on a working electrode in an electrochemical cell filled with an aqueous HNO_3 electrolyte is used to mimic SX conditions.

RESULTS AND DISCUSSION

Redox state manipulations enable selective metal separations that underpin key systems, such as PUREX (Plutonium Uranium Reduction EXtraction) and the partitioning of cerium from the other rare earths. We have exploited the core phenomena of electrochemistry, which involve electron- and ion-transfer processes at phase boundaries. Of these, liquid-liquid interfaces are the crux of SX, in which cation transfer between phases must be a charge-neutral process. This occurs either by the simultaneous transfer of anions in same direction or, alternatively, cations in opposite direction. We have probed the chemistry of the Ce(IV)/Ce(III) redox couple in organic phases—containing the extractants shown in Fig. 2—that are in contact with aqueous nitric acid electrolytes. We demonstrate that the Ce redox activities in the organic phases are correlated with the cerium coordination environments and nitrate ion transfer between the aqueous and organic phases.

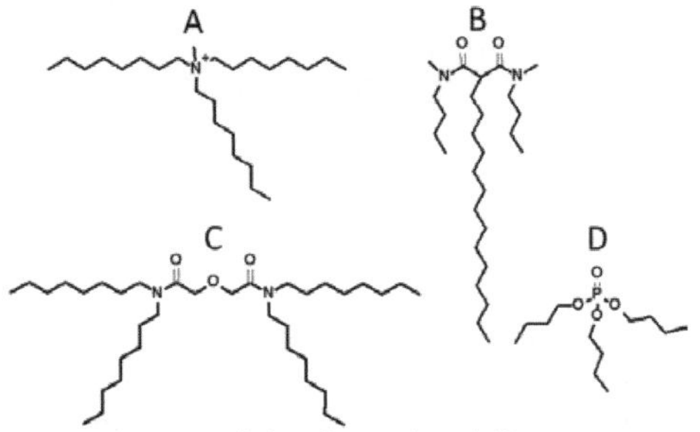

Fig. 2. Structures of the four oleophilic extractants for Ce extraction, including (A) tri-*n*-octyl methyl ammonium (TOMA) nitrate; (B) N,N'-dimethyl *N,N'*-dibutyl tetradecylmalonamide (DMDBTDMA); (C) *N,N,N',N'*-tetra(*n*-octyl)diglycolamide (TODGA); (D) tri-*n*-butyl phosphate (TBP).

CONCLUSIONS

Cerium redox chemistry has myriad applications in laboratory- and industrial-scale electrochemical reactions and catalytic processes.[5] Our use of the method of three-phase-electrode voltammetry (Fig. 1r) provides a means to decouple the energetics of cerium cation extraction and anion transfer in separation science applications.

REFERENCES

1. B. Weaver, F. A. Kappelmann, and A. C. Topp, *J. Am. Chem. Soc.* **75**, 3943, 1953
2. R. J. Ellis, M. K. Bera, B. Reinhart, and M. R. Antonio, *Phys. Chem. Chem. Phys.* **18**, 31254, 2016
3. R. J. Ellis and M. R. Antonio, *ChemPlusChem* **77**, 41, 2012
4. M. K. Bera, T. H. Bray, R. J. Ellis, and M. R. Antonio, *ChemElectroChem* **1**, 1173, 2014
5. L. F. Arenas, C. P. de León, and F. C. Walsh, *Electrochim. Acta* **205**, 226, 2016

ACKNOWLEDGMENTS

This work is supported by the U. S. Department of Energy, Office of Science, Office of Basic Energy Sciences, Division of Chemical Sciences, Biosciences and Geosciences, under contract No DE-AC02-06CH11357. The author warmly acknowledges the assistance of Mrinal Bera, Ross Ellis, and Shanna Estes.

LA-UR-17-21907

Chemical and Mineralogical Contents in New Mexico Coals and Coal By-products – Assessing the REE Resources

Giday WoldeGabriel[1], Hongwu Xu[1], Emily Kluk[1], Rose Harris[1], and Frank Tamakloe[1]

[1]Earth Environmental Sciences Division
Los Alamos National Laboratory
Los Alamos, NM 87545, USA, wgiday@lanl.gov

INTRODUCTION

Advancing the ongoing national renewable energy expansion and high-tech manufacturing in the US requires sustainable domestic supply of strategic mineral resources like rare earth elements (REE) and other high-value metals from unconventional resources like coals and associated sediments, coal by-products, and other sources. New Mexico is the 12th major coal producer in the US and currently, its coal deposits power three electric plants in the NW part of the State. These coals produce Class F type fly ash that is characterized by high-silica and low-calcium and low-Loss on Ignition (LOI). Significant amounts of fly ashes (44%) are shipped to Arizona for Portland cement production[1]. Preliminary mineralogical and chemical analyses indicate that some of the coals and fly ashes contain higher contents (>1000 ppm) of REE+Y. The coals and fly ashes represent untapped reliable and sustainable strategic mineral resources for domestic supply.

EXPERIMENTAL METHODS

Coals and fly ashes were collected from the three power plants and from deposits in the NW, NE, and central parts of the State. The samples were optically characterized and mineralogical and chemical compositions were determined using X-ray diffraction (XRD) and X-ray fluorescence (XRF) methods.

RESULTS AND DISCUSSION

Preliminary results from XRD and XRF analyses are given in Table 1. Generally, the coals and fly ashes from the three power plants indicate similar contents. The coals from the Raton coal field contain higher light REE concentrations. Synthetic glass and mullite and natural minerals such as smectite, illite, calcite, quartz, feldspars, and minor hematite are present in the coals and fly ashes. In most cases, the REE contents of the coal deposits in the San Juan and Raton Basins are higher than background concentration measured in soils and surficial materials[2]. The light REE concentrations of the sedimentary rocks interbedded with the Raton coal seams are also slightly higher than background concentrations but generally lower than the coal beds.

Leaching experiments on fly ashes using deionized water indicated variable amounts of sulphate, bicarbonate, calcium, sodium, and minor amounts of fluorine, potassium, and nitrate. The leaching solutions contained no heavy metals.

Table 1. Preliminary REEs and trace metals (ppm).

1. Four Corners	REE	Y	Er
a) Coal	190	25	
b) Fly ash	183	41	
2. San Juan			
a) Coal	221	26	
b) Fly ash	204	42	
3. Escalante			
a) Coal	184	26	
b) Fly ash	163	45	
4. a) Raton coals	1093	425	
b) Coals	318	90-353	90-207
c) Coal/sediments	238	53	<42
5. Cerrillos coals	138	86	22-62

CONCLUSIONS

Apart from being a major source of energy, the coal deposits in New Mexico as well as the fly ashes contain elevated light REE concentrations (e.g., La, Ce, Nd, etc.). The strategic REE and high-value metal contents in some of the coals and fly ashes represent major economic resources for New Mexico and the Nation. Moreover, the beneficial use of fly ashes in concrete and flue gas desulfurization gypsum for wallboard productions represent significant opportunities to advance sustainable raw material resources and management for manufacturing engineered building products.

REFERENCES

1. Hoffman, G.H., NM Geology, 22: 25-36, 2000.
2. Shacklette, H.T., and Boerngen, J.G., USGS Professional Paper 1270, 63 p., 1984.

ACKNOWLEDGMENTS

The study was supported by the New Mexico Small Business Assistance program at Los Alamos National Laboratory.

Chemical Bonding in Gallides of the Rare-Earth Metals: from Zintl Concept to Positional-Space Approach

Yuri Grin[1*], Anatolii Fedorchuk[2]

[1]Max-Planck-Institut für Chemische Physik fester Stoffe, Dresden, Germany, grin@cpfs.mpg.de
[2] Lviv National University of Veterinary Medicine and Biotechnologies, Lviv, Ukraine

The composition and crystal structures of numerous compounds of the alkaline and alkaline-earth metals with tetrels can be understood within the framework of the Zintl concept[1] applying the generalized Pearson rule[2]. E.g., the interpretation of the structural features of $CaSi_2$, CaSi and Ca_2Si by means of charge transfer from calcium to the silicon which leads to the formation of the polyanions with three-, two-bonded or isolated Si anions, can be found already in the handbooks[3,4]. The basis for such understanding is - among others - the sufficient number of valence electrons in the according system (cf., e.g., Ref. 5).

Substituting tetrels by triels reduces this number which has a consequence in a very different behaviour of the triels. They tend to form crystal structures which organisation and composition only exceptionally can be understood within the Zintl model[3]. Its extensions, like the Wade-Mingos rules[6,7], have to be involved for description of the tetrels' cluster polyanions[8].

The substitution of the alkaline and alkaline-earth metals by their rare-earth homologous causes the next complication connected with the electron counting. Therefore, systematic descriptions were developed mostly for the divalent rare-earth metals[8].

Further reduction of the number of valence electrons in the gallides of the rare-earth metals may lead to the multicentre interaction even with more than three centres per bond. In terms of the electronic band structure, the stabilization of a structural pattern is usually attributed to the Hume-Rothery interaction of the Fermi surface with the Brillouin zone[9,10]. Such model yields electron numbers which should be sufficient for the stabilization of structural patterns based on the packing of spheres (e.g. in trigallides of heavy rare-earth metals). However, the band-structure based consideration has difficulties in interpretation of the composition and atomic coordinations.

Therefore, further information on atomic interactions in rare-earth gallides can be obtained by quantum mechanical techniques. A suitable way is opened by application of bonding analysis in position space. The quantum theory of atoms in molecules (QTAIM[11]) allows for analysis of the electron density distribution in terms the basins representing the individual atoms in the system. The QTAIM charges reflect the ionic component of atomic interactions in intermetallic compounds as it depends on the composition and/or on the difference of the electronegativities of the components. The next step in the study of the spatial organization of a crystal structure can be realized applying quantum chemical tools describing the behaviour of electron pairs in position space. One of such tools - electron localizability indicator (ELI)[12] describes the correlation of electronic motion in real space. ELI reveals regions of space (basins) that can be used as descriptors of atomic shells, bonds, and lone pairs. Therefore, in analogy to the QTAIM approach, ELI basins can be determined and used as meaningful descriptors of chemical bonding[13]. Recently the energetic considerations for intermetallic compounds and their prediction were realized on the base of a combined ELI/QTAIM application[14].

Application of the position-space tools on the bonding analysis of the rare-earth-metals' gallides reveals a wide spectrum of the bonding patterns. Whereas the bonding in $EuGa_2$ with four-bonded gallium network can be interpreted as a variant of the Zintl phase, the further europium gallides can be understood applying only a developed from the ELI/QTAIM techniques Zintl-like electron counting which yields the lower effective charges for the differently bonded gallium species in comparison with the Zintl model. This kind of counting allows also to understand the compositions and structure of the gallides of ytterbium, especially in the gallium-rich part of the Yb-Ga system. An extension of such approach on the gallides of other rare-earth metals requires quantum chemical techniques which may account for the electrons of the inner shells.

REFERENCES

1. Zintl, E. *Naturwiss.* 17:782 1929.
2. Pearson, W. B. *Acta Crystallogr.* 17:1 1964.
3. Corbett, J. D. In: *Chemistry, Structure and Bonding of Zintl Phases and Ions*, VCH Publ. 139-181 1996.
4. Müller, U. *Inorganic Structural Chemistry*, Wiley& Sons, 128-149 2007.
5. Grin, Yu. In: *Comprehensive Inorganic Chemistry II*, Elsevier, 359-373 2013.
6. Wade, K. *Adv. Inorg. Radiochem.* 1:18 1976.
7. Mingos, D. M. P. *Acc. Chem. Res.* 17:311 1984.
8. Eisenmann, B., Cordier, G. In: *Chemistry, Structure and Bonding of Zintl Phases and Ions*, VCH Publ. 61-137 1996.
9. Hume-Rothery, W. *Materials Science and Engineering*; McGraw Hill, 3-23 1967.
10. Mizutani, U. *Hume-Rothery Rules for Complex Metallic Alloys*; T&F, 2010.
11. Bader, R. F. W. *Atoms in Molecules*; Oxford University Press, 1990.
12. Kohout, M. Int. J. Quantum Chem. 97:651 (2004).
13. Wagner, F. R., et al. Chem. Eur. J. 13:5724 (2007).
14. Bende, D. et al. Angew. Chem. 56:1313 (2017).

Chemical Speciation of Rare Earths in Wyoming Powder River Basin Coal Fly Ashes

De-Tong Jiang[1], Ning Chen[2], Zachary Arthur[2], Wendong Wang[3], Gaurab Rimal[3], Zaixing Huang[4], Maohong Fan[4], and Jinke Tang[3]
[1]Department of Physics, University of Guelph, 50 Stone Road East, Guelph, ON N1G 2W1, Canada
[2]Canadian Light Source, Inc., 44 Innovation Boulevard, Saskatoon, SK, S7N 2V3, Canada
[3]Department of Physics & Astronomy, University of Wyoming, Laramie WY 82071, USA
[4]Department of Chemical and Petroleum Engineering, University of Wyoming, Laramie WY 82071, USA

INTRODUCTION

Despite the widespread use of rare earth elements (REEs), production in Western countries remains very low due to the significant environmental impact associated with their extraction and processing. As a matter of fact, with the huge sacrifices in environmental quality and human health, China supplies more than 90-95% of the world's REE consumption over the past several decades [1]. On the other hand, coal by-products such as fly ash has emerged as a viable source for REEs [2]. Determination of the chemical speciation of rare earths in coal fly ash is of great importance to providing guidance for environmental friendly strategy and methods that will be employed in extraction and processing of rare earths. Such knowledge is largely unknown at the present due to the low concentrations of REEs and complexity in mineralogy in coal ashes although a couple reports have emerged within the past year [3-5].

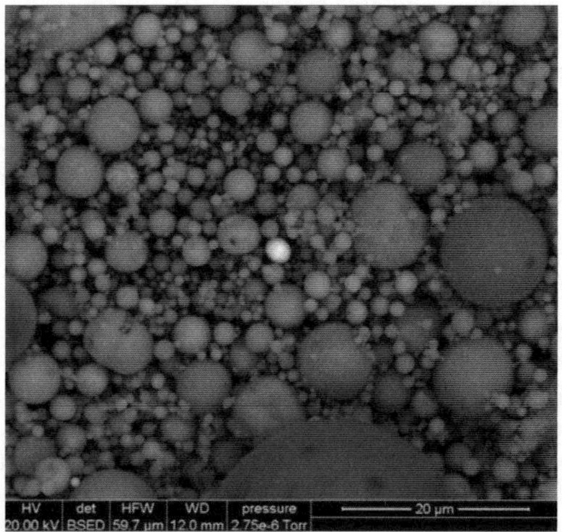

Figure 1. A spherical monazite granule (bright spot near the center) in coal fly ash.

EXPERIMENTAL METHODS

In this study, the chemical speciation of several representative rare earth elements (Y, Ce, Nd) in Wyoming Powder River Basin coal fly ashes have been carried out using synchrotron-radiation (SR) x-ray absorption spectroscopy at the Canadian Light Source. The morphology characterized by both scanning electron microscopy (SEM) and SR microprobe is combined with REE concentration studies using inductively coupled plasma mass spectroscopy (ICP-MS).

RESULTS AND DISCUSSION

The primary results obtained in this study include that Ce exists in these samples as a mixed Ce^{3+} in monazite phase and Ce^{4+} in CeO_2 phase; in contrast, the Nd and Y are both in 3+ oxidation state and possessing predominantly the monazite and xenotime mineral phases, respectively. The morphology of these REEs in two types of Wyoming Powder River Basin coal fly ashes possesses a REE concentrated granular feature with dimensions typically in micrometers (see Figure 1). They appear to largely survive the combustion process due to the high melting points of monazite and xenotime. The details of the investigation and results will be discussed.

CONCLUSIONS

Understanding the chemical speciation of rare earths in coal fly ash is of great importance to providing guidance for environmental friendly extraction and processing of rare earths. X-ray absorption near edge structure (XANES) has been used to study REE binding environment and electronic structure and to establish the mineralogy content of the REE in the coal ash. Combined with SEM, the REE phases and morphology in coal fly ash have been identified and revealed.

REFERENCES

[1] China's Continuing Monopoly over Rare Earth Minerals, J. Nesbit, U.S. News and World Report, April 2, (2013). http://www.usnews.com/news/blogs/at-the-edge/2013/04/02/chinas-continuing-monopoly-over-rare-earth-minerals

[2] Lucid Insight, (2015), Commercial Recovery of Metals from Coal Ash – Global Review, http://cornerstonemag.net/commercial-recovery-of-metals-from-coal-fly-ash/

[3] R.K. Taggart, J.C. Hower, G.S. Dwyer, and H. Hsu-Kim, (2016) Environ. Sci. Technol. **50**, 5919.

[4] C.L. Lopano, M. Stuckman, C.L. Thomas, J. Jain, E. Granite, (2016), Geological Society of America Abstracts with Programs. **47**(7), 391.

[5] S. Montross, C. Verba, C.L. Lopano, M. Stuckman, and K. Collins, (2016) https://www.microscopy.org/MandM/2016/program/abstracts/PDP-18.pdf

ACKNOWLEDGMENTS

We acknowledge the support from University of Wyoming School of Energy Resources via the Carbon Engineering Initiative.

Chemically and Structurally Flexible Hosts for Yb–Er Sensitizer–Activator Pairs

Federico A. Rabuffetti, K. Tauni Dissanayake, and B. Dulani Dhanapala

Department of Chemistry, Wayne State University, far@chem.wayne.edu

INTRODUCTION

Upconverting nanocrystals, which sequentially absorb multiple NIR photons and emit higher-energy NIR or visible photons, offer several potential advantages over molecular downconversion fluorophores traditionally employed in biophotonic applications such as imaging and sensing. However, their implementation as optical probes has been limited by their low quantum efficiency and lack of full-spectrum color tunability; these deficiencies stem from the occurrence of deleterious energy-transfer processes. Despite the key role played by the host lattice in mediating energy-transfer processes relevant to light upconversion, the ability to rationally manipulate its chemical composition and crystal structure to direct the flow of energy remains limited. Our group's research aims bridging this fundamental knowledge gap by using upconverting nanocrystals that couple rare-earth activators (Yb-Er, Yb-Tm) to a family of host materials (MFX; M: Ca, Sr, Ba; X: Cl, Br, I) whose crystal structure can be rationally manipulated through chemical composition.[1,2]

RESULTS AND DISCUSSION

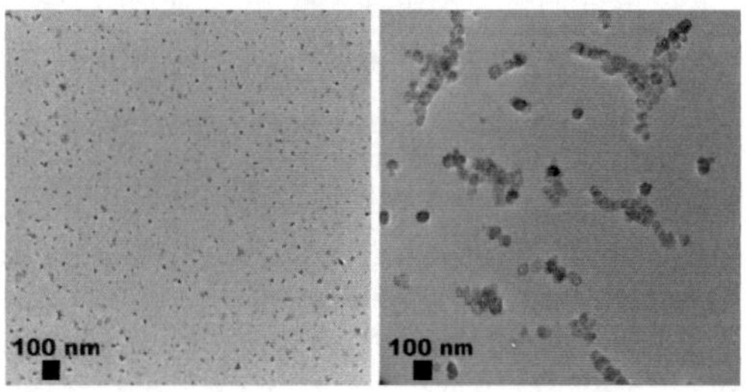

Figure 1. Electron microscopy images of fluorohalide upconverting nanocrystals of two different chemical compositions.

In this work, we present the synthesis, structural characterization, and infrared-to-visible light upconversion properties of semiconducting MFX (X: Cl, Br, I) nanocrystals codoped with Er and Yb (Figure 1). According to Rietveld and pair distribution function analyses, the average and local crystal structure conform to that expected on the basis of the *P4/nmm* tetragonal space-group. Excitation of Yb at 980 nm results in two-photon upconversion covering the 520-700 nm spectral range. The excited states of the Er activator in the MFX hosts exhibit a biexponential decay with lifetimes in the 0.010-0.2 and 0.3-0.4 ms ranges. Both, the chromaticity of the upconverted visible light and the excited-state dynamics can be tuned via chemical composition. Our findings prompt for an expansion of the library of upconverting nanocrystals into novel materials containing heavy halide anions (i.e., Cl^-, Br^-, and I^-).

CONCLUSIONS

In summary, the preparation of compositionally flexible rare-earth-doped alkaline-earth fluorohalide nanocrystals was accomplished via colloidal synthesis. Yb-Er sensitizer-activator pairs show color-tunable NIR-to-visible upconversion luminescence when incorporated in alkaline-earth fluorohalide hosts.

REFERENCES

1. Dissanayake, K.T. and Rabuffetti, F.A. *J. Mater. Chem. C* **2016**, 4, 2447.
2. Dissanayake, K.T. and Rabuffetti, F.A. *Submitted* **2017**.

ACKNOWLEDGMENTS

The authors would like to acknowledge the financial support of the National Science Foundation (Grant No. DMR-1606917) and of the Department of Chemistry at Wayne State University. Use of the Advanced Photon Source at Argonne National Laboratory was supported by the U.S. Department of Energy, Office of Science, Office of Basic Energy Sciences, under Contract No. DE-AC02-06CH11357.

Important:
File size is limited to 3Megabyte
Reduce the size of images, graphs etc. if your file size exceeds this limit.

The abstract must fit within a single page

Chemistry at the edge: actinide interactions with (poly)aminopolycarboxylates

Jenifer Braley[1], Matt Urban[1], Nathan Bessen[1], Morgan Luckey[1], Morgan Kelley[2], Ping Yang[2]

[1]Deparement of Chemistry, Nuclear Science & Engineering Program, Colorado School of Mines, U.S.A,
[2*]Los Alamos National Laboratory, Theoretical Division, U.S.A., jbraley@mines.edu

INTRODUCTION

Recent work has suggested that covalent interactions may be increasingly relevant for actinides with increasing atomic number starting around berkelium.[1-3] The reason for this, or the role increases in covalent interactions might have in metal-ligand complexes has not been fully understood. This work considers the complexation of actinides with dipicolinic acid and other aminopolycarboxylates to understand changes in complexation strength that might arise from increases in covalent metal-ligand interactions.

EXPERIMENTAL METHODS

Stability constants are determined by radiotracer, solvent extraction competition measurements under well specified solution conditions. The metal extraction constants (K_{Ex}) of bis-2-ethyl hexyl phosphoric acid (HDEHP) or 2-ethylhexylphosphonic acid mono-2-ethylhexyl ester (HEH[EHP]) are assessed in the absence of an aqueous complexing reagent. The partitioning of actinides between aqueous and organic phases in the presence of the extractant and appropriate complexing reagent are assessed. The stability constants are minimized by using mass balance relationships and non-linear least squares fitting. Enthalpic and entropic contributions giving rise to a given stability constant are assessed using van't Hoff relationships. Gamma emissions are used to quantify Am-241 and Cf-249 partitioning. Alpha emissions are used to assess Cm-244 and Es-253 partitioning. Beta emissions are used to assess Bk-249 partitioning.

RESULTS AND DISCUSSION

Preliminary data suggests the 1:3 metal:ligand stability constants of americium and curium complexation with dipicolinic acid is comparable and berkelium and californium complexation is comparable, Table 1.

Figure 1. Stability constants for $M(DPA)_3^{3-}$ species in 1 M $HClO_4$ media.

Am^{3+}	Cm^{3+}	$Bk^{3+,1}$	Cf^{3+}
22.65 ± 0.1	22.67 ± 0.5	23.14 ± 0.01	23.29 ± 0.03

The difference in stability constants observed between Cm and Bk is consistent with increased covalency in the Bk and californium complexes. Data with einsteinium and other aminopolycarboxylates is pending.

CONCLUSIONS

The increase in stability constants observed between curium and berkelium actinide:dipicolinates suggests that increased covalency may be relevant for the heavier actinides in this system starting at berkelium. Additional systems should be considered and computational modelling efforts should be used to understand the origins of covalency in these actinide-ligand interactions.

REFERENCES

1. Silver, Mark A., et al. "Characterization of berkelium (III) dipicolinate and borate compounds in solution and the solid state." Science 353.6302 (2016): aaf3762.
2. Cary, Samantha K., et al. "Emergence of californium as the second transitional element in the actinide series." Nature Comm. 6 (2015).
3. Polinski, Matthew J., et al. "Unusual structure, bonding and properties in a californium borate." Nature chemistry 6.5 (2014): 387-392.

ACKNOWLEDGMENTS

This work was supported by the U.S. Department of Energy, Office of Science, Office of Basic Energy Sciences, Heavy Elements Chemistry Program at Colorado School of Mines (under Award Number DE-SC0012039) and at Los Alamos National Laboratory. Los Alamos National Laboratory is operated by Los Alamos National Security, LLC, for the National Nuclear Security Administration of U.S. Department of Energy (contract DE-AC52-06NA25396). The Bk-249 and Cf-249 used in this and prior research by this PI was supplied by the United States Department of Energy Office of Science by the Isotope Program in the Office of Nuclear Physics. Electronic structure calculations were completed using the Molecular Science Computing Facilities in the William R. Wiley Environmental Molecular Sciences Laboratory, a national scientific user facility sponsored by the U.S. DOE BER and located at Pacific Northwest National Laboratory.

Chiroptical and magneto-optical properties of chiral nona-nuclear Tb(III) clusters

Satoshi Wada[1], Yuichi Kitagawa[2], Takayuki Nakanishi[2], Koji Fushimi[2], Yasuhiro Morisaki[3], Koji Fujita[4], Katsuaki Konishi[5], Katsuhisa Tanaka[4], Yoshiki Chujo[4], Yasuchika Hasegawa[2*]

[1]Graduate School of Chemical Sciences and Engineering, Hokkaido University, Japan
[2*]Faculty of Engineering, Hokkaido University, Japan, hasegaway@eng.hokudai.ac.jp
[3]School of Science and Technology, Kwansei Gakuin University, Japan
[4]Graduate School of Engineering, Kyoto University, Japan
[5]Faculty of Environmental Earth Science and Graduate School of Environmental Science, Hokkaido University, Japan

INTRODUCTION

The Faraday effect causes a rotation of polarization by a material as the result of the application of an external magnetic field.[1] We recently described nona-nuclear Tb(III) clusters coordinated with salicylate ligands that were a new type of lanthanide (Ln(III)) material exhibiting a large Faraday effect.[2] We also found that the Faraday rotation angles were different depending on the organic ligands of these clusters. To maximize the unique characteristics of the inorganic-organic hybrid clusters, we focused on introducing chiral ligands to the Tb(III) cluster. Ln(III) complexes containing chiral ligands show circular dichroism (CD) and circularly polarized luminescence (CPL), which are derived from the chirality. The origin is different from that of the Faraday effect.[1,3] The dissymmetry factors (g_{CD} or g_{CPL}) of chiral Ln(III) complexes are known to be large, which is attributed to the large transition magnetic dipole moment of the Ln(III) ion.[4] In this work, the effect of large transition magnetic dipole moments on the Faraday rotation was investigated for the chiral Tb(III) clusters.

EXPERIMENTAL METHODS

Chiral Tb(III) clusters $[Tb_9(\text{sal-}(R)\text{-Bt})_{16}(\mu\text{-OH})_{10}]^+[NO_3]^-$ (Tb-(*R*)-Bt: sal-(*R*)-Bt = (*R*)-2-butyl salicylate) and $[Tb_9(\text{sal-}(S)\text{-Bt})_{16}(\mu\text{-OH})_{10}]^+[NO_3]^-$ (Tb-(*S*)-Bt: sal-(*S*)-Bt = (*S*)-2-butyl salicylate) were synthesized by the complexation of $Tb(NO_3)\cdot 6H_2O$ with chiral organic ligands. The chiroptical properties of these clusters were characterized using CD and CPL measurements. We also performed Faraday effect measurements for these chiral Tb(III) clusters in poly(methyl methacrylate) (PMMA) films under external magnetic field (H = 15 kOe).

RESULTS AND DISCUSSION

The Tb(III) clusters Tb-(*R*)-Bt and Tb-(*S*)-Bt were composed of 9 Tb(III) ions bridged by 10 μ-OHs and 16 chiral salicylate ligands (Fig. 1). The clusters exhibited bisignate CD bands because of π–π* transitions of the salicylate ligands. CPL measurements were performed for these clusters, and the resulting signals were observed corresponding to the transitions of the Tb(III) ions. The largest g_{CPL} value was observed at the $^5D_4 \rightarrow {}^7F_5$ transitions (g_{CPL} = 0.04). The large g_{CPL} consequently leads to a large g_{CD},[4] and this is expected to give large transition magnetic dipole moments around the Faraday-active wavelength based on the absorption of the Tb(III) clusters. Finally, the Faraday effect was measured using PMMA films containing Tb-(*R*)-Bt or Tb-(*S*)-Bt. Negative Faraday rotation angles are observed at around 450 nm. It was found that the rotation angle of the Tb-(*R*)-Bt was larger than that of the Tb-(*S*)-Bt. This result represents the first observation of a Faraday effect dependent on the chirality of the clusters, which might be attributable to the large transition magnetic dipole moments of the Tb(III) clusters.[5]

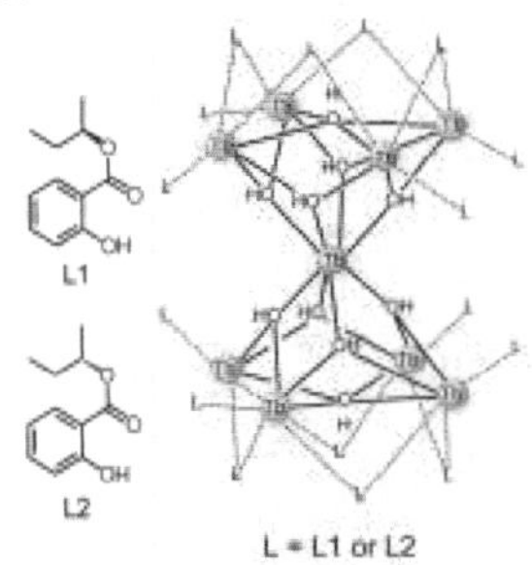

Fig. 1 Chiral Tb(III) clusters

CONCLUSIONS

Chiroptical and magneto-optical properties of novel chiral nona-nuclear Tb(III) clusters were assessed. Faraday effect measurements indicated that the Faraday rotation angles were dependent on the chirality of the ligands in the Tb(III) clusters.

REFERENCES

1. Schatz P. N. *et al.*, Q. Rev. Chem. Soc. 23:552-584, 1969
2. Nakanishi T. *et al.*, Inorg. Chem. 53:7635-7641, 2014
3. Harada N. *et al.*, Circular Dichroic Spectroscopy: Exciton Coupling in Organic Stereochemistry, Tokyo Kagaku Dojin, Tokyo, 1982
4. Richardson F. S., Inorg. Chem. 48:2806-2812, 1980
5. Wada S. *et al.*, NPG Asia Mater. 8:e251, 2016

ACKNOWLEDGMENTS

This work was partly supported by (i) Grants-in-Aid for Scientific Research on Innovative Areas of 'New Polymeric Materials Based on Element-Blocks (no. 2401)' (no. 24102012) of the Ministry of Education, Culture, Sports, Science and Technology (MEXT), Japan, (ii) the Frontier Chemistry Center Akira Suzuki 'Laboratories for Future Creation' Project for their support, and (iii) The Ministry of Education, Culture, Sports Science and Technology through Program for Leading Graduate Schools (Hokkaido University 'Ambitious Leader's Program').

Competitive Solvation Effects Within Multicomponent Solutions of Trivalent Ions

Morgan Kelley,[1] Ping Yang[2] Aurora E. Clark[1*]

[1*]Department of Chemistry and the Materials Science and Engineering Program, Washington State University, USA, auclark@wsu.edu
[2]Theory Division, Los Alamos National Laboratory

INTRODUCTION

Binary water/methanol mixtures are used in a wide variety of industrial and analytical applications, including separations, methanol based fuel cells, and electrospray mass spectrometry. Simple alcohols have been found as cosolvents in wastes at Department of Energy sites, and cosolvation is known to significantly alter the behavior of environmental contaminant metals. Water/methanol mixtures have therefore seen extensive study, with the bulk of the work involving ion solvation limited to singly charged ions. Significant work remains in understanding the effect of water/methanol solvation on more highly charged systems. Herein, we focus upon trivalent Ln and An ions due to the potential impact of cosolvation in nuclear wastes and reprocessing schemes.

EXPERIMENTAL METHODS

Structures of each ion at each water/methanol 1st shell composition – from $M(H_2O)_9(CH_3OH)_0^{3+}$ to $M(H_2O)_0(CH_3OH)_9^{3+}$ for M = La, Gd, Cm and from $M(H_2O)_8(CH_3OH)_0^{3+}$ to $M(H_2O)_0(CH_3OH)_8^{3+}$ for M = Lu – were optimized via the hybrid B3LYP functional. Single point MP2 calculations determined the solvent dissociation energy of water and methanol molecules from the 1st solvation shell of each ion. In addition to the solvent binding energy calculations discussed above, MP2 calculations were used in the development of classical inter-molecular potentials between the metal ion and solvent. Classical molecular dynamics and Carr-Parinello dynamics were performed in pure water and methanol solutions. Portions of trajectories were analyzed via intermolecular network theory (INT) using the ChemNetworks software package.

RESULTS AND DISCUSSION

In all water/methanol compositions Cm^{3+}, La^{3+}, and Gd^{3+} have average coordination numbers (CN) of nine, while Lu^{3+} has an average CN of eight. Preferential solvation by water over methanol exists for all ions. The 1st solvation shell for the CN=9 ions maintains the expected tri-capped trigonal prismatic geometry. Preferential solvation by water exists in the 1st solvation shell and also extends out into the 2nd solvation shell of each ion. Analysis of the exchange mechanisms of water and methanol entering the 1st solvation shell reveals that all three exchange mechanisms are observed, though the intermediate exchange events are the most common for water with both associative and dissociative mechanisms occurring in less than 12% of observed exchange events. Unlike water, methanol is nearly always added to the 1st solvation shell of Cm^{3+} via an associative mechanism.

MP2 dissociation energy calculations reported previously for water and methanol with the Cm^{3+} ion and its 1st solvation shell predict that methanol has a consistently higher dissociation energy than water irrespective of 1st shell water/methanol composition. Furthermore, the dissociation energy of both solvents decreases as more methanol molecules are added to the 1st solvation shell. This suggests that for the trivalent ions, the ion-solvent dynamics should be faster as the methanol concentration in solution increases. However, the observed ion-solvent lifetimes do not follow this prediction; coordination states with greater methanol content generally have longer lifetimes and as methanol is added to the bulk solution both the ion-water and ion-methanol lifetimes increase. This increase in ion-water and ion-methanol lifetimes as the concentration of methanol in solution increases has been observed previously in singly charged ions[28] (Na^+ and Cl^-) and is correlated with the increase in the lifetimes of the hydrogen bonds despite the opposing effect from the solvent dissociation energy.

CONCLUSIONS

Electronic structure calculations demonstrate that as more methanol is introduced into the 1st shell, solvent molecules become easier to remove. This suggests that coordination states with greater methanol content should have shorter durations, which is the opposite of what is observed in the simulations. The effect of the solvent dissociation energy is overridden by changes in the hydrogen bond network of the solution as the amount of methanol in bulk solution increases, causing the network to become more static.

ACKNOWLEDGMENTS

M. Kelley acknowledges support from the Los Alamos National Laboratory G.T. Seaborg Institute and the DOE Office of Science Graduate Student Research Fellowship. P. Yang was supported by the U.S. DOE, CSGB, Heavy Element Chemistry Program at Los Alamos National Laboratory under contract No. DE-AC52-06NA25396. Sue Clark acknowledges support from the US Department of Energy, Office of Science, Heavy Elements program, DE-SC-000-4102. Aurora Clark acknowledges support from the US Department of Energy, Office of Science, Heavy Elements program, DE-SC-000-1815.

Complex Intermetallics Grown from Rare Earth / Transition Metal Eutectic Fluxes

Susan Latturner[1]

[1]Department of Chemistry, Florida State University, USA
latturner@chem.fsu.edu

INTRODUCTION

Intermetallics containing both rare earth and transition metals often exhibit complex structures and magnetic behaviour that is dependent the combination of localized f-electron moments and the itinerant magnetism of the transition metals. Our research group has explored the use of eutectic melts of early rare earths combined with late first row transition metals as fluxes for the synthesis of new complex R/T/M/M' compounds.[1] In our most recent work, we have found two closely related competing Pr/Fe/Si/C products forming from the same reaction. Size effects were explored by replacing Si with other main group elements, and thermal effects may also play a role in determining which product forms.

EXPERIMENTAL METHODS

Pr/Ni flux was made by arc melting a 78:22 mole ratio mixture of Pr and Ni. Pieces of this alloy were used in (Pr/Ni)/Fe/M/C reactions carried out in alumina crucibles. Iron, carbon, and M = Si, P, Ge, or Bi powders in 1-2 mmol quantities were sandwiched between 1.5 g of the Pr/Ni flux. The crucibles were sealed in quartz ampoules under vacuum. A variety of heating profiles were explored to optimize crystal growth and yield, with all reactions being centrifuged at 625°C to remove excess flux. Product crystals were characterized using SEM-EDS, single crystal and powder X-ray diffraction, magnetometry, and density of states calculations.

RESULTS AND DISCUSSION

$Pr_{62}Fe_{22}Si_{15}C_{32}$ (with a new tetragonal structure shown in Figure 1) and $Pr_{21}Fe_8Si_7C_{12}$ (an analog of the cubic $La_{21}Fe_8Sn_7C_{12}$ structure type)[2] grow from reactions of iron, carbon, and silicon in Pr/Ni eutectic (m.p. 500°C). Both structures feature iron sites coordinated by carbon to form trigonal planar FeC_3 units. These units share corners, forming a zeolite-like framework in the tetragonal structure, and Fe_4C_6 clusters in the cubic structure. The silicon sites are surrounded by Pr cations, forming either Si@Pr_9 monocapped square antiprisms, or larger Si@Pr_{12} cuboctohedra. These polyhedra share faces to form a Pr/Si network that fills the Fe/C network of the tetragonal phase, or surrounds the Fe_4C_6 clusters of the cubic phase.

An analog of the tetragonal phase is formed when silicon is replaced by phosphorus. Conversely, only cubic $Pr_{21}Fe_8M_7C_{12}$ is formed if larger main group elements (M = Ge, Bi) are used in the synthesis. This size effect may stem from the fact that the cubic phase contains a higher ratio of large M@Pr_{12} to small M@Pr_9 building blocks compared to the tetragonal structure.

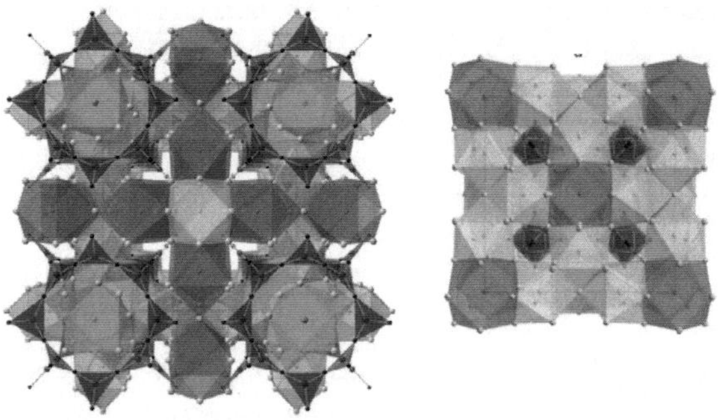

Figure 1. *Left:* tetragonal $Pr_{62}Fe_{22}Si_{15}C_{32}$ viewed down the *c*-axis. *Right*: cubic $Pr_{21}Fe_8Si_7C_{12}$. FeC_3 trigonal planar units are shown in red; silicon-centered Pr polyhedra are shown in blue and green. Pr and carbon atoms are yellow and black spheres, respectively.

CONCLUSIONS

The competition between the two products $Pr_{62}Fe_{22}Si_{15}C_{32}$ and $Pr_{21}Fe_8Si_7C_{12}$ is likely affected by the size of the silicon atoms; replacing the silicon with a larger or smaller atom results in formation of only one of the two structures. Thermodynamic considerations may also affect the outcome; longer heating times promote formation of the cubic phase, indicating it might be more thermodynamically stable relative to the tetragonal phase.

REFERENCES

1. Zhou, S.; Latturner, S.E. *J. Solid State Chem.* **2016**, *236*, 159-165.
2. Benbow, E.; Dalal, N.; Latturner, S.E. *J. Am. Chem. Soc.* **2009**, *131*, 3349-3354.

ACKNOWLEDGMENTS

This research is supported by the National Science Foundation (Division of Materials Research, grant no. DMR-14-10214).

Computational study of Pacman complexes of Th(IV) and U(IV) in comparison to their U(VI) analogues

Gurpreet-Kour[1] and Georg Schreckenbach[1]
[1]Department of Chemistry, University of Manitoba, Canada,
kourg@myumanitoba.ca

INTRODUCTION

Studies of complexes of actinides with aromatic **π**-electron systems are important in understanding the physical and chemical properties of actinides as they give insight into the covalent contributions to the metal-ligand bonding. On the basis of known uranyl complexes with the binucleating Schiff-base polypyrrolic macrocyclic ligand[1-2] (H_4L) (Fig.1a) that resembles a Pacman structure, Density Functional Theory (DFT) has been applied to study Th(IV) complexes of L as well as U(IV) analogues. Mono and binuclear complexes of Th(IV) and U(IV) with the Pacman ligand have been studied computationally.

COMPUTATIONAL DETAILS

All calculations were done using Priroda code. We performed the calculations using approximate DFT in the form of the PBE exchange-correlation functional, which is a generalized gradient approximation (GGA) version of DFT. Scalar relativistic effects were also incorporated.

RESULTS AND DISCUSSION

On the basis of previously studied uranyl structures with the Pac-Man cofacial diporphyrins ligand[1], we studied their Th(IV) and U(IV) analogues and found various interesting structures. All these complexes exhibit cation-cation interactions (CCI) and show unusual coordination of thorium similar to those in the known uranyl U(VI) structures.

In case of mononuclear complexes, Th(IV) $[(ThO_2)(H_2L)(THF)]^{-2}$ and U(IV) $[(UO_2)(H_2L)(THF)]^{-2}$ both have very similar structures to those of the corresponding U(VI) complex but the Th-O bond length is significantly larger than the U-O bond, and O_{endo} is abstracting two hydrogens from the two nitrogen donors of the lower ligand cavity. Moreover the O-Th-O bond angle is quite bent as compared to the O-U-O bond angle, which is almost linear. In case of the mononuclear U(IV) complex $[(UO_2)(H_2L)(THF)]^{-2}$ the spin density on uranium is found to be 1, and the rest is ditributed on the ligand; this clearly corresponds to an interanl redox reaction resulting in a U(V) complex.

Among binuclear complexes of thorium with the Pacman ligand, in addition to the known Butterfly structure B, we explored two new geometries termed as Diamond D and Zigzag structures (Fig.1b). These complexes have been named based on the shape of actinide-oxygen Skelton.

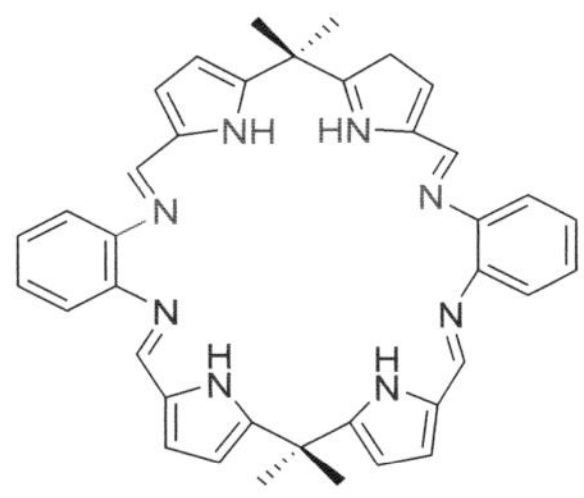

H4L

(a)

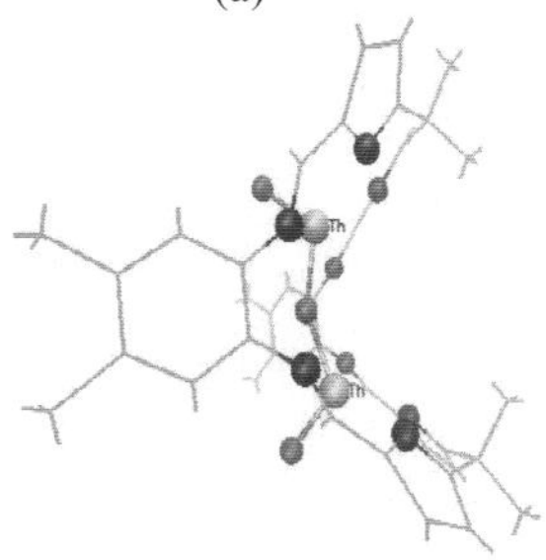

(b)

Figure 1(a): Structure of polypyrrolic macrocyclic ligand; (b): Zigzag structure of the binuclear complex of Th(IV)[pink Th, Blue B, Red O]

CONCLUSION

On the basis of the DFT calculations, we were able to analyse the geometrical parameters of the various structures of thorium and uranium in +4 oxidation state and also we compared these structures to the known U(VI) complexes. We are further looking for the detailed bonding analysis of these complexes using QTAIM approach.

REFERENCES

1. Pan Q. J., Schreckenbach G., *et al.*, Chem. Eur., 16:2282-2290, 2010
2. Berard J.J *et al.*, Chem. Eur. 47:11583-11592, 2008

ACKNOWLEDGMENTS

The authors would like to thank Xiaobin Zhang and Jeffery Perkins for useful discussions. We are grateful to WestGrid for using its computing resources. Financial support from the Natural Sciences and Engineering Research Council of Canada (NSERC) is also gratefully acknowledged.

Concentration and Temperature Dependent Upconversion Luminescence of $CaWO_4:Er^{3+},Yb^{3+}$ 3D Microstructure Materials

Jing Liu[1], Anna M. Kaczmarek[1] and Rik Van Deun[1*]

[1]Department of Inorganic and Physical Chemistry, Ghent University, Belgium, Jing.Liu@ugent.be

INTRODUCTION

Near-infrared (NIR) to visible (VIS) photon upconversion is a process where the absorption of (at least) two NIR photons excited at a certain wavelength results in luminescence at a shorter wavelength in the VIS range. Many reported upconversion materials have good chemical stability, strong and stable luminescence properties, a large stoke shift, long luminescence lifetimes, high penetration depth and so on[1,2], which accounts for their popularity in materials chemistry research. The influence of the precursor concentration on the materials' morphology has been much less investigated and was therefore one of the aims of the undertaken work. The most suitable heat-treatment temperature for tungstate and molybdate materials has been discussed quite extensively in literature, usually though only to compare the intensity of luminescence spectra, not to investigate the changes of the ratio of different emission bands. In addition, in order to change the emission colour (and thus the CIE coordinates), the most common practice is to change the ratio of the doped rare-earth ions in order to influence the intensity of the corresponding emission bands.

In this paper, we focus on two factors that can influence the upconversion emission and morphology of $CaWO_4:Er^{3+},Yb^{3+}$ materials: the precursor concentration and the heat-treatment temperature.

RESULTS AND DISCUSSION

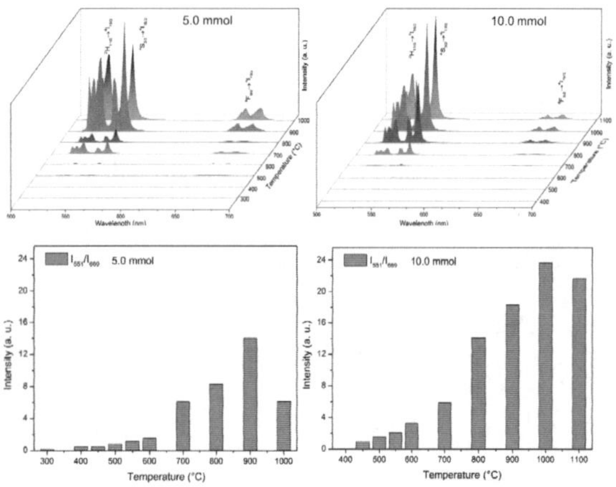

Top left, right: Upconversion luminescence spectra of sample **IV**, **V** heat treated at temperatures from 300 °C to 1000 °C, recorded under excitation at 975 nm. Bottom left: histogram of sample **IV**, **V** displaying the ratio of the luminescence intensities of the peaks at 551 nm and 669 nm at various temperatures.

Table 1. Overview of prepared materials (the total volume for all samples was 40 mL)

Label	Concentration (mmol/mL)	$CaCl_2$ (mmol)	Na_2WO_4 (mmol)	$ErCl_3·6H_2O$ (mmol)	$YbCl_3·6H_2O$ (mmol)
I (0.5 mmol)	0.0125	0.4	0.5	0.01	0.09
II (1.0 mmol)	0.025	0.8	1.0	0.02	0.18
III (2.5 mmol)	0.0625	2	2.5	0.05	0.45
IV (5.0 mmol)	0.125	4	5	0.1	0.9
V (10.0 mmol)	0.250	8	10	0.2	1.8

CONCLUSIONS

In conclusion, the possibility of employing different concentrations of the reaction precursors in order to yield different morphologies of $CaWO_4:Er^{3+},Yb^{3+}$ materials has been investigated. After heat treatment of the samples which were synthesized at different concentrations of the precursors, the samples showed similar luminescence behavior, the difference being that for the 5.0 mmol (0.125 mmol/mL) $CaWO_4:Er^{3+},Yb^{3+}$ sample the highest luminescence intensity is obtained when the sample is heat treated at 900 °C. For this sample at 550 °C a significant change in the I_{551} (green) to I_{669} (red) intensity ratio is observed. The 669 nm peak starts to be dominant at this heat-treatment temperature. For the 10.0 mmol (0.250 mmol/mL) $CaWO_4:Er^{3+},Yb^{3+}$ sample, the highest intensity is recorded after heat treatment at 1000 °C and the 551 nm peak becomes relatively most intense compared to the 669 nm peak after heat treatment at 500 °C.

REFERENCES

1. Liu Y. *et al.*, Chem. Soc. Rev. 42:6924-6958, 2013.
2. Chen G. *et al.*, Chem. Soc. Rev. 114:5161-5214, 2014.

ACKNOWLEDGMENTS

This work was funded by the China Scholarship Council. AMK acknowledges Ghent University's Special Research Fund (BOF) for a Postdoctoral Mandate (project BOF15/PDO/091). RVD thanks the Hercules Foundation (project AUGE/09/024 "Advanced Luminescence Setup") for funding.

Construction of Energy Diagrams for Ce^{3+} Coordinated with Six O^{2-} Ions Based on First-Principles Calculations

Kazuyoshi Ogasawara[1]

[1]Department of Chemistry, Kwansei Gakuin University, Japan, ogasawara@kwansei.ac.jp

INTRODUCTION

In order to improve the color rendering properties of white LEDs, (Sr,Ca)$AlSiN_3$:Eu phosphor is usually used as a red phosphor. However, both the luminescent ion (Eu) and the host crystal (nitride) are expensive. Therefore, cheaper alternative of (Sr,Ca)$AlSiN_3$:Eu phosphor is strongly desired. One candidate is Mn^{4+} phosphor. However, the shortest emission wavelength reported for Mn^{4+}-doped oxide phosphor is ca. 650 nm,[1] which is still longer than 630 nm required for the red phosphor for white LEDs. Therefore, establishment of a design guideline for Mn^{4+}-doped oxide phosphor with desired emission wavelength is an important issue. In order to provide such information, we have recently constructed the energy level diagrams and maps for Mn^{4+} in oxides based on first-principles calculations.[2] The other candidate for the red phosphor for white LEDs is Ce^{3+} phosphor. In this case, the emission wavelength is always shorter than 630 nm. Therefore, a design guideline to increase the wavelength is required. Since the emissions at ca. 600 nm was recently reported for Ce^{3+}-doped oxide where Ce^{3+} occupies a six coordination site[3], the relationship between the local structure and the energy levels of CeO_6^{9-} cluster has been investigated in this work. The energy level diagram and maps for CeO_6^{9-} clusters with O_h and D_{4h} symmetries were constructed by performing first-principles calculations. Especially the relationship between the local structure and the $4f$-$5d$ transition energy was investigated in detail.

COMPUTATIONAL METHODS

The first-principles relativistic molecular orbital calculations were performed using the relativistic SCAT code[4] for CeO_6^{9-} clusters with O_h and D_{4h} symmetries. In the case of O_h symmetry, the calculations were performed for the clusters with gradually changed bond lengths and the energy level diagram depending on the bond length was constructed by connecting the obtained orbital energies. In the case of D_{4h} symmetry, two types of bond lengths d_{xy} and d_z were considered and the calculations were performed for clusters with various combinations of d_{xy} and d_z. Then the contour maps representing the variation of energy levels depending on d_{xy} and d_z were constructed.

RESULTS AND DISCUSSION

The energy level diagram for CeO_6^{9-} cluster with O_h symmetry is shown in Fig. 1 In this diagram, the lowest $4f$ level is set at zero so as to represent the variation of the $4f$-$5d$ transition energy. The $5d$ level splits into t_{2g} and e_g due to the crystal field and the splitting increases as the bond length decreases. The crystal field splitting of $4f$ level also increases as the bond length decreases. As a result, the $4f$ level and $5d$ level begin to overlap at ca. 1.9 Å.

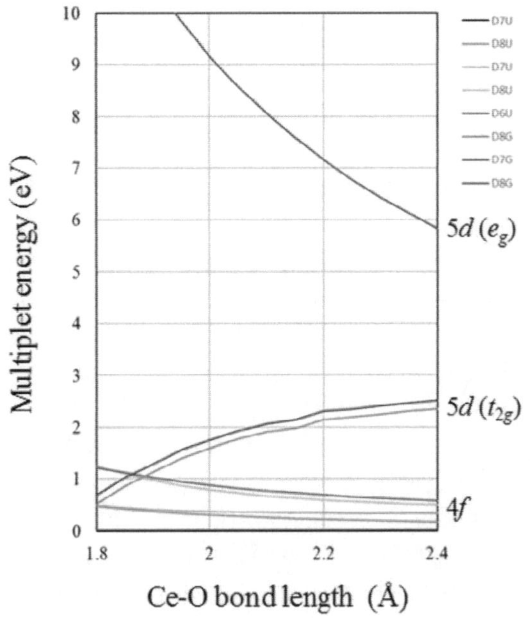

Fig. 1. The energy diagram for CeO_6^{9-} cluster with O_h symmetry. The lowest $4f$ level is set at zero.

CONCLUSIONS

The energy diagram for CeO_6^{9-} cluster with O_h symmetry and the energy maps for CeO_6^{9-} cluster with D_{4h} symmetry were constructed based on first-principles calculations. The relationship between the $4f$-$5d$ transition energy and the local structure was clarified. Such information would be useful for the theoretical search of the novel Ce^{3+}-doped oxide red phosphor for white LEDs.

REFERENCES

1. M.H. Du, *J. Mater. Chem.*, *C***2**, 2475 (2014).
2. K. Ogasawara *et al.*, *ECS J. Solid State Sci. Technol.*, **5**, R3191 (2016).
3. Y. Kawano *et al.*, *Opt. Mater. Express*, **4**, 1770 (2014).
4. A. Rosen *et al.*, *J. Chem. Phys.*, **65**, 3629 (1976).

Controlling lanthanide luminescence in core-shell nanoparticles

Feng Wang

Department of Physics and Materials Science, City University of Hong Kong, 83 Tat Chee Avenue, Hong Kong SAR, China and City University of Hong Kong Shenzhen Research Institute, Shenzhen 518057, China

Abstract: Spectral conversion of light radiation is essential for a variety of modern-day technologies including displays, photovoltaics, data storage, and cancer therapy. In the past decades, a variety of luminescent materials, including organic dyes and semiconductor quantum dots, were developed to realize photon energy conversion in the ultraviolet (UV) to near infrared (NIR) spectral range. However, practical application of these techniques is typically limited by constraints such as toxicity and/or poor stability. In this regard, lanthanide-doped upconversion nanoparticles have been developed as a promising alternative to conventional luminescent nanomaterials. By comparison, these nanoparticles offer low cytotoxicity, sharp emission peaks, long-lived excited electronic states, as well as high stability against photo bleaching and chemical degradation. In this talk, I focus on our recent efforts on lanthanide-doped core-shell nanoparticles that are generally composed of a layer-by-layer structure with a set of lanthanide ions incorporated into separate layer. Examples will be given to demonstrate how to tune the luminescence process and enhance multiphoton upconversion emission for exciting new technological applications.

References

[1] X. Chen, D. Peng, Q. Ju, F. Wang, Photon upconversion in core-shell nanoparticles. Chem. Soc. Rev. **2015**, 44, 1318.

[2] Wen, H.; Zhu, H.; Chen, X.; Hung, T. F.; Wang, B.; Zhu, G.; Yu, S. F.; Wang, F. Upconverting near-infrared light through energy management in core–shell–shell nanoparticles, *Angew. Chem. Int. Ed.* **2013**, *52*, 13419.

[3] X. Chen, L. Jin, W. Kong, T. Sun, W. Zhang, X. Liu, J. Fan, S. F. Yu, Feng Wang, Confining energy migration in upconversion nanoparticles towards deep ultraviolet lasing, Nature Communications **2016**, 7, 10304, doi: 10.1038/ncomms10304.

[4] Chen, B.; Liu, Y.; Xiao, Y.; Chen, X.; Li, Y.; Li, M.; Qiao, X.; Fan, X.; Wang, F. Amplifying Excitation-Power Sensitivity of Photon Upconversion in a $NaYbF_4$:Ho Nanostructure for Direct Visualization of Electromagnetic Hotspots, *J. Phys. Chem. Lett.* **2016**, *7*, 4916.

Coordination complexes of d^0 vs. d^{10} metals: what can we learn about bonding?

David S. Tatum[1,2] and Kenneth N. Raymond[2]

[1]Chemical Sciences Division, Lawrence Berkeley National Laboratories, Berkeley, California 94720, United States
[2]Department of Chemistry, University of California, Berkeley, California 94720, United States
dtatum@berkeley.edu

INTRODUCTION

The oxophilicity of trivalent rare earth ions (REs) is well understood by hard-soft acid-base (HSAB) theory. The bonding interactions between hard, positively charged REs and hard, negatively charged oxygen coordinating ligands is mostly ionic and non-directional. However, we have recently been interested in the covalency of RE-oxygen bonds, since these covalent interactions may help mediate energy transfer in photoluminescent lanthanide complexes. Although the valence *4f* orbitals of lanthanides do not participate much in bonding due to poor spatial overlap with ligand donor orbitals, the Ln(III) ions have empty *d* orbitals available for bonding, just like Y(III) and Sc(III). The covalent interactions between filled ligand orbitals and empty RE *d* orbitals are expected to be small due to a large energy mismatch, but we find evidence that they are chemically significant nonetheless.

RESULTS AND DISCUSSION

We have synthesized and characterized 23 eight-coordinate ML_2 complexes using the novel tetradentate ligand *m*-xylylene-bis-1,2-hydroxypyridinone (*m*Xy-1,2-HOPO, Fig. 1), including all 16 of the non-radioactive REs.

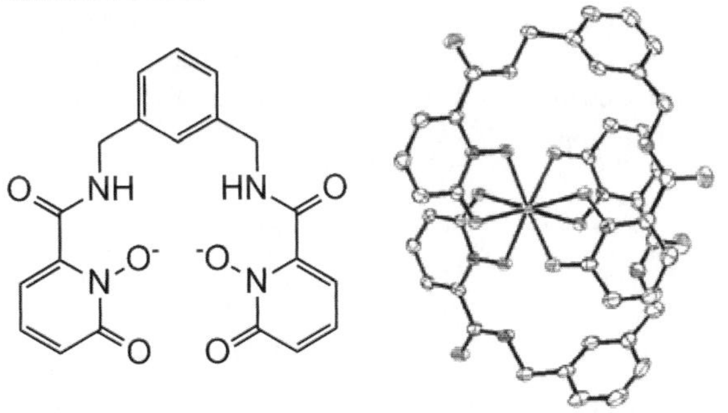

Figure 1. *m*Xy-1,2-HOPO ligand (left) and $[Lu(m\text{Xy-1,2-HOPO})_2]^-$ XRD structure (right)

We compare the stereochemistry of the group 3 and 4 d^0 metal complexes to the stereochemistry of the group 13 and 14 d^{10} metal complexes using structures derived from single crystal XRD and DFT geometry optimizations. We find that the d^0 metal complexes tend toward an *mmmm*-edge dodecahedral (*mmmm*-Dod) geometry, while the d^{10} metal complexes tend toward a *ssss*-edge square antiprismatic (*ssss*-SA) geometry. These changes in geometry are also apparent from differences in the unit cell constants. In solution, the complexes undergo an intramolecular racemization process that is the 8-coordinate analog of the 4-coordinate tetrahedral inversion and the 6-coordinate Bailar twist. The transition state geometry of this 8-coordinate twist inversion process has also been optimized by DFT (Fig. 2).

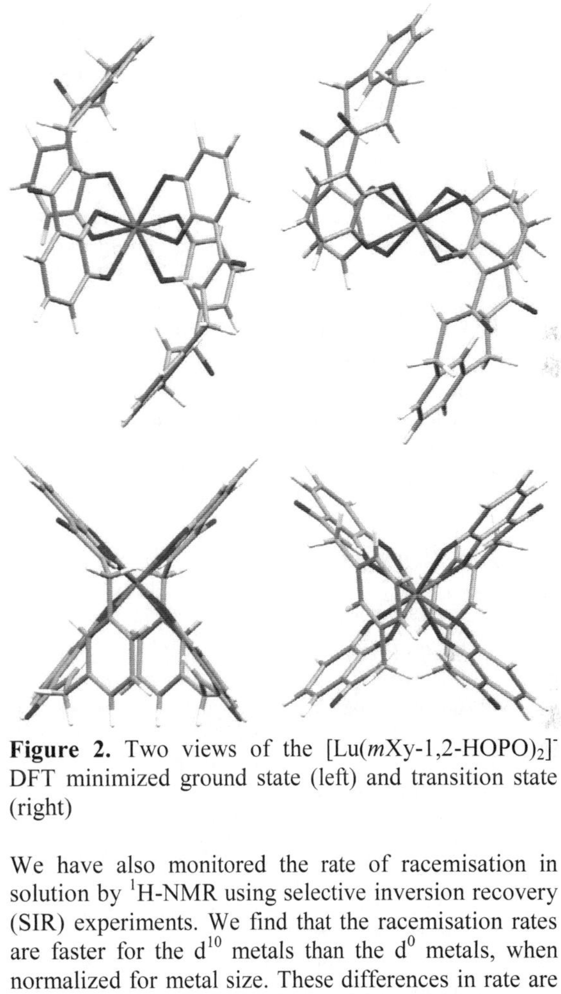

Figure 2. Two views of the $[Lu(m\text{Xy-1,2-HOPO})_2]^-$ DFT minimized ground state (left) and transition state (right)

We have also monitored the rate of racemisation in solution by ^{1}H-NMR using selective inversion recovery (SIR) experiments. We find that the racemisation rates are faster for the d^{10} metals than the d^0 metals, when normalized for metal size. These differences in rate are rationalized with the aid of natural bond orbital (NBO) calculations, which show a larger change in partial charge at the metal center for the d^0 metals. If there were no covalent interactions between the RE ions and the HOPO ligands, then we would expect the d^0 and d^{10} metal complexes properties to be similar. That they are significantly different leads us to conclude that the covalent portion of the RE-HOPO ligand bond is chemically significant and important.

ACKNOWLEDGMENTS

The authors would like to thank the U.S. Department of Energy (Contract No. DE-AC02-05CH11231) for providing financial support to this project.

Coordination of Macrocyclic Ligands to the Uranyl Ion

Trevor W. Hayton

Department of Chemistry and Biochemistry, University of California Santa Barbara, Santa Barbara CA 93106, USA, hayton@chem.ucsb.edu

INTRODUCTION

The uranyl ion, UO_2^{2+}, is one of the most common fragments in actinide chemistry. It features an exclusively *trans* arrangement of its two oxo ligands. Indeed, there are no known examples of an authentic *cis*-uranyl complex in the literature.[1] Nonetheless, there is considerable interest in the synthesis of an authentic *cis*-uranyl complex, as this fragment should provide unique insights into actinide covalency and f-orbital participation in bonding. Previously, Clark and co-workers attempted to enforce *cis*-oxo stereochemistry by ligating a tripodal ligand to the uranyl framework.[2] However, this resulted in ligand decomposition rather than the intended *trans*/*cis* isomerization. Drawing inspiration from the results of Clark and co-workers, we have begun to explore the ligation of 14-membered macrocyclic ligands to the uranyl fragment. We hypothesized that these macrocycles would be unable to accommodate the *trans*-UO_2^{2+} fragment within their small ring cavities, thereby forcing the desired *trans* to *cis* isomerization.

RESULTS AND DISCUSSION

This presentation will describe our attempts to ligate the 14-membered macrocyclic ligands, cyclamMe$_2$ (1,8-dimethyl-1,4,8,11-tetraazacyclotetradecane) and tmtaaH$_2$ (dibenzotetramethylaza[14]annulene), to the uranyl fragment, along with the characterization of the resulting complexes. Specifically, reaction of $[UO_2Cl_2(THF)_2]_2$ with Li$_2$(cyclamMe$_2$) results in formation of a complex with an undesired 2:1 ligand-to-metal ratio, e.g., $[Li]_2[UO_2(cyclamMe_2)_2]$ (**1**), regardless of the stoichiometry using in the reaction. Complex **1** was characterized by X-ray crystallography and found to exhibit an O-U-O angle of 180°. Additionally, each cyclamMe$_2$ ligand coordinates to uranyl with only two of its nitrogen atoms, giving the uranium centre an overall octahedral geometry.

CONCLUSIONS

The isolation of **1** highlights several major challenges in isolating a *cis*-uranyl complex, namely that ligand flexibility, along with unanticipated ligand binding modes, can result in unwanted reaction products. We suggest that further work in this area will be necessary before these factors can be fully identified and mitigated.

REFERENCES

1. For leading references, see: Pedrick, E. A.; Schultz, J. W.; Wu, G.; Mirica, L. M.; Hayton, T. W. *Inorg. Chem.* **2016**, *55*, 5693−5701.
2. Duval, P. B.; Burns, C. J.; Buschmann, W. E.; Clark, D. L.; Morris, D. E.; Scott, B. L. *Inorg. Chem.* **2001**, *40*, 5491−5496.

ACKNOWLEDGMENTS

This work was supported by the U.S. Department of Energy, Office of Basic Energy Sciences, Chemical Sciences, Biosciences, and Geosciences Division under Contract No. DE-SC-0001861

Covalency of Actinides with Chelating Agents Across the Actinide Series

Morgan Kelley[1], Jing Su[1], Jennifer Braley[2], Enrique Batista[1], Ping Yang[1]

[1]Theoretical Division, Los Alamos National Laboratory, New Mexico, 87544
[2]Department of Chemistry, Colorado School of Mines, Golden, CO, 80401
pyang@lanl.gov

INTRODUCTION

Actinide-ligand bonding has a long history of investigation due to its importance to the nuclear fuel cycle, environmental remediation, nuclear medicine, and other concerns. Early actinides (up to plutonium) generally mimic some aspects of transition metal chemistry, including the formation of covalent An-ligand bonds through the 6d orbitals.[1] Whereas the early actinides have seen extensive study, studies on the actinides heavier than plutonium are sparse. As the actinide series is traversed, the contraction of the 5f orbitals stabilizes the trivalent oxidation state, potentially limiting covalent interactions and enforcing lanthanide-like behavior. It is generally thought, therefore, that the heavier actinides do not form significant covalent interactions,[2] though some evidence suggests this is not the case.[3]

This work investigates covalent interactions across the actinide series using two model systems: the biological decorporation ligand 3,4,3-LI(1,2-HOPO) and the common dipicolinate ligand.[4,5] Changes in covalent interactions across the actinide series in both An(III) and An(IV) ions are investigated, offering new insights into the complexation of the late actinides and guiding principles for the design of new chelating ligands.

METHODS

Density functional theory calculations were carried out using the PBE functional, relativistic ZORA small core basis sets for the actinide ions, and triple zeta basis sets for atoms comprising the ligands. Solvation by water was considered using the COSMO implicit solvation model.

RESULTS AND DISCUSSION

Despite the contraction of the 5f orbitals across the actinide series, actinide-ligand interactions rely more upon the 5f and less upon the 6d orbitals as the series is traversed. Figure 1 shows increased electron donation (in terms of Mulliken population) into the actinide 5f orbitals by the 3,4,3-LI(1,2-HOPO) ligand. In complexes with An(IV) ions (right panel of Figure 1), 5f orbitals actually accept more electron density than the 6d orbitals for Cm, Cf, and Es. While counter to the common consensus on bonding in the later actinides described above, these finding are consistent with further electronic structure calculations and thermodynamic measurements on the dipicolinate system. The covalent interactions of the early actinides, with electron density primarily donated into the 6d orbitals, change across the series. An(IV) ions are shown to form stronger bonds with the HOPO ligand than An(III) ions, correlating with the increased participation of the 5f orbitals.

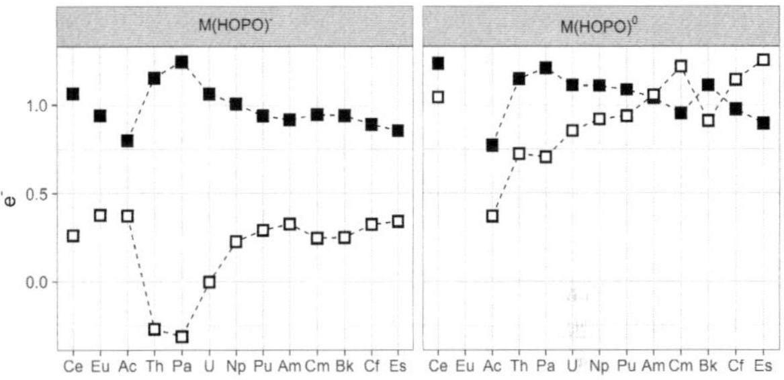

Figure 1. Electron donation into the valence d (■) and f (□) orbitals by the chelating HOPO ligand.

CONCLUSIONS

The results clearly indicate the importance of the 5f orbitals in covalent actinide-ligand interactions across the actinide series. The changing nature of these covalent interactions as the series is traversed offers new insights into the complexation of the late actinides, and may offer guiding principles for the design of new chelating ligands for this part of the periodic table.

REFERENCES

1. G. R. Choppin, *J. Alloys Compd.*, 2002, 344, 55–59.
2. J. A. Drader, M. Luckey and J. C. Braley, *Solvent Extr. Ion Exch.*, 2016, 34, 114–125.
3. S. G. Minasian et. al., *J. Am. Chem. Soc.*, 2012, 134, 5586–97.
4. Kelley, M.P., Batista, E.R., Yang, P. *Manuscript Submitted April 2017*
5. Kelley, M.P., Su, J., Urban, M., Luckey, M., Batista, E.R., Yang, P., Braley, J.C. *Manuscript Submitted April 2017.*

ACKNOWLEDGMENTS

This work was supported by the U.S. Department of Energy, Office of Science, Office of Basic Energy Sciences, Heavy Elements Chemistry Program at Los Alamos National Laboratory under contract number DE-AC52-06NA25396 and at the Colorado School of Mines under Award Number DE-SC0012039.

Covalent Stabilisation of Bonding in $M(BTP)_3$ and $M(BTPhen)_2$ (M = Ce-Lu, Cm, Am; BTP = bis(triazinyl)pyridine; BTPhen = bis(triazinyl)phenanthroline)

Izaak Fyer-Kanssen[1], Jonathan Austin[2] and Andy Kerridge[1*]

[1*] Department of Chemistry, Lancaster University, Bailrigg, Lancaster LA1 4YB, U.K., a.kerridge@lancaster.ac.uk
[2] National Nuclear Laboratory, Chadwick House, Birchwood Park, Warrington WA3 6AE, U.K.

INTRODUCTION

The presence of lanthanides in spent nuclear fuel increases the difficulty of both its disposal and its potential recyclability. In the former, the relatively short lanthanide (Ln) half-lives, compared to those of the actinides (An), mean that greater volumes than necessary currently need to be disposed of in long-term storage facilities, while in the latter, a hurdle due to the strong neutron absorption cross sections of lanthanides arises. These issues can be effectively surmounted by the separation of lanthanides from actinides in the spent fuel. However, this becomes technically challenging when considering the later actinides such as americium (Am) and curium (Cm), whose chemistry strongly resembles that of the trivalent lanthanides.

In this contribution, we present the first systematic study of the topological and integrated electronic properties of aquo and BTP complexes of the lanthanides (Ce–Lu)[1] and compare them to complexes of two technologically relevant actinides (Am and Cm) in order to investigate the relationship between bond covalency and complex stability. Nitrate and BTPhen complexes of Eu, Gd, Am and Cm are also considered and analysed.

RESULTS AND DISCUSSION

Scalar-relativistic DFT simulations of all complexes were performed and successfully compared to literature XRD/EXAFS structural data. Resulting electron density distributions were subjected to analysis via the Quantum Theory of Atoms in Molecules (QTAIM), which can estimate contributions to both overlap-driven and degeneracy-driven covalency[2], to obtain topological and integrated properties of the electron density.

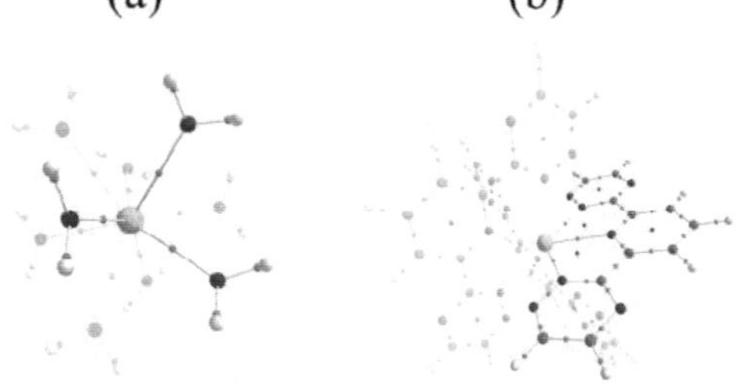

Figure 1. Representative molecular graphs of (a) $[Ln(H_2O)_9]^{3+}$ and (b) $[Ln(BTP)_3]^{3+}$.

As expected, QTAIM analysis revealed only weak variation in covalency across the lanthanide series. However, comparison between Ln and An analogues (Gd/Cm & Eu/Am) revealed significantly enhanced covalency in the latter. Evaluation of the electronic energies associated with the exchange reactions between aquo/nitrate and BTP/BTPhen ligands revealed a corresponding stability of the latter to actinide species.

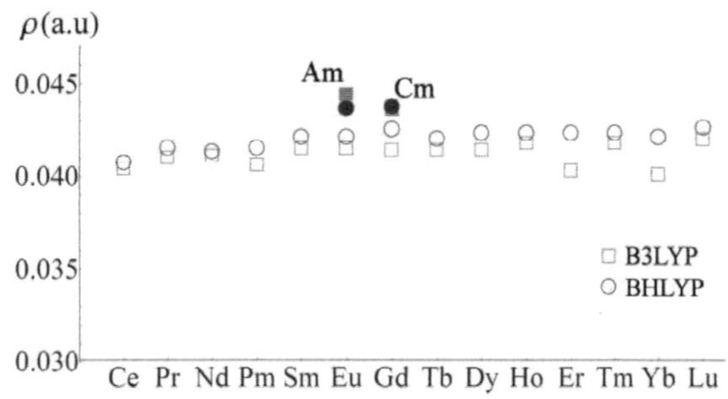

Figure 2. Calculated electron densities at the M-N bond critical points of $[Ln(BTP)_3]^{3+}$ complexes.

CONCLUSIONS

Analysis of DFT-generated electron densities establishes that enhanced An-N bond stability is accompanied by enhanced An-N bond covalency, indicating that there is an electronic component to the selectivity of BTP and BTPhen ligands towards minor actinides over lanthanides.

REFERENCES

References must be numbered. Keep the same style.
1. Fryer-Kanssen, I, Austin, J. and Kerridge, A., Inorg. Chem. 55:10034-10042, 2016
2. Kerridge, A., Chem. Comm., accepted, 2017

ACKNOWLEDGMENTS

A.K. thanks the EPSRC for the award of a career acceleration fellowship (Grant EP/J002008/2). We thank the Nuclear Decommissioning Authority for financial support and the National Nuclear Laboratory for industrial supervision and acknowledge both the High End Computing facility at Lancaster University for access to HPC resources and the National Service for Computational Chemistry Software for access to the Slater HPC facility.

Crystal Chemistry of New Rare Earth Nickel Silicides

Marcella Pani[1*], Alessia Provino[1], Volodymyr Smetana[2], Pietro Manfrinetti[1]

[1]Department of Chemistry, University of Genova, Italy, marcella.pani@unige.it
[2]Ames Laboratory, Iowa State University, Ames, IA, USA

INTRODUCTION

Among intermetallic compounds, silicides form a numerous class, extensively studied in both fundamental and applied research. In particular, ternary compounds formed by Si in combination with a rare earth element and a transition metal are interesting due to the richness of their compositions and structural arrangements, as well as to their magnetic behavior.[1,2] Despite the large number of literature data on both the formation and crystal structure of these systems,[3] their knowledge results still incomplete, mainly due to the complexity of the ternary phase diagrams and the objective difficulty in obtaining monophasic samples. In this work, we present some crystallographic results on new ternary phases identified in the La-Ni-Si, Ce-Ni-Si, Pr-Ni-Si and Nd-Ni-Si systems.

EXPERIMENTAL METHODS

About twenty different samples were synthesized in the ternary systems RE-Ni-Si (RE=La,Ce,Pr,Nd), within the composition ranges 25-35 at. % RE, 40-50 at. % Ni and 25-33 at. % Si. The alloys were prepared by arc-melting, starting from the pure elements, followed by annealing at 750-900°C for 10-20 days. They were then characterized by means of microscopic, X-ray diffraction and DTA techniques.

RESULTS AND DISCUSSION

According to the preliminary results obtained for the four investigated systems, eleven new compounds have been identified: five in the La system, and two in each of the other systems. Among the La phases, three have been fully characterized: $La_6Ni_7Si_4$ ($Pr_6Ni_7Si_4$ type[4], *oP*68, *Pbcm*), $LaNi_2Si$ ($SrCu_2Ga$ type[5], *hR*12, *R*-3*m*) and $La_2Ni_3Si_2$, which crystallizes with a new monoclinic type (*mP*28, $P2_1/c$, Figure 1). The other new phases, having stoichiometry 3:4:2 and ~2:4:3, are common also to the Ce, Pr and Nd systems.

Concerning the former composition, it corresponds to a new structure type, successfully elucidated for $Ce_3Ni_4Si_2$: its monoclinic cell (*mS*36, *C*2/*c*) is strictly related to that of the adjacent 6:7:4 phase. In fact, a geometrical relationship between the lattice parameters of the two types exists, and, accordingly, the two structures contain identical blocks which are stacked, in a different way, along the respective longest lattice parameter.

Regarding the ~2:4:3 phase, its crystal structure has not been solved yet; however, a hexagonal unit cell has been obtained by single crystal data and confirmed by X-ray powder diffraction.

A crystal chemical approach will be used to describe the structural relationships among the different prototypes adopted by these compounds.

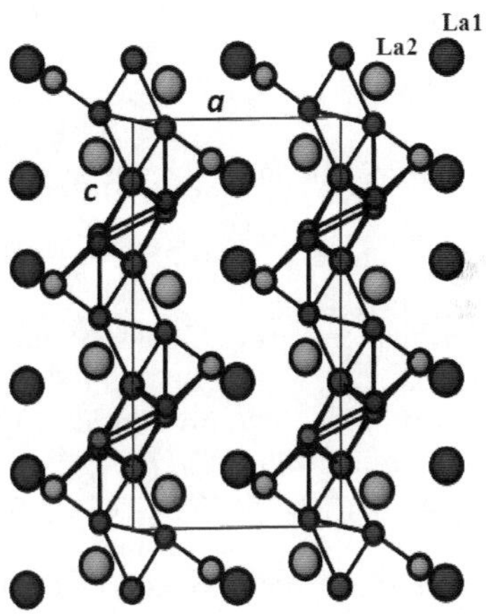

Figure 1. A projection of the crystal structure of $La_2Ni_3Si_2$, viewed along the *b* axis. Bond contacts within the two-dimensional [Ni-Si] network are highlighted.

REFERENCES

1. Rogl P., in: Handbook on the Physics and Chemistry of Rare Earths, K.A. Gschneidner, Jr. and L. Eyring Editors (1984) Vol. 7, Ch. 51.

2. Gladyshevsky I. *et al.*, in: Handbook on the Physics and Chemistry of Rare Earths, K. A. Gschneidner, Jr. and L. Eyring Editors (1990) Vol. 13, Ch. 88.

3. Villars P. *et al.*, Pearson's Crystal Data - Release 2015/16, ASM International, Materials Park, Ohio, USA.

4. Hovestreydt E. *et al.*, Acta Crystallogr. C. 40:1992-1995, 1984.

5. Fornasini M.L. *et al.*, J. Less-Common Met. 142: 289-294, 1988.

Crystal structure and thermoelectric properties of layered compound NdOZnSb

Meng Pan, Xin-Xin Yang, Kai Guo,Jing-Tai Zhao*

School of Materials Science and Engineering, Shanghai University, Shanghai, 200444

Quaternary layered compound NdOZnSb has the tetragonal symmetry of the space group *P4/nmm*, adopting the ZrCuSiAs structure type. This crystal structure composes of [NdO] layers and [ZnSb] layers alternately stacked along the *c* axis. In the [ZnSb] layer, Zn is tetrahedrally coordinated by four Sb atoms, forming edge-sharing tetrahedrons [$ZnSb_{4/4}$]. Similarly, [NdO] layers are formed by edge-sharing tetrahedrons [$ONd_{4/4}$] [1]. As a narrow band-gap *p*-type semiconductor, NdOZnSb has lower electrical conductivity, resulting in poor thermoelectric performance despite considerably high Seebeck coefficient and low thermal conductivity. Polycrystalline samples $Nd_{1-x}Sr_xOZnSb$ [2] and $NdOZn_{1-x}Ag_xSb$ [3] have been prepared by conventional solid state reaction with the aim to increase the hole concentration and improve the electrical properties via *p*-type doping. It is found that the maximum thermoelectric figure of merit zT ($zT = S\sigma^2T/\kappa$) reaches 0.18 for $Nd_{0.94}Sr_{0.06}OZnSb$ and 0.44 for $NdOZn_{0.96}Ag_{0.04}Sb$ at 725 K.

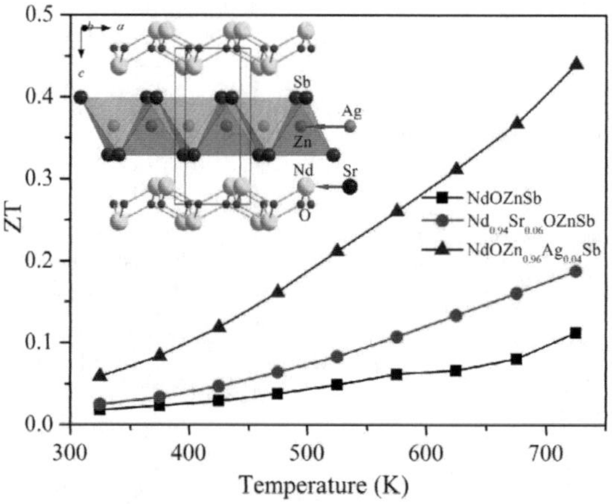

References:

[1] Guo K, Man ZY, Wang XJ, Chen HH, Tang MB, Zhang ZJ, et al. Chemical bonding and properties of "layered" quaternary antimonide oxide REOZnSb (RE = La, Ce, Pr, Nd). Dalton transactions. 2011;40:10007-13.

[2] Pan M, Wang K, Lv W, Hu D, Guo K, Yang X, et al. Structure and thermoelectric performance of layered compounds Nd1−xSrxOZnSb. Journal of Alloys and Compounds. 2016;688:153-7.

[3] Pan M, Guo K, Lv W, Zhang J, Hu D, Yang X, et al. Influence of Ag doping on the thermoelectric properties of layered compound NdOZnSb. Materials Letters. 2017;189:126-30.

* Corresponding author: jtzhao@shu.edu.cn

Crystallization behavior and component optimization study on cerium doped $(Gd_{1-x}Y_x)_2Si_2O_7$ single crystal

WAN Huanhuan[1], Wang Yong[2],MAO Rihua[2], ZHANG Zhijun[1], XU Zhan[1], ZHAO Jingtai[1], FENG He[1*]

[1]School of Materials Science and Engineering, Shanghai University, China, email:fh117@shu.edu.cn
[2]Shanghai Institute of Ceramics, Chinese Academy of Science, China

Abstract
Cerium-doped gadolinium pyrosilicate crystal attracts a broad attention as a novel scintillation crystal. Due to the pyrosilicate phase melts incongruently in the Gd_2O_3-SiO_2 system, it resulted in great difficulties in the large size crystal preparation. In this experiment, A series of 1at% cerium doped $(Gd_{1-x}Y_x)_2Si_2O_7$(x=0,0.1,0.2,...0.99) crystals were grown by the optical floating zone method. The effect of Y on the crystallization behavior of GPS was studied systematically. We investigated the crystallization behavior of $Gd_2Si_2O_7$ with slow cooling floating zone method. Through quenching method, we obtained information of the peritectic point and melt composition during crystal growth. The mechanism of Y influence on the crystallization behavior of GPS:Ce was proposed based on above result. The structure of $(Gd_{0.99-x}Y_xCe_{0.01})_2Si_2O_7$ samples was tested by X-ray diffraction. Their optical and scintillation properties were characterized through X-ray excitation luminescence spectra, fluorescence excitation and emission spectra, fluorescence decay curves and multi-channel energy spectra. The optimal Y concentration was determined, in which its best scintillation performance and growth stability were obtained.

Keywords
$(Gd_{0.99-x}Y_xCe_{0.01})_2Si_2O_7$; floating zone method; crystallization behavior; scintillation performance

Crystallographic ordering and magnetic structure in the next-to-ternary $RCo_{1-x}Ni_x$ compounds (R = Gd-Lu)

Alessia Provino[1*], Vitalij K. Pecharsky[2], Volodymyr Smetana[2], Clemens Ritter[3], Anja-Verena Mudring[2], Pietro Manfrinetti[1]

[1*]Department of Chemistry, University of Genova, Italy, alessia.provino@spin.cnr.it
[2]Ames Laboratory, Iowa State University, Ames, IA, USA
[3]Institut Laue-Langevin, Grenoble, France

INTRODUCTION

The equiatomic RT phases (R = rare earth, T = Co, Ni) are formed only by Ni and not with Co; the compounds containing light lanthanides and Gd crystallize in the orthorhombic CrB-type, while those formed by heavier R adopt the orthorhombic FeB-type structures[1]. The solubility of Co in the CrB-type GdNi intermetallic has been previously reported to reach about 20 at.%[2,3]. On the other hand, for some of the heavy R, earlier literature data reported the existence of a ternary $R_5Co_3Ni_2$ phase adopting the MoB-type[4]. During our previous work on the Gd_3Ni_2-to-Gd_3CoNi solid solution system[5], we also faced with a new phase $GdCo_{1-x}Ni_x$, no longer crystallizing in the CrB-type but in a tetragonal cell of the MoB-type or, most likely, a derivative of this prototype. Therefore, we have investigated the formation, crystal structure and physical properties of the $RCo_{1-x}Ni_x$ compounds (with R = Gd and some of the heavy R) in view of their potential for a strong magnetocaloric effect.

RESULTS AND DISCUSSION

Our results show that in the $RCo_{1-x}Ni_x$ compounds a higher Co content determines a structural change from the CrB-type to a tetragonal superstructure, ternary derivative of the MoB-type[4]. We also ascertained that in $GdCo_xNi_{1-x}$ the CrB-type structure is preserved up to only 21 at.% Co, while the Co solid solubility in the new tetragonal structure, as determined by accurate microprobe analysis (EDX) and single crystal X-ray diffraction, was found to range from 24 to 36 at.% Co. For this series, the trend of the lattice parameters, as a function of Co content, is shown in **Figure 1**. Preliminary single crystal data indicate a space group less symmetric than the one of the MoB-type ($I4_1/amd$), with preferential Co/Ni site occupation. Further crystal structure refinement by neutron diffraction data for the Tb and Ho compounds is now underway to confirm the site ordering and to investigate their magnetic structure. Physical properties have been measured on some representative compounds (R = Gd, Tb, Dy and Ho). Paramagnetic susceptibility data are consistent with the trivalent state of R ions, with negligible magnetic contribution from the 3*d* electrons of Co and Ni. Ferromagnetic-like transitions are generally observed in these compounds, with Curie temperatures, T_C, between 130-160 K for the $GdCo_{1-x}Ni_x$ series, ≈ 90 K and ≈ 50 K for $TbCo_{0.6}Ni_{0.4}$ and $DyCo_{0.6}Ni_{0.4}$, respectively (**Figures 2** and **3**). Magnetocaloric effect has been measured for the Gd phases on two compositions close to the solubility limits ($GdCo_{0.5}Ni_{0.5}$ and $GdCo_{0.7}Ni_{0.3}$).

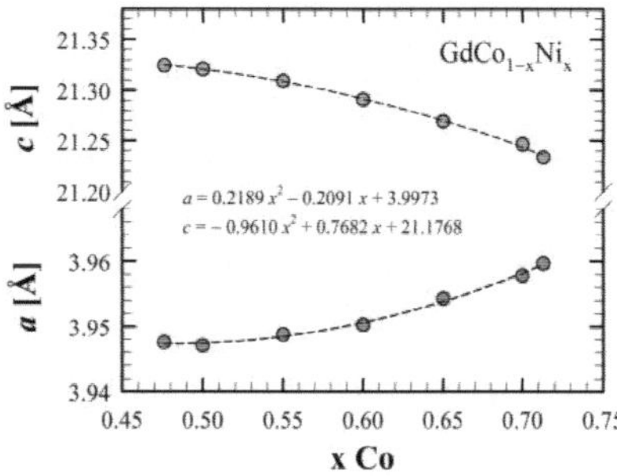

Figure 1. Trend of the lattice parameters *vs*. Co content in the $GdCo_{1-x}Ni_x$ series.

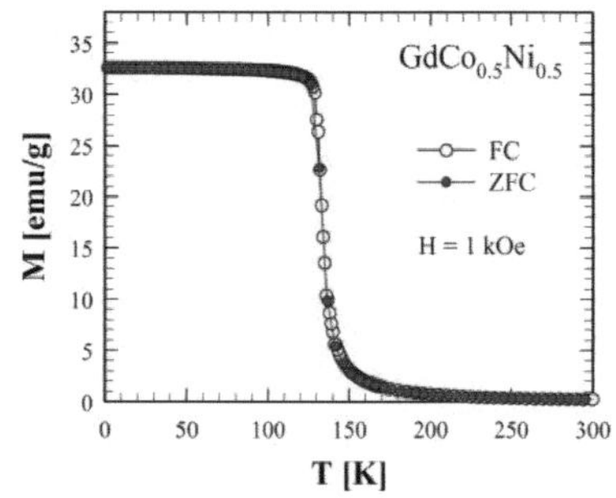

Figure 2. Magnetization *vs*. T for $GdCo_{0.5}Ni_{0.5}$ at H = 1kOe.

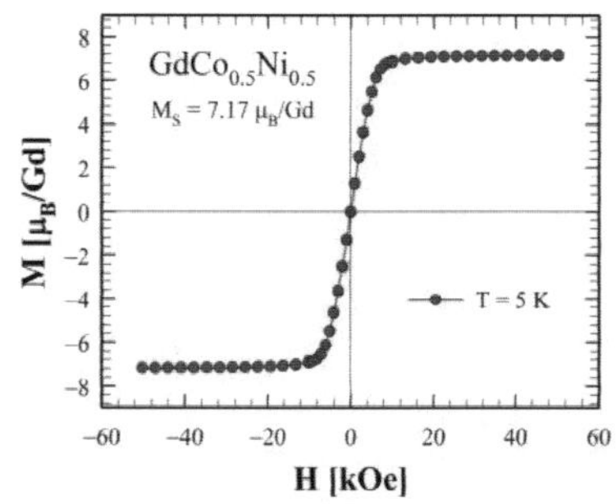

Figure 3. Isothermal magnetization at 5 K for $GdCo_{0.5}Ni_{0.5}$.

REFERENCES

1. P. Villars, K. Cenzual, *Pearson's Crystal Structure Database for Inorganic Compounds*, 2015/16, ASM International, Materials Park, OH, USA.
2. Y. Mudryk *et al.*, Acta Mater. 92,18-24 (2015).
3. T. E. Prost *et al.*, J. Alloys Compd. 696, 382-390 (2017).
4. O. I. Kharchenko, *et al.*, Russ. Metall. 1, 170-176, (1977).
5. A. Provino *et al.*, J. Mater. Chem. C 4, 6078-6089 (2016).

ACKOWLEDGMENTS

This work was supported by the Division of Materials Science and Engineering Programs of the U.S. DOE under contract No. DE-AC02-07CH11358 with Iowa State University.
A. Provino and P. Manfrinetti would like to thank the Ames Laboratory for hosting their visits during last years.

CyTTHA: a versatile scaffold for designing lanthanide imaging probes.

Ali Mohamadi[1], Lawrence W. Miller*[1]

[1]Department of Chemistry, University of Illinois at Chicago, USA, *lwm2006@uic.edu

INTRODUCTION

Bifunctional chelators (BFCs) are a critical component of metal-based contrast agents used in magnetic resonance imagine (MRI), time-gated luminescence microscopy and positron emission tomography (PET).[1-4] Effective BFCs must meet a number of requirements. First, they must be unimolecular constructs that can be easily conjugated to antibodies, peptides or receptor-targeting ligands. Second, they must bind to a given ion with high thermodynamic stability and kinetic inertness. While stability and inertness are paramount, the ability to form metal complexes under mild physiological conditions is desirable so that antibody conjugates or other heat- or pH-sensitive preparations can be used. Third, the synthetic route to preparation must be efficient and adaptable so that ligands' coordination profiles or functional substitutions can be easily varied. For luminescent lanthanide complexes, the synthesis must allow incorporation of sensitizing chromophores.

RESULTS AND DISCUSSION

We have developed an effective synthetic route to cyclohexanyl triethylene tetramine hexaacetic acid CyTTHA analogs.[5] Synthetic tractability made it possible to prepare rationally designed emissive Tb(III) and Eu(III) complexes with exceptional brightness and enhanced kinetic inertness.[6] Eu(III) complexes of ligand cs124CF_3-cyTTHA (**Figure 1**) and its variants exhibit overall quantum yields of emission in water up to 40% and absorption out to 400 nm. Moreover, Ln(III) complexes of cs124CF_3-cyTTHA are highly resistant to transchelation, remaining intact for days at high dilution in mM EDTA solution. The high brightness is due to careful positioning of a pendant quinolinone sensitizer at a central nitrogen. The pre-organized cyclohexanyl backbone yields high kinetic inertness while preserving fast metal-binding kinetics.

In more recent work, we have prepared an 8-dentate CyTTHA analog, Gd-39-6-3 that exhibits specific relaxivities comparable to Gd-DTPA and high resistance to anions including phosphate (**Figure 2**).

CONCLUSIONS

Pre-organized, acyclic CyTTHA is a versatile scaffold for designing functional metal complexes with favourable metal binding characteristics. An efficient and flexible synthetic strategy permits access to analogs with varied pendant substitutions and denticities. Quinolinone-sensitized CyTTHA Tb(III) and Eu(III) complexes are bright and kinetically inert. Ongoing studies are focused on developing backbone-functionalized ligands for bioconjugation as well as continued development of optical, MRI and radiotracer imaging agents.

REFERENCES

1. Price, E. W.; Orvig, C., *Chem Soc Rev* 2014, 43, 260-90.
2. Heffern, M. C.; Matosziuk, L. M.; Meade, T. J., *Chem Rev* 2014, 114, 4496-539.
3. Pierre, V. C.; Allen, M. J.; Caravan, P., *J Biol Inorg Chem* 2014, 19, 127-31.
4. Bunzli, J. C., *Coord Chem Rev* 2015, 293, 19-47.
5. Mohamadi, A., Miller L.W. *Tetrahedron Lett* 2017 http://dx.doi.org/10.1016/j.tetlet.2017.02.056
6. Mohamadi, A., Miller L.W. *Bioconjug Chem* 2016, 27, 2540-2548.

ACKNOWLEDGMENTS

The authors would like to thank the National Instituties of Health (Grant no: GM081030) for providing financial support to this project and Valerie Pierre for providing valuable input.

Figure 1. Excitation and emission spectra of Eu-cs124CF_3-cyTTHA.

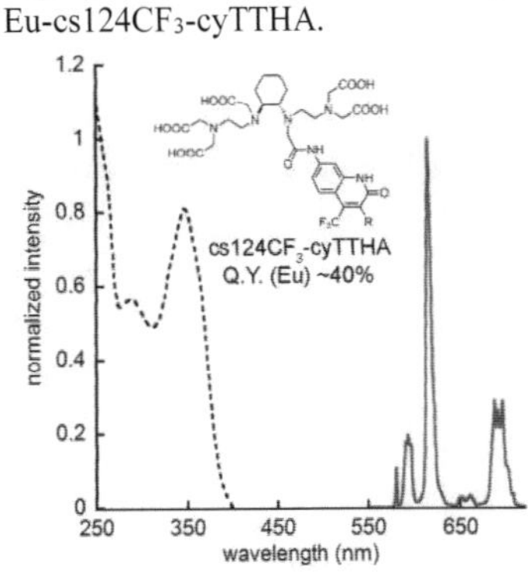

Figure 2. Effect of ions on the inverse longitudinal relaxation times of 0.025 mM Gd-39-6-3 in the presence of 5 mM anions (solid) and 100 mM (patterned) in 100 mM HEPES pH 7.4 at 25 °C. (n = 3)

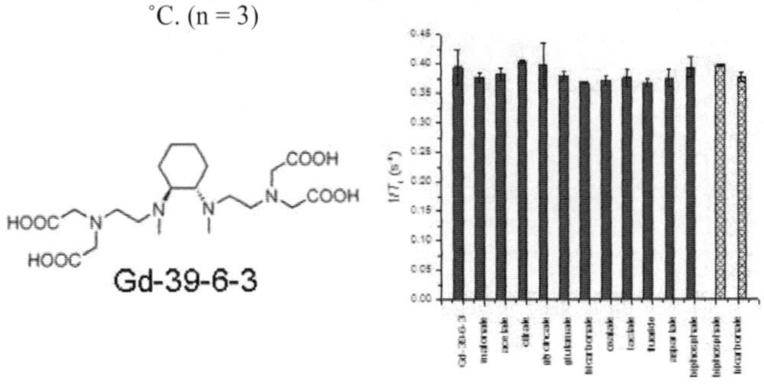

Designer materials: Molecular nano-magnets based on lanthanides

Muralee Murugesu[1]

[1]Department of Chemistry and Biomolecular Sciences, University of Ottawa, Canada, m.murugesu@uottawa.ca

Multimetallic rare-earth systems and especially Dy^{III}, Er^{III}-based materials have sparked much interest in the area of molecular magnetism due to the large intrinsic magnetic anisotropy of the lanthanide ions. When such a unique property is combined with a high-spin ground state (*S*) in a molecular complex, it causes slow relaxation of the magnetization as seen for Single-Molecule Magnets (SMMs). Lanthanide only SMMs are rare due to the difficulty in promoting the magnetic interactions *via* the overlapping bridging ligand orbitals with the contracted 4*f* orbital of the ions. Toward the goal of inducing significant magnetic interaction between lanthanide ions and subsequently isolating high-energy barrier SMMs, our research is currently focused on the use of phenoxide bridges as superexchange pathways between spin carriers.[1-4] In addition, organometallic lanthanide sandwich complexes provided an alternative approach for understanding intriguing quantum mechanical properties of SMMs.[5-8]

REFERENCES

1. Lin, P.-H.; Murugesu, M.* and co-workers, *Angew. Chem. Int. Ed.*, 2008, *47*, 8848. 2. Lin, P.-H.; Murugesu, M.* and co-workers, *Angew. Chem. Int. Ed.*, 2009, *48*, 9489. 3. Long, J.; Murugesu, M.* and co-workers, *J. Am. Chem. Soc.*, 2011, *133*, 5319. 4. Habib, F.; Murugesu, M.* and co-workers, *J. Am. Chem. Soc.*, 2011, *133*, 8830. 5. Jeletic, M.; Murugesu. M.* and co-workers, *J. Am. Chem. Soc.*, 2011, *133*, 19286. 6. Le Roy, J.; Murugesu. M.* and co-workers, *J. Am. Chem. Soc.*, 2013, *135*, 3502. 7. Le Roy, J.; Murugesu. M.* and co-workers, *J. Am. Chem. Soc.*, 2014, *136*, 8003. 8. Le Roy, J.; Murugesu. M.* and co-workers, *Angew. Chem. Int. Ed.* 2014, *53*, 4413.

ACKNOWLEDGMENTS

The authors would like to thank the NSERC-DG Canada program for providing financial support to this project.

Determining Computational Methods to Aid in Rare Earth Element Extractant Design

Marilu G. Perez[1*], Federico Zahariev[1], Nuwan De Silva[1], Benjamin P. Hay[2], Vyacheslav S. Bryantsev[3], Mark S. Gordon[1], and Theresa L. Windus[1]

[1]Ames Laboratory, Iowa State University, United States of America, marilu@iastate.edu
[2] Supramolecular Design Institute, United States of America
[3] Oak Ridge National Laboratory, United States of America

INTRODUCTION

One mining process involves solvent extraction to purify rare earth (RE) metal ions by selectively increasing the concentration of RE ions in organic diluents by extraction from aqueous solution. Computational methods can provide a safe and fast way to screen new extractants thus decreasing the overhead associated with the search, synthesis, and testing. Determining a proper computational method that is both fast and accurately predictive of experimental observables is presented.

Metal ligand binding is one factor affecting the extraction capability of a compound; that is, stronger binding ligands should have a greater ability to bind to a metal and thus transfer it to the organic phase. Using a modified thermodynamic cycle, Fig. 1, related to one previously used to determine uranyl and lanthanide critical stability constants,[1] the relative aqueous phase binding energies for different ligands can be approximated. These ligand binding energy trends are then compared to experimental distribution coefficients – the ratio of metal ion in organic versus aqueous phase at equilibrium.

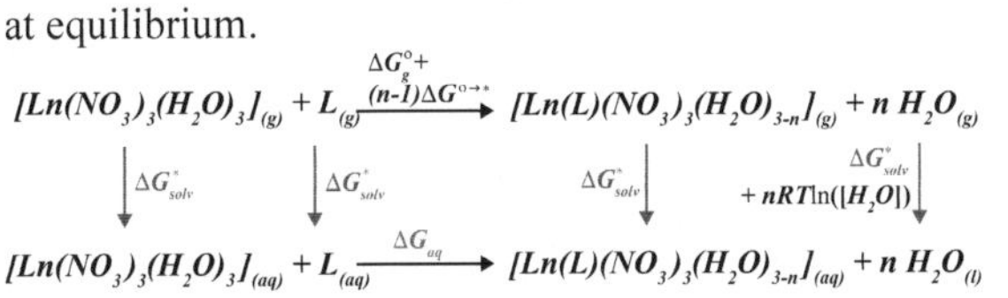

Figure 1. Thermodynamic cycle used to calculate the energy, ΔG_{aq}, of water displaced from the first solvation sphere around a lanthanide-nitrate complex, $Ln(NO_3)_3$, by a ligand, L.

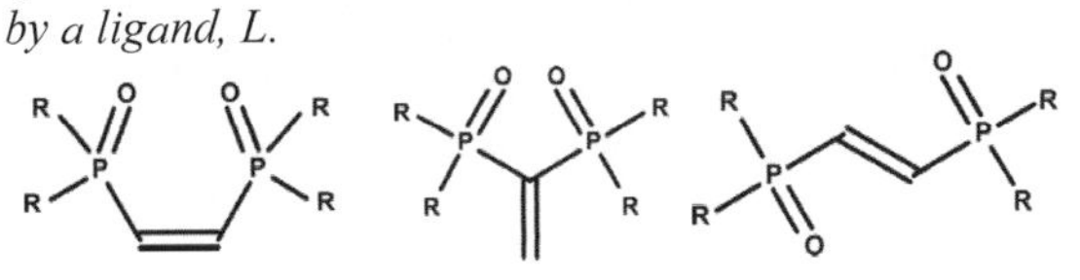

I II III

Figure 2. Diphosphine dioxides used for computing energies. R is methyl and phenyl in computations and experiment, respectively.

Three ligands, Fig. 2, were chosen based on experimental distribution coefficient data availability for metals with similar behaviour to lanthanide.[2] These compounds were also computationally predicted to be good extractants.[3]

COMPUTATIONAL METHODS

After preliminary tests, B3LYP with 6-31+G* basis for light elements and Stuttgart basis and segmented effective core potentials for the metal were chosen for the calculations. First, gas phase structures were optimized for each of the compounds in Fig. 1 with Ln=lanthanide and L=**I**, **II**, and **III**, Fig. 2. Minimum energy structure confirmation, zero-point, and thermal energy corrections were obtained from Hessian calculations. The energy of solvation, ΔG_{solv} was calculated using the SMD model, and all calculations were run using NWChem.[4]

RESULTS AND DISCUSSION

The experimental distribution coefficients[2] for **I**, **II**, and **III** were compared to the relative aqueous binding energies, ΔG_{aq}, and found to reproduce the same trend, Table 1. The energy for the systems increase from **I** to **III** which indicates that the most stable metal ligand complex is Ln-**I**. This is consistent with the observed distribution coefficient trend, where Ln-**I** has the highest distribution coefficient, the greatest fraction of metal in the organic phase, followed by Ln-**II**, and then Ln-**III**.

Table 1. Comparison of computationally derived relative binding energies and experimental distribution coefficients.[2]

Ligand	Comp. ΔG_{aq} in kcal/mol (Metal Ion)	Expt. Dist. Coeff (Metal Ion)
I	-32.76 (La)	1.23 (Eu)
II	-28.42 (La)	0.79 (Eu)
III	-23.15 (La)	0.01 (Am)

CONCLUSIONS

The distribution coefficients correlate with the binding energy determined using computational methods. Though further testing is needed to determine if trends using other metals reproduces the experimental trends, this finding shows promise in predicting relative distribution coefficients.

REFERENCES

1. Vukovic, S. et al. Inorg. Chem. 54(8):3995-4001 2015
2. Chmutova, M. K. et al. J. Inorg. Nucl. Chem. 42(6):897-903 1980
3. McCann, B. W. et al., Inorg. Chem. 55(12):5787-5803 2016
4. Valiev, M. *et al.*, Comput. Phys. Commun. 181:1477-1489, 2010

ACKNOWLEDGMENTS

This work is supported by the Critical Materials Institute, an Energy Innovation Hub funded by the U.S. Department of Energy, Office of Energy Efficiency and Renewable Energy, Advanced Manufacturing Office.

Developing Best Practices for the United States Department of Defense Rare Earth Mining and Recovery Efforts

Gregory Nichols[1] and Jamie Glover[1]

[1]Homeland Defense and Security Information Analysis Center, USA, gnichols@hdiac.org

INTRODUCTION

Rare earth elements (REEs) are necessary for many components of electronics, including hard drives, televisions, cell phones, and critical instrumentation for missile guidance systems and aircraft, as well as lasers and magnets. Although REEs are central to modern life, including defense applications, in the United States (US), no rare earth mining currently takes place in the US. The last rare earth mining operation in the US filed for bankruptcy in 2015, and nearly all REE needs are filled though foreign imports, primarily from China. This reliance on imports for REEs has become an issue of national security, as the threat of the REE supply being shut off at any time would immediately harm US defense strategy.

Reports from the US Government Accountability Office[1] and Congressional Research Service[2] have highlighted the urgency for the Department of Defense (DoD) to maintain uninterrupted access to critical materials, including certain REEs, and other sources suggest adding those REEs to the National Defense Stockpile. In addition, proposed legislation would mandate the DoD obtain REEs produced exclusively in the US. Given that no rare earth mining, recovery, or recycling operations exist in the US, and these operations have historically been spotty, best practices for the industry are underdeveloped. Furthermore, the DoD has no history of directly overseeing REE extraction of any kind; therefore, a standard set of guidelines for developing a sustainable REE enterprise within the DoD is needed to ensure a successful endeavour.

METHODS

The Homeland Defense and Security Information Analysis Center (HDIAC) conducted a literature review to identify existing frameworks, guidelines, best practices, and risk management strategies specifically for rare earths mining. Key words used in the search of peer-reviewed literature, grey literature, and US government publications, included "rare earth element," "rare earth mining," and names of individual rare earth elements, with one or more of the following terms: framework, guidance, best practice, recommendation, and risk management. HDIAC searched US federal agency websites individually to ensure thorough and complete results; websites included the DoD, National Institute of Environmental Health Sciences, Centers for Disease Control and Prevention, National Institute for Occupational Safety and Health, Occupational Safety and Health Administration, Environmental Protection Agency, and Agency for Toxic Substances and Disease Registry.

RESULTS AND DISCUSSION

The search revealed no comprehensive frameworks or guidance specific to the US; however, our review identified the relevant Framework for Sustainable Mining of Rare Earth Elements presented by Kamenopoulos, Agioutantis, & Komnitsas[3] and best practices presented by Kamenopoulos & Agioutantis[4] aimed at multinational and European Union efforts, respectively.

The published framework and best practices provide a comprehensive foundation upon which a US framework for sustainable REE mining can be developed. HDIAC proposes to adapt these guidelines to accommodate the complexities of US federal and state regulations, as well as DoD directives, to create necessary best practices as outlined below.

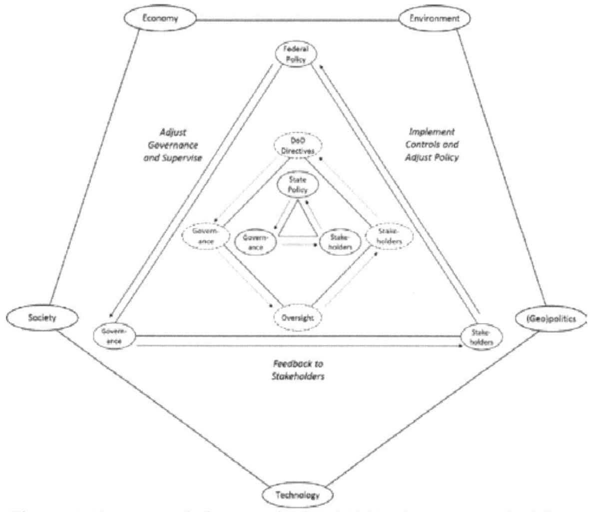

Figure 1. Framework for sustainable US/DoD-sponsored REE mining development, adapted from Kamenopoulos, *et al.*, 2015

CONCLUSIONS

HDIAC's research has exclusively focused on mining as this will likely be the most viable method to secure REEs in the quantities needed utilizing available technologies. Establishing a comprehensive framework and best practices specific to the US/DoD will ensure effective and sustainable practices as the US ensures it has continuous access to REEs.

REFERENCES

1. Government Accountability Office, GAO-16-161, 2016
2. Congressional Research Service, R41744, 2011
3. Kamenopoulos S. *et al.*, Rare Earths Industry. Elsevier, 2016
4. Kamenopoulos S. *et al.*, International Conference on Sustainable Development in the Minerals Industry, 2013

Development of "GAP MAGNET" in Ce – TM – TM' systems.

Andriy Palasyuk[1*], Tej Lamichhane,[1,2] Olena Palasyuk,[1,3] Michael Onyszczak,[2] Daphne Gates,[2] Valentin Taufour,[4] Vladimir Antropov,[1,3] Sergey L. Bud'ko[1,2] and Paul C. Canfield[1,2]

[1] Ames Laboratory, Iowa State University, Ames, IA 50011, U.S.A.
[2] Department of Physics and Astronomy, Iowa State University, Ames, IA 50011, U.S.A.
[3] Department of Material Science and Engeneering, Iowa State University, Ames, IA 50011, U.S.A.
[4] Department of Physics, University of California, Davis, CA 95616, U.S.A.

INTRODUCTION

Recent rapid development of alternative energy sources created increased demand on high-energy permanent magnets. Particular interest is attracted to the niche of a so called "GAP MAGNETS" that could reach the Energy Products of ~10 – 20 MGOe and be both reliable and easily approachable with respect to the cost of the components as well as the processing. This may be realized in Ce-containing transition metal (TM) systems as more economically affordable alternative to magnets containing expensive and supply dependent critical rare earths, e.g., Sm, Nd, Dy, etc.

EXPERIMENTAL METHODS

We use both high temperature flux techniques for growths of single crystals and arc-melting, drop casting approaches for polycrystalline bulk samples. The characterization was done by means of powder XRD, Magnetic-DSC and VSM, SEM and TEM methods.

RESULTS AND DISCUSSION

In the project we use both theoretical and experimental efforts to evaluate the potential of Ce intermetallic compounds for use as primer materials for high flux permanent magnets for electric drive motors. The theoretical work is focused on determination of the degree of hybridization which occurs in known systems and evaluates the effect on the *magnetization, anisotropy and Curie temperature*, while maintaining *coercivity*.

Experimentally, the attempts is focused $CeTM_{5-x}TM'_x$ (TM = Co, TM' = Fe, Ni, Cu, Zn) systems with small additions of refractory metals (RM).

We have demonstrated that Sm, which is most expensive component in 1:5-type magnets, can be fully substituted by less expensive Ce, and yet the surrogates, deliver magnetic characteristics acceptable for the "GAP MAGNET", i.e., $T_{c.} > 300$ °C, $H_{cj.} = 0.5 – 1.0$ T, $B_r = 0.4 – 0.8$ kG and $(BH)_{max.} = 7 – 15$MGOe.

* - palasyuk@iastate.edu

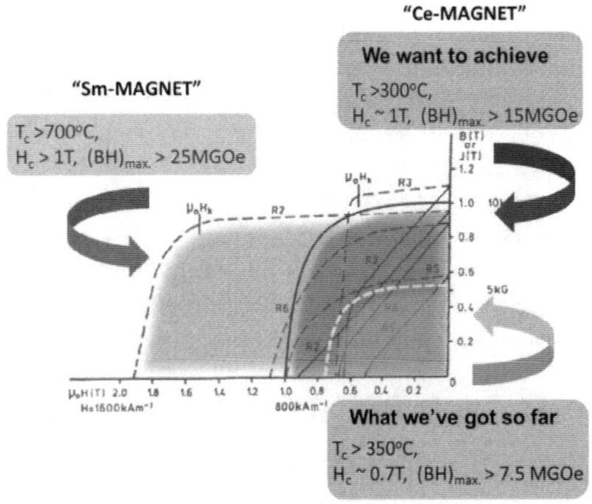

Figure shows typical demagnetization curves of Sm-containig 1:5 and 2:17 magnets (dashed black lines)[1] in comparison to our objectives with partially or/and completely substituted Ce alternative (solid red line) and the actual data on 1:5-type $Ce(Co,Fe,Cu,RM)_5$ magnet (dashed yellow line).

CONCLUSIONS

Due to an excess of Ce on the market, the development of a high magnetic flux density Ce–TM based permanent magnets would facilitate an increase of the supply by a factor of 2 to 3 as well as its domestic nature., i.e., ~50% of the rare-earth (RE) content of US-mined Bastnasite concentrate. Our preliminary testing on $CeTM_{5-x}TM'_x$ (TM = Co, TM' = Fe, Ni, Cu, Zn) clearly indicate the potential for "GAP MAGNET" development in these Ce-containing systems.

REFERENCES

1. Jiles D. Introduction to Magnetism and Magnetic Materials J. Taylor & Francis Group. 1998.

ACKNOWLEDGMENTS

This research was supported by the Critical Materials Institute, and Energy Innovation Hub funded by the U.S. Department of Energy, Office of Energy Efficiency and Renewable Energy, Advanced Manufacturing Office.

Development of Aromatic Macrocyclic Bi- and Tri-Functional Chelators for Lab and Clinic

Kenneth N. Raymond, David Tatum, Jide Xu and Darren Magda*

Lumiphore, Inc., Berkeley, CA, USA, dmagda@lumiphore.com

INTRODUCTION

Lumiphore, Inc.[1] was founded in 2001 to develop some of the discoveries in lanthanide and actinide chelation made by the Raymond laboratory at the University of California, Berkeley.[2-4] With the assistance of private, institutional, and governmental funding, Lumiphore has built upon this core technology to optimize product characteristics and expand market opportunities. An overview of selected R&D programs will be presented.

RESULTS AND DISCUSSION

The initial focus of our research was the optimization of the 2-hydroxyisophthalamide (IAM) chelate for Tb(III) as a potential commercial agent for use in Homogeneous Time Resolved Fluorescence (HTRF) technology. Due to the long-lived luminescence of a lanthanide energy donor, the sensitivity of the Förster Resonance Energy Transfer (FRET) assay format can be vastly improved by using time gated excitation and detection (Figure 1). These assays have become widespread in pharmacology and biotechnology. Our resulting flagship product, Lumi4®-Tb (Figure 2) is now used in commercial products by four different companies.

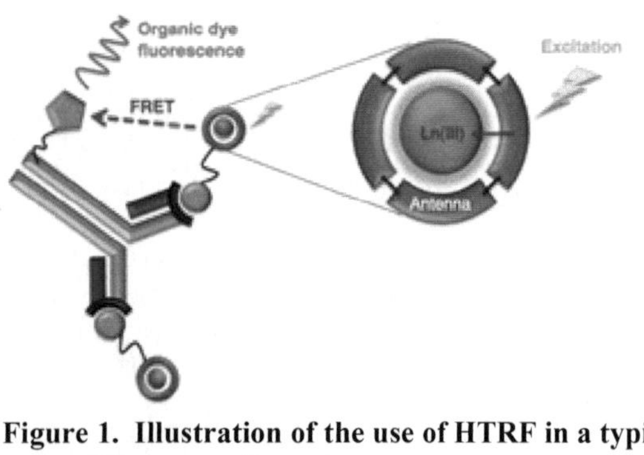

Figure 1. Illustration of the use of HTRF in a typical competition assay format.

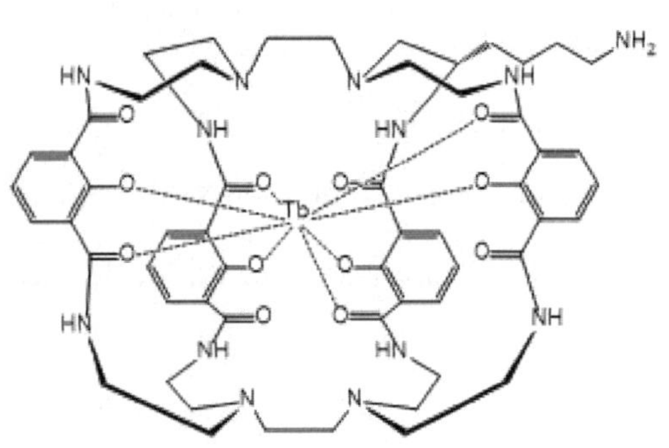

Figure 2. Lumi4-Tb-NH_2

Recent efforts have also involved extending the use of IAM–based chelators to radiopharmaceutical applications. For example, we conjugated an antibody targeting HER2 receptor with the Lumi4 chelator. The resulting antibody-Lumi4 conjugate was labelled under ambient conditions with ^{177}Lu ($T_{1/2}$ = 6.7 d; 133 keV β-emission) in 95% radiochemical yield. Administration of a single 200 μCi dose of the anti-HER2-Lumi4-^{177}Lu conjugate to nude mice bearing SKOV-3 tumor xenografts led to tumor growth inhibition. Specific localization to tumor was observed using SPECT/CT 72 hr following a single 1 mCi dose relative to an antibody blocking control group. Lumiphore is also studying Lumi4 and related chelating systems for use with ^{89}Zr for PET imaging and has licensed technology for use in targeted thorium conjugates (TTC's). A Phase I clinical trial investigating a TTC is currently in progress.[5]

CONCLUSIONS

We have shown that aromatic macrocyclic chelators based on IAM coordinating units display facile coordination with lanthanide cations to form kinetically and thermodynamically stable complexes. In the case of Tb(III), efficient (QY= ca. 60%) luminescent sensitization is facilitating development of HTRF assays for use in research and in vitro diagnostics. IAM chelators are also being developed for use in radiopharmaceutical agents, where they serve to: 1. Form a covalent attachment to a site-directing molecule. 2. Bind a metallic radionuclide. 3. Provide a luminescent signal upon terbium treatment to quantify the presence of unbound chelator.

REFERENCES

1. http://www.lumiphore.com
2. Petoud, S. *et al.*, JACS, 2003, 125, 13324.
3. Samuel, A.P.S., *et al.*, Inorg. Chem., 2008, 47, 7535.
4. J. Xu, *et al.*, J. Am. Chem. Soc. 2011, 133, 19900.
5. https://clinicaltrials.gov/ct2/show/NCT02581878

ACKNOWLEDGMENTS

"The authors would like to thank the National Science Foundation (Grant no: IIP-1215462) for providing financial support to this project".

Development of CMPO-Based Ligands for Selective Lanthanide Extraction

Eric J. Werner

Department of Chemistry, Biochemistry and Physics, The University of Tampa, U.S.A.
ewerner@ut.edu

INTRODUCTION

The wide range of modern applications of the lanthanides has driven a resurgence in research devoted to the selective binding and extraction of these valuable metals. Given recent concerns related to the lanthanide supply chain, coupled with a growing interest in recycling Ln metals from consumer products (cell phones, batteries, etc.),[1] development of efficient, scalable technologies to accomplish selective extraction remains worthwhile. As a contribution to this field, our group has been studying multipodal ligands possessing carbamoylmethylphosphine oxide (CMPO), a chelating group well-known in the areas of *f*-element coordination and liquid-liquid extraction. In generating tripodal CMPO-based ligands, we have seen promising actinide and lanthanide extraction selectivity as a function of ligand capping architecture.[2,3] The focus of this presentation will be on the further development of a tris-(2-aminoethyl)amine (TREN) capped CMPO ligand, TREN-CMPO-OEt (**1**, Figure 1), as well as variation of extraction conditions to further probe the unique terbium(III) selectivity observed previously.

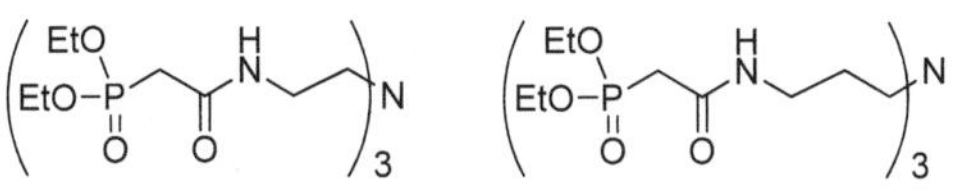

TREN-CMPO-OEt **1** **TRPN-CMPO-OEt** **2**

Figure 1. Tripodal CMPO-based ligands.

EXPERIMENTAL METHODS

Synthesis of a new, tris-(3-aminopropyl)amine (TRPN) capped ligand, TRPN-CMPO-OEt (**2**, Figure 1), was accomplished using the same reaction conditions as reported for the TREN-capped ligand.[2] Extraction experiments were carried out by combining 2 mL of TREN or TRPN ligand solutions (10^{-3} M in CH_2Cl_2 for the 10:1, ligand:metal studies) with 2 mL of Ln metal solutions prepared from nitrate salts for La - Lu (except Pm; 10^{-4} M in 1 M HNO_3). Extractions were performed by mixing the ligand solution with individual metal solutions and with an ICP-AES rare earth standard mixture of all Ln metals (100 ppm/Ln(III)). The mixtures were stirred vigorously for 24 h before phase separation and analysis of each extracted aqueous layer by ICP-AES to determine percent extraction values.

RESULTS AND DISCUSSION

Noting the unique Tb(III) selectivity revealed in our initial studies of the TREN-CMPO-OEt ligand, we were motived to further explore the importance of the TREN scaffold and its impact on extraction selectivity. To help investigate this, the larger TRPN capped ligand (**2**, Figure 1) was synthesized. The larger cap may increase extraction efficiency for the larger Ln ions, while also revealing the importance of the TREN cap in influencing the original Tb(III) preference. As indicated by extraction experiments, tethering three CMPO units with the larger cap results in a loss in Tb(III) selectivity. Also, overall Ln extraction percentage values are lower for the TRPN ligand presumably due to the more flexible scaffold and loss in preorganization of the metal binding cavity.

Noting the importance of the TREN cap with regard to Ln extraction, variation of the metal extraction conditions was also undertaken in an effort to further augment selectivity. The ratio of ligand to metal was varied and indicated that a 10:1 L:M ratio was optimal for promoting Tb(III) selectivity but, surprisingly, that a smaller L:M ratio of 1:1 led to overall higher Ln extraction across the series. Finally, utilizing similar extraction conditions but instead varying the CMPO substituent with more hydrophobic groups was also accomplished to test the effect of ligand/complex solubility on extraction efficiency.

CONCLUSIONS

The unique Tb(III) extraction selectivity of TREN-CMPO-OEt has been confirmed and further explored in this work, and variation of extraction conditions have revealed optimal conditions to promote selective Ln extraction utilizing this ligand system.

REFERENCES

1. Zhang, H. *et al.*, ACS Appl. Mater. Interfaces 8:9523–9531, 2016
2. Sartain H.T. *et al.*, Inorg. Chim. Acta 426:126–135, 2015
3. Coburn, K.M. *et al.*, Inorg. Chim. Acta 449:96–106, 2016

ACKNOWLEDGMENTS

We would like to thank Grand Valley State University (Weldon Fund, OURS, CSCE) and The University of Tampa (Dept. of Chemistry, Biochemistry and Physics) for providing financial support for this work.

Distribution of Yttrium and Europium In Organic-Aqueous Phases Containing Sulphuric Acid and HDEHP

Mark D. Williams-Wynn [1,2], Thandiwe Mhlongo [2], Milan Carsky [1,2], Paramespri Naidoo [1], Deresh Ramjugernath [1,2]

[1] Thermodynamics Research Unit, School of Engineering, University of KwaZulu-Natal, Howard College Campus, King George V Avenue, Durban, 4041, South Africa;
[2] Rare Earth Recycling Technologies., UKZN InQubate, University of KwaZulu-Natal, University Road, Chiltern Hills, Westville, 3629, williams-wynn@ukzn.ac.za

The distribution ratios of yttrium and europium between an aqueous phase consisting of various concentrations of sulphuric acid and an organic phase consisting of di-2-ethylhexylphosphoric acid (HDEHP), diluted with n-dodecane, were measured at 298.2 K.

INTRODUCTION

Rare earth metals (rare earths) are used in a wide variety of technologies, many of which require very high purity rare earths. The most common purification technique for obtaining these high purities is solvent extraction [1-2]. Although many different solvents have been proposed for these processes [3], there is little equilibrium data available in open literature. This equilibrium data would be useful for the optimisation of process parameters for the solvent extraction of rare earths.

In this study, we report on the equilibrium distribution ratios of the rare earths, yttrium and europium, between sulphuric acid solutions of varied concentrations and HDEHP, diluted with n-dodecane.

EXPERIMENTAL METHODS

Modified versions of the liquid-liquid equilibrium (LLE) cell used by Narasigadu et al. [4] were used to attain equilibrium between the phases. The phases were analysed using an inductively coupled plasma optical emission spectrometer (ICP OES). Fluid flowing through the jackets of the equilibrium cells was used to maintain the cell contents at 298.2 K.

RESULTS AND DISCUSSION

The distribution ratios of both yttrium and europium were measured. A graphical representation of the distribution ratios of yttrium, between sulphuric acid and HDEHP/n-dodecane, at varied sulphuric acid concentrations, calculated as described by Equation 1 are given in Figure 1.

$$Distribution\ ratio\ (i) = \frac{[C_{i,organic}]}{[C_{i,aqueous}]} \quad [1]$$

where $C_{i,j}$ is the concentration of component i in phase j.

The results for yttrium and europium both follow a similar logarithmic trend, with more of the rare earths present in the organic phase at lower acid concentrations. The results can be used to optimise the solvent extraction of yttrium or europium from a sulphuric acid solution, using HDEHP.

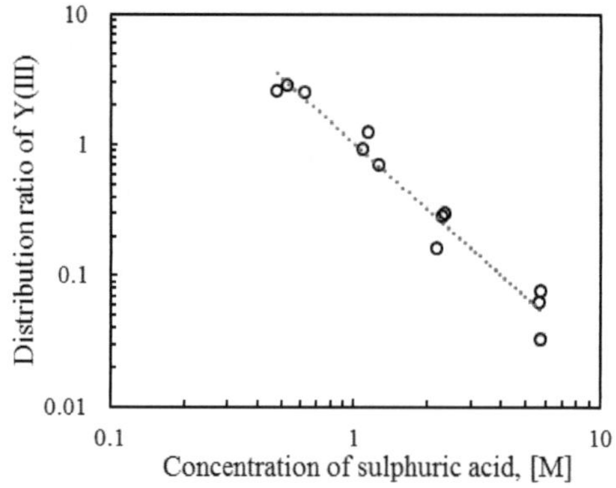

Figure 1. Distribution ratio of Y(III) as a function of sulphuric acid concentration with an organic phase of 0.25 M HDEHP in n-dodecane

CONCLUSIONS

The distribution coefficients of yttrium and europium at differing concentrations of sulphuric acid, between an aqueous sulphuric acid phase and an organic HDEHP/n-dodecane phase were measured using LLE cells. These data can be used to determine the optimal operating conditions for a solvent extraction process.

REFERENCES

1. Peppard, D. F.; Driscoll, W. J.; Sironen, R. J.; McCarty, S. *J. Inorg. Nucl. Chem.* 1957, *4* (5), 326-333.
2. Peppard, D. F.; Mason, G. W.; Maier, J. L.; Driscoll, W. J. *J. Inorg. Nucl. Chem.* 1957, *4* (5), 334-343.
3. Xie, F.; Zhang, T. A.; Dreisinger, D.; Doyle, F. *Miner. Eng.* 2014, *56*, 10-28.
4. Narasigadu, C.; Naidoo, M.; Ramjugernath, D. *J. Chem. Eng. Data* 2014, *59* (11), 3820-3824.

ACKNOWLEDGMENTS

The financial assistance of the National Research Foundation (NRF) and the Department of Economic Development, Tourism and Environmental Affairs of the Province of KwaZulu-Natal (EDTEA KZN) towards this research are hereby acknowledged. Opinions expressed and conclusions arrived at, are those of the authors, and are not necessarily to be attributed to the NRF or EDTEA KZN.

Divalent Europium in Magnetic Resonance Imaging

Matthew J. Allen* and Lina A. Basal

Department of Chemistry, Wayne State University, USA, mallen@chem.wayne.edu

INTRODUCTION

Eu^{II} is isoelectronic with Gd^{III}, resulting in both ions having desirable properties with respect to enhancing contrast in magnetic resonance imaging. Recently, the use of Eu^{II} as an in vivo contrast agent was reported.[1] In this context, the contrast enhancement ceased upon oxidation, enabling the hypoxic core of a tumor to be imaged. However, the oxidized europium species did not produce enhancement in T_1-weighted images, leading to a desire to couple a detection method for Eu^{III} with Eu^{II}-based contrast agents.

RESULTS AND DISCUSSION

To detect Eu in both the +2 and +3 oxidation state, we used a variety of ligands to access properties amenable to imaging techniques other than T_1-weighted ^1H-magnetic resonance imaging. These methods include imaging other nuclei or using chemical exchange saturation transfer with magnetic resonance imaging and using other modalities such as photoacoustic imaging. Our results indicate that it is possible to detect both oxidation states of europium, but that the best method of detection depends on the ligand.

CONCLUSIONS

Different strategies have been tried to detect Eu^{III} after the oxidation of Eu^{II}-based contrast agents. Some of the strategies are promising for enabling the detection of Eu in both the +2 and +3 oxidation states.

REFERENCES

1. Ekanger L. A. *et al.*, Angew. Chem. Int. Ed. 54:14398-14401, 2015

ACKNOWLEDGMENTS

The National Institutes of Health (Grant number EB013663) is acknowledged for providing financial support to this project.

Early Actinide Superlatives

Han-Shi Hu and Nikolas Kaltsoyannis[1]

[1]Centre for Radiochemistry Research, School of Chemistry, The University of Manchester, Manchester M13 9PL, UK
nikolas.kaltsoyannis@manchester.ac.uk

Compounds featuring unsupported metal–metal bonds between actinide elements remain highly sought after yet confined to inert gas matrix studies. Notwithstanding this paucity, actinide–actinide bonding has been the subject of significant computational effort. In this contribution, density functional theory (DFT), CASPT2 and CCSD(T) calculations at both the scalar and spin orbit levels are used to survey the Th–Th bonding in a range of systems of general formula LTh–ThL, where L is one of a range of neutral p-block donor ligands. Several of these are found to have very short Th–Th distances arising from a new type of Th–Th quadruple bond with a previously unreported electronic configuration featuring two unpaired electrons in 6d-based δ bonding orbitals. $H_3AsThThAsH_3$ is found to have the shortest Th–Th bond yet reported (2.590 Å). The Th_2 unit is a highly sensitive probe of the L ligand's electron donor/acceptor ability; we can tune the Th–Th bond from quadruple to triple, double and single by judicious choice of L, up to 2.888 Å for singly bonded ONThThNO.[1]

The current highest coordination number in molecular chemistry is 16. This has been seen very recently in the experimental work of Pollak *et al.*,[2] who found 16 Cs···F contacts in $Cs[H_2NB_2(C_6F_5)_6]$, and that by Popov *et al.* in CoB_{16}^-,[3] and has also been predicted theoretically by Gu *et al.*[4] for M@Li16 (M = Ca, Sr, Ba, Ti, Zr and Hf). The record for the actinide elements is 15, established experimentally by Daly *et al.* for $Th(H_3BNMe_2BH_3)_4$,[5] although the same authors used DFT to predict that in the gas phase the number of Th···H contacts could rise to 16. All of these contributions make reference to the work of Hermann *et al.*[6] who, in 2007, published a DFT survey of the geometric structures of Pb^{2+} coordinated by He atoms. The maximum value of n in $PbHe_n^{2+}$ was found to be 15, with no true minimum structures being located for n > 15 in which all the He atoms are located in the first shell around the Pb. Intriguingly, buried at the end of the Methods section, Hermann *et al.* write "We note that higher coordination number might be expected for charged actinide–He interactions". In this contribution, I will reveal if this expectation is correct.

REFERENCES

1. Hu H.-S. and N. Kaltsoyannis, Phys. Chem. Chem. Phys. 19:5070, 2017
2. Pollak D. *et al.*, J. Am. Chem. Soc. 138:9444, 2016
3. Popov I.A. *et al.*, Nat. Comm. 6, 2015
4. Gu X., *et al.*, Nanoscale 4:2567, 2012
5. Daly S.R. *et al.*, Angew. Chem. Int. Ed. 49:3379, 2010
6. Hermann A. *et al.*, Angew. Chem. Int. Ed. 46:2444, 2007

ACKNOWLEDGMENTS

We are grateful to the University of Manchester for a postdoctoral research fellowship (to HSH) and for computational resources from the University's Computational Shared Facility.

Effect of Mn^{2+} in β-$NaYF_4$: Eu^{3+} on the morphologies and enhanced luminescence of Eu^{3+} by inhibition the luminescence from Eu^{2+}

Guang-Xin Xiang [1], Zhi-Jun Zhang [2,*], Jing-Tai Zhao [3,*]

[1] School of Materials Science and Engineering, Shanghai University, Shanghai 200072, P. R. China

[2*] School of Materials Science and Engineering, Shanghai University, Shanghai 200072, P. R. China

Email: zhangzhijun@shu.edu.cn

[3*] School of Materials Science and Engineering, Shanghai University, Shanghai 200072, P. R. China

Email: jtzhao@shu.edu.cn

Abstract

Eu^{3+}-Mn^{2+} co-doped $NaYF_4$ micro/nano-crystals with multiple morphologies, such as microrod, hexagonal nanoprisms and nanoplates have been prepared via a facile hydrothermal route. X-ray diffraction (XRD), scanning electron microscopy (SEM) exhibits the uniform and monodispersed hexagonal nano-prism with a size of about 500 nm in height and about 1.0 μm in longest diameter. It is found that trisodium citrate (Cit^{3-}) and pH values rather than Mn^{2+} play important roles in fine-tuning the morphologies. However, broad band emission from the *5d*-*4f* transition of Eu^{2+} and line emissions from *4f*-*4f* transitions of Eu^{3+} are observed in the mixed-valent Eu^{3+}/Eu^{2+}, Mn co-doped $NaYF_4$ with further discussion and confirmed. The internal quantum efficiency of Eu^{3+} increased from 6.31 to 15.13% by inhibition the luminescence of Eu^{2+}. Therefore, Mn^{2+} ions play a key role in controllable emissions from Eu^{3+}/Eu^{2+} co-doped $NaYF_4$.

Key word: hexagonal, $NaYF_4$, Mn^{2+}, Eu, luminescence

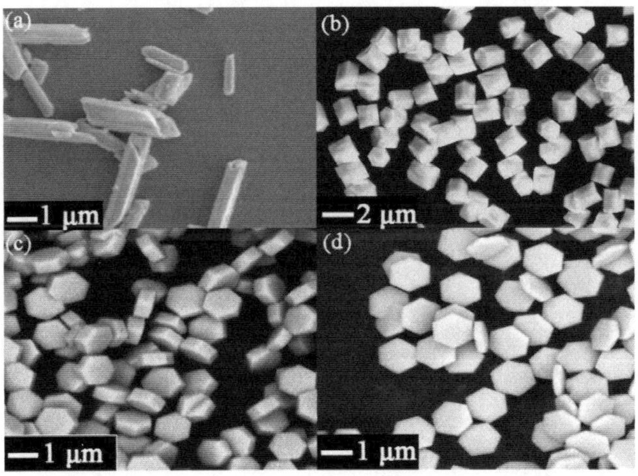

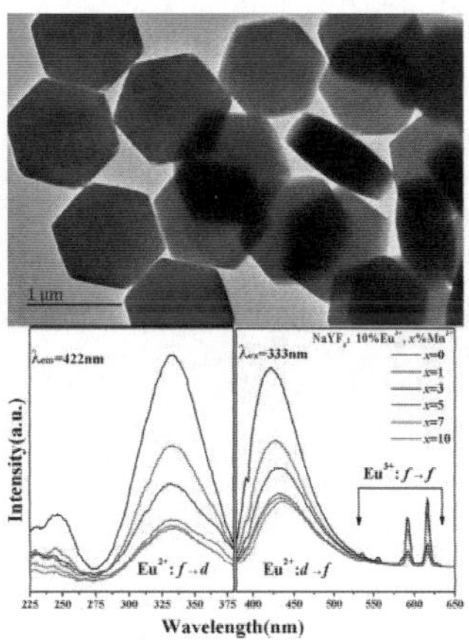

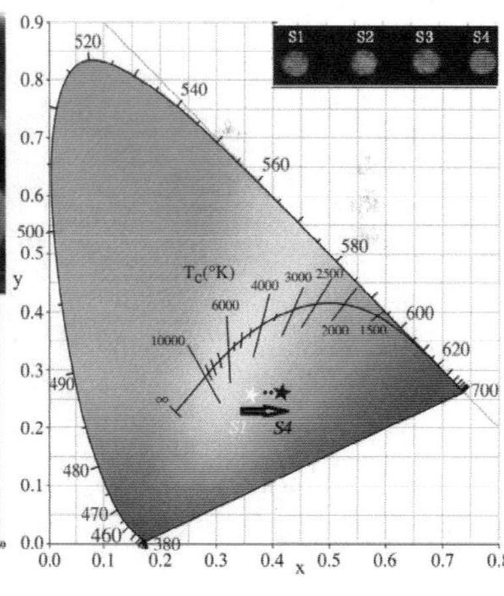

Effective Photo- and Triboluminescent Eu(III) Coordination Polymer with Triangular Spacers

Yasuchika Hasegawa

Faculty of Engineering, Hokkaido University, Japan, hasegaway@eng.hokudai.ac.jp

INTRODUCTION

Lanthanide coordination polymers are generally constructed from lanthanide ions and small organic joining parts with two coordination sites. We have reported luminescent lanthanide coordination polymers linked with bidentate phosphine oxide groups. Combination of bidentate phosphine oxide and low vibrationa frequency $Eu(Hfa)_3$ (hfa: hexafluoroacetyl-acetonate) units are effective for the preparation of strongly luminescent lanthanide coordination polymers that exhibit large emission quantum yields up to 70% under 4f-4f excitation.[1-3]

In this study, we designed novel Eu(III) coordination polymers with rigid triangular spacer ligands tppb (tris(4-diphenylphosphorylphenyl)benzene) for effective photo- and triboluminescence (Figure 1). Tribolumninescence is the emission of light originating from mechanical stress on bulk solid materials. The Eu(III) coordination polymers with triangular spacers are expected to provide close packed structure and effective crystal cracking under mechanical stress. In the present study, photo- and triboluminescence properties of Eu(III) coordination polymer with triangular spacers are demonstrated.[4]

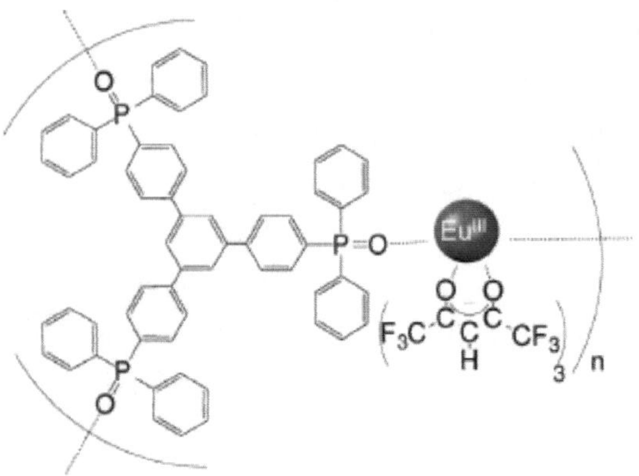

Figure 1 Chemical structure of Eu(III) coordination polymers with rigid triangular spacer ligands

EXPERIMENTAL METHODS

Ratio of $Eu(hfa)_3$ and triangular spacer was estimated by elemental analyses. The emission quantum yield is estimated calculated from the emission spectrum and lifetime. Triboluminescence properties are estinated under shockwave irradiation using Nd:YAG nanosecond pulse laser.

RESULTS AND DISCUSSION

The Eu(III) coordination polymer with triangular spacers are composed of luminescent $Eu(hfa)_3$ units and triangular spacer. Based on the elemental analysis, ratio of $Eu(hfa)_3$ and triangular spacer was estimated to be 2:3. The polymeric structure of $[Eu_2(hfa)_9(dpbp)_2]_n$ were characterized using ESI-MS.

$[Eu_2(hfa)_9(dpbp)_2]_n$ shows high termo-stability (decomposition temperature = 354 ^oC) and photoluminescence quantum yield ($\Phi_{4f\text{-}4f}$ = 82%, photosensitized energy transfer efficiency = 78%). The triboluminescence efficieny of $[Eu_2(hfa)_9(dpbp)_2]_n$ under laser pulse irradiation is caluculated to be 49%. These results indicate that mechanical crushing of molecular scystals leads to enhancement of the non-radiative transition in the quantum process. We consider that the shockwave irradiation promotes effective mechanical crushing of organic crystals and the formation of the excited state of the ligands such as in the X-ray scintillation effect.

CONCLUSIONS

In this study, Eu(III) coordination polymer $[Eu_2(hfa)_9(dpbp)_2]_n$ with effective photo- and triboluminescence properties was performed for bthe first time. Successful observations were achieved using close packing structures of Eu(III) coordination polymer. Highly photosensitized luminescence efficiency for Eu(III) coordination compound should be also key point for observation of tribolunimecence. The effective molecular design and experimental analysis of two-dimensional luminescent lanthanide coordination polymers are expected to open-uo a new frontier between coordination, materials and photophysical chemistry.

REFERENCES

1. Hasegawa Y. *et al.,* Angew. Chem. Int. Ed. 52, 6413-6416, 2013.
2. Hasegawa Y. *et al*., RSC adv. 5:338-353, 2015
3. Hirai Y. and Hasegawa Y. *et al,* Angew. Chem. Int, Ed. 55, 12059-12062, 2016.
4. Hasegawa Y. *et al.,* Chem. Eur. J. in press.

ACKNOWLEDGMENTS

This work was partly supported by Grabts-in-Aid for scientific Research on Innovative Areas of "New Polymeric Materials Based on Element-Blocks (No.2401)" (24102012) of the Ministry of Education, Culture, Sports, Science and Technology (MEXT) of Japan.

Electrochemical Performance of $Ni_{1-x}Fe_x$-SDC Composite Anodes for Intermediate Temperature Solid Oxide Fuel Cell

Lili Zhu, Xiaomei Liu*, Hailin Bi, ShenglongYu, Jialing Sun, Huamin Yu, Li Pei
Key Laboratory of Physics and Technology for Advanced Batteries, College of Physics, Jilin University, Changchun 130012, P. R. China, xiaomeiliu58@sina.com

INTRODUCTION

Intermediate Temperature Solid Oxide Fuel Cells (IT-SOFCs) have recently received considerable attention due to their expected improved stability, reliability and reduced cost. IT-SOFCs operation at lower temperatures could enable the use of inexpensive metal components as interconnects, improve the stability and durability of all components and provide faster start-up and shutdown procedures. As for high performance anodes, doped ceria, particularly samaria-doped ceria SDC ($Ce_{0.85}Sm_{0.15}O_{1.925}$,) together with Ni is most commonly used due to the high ionic conductivity of SDC at intermediate temperature. Since the performance of an anode depends strongly on its composition, fabrication method, and resulting microstructure, it is necessary to investigate how the performance of a Ni-SDC composite anode is infuluenced by its microstructure on a doped ceria electrolyte-based fuel cell. The electrochemical and catalytic properties of Ni might be improved by partially replacing Ni with other metals such as Fe, Co and Cu. We studied the Ni1-xCux -SDC composite anodes [1]. The $Ni_{1-x}Cu_x$ –SDC(x = 0.0–0.15) cermets have been evaluated as anode materials for its potential application in IT-SOFCs. The electrical conductivity of the $Ni_{0.90}Cu_{0.10}$ is higher than Ni. The $Ni_{0.90}Cu_{0.10}$ –SDC anode exhibits the maximum electrical conductivity. And the $Ni_{0.90}Cu_{0.10}$ –SDC cell shows the highest power density. In the present work, Ni1-xFex-SDC powders were prepared using the glycine – nitrate process. The results of XRD phase analysis, XPS valence analysis and SEM of the $Ni_{1-x}Fe_x$ O are herein reported. The electrical conductivities of the $Ni_{1-x}Fe_x$ O under H 2 at 400–800℃ are studied in detail by the DC four-probe method. The effect of the $Ni_{1-x}Fe_x$ -SDC composition is also investigated. In addition, the electrochemical properties of the $Ni_{1-x}Fe_x$ -SDC composite anode cells are studied.

EXPERIMENTAL METHODS (if relevant)

$Ni_{1-x}Fe_xO$ (x=0, 0.1, 0.2, 0.25, 0.3) and SDC were prepared by glycine- nitrate process. $Ni_{1-x}Fe_xO$ and SDC were mixed in the weight ratio of 6:4 to make composite anodes. Current–Voltage measurements of the single cells with a Ba 0.5 Sr 0.5 Co 0.8 Fe 0.2 O 3 – δ –30 wt. % SDC cathode andthe SDC electrolyte were measured.

RESULTS AND DISCUSSION

The surface microstructures of the samples sintered in air at 1000℃ for 5h and reduced in H2 at 800℃for 2h indicated that although no pore formers were used, all samples had a porous structure. The electrical conductivities of the $Ni_{1-}xFe_xO$-SDC under H2 at 400–800 ℃ are studied in detail by the DC four-probe method. The electrical conductivities of the $Ni_{1-x}Fe_x$-SDC increase with increasing Fe content (from x= 0 to x=0.25) and then decrease at x=0.30. To get insight into the effect of Fe-doping on the electronic structure of the $Ni_{1-x}Fe_x$ samples, first-principles calculation was performed to get the band structure and the density of states of the material based on density functional theory using Materials Studio software. The band diagrams of Ni and $Ni_{0.75}Fe_{0.25}$ indicate that the d-band of $Ni_{0.75}Fe_{0.25}$ is wider than Ni and the d-band fluctuations of $Ni_{0.75}Fe_{0.25}$ is larger than Ni .The results show that the electron localization of $Ni_{0.75}Fe_{0.25}$ decreases, the effective mass of $Ni_{0.75}Fe_{0.25}$ decreases, and the conductivity of $Ni_{0.75}Fe_{0.25}$ is enhanced. The cells were operated at 800 ℃ with hydrogen as the fuel and stationary air as the oxidant. The $Ni_{0.75}Fe_{0.25}$-SDC cell shows the highest power density and yields improved performance compared with the Ni-SDC cell.

CONCLUSIONS

The Ni1−xFexO-SDC (x = 0.0–0.30) have been evaluated as anode materials for its potential application in SOFCs. $Ni_{1-x}Fe_xO$-SDC form a well-cubic fluorite structure of SDC and a good $Ni_{1-x}Fe_x$ alloy phase at x = 0.0–0.30. The electrical conductivity of the $Ni_{0.75}Fe_{0.25}$ is higher than Ni. The $Ni_{0.75}Fe_{0.25}$ –SDC anode exhibits the maximum electrical conductivity. And the $Ni_{0.75}Fe_{0.25}$ –SDC cell shows the highest power density.

REFERENCES

[1] L. L. Zhu, X. M. Liu, F. Han, J. L. Sun, H. L. Bi, H. P. Wang, S. L. Yu, L. Pei. Ni1-xCux-SDC anodes for intermediate temperature solid oxide fuel cell. Solid State Ionics 288 (2016) 115-119.

ACKNOWLEDGMENTS

This work was supported by the National Natural Science Foundation of China (Nos. 51272087 and 50872041) and the National Foundation for Fostering Talent in Basic Science of China (No. J1103202)

Electronic Structure in Lanthanide and Actinide Dialuminides from Al K-edge XAS

Stefan Minasian[1*], Alison Altman[1], Eric Bauer[2*], Corwin Booth[1*], Joseph Pacold[1], David Prendergast[3], David Shuh[1*], Tolek Tyliszczak[3], and Jian Wang[4]

[1*]Chemical Sciences Division, Lawrence Berkeley National Laboratory, United States, sgminasian@lbl.gov
[2]Materials Physics and Applications Division, Los Alamos National Laboratory, United States
[3]Molecular Foundry and Advanced Light Source, Lawrence Berkeley National Laboratory, United States
[4]Canadian Light Source, Canada

INTRODUCTION

Several recent studies have challenged the traditional viewpoint that the 4f and 5f orbitals of lanthanides and actinides are core-like and that f electrons are localized to the metal.[1] In some cases, indirect exchange processes mediated by the d orbitals have been invoked to explain unusual electronic, magnetic, and structural phenomena.[2] Our recent oxygen K-edge X-ray absorption spectroscopy (XAS) studies of lanthanide sesquioxides showed that changes in O 2p mixing with the 5d states were directly related to established trends in the chemical behavior of the 4f electrons.[3] However, the interplay between f and d orbitals is still difficult to probe experimentally.

For more than 60 years, lanthanide and actinide dialuminides ($LnAl_2$ and $AnAl_2$) have been employed as prototypical correlated electron systems because they all have the same cubic, Cu_2Mg structure.[2] In this study, aluminum K-edge XAS and density functional theory (DFT) calculations were used to evaluate the electronic structure of the dialuminides, where Ln = Ce, Sm, Eu, Yb, Lu and An = U and Pu. Because the d orbitals are more diffuse than the f orbitals, we hypothesized that enhancements in d orbital mixing would be closely tied to energetic stabilization due to orbital mixing.[4]

EXPERIMENTAL METHODS

Samples of the dialuminides were prepared by arc- or induction-melting of the elements. Al K-edge XAS measurements were conducted with the scanning transmission X-ray microscopes (STXMs) at the Advanced Light Source (11.0.2) and Canadian Light Source (10-ID1). Al K-edge XAS were simulated using the eXcited-electron Core-Hole XCH approach.

RESULTS AND DISCUSSION

Fig. 1 shows the background-subtracted and normalized Al K-edge XAS for selected dialuminides $YbAl_2$, $LuAl_2$, and $PuAl_2$. The edge onset for each alloy is 0.4 to 0.8 eV lower in energy than observed for pure Al metal and – in the case of $YbAl_2$ and $LuAl_2$ – well-resolved features are observed. XCH-DFT calculations were used to develop spectral assignments, which showed that the first low energy features near 1558.5 eV involved directional, σ -type bonding interactions between the Al 3p and Ln or An d orbitals. At higher energy, the broad features observed at 1565-1566 eV involved states derived from a σ-antibonding interaction between the Al 3p and Ln or An d orbitals. A simple theoretical framework was developed to rationalize these results, which suggested that the increases in mixing with the high energy Al 3p orbitals were correlated with both the energy and radial extension of the 5d and 6d orbitals.

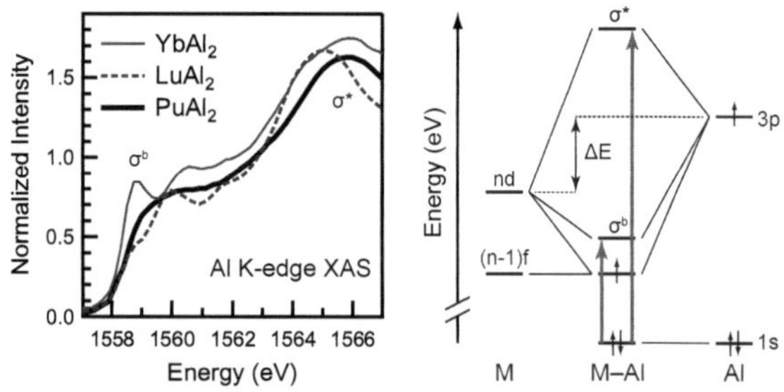

Fig 1. Al K-edge XAS of selected MAl_2 alloys (M = Yb, Lu, Pu) and proposed molecular orbital model.

CONCLUSIONS

These findings may have important implications for controlling bond energies, phase stabilities, and mechanical properties of actinide intermetallics by enhancing overlap with the diffuse 6d orbitals.[5]

REFERENCES

1. Bauer, E. D., Thompson, J. D., Annu. Rev. Condens. Matter Phys.; J. S. Langer, Ed. 6:137-153, 2015.
2. Paudyal, D., Pecharsky, V. K., Gschneidner, K. A., Jr., J. Appl. Phys. 115:17E127, 2014.
3. Altman, A. B., Pacold, J. I., Wang, J., Lukens, W. W., Minasian, S. G., Dalton Trans. 45:9948-9961, 2016.
4. Bursten, B. E., Strittmatter, R. J., Angew. Chem. Int. Ed. 30:1069-1085, 1991.
5. Hecker, S. S., Metall. Mater. Trans. A 35A:2207-2222, 2004.

ACKNOWLEDGMENTS

This work was supported by the Director, Office of Science, Office of Basic Energy Sciences, Division of Chemical Sciences, Geosciences, and Biosciences Heavy Element Chemistry Program of the U.S. Department of Energy (DOE) at LBNL under Contract No. DE-AC02-05CH11231. Work at Los Alamos National Laboratory was performed under the auspices of the US Department of Energy, Office of Basic Energy Sciences, Division of Materials Sciences and Engineering.

Electronic Structure, Magnetic Properties, and Magnetocrystalline Anisotropy of Lanthanide Dialuminides

Timothy A. Hackett,[1*] Durga Paudyal,[1] and Vitalij K. Pecharsky[1,2]
[1*]Ames Laboratory, Iowa State University, Ames, IA 50011, USA, thackett@ameslab.gov
[2]Department of Materials Science and Engineering, Iowa State University, Ames, Iowa, 50011-2300, USA

INTRODUCTION

The theoretical understanding of interactions between localized 4*f* and mobile 5*d* electrons is important because of their key role in the model magnetostructural materials for magnetocaloric applications.[1,2] Lanthanide dialumindes crystalize in the cubic Laves phase at room temperature. The R atoms have tetrahedral (Td) site symmetry and aluminum atoms are in the D3d local symmetry.[3] The orbital moment of Ce in $CeAl_2$ is quenched.[4] $PrAl_2$ shows 4*f* crystal field (CF) splitting and undergoes tetragonal distortion.[2] In contrast, $NdAl_2$ remains cubic at low temperature. $SmAl_2$ has been found to have 4*f* spin and 4*f* orbital magnetic moments which cancel each other.[5] Earlier calculations have shown $ErAl_2$ to have weak 5*d* exchange and strong 5*d* crystal field splitting.[6] The direction dependent magnetic moments, magnetocrystalline anisotropy, and the effect of 4*f* and 5*d* electrons on the crystal lattice structure in lanthanide dialumindes have not been investigated. Therefore, in this study we focus on the direction dependent magnetism and magnetic anisotropies within the observed crystal structures.

THEORETICAL METHODS

We have employed local spin density approximation including onsite electron correlation and spin orbit coupling (LSDA+U+SOC) to calculate magnetic and magnetocrystalline anisotropy properties of lanthanide dialuminides. The k-space integrations have been performed with at least 16×16×16 Brillouin zone mesh which was sufficient for the convergence of total energies (10^{-6} Ryd.), charges, and magnetic moments. We calculated magnetocrystalline anisotropy energy (MAE) as the total energy difference with magnetic moment aligned along the planar and c-axis directions.

RESULTS AND DISCUSSION

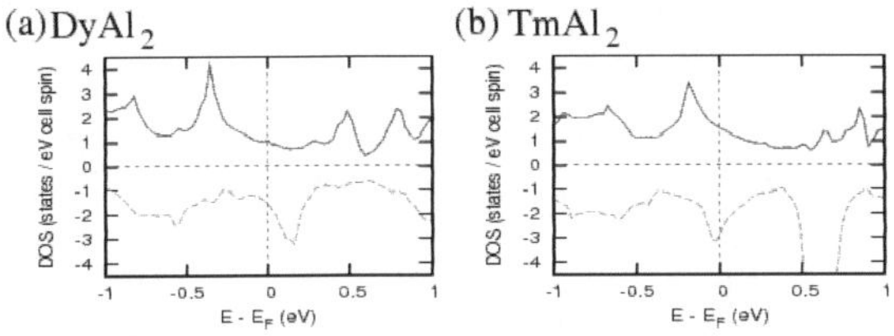

FIG. 1. Density of States (DOS) of DyAl2 (a) and TmAl2 (b).

The calculated spin and orbital moments follow the trend of Hund's rule, however, the orbital moments are underestimated indicating partial quenching in the light lanthanide dialuminides. The total DOS around the Fermi level shows the 5*d* exchange splitting and 5*d* crystal field splitting that are responsible for magnetic and structural transformations (Fig. 1). The asphericity of electron density indicates the anisotropic nature of dialumindes in their valence differences in RAl_2. The calculated magnetocrystalline anisotropy considering the cubic Laves phase indicates possibilities of structural deformations. As for example, $DyAl_2$ and $TmAl_2$ undergo tetragonal and rhombohedral deformation.

In lanthanide dialuminides, complicated Fermi surface structures represent the velocities of electrons that are related to the metallic conductivity.[7] We present how the 4*f* electrons have an effect on the occupation of electronic bands, and thus the magnetism and magnetic anisotropy.

CONCLUSIONS

We have presented anisotropic and magnetic calculations with the all-electron APW method. The lanthanide dialuminides exhibit orbital quenching and follow Hund spin rule. The reported magnetocrystalline anisotropy indicated competing c-axis and in-plane anisotropies in each lanthanide dialuminides. The charge densities, onsite electron correlation, and spin orbit coupling are highly dependent on the 4*f* states. Partial orbital quenching gave rise to crystalline electric field splitting.

REFERENCES

[1]H. -G. Purwins, A. Leson, Adv. in Phys. **39**, 309 1990.
[2]V. K. Pecharsky *et al.*, J. Magn. Magn. Mater. 2009, **321**, 3541.
[3]J. R. Cullen, A. E. Clark, Phys. Rev. B **15**, 4510 1977.
[4]M. Sagawa *et al.*, IEEE Trans. Magn. 1984, **20**, 1584; J. F. Herbst, Rev. Mod. Phys. 1991, **63**, 819; R. Skomski and J. M. D. Coey, Phys. Rev. B, 1993, **48**, 15812.
[5]H. J. Gotsis and I. I. Mazin, 68, 224427-1 2003.
[6]D. Paudyal *et al.*, J. Phys.: Condens. Matter 25, 396002 2013.
[7]A. H. Reshak, S. A. Khan, Journal of Magnetism and Magnetic Materials 354 (2014) 216-221.

ACKNOWLEDGMENTS

The Ames Laboratory is operated for the U. S. Department of Energy by Iowa State University of Science and Technology under contract No. DE-AC02-07CH11358. This work was supported by the Department of Energy, Office of Basic Energy Sciences, Materials Sciences Division.
T. H. would like to strongly thank Ed Moxley for all of the technical support and Manish K. Kashyap for WIEN2k and other needed guidance.

Engaging the terminal: promoting supramolecular assembly with the uranyl oxo atoms

Korey P. Carter[1], Mark Kalaj[1], Robert G. Surbella III[1], Lucas C. Ducati[2], Jochen Autschbach[3], and Christopher L. Cahill[1*]

[1*]Department of Chemistry, The George Washington University, Washington, DC, 20052, USA
koreypcarter@gwmail.gwu.edu
[2]Department of Fundamental Chemistry, Institute of Chemistry, University of São Paulo, P.O. Box 26077, 05513-970, São Paulo, SP, Brazil
[3]Department of Chemistry, University at Buffalo, State University of New York, Buffalo, New York, 14260, USA

INTRODUCTION

Engaging the nominally terminal oxo atoms of the linear uranyl (UO_2^{2+}) cation in non-covalent interactions represents both a significant challenge and opportunity within the field of f-element hybrid materials. We have developed an approach for promoting oxo atom participation in a range of such interactions through judicious choice of appropriately polarizable halogen-donor atoms and strongly electron donating equatorial ligands,[1] and herein we aimed to identify means to *systematically promote non-covalent assembly* with the uranyl oxo atoms.

EXPERIMENTAL METHODS

Materials have been prepared via hydrothermal syntheses and characterized via single crystal X-ray diffraction, Raman, Infrared (IR), and luminescence spectroscopy, as well as with electrostatic surface potential calculations.

RESULTS AND DISCUSSION

Delineation of criteria for oxo participation in halogen bonding interactions has been achieved by preparing materials containing 2,5-dichloro-, 2,5-dibromo-, or 2,5-diiodobenzoic acid coupled with 2,2'-bipyridine (bipy), 1,10-phenanthroline (phen), 2,2':6',2''-terpyridine (terpy), or 4'-chloro-2,2':6',2''-terpyridine (Cl-terpy)). In organic systems it has been shown that single atom modifications to a molecular structure can determine the primary interaction(s),[2] and we observed similar trends in uranyl materials featuring terpy and Cl-terpy as demonstrated by XRD, vibrational spectroscopy, and sigma hole calculations (Table 1).

Looking at the modes of supramolecular assembly in 2,5 dihalobenzoic acid materials we note the importance of *halogen polarizability* in promoting halogen bonding interactions with the uranyl oxo atoms. Whereas we noted that with 2,5-*diiodo*benzoic acid occurrences of oxo interactions appeared to be the norm,[1] we only observe one example (out of six complexes) with 2,5-*dichloro*benzoic acid and two instances (out of four complexes) with 2,5-*dibromo*benzoic acid.

Table 1 Symmetric (ν_1) and asymmetric (ν_3) stretches for 2,5-dihalobenzoic acid TPY and Cl-TPY materials along with electrostatic surface potentials of oxo atoms (kJ/mol) for all six complexes.

Complex	**υ_1 (cm^{-1})**	**υ_3 (cm^{-1})**	**Sup. Int.**	**ESP of Oxo Atoms (kJ/mol)**
[UO_2 ($C_{15}H_{11}N_3$) ($C_7H_3\mathbf{Cl_2}O_2)_2$]	833	919	Cl-Cl	-135.7 -127.8
[UO_2 ($C_{15}H_{11}N_3$) ($C_7H_3\mathbf{Br_2}O_2)_2$]	834	920	Br-Br	-136.8 -123.4
[UO_2 ($C_{15}H_{11}N_3$) ($C_7H_3\mathbf{I_2}O_2)_2$][1]	829	910	I-oxo (x2)	-118.9 -150.9
[UO_2 ($C_{15}H_{10}ClN_3$) ($C_7H_3\mathbf{Cl_2}O_2)_2$]	836	921	H-Bonding	-133.8 -120.2
[UO_2 ($C_{15}H_{10}ClN_3$) ($C_7H_3\mathbf{Br_2}O_2)_2$]	836	921	Br-Br	-133.0 -115.5
[UO_2 ($C_{15}H_{10}ClN_3$) ($C_7H_3\mathbf{I_2}O_2)_2$][1]	826.5	910	I-oxo	-140.9 -112.6

CONCLUSIONS

Our findings indicate that there is a synergistic interplay between electron donating equatorial ligands and benzoic acid groups featuring multiple large, polarizable halogen atoms that stimulates oxo participation in non-covalent assembly, yet it is the *polarizability* of the selected halogen atom that ultimately promotes halogen bonding interactions with the uranyl oxo atoms.

REFERENCES

1. Carter K. P. *et al.*, Inorg. Chem. Front. 4:165-178, 2017.
2. Shirman T. *et al.*, Cryst. Growth Des. 15: 4756-4759, 2015.

ACKNOWLEDGMENTS

This material was supported by the U. S. Department of Energy—Chemical Sciences, Geosciences and Biosciences Division, Office of Basic Sciences, Office of Science, Heavy Elements Program, under grant number DE-FG02-05ER15736. K. P. C. would also like to acknowledge the George Washington University for a Presidential Merit Fellowship award.

Enhanced cooling power of ball-milled Gd_5Si_4

S. M. Harstad,[1] A. A. El-Gendy,[1,2] S. Gupta,[3] V. K. Pecharsky,[3,4] and R. L. Hadimani[1]

[1]Department of Mechanical and Nuclear Engineering, Virginia Commonwealth University, USA
[2]Nanotechnology and Nanometrology Lab, National Institute for Standards, Egypt
[2]Ames Laboratory, Division of Materials Science and Engineering, U.S. Department of Energy, USA
[3]Department of Materials Science and Engineering, Iowa State University, USA, smharstad@vcu.edu

INTRODUCTION

Recently there has been much effort in research directed at materials with a magnetocaloric effect (MCE) near room temperature for refrigeration applications.[1] The MCE is described by a change in entropy (ΔS_M) and/or temperature (ΔT_{ad}) of a material in response to an applied magnetic field. Ball milled samples generally exhibit smaller maximum ΔS_M values than their bulk material counterparts but the MCE occurs over a broader range of temperatures, potentially increasing their relative cooling power.[2] Gd_5Si_4 is an attractive option due to its magnetostructural transition at ~335K yielding near room temperature cooling potential.

EXPERIMENTAL METHODS

Polycrystalline sample of Gd_5Si_4 was synthesized by arc-melting of the stoichiometric mixture of commercial grade Gd and Si in an argon atmosphere. The sample was remelted six times turning over each time to ensure homogeneity. Using commercial grade Gd leads to the formation of a small amount of Gd_5Si_3 impurity in the predominantly Gd_5Si_4 matrix. To obtain sub-micrometer particles of Gd_5Si_4, the as-cast bulk material was first ground in an agate mortar and screened to eliminate particles larger than 53 μm. Further reduction in the particle size was achieved by high-energy ball-milling of the powder in a SPEX 8000M mill. The milling procedure used a powder to ball weight ratio of 5:1 with 2 balls of 11 mm diameter and 4 balls of 4 mm diameter. 10 wt% poly(ethylene glycol) (PEG) (molecular weight-8000 Da) was added to the Gd_5Si_4 powder to act as a surfactant during milling. A two-step milling process was used in which the powder and PEG mixture was first milled for 1 h at which point the milling was interrupted and 5 ml of heptane was added to the vial. The milling was then continued for another 1h for sample S1 and 2 h for S2. After milling, the PEG was removed by washing with deionized water.

RESULTS AND DISCUSSION

After the first hour of milling, the sample consisted of flakes that coated the walls of the vial indicating that the PEG was well integrated with the Gd_5Si_4 powder. Upon milling with heptane and drying, the powder appeared to have very large surface area and fine particle size. Magnetization (M) isotherms of 0-3 T were collected from 280K-350K in order to calculate the magnetic entropy change via the Maxwell relation[2]

$$\Delta S_M = \int_0^H \frac{\delta M}{\delta T} dH$$

where T is the absolute temperature and H is the applied magnetic field. Figure 1 shows the magnetic entropy change and the magnetization with respect to temperature for the milled samples, S1 and S2, as well as bulk Gd_5Si_4. The MCE of the milled samples has decreased in magnitude and broadened over a larger temperature range. The peak ΔS_M value has also shifted to lower temperature in the milled samples. The shift follows the same trend as the magnetic transition temperature in ball milled Gd_5Si_4.[3] Increased milling time has decreased ΔS_M with little impact on broadening the response peak.

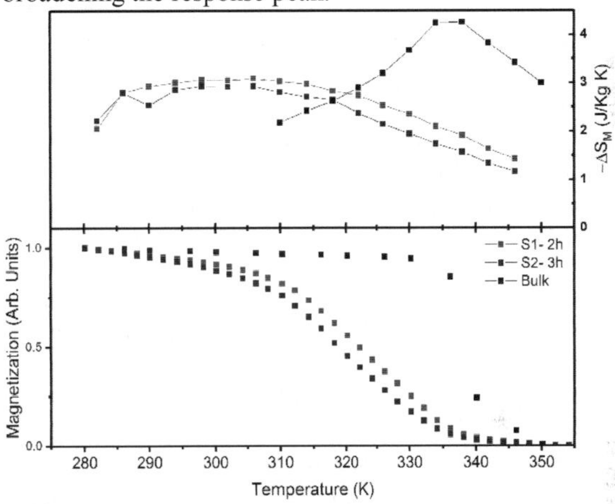

CONCLUSIONS

Gd_5Si_4 was ball-milled using PEG as a surfactant and heptane as a liquid processing agent. The resultant powder was sub-micron, maintained crystallinity and the magnetic transition temperature was reduced to 325K from the bulk value of ~340K. The MCE was decreased as expected but it remains substantial and nearly constant over a wide temperature range between ~280K and ~320K.

REFERENCES

1. Mathur N. and Moya X., Phil. Trans. Royal Soc. 374A: 20150314, 2016.
2. Gschneidner, K.A., Jr. and Pecharsky, V.K., Annu. Rev. Mater Sci. 30:387-429, 2000.
3. Hadimani R. L., *et al*, IEEE Trans. Mag. 51, 2015

ACKNOWLEDGMENTS

Synthesis and materials processing were done at the Ames Laboratory. The Ames Laboratory is operated for the U. S. Department of Energy by Iowa State University of Science and Technology under contract No. DE-AC02-07CH11358.

Enhancement in Nd-Fe-B Coercivity: Dy Diffusion or Heat Treatment?

Kinjal Gandha[1], Matthew J. Kramer[1, 2] and Ikenna C. Nlebedim[1]

1. Critical Materials Institute 2. Division of Materials Science and Engineering
Ames Laboratory, United States DOE, Ames, IA 50011, USA
kinjalgandha@ameslab.gov

INTRODUCTION

A means of conserving critical REE dysprosium for magnet application, coercivity and its temperature dependence can be improved by diffusing Dy into sintered Nd-Fe-B. This requires coating sintered Nd-Fe-B with DyF_3 or Dy_2O_3 and facilitating diffusion by heat treatment. Coercivity of Dy-diffused Nd-Fe-B is usually improved compared to the starting materials.[1] This has been shown to be due to effect of Dy in limiting demagnetization at surfaces and edges where demagnetization is highest.[2]

In this study, Dy was diffused into sintered commercial Nd-Fe-B magnets at temperatures 600 – 900 °C. 2nd quadrant magnetic hysteresis measured at 300 K is shown Fig. 1. Fig. 2 shows the variation of coercivity and the energy product with dysprosium diffusion temperature. Both figures show enhancement in coercivity for dysprosium treated materials as others have observed.

However, the assumption is often that magnetic properties improvements are only due to the diffused Dy. This presentation will show that such assumption may be over-quantifying the effect of Dy in enhancing coercivity. This study provides insight into quantifying the true gain in magnetic properties via Dy diffusion into sintered Nd-Fe-B magnets.

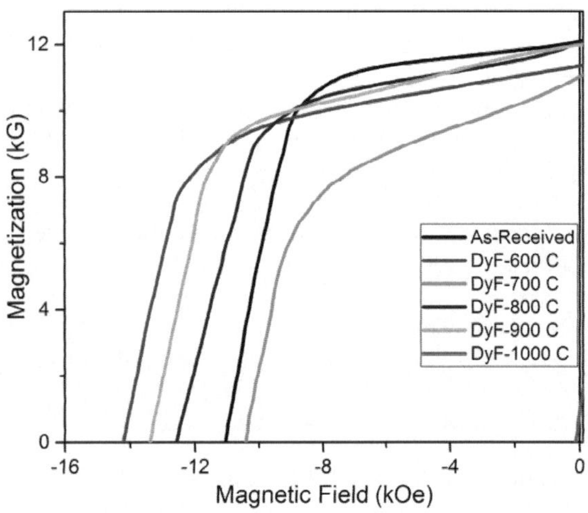

Fig. 1: Second quadrant demagnetization plots of samples used for study

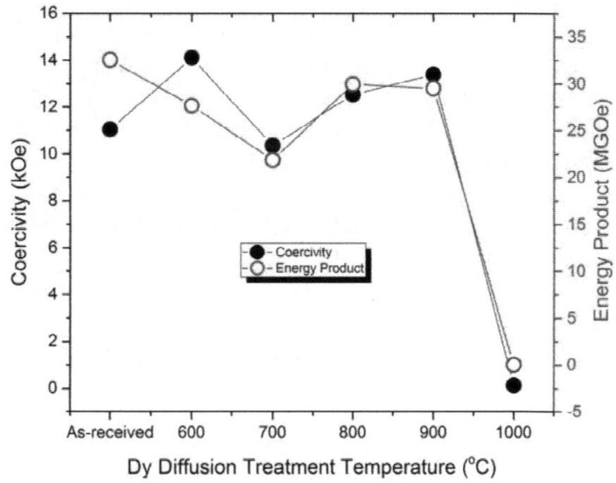

Fig. 2: Coercivity and energy product of samples before and after dysprosium diffusion at different temperatures.

REFERENCES

1. W. Tang, K. W. Dennis, M. J. Kramer, I. E. Anderson, and R. W. McCallum, J. Appl. Phys. 111, 07A736 (2012).

2. Ucar, Huseyin, David S. Parker, I. C. Nlebedim, R. W. McCallum, S. K. McCall, M. Parans Paranthaman, Adv. Mater. Res., Vol 4.

ACKNOWLEDGMENTS

This work was supported by the Critical Materials Institute, an Energy Innovation Hub funded by the U.S. Department of Energy, Office of Energy Efficiency and Renewable Energy, Advanced Manufacturing Office. The work was supported by the U.S. Department of Energy (DOE), Office of Science, Basic Energy Sciences, Materials Science and Engineering Division. The research was performed at the Ames Laboratory, which is operated for the U.S. DOE by Iowa State University under contract # DE-AC02-07CH11358.

Enhancing lanthanide upconversion with molecular concentrators, micro pillars and energy-looping nanoparticle-based lasers

P. James Schuck[1*], David Garfield[1,2], Angel F. Bravo[1], Elizabeth Levy[1], Bining Tian[1], Cheryl Tajon[1], Edward S. Barnard[1], Maysamreza Chamanzar[2,3], Jillian Iafrati[4], Michel M. Maharbiz[2], Vikaas S. Sohal[4], Nicholas J. Borys[1], Emory M. Chan[1], Bruce E. Cohen[1]

[1]The Molecular Foundry, Lawrence Berkeley National Lab, USA
[2]University of California, Berkeley, USA
[3]Carnegie Mellon University, USA
[4]University of California San Francisco, USA
[*]pjschuck@lbl.gov

ABSTRACT
Though their unique optical properties are proving advantageous for numerous applications, an ongoing challenge facing uponverting nanoparticles remains achieving satisfactory upconverted signal and quantum yield while illuminating at low fluences. Here, I will describe the pros and cons of various strategies used to improve UCNP signals and efficiencies. I will follow this with a discussion of current efforts at our Facility aimed at applying these strategies for deep-tissue imaging, biosensing, optogenetics, and solar light harvesting. This will include demonstrations of imaging of ucpn-based micropillars and microbeads through >2mm of tissue, continuous-wave multicolor upconverting microlasers, and sub-20nm UCNPs with >5% quantum yield.

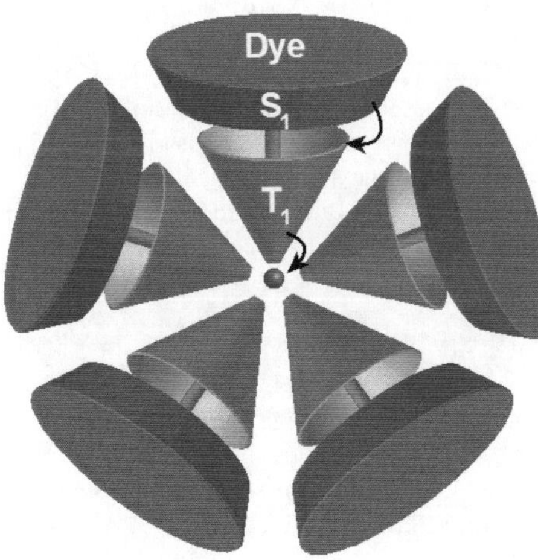

Figure 1 is a schematic of the concept behind dye-sensitized upconverting nanoparticles. Here, the dye acts as an antenna for capturing the near-infrared excitation light, and energy transfer between the dye and the UCNP is mediated by the molecule's triplet state. The relative sizes of the dye antennas vs. the UCNP are scaled to roughly match their relative absorption cross-sections.

Figure 2 is an illustration of the 5 μm sized whispering gallery mode (WGM) upconverting microlasers.

ACKNOWLEDGMENTS
Work at the Molecular Foundry was supported by the Office of Science, Office of Basic Energy Sciences, of the U.S. Department of Energy under Contract No. DE-AC02-05CH11231. S. T. acknowledges funding from NSF DMR-1552220..

Enhancing Magnetic Functionality with Scandium in Gadolinium-based Rare-Earth Alloys

Yaroslav Mudryk,[1*] Durga Paudyal,[1] Jing Liu,[1,2,3] Vitalij K. Pecharsky[1,2]

[1*]Division of Materials Science and Engineering, Ames Laboratory of US DOE, Iowa State University, Ames, Iowa, 50011-3020, USA, slavkomk@ameslab.gov
[2]Department of Materials Science and Engineering, Iowa State University, Ames, Iowa, 50011-2300, USA
[3]Current Affiliation: Seagate Technology, Bloomington, MN, USA

INTRODUCTION

In magnetostructural transformations (MSTs) a strong coupling between magnetic and crystal lattices leads to large simultaneously occurring changes in both magnetic state and crystal structure. Such changes can be controlled by external thermodynamic stimuli such as temperature, hydrostatic pressure, and applied magnetic field as well as by tuning the chemical composition. The MSTs commonly occur in R_5T_4 family of intermetallic compounds (R= rare earth element, T= group 13-15 element) resulting in a wealth of interesting physical phenomena such as giant magnetocaloric effect, giant magnetostriction, giant magnetoresistance, spontaneous generation of voltage, acoustic emissions, kinetic arrest, and magnetic deflagration.[1] This magnetoresponsive behavior is usually observed in systems where R is a magnetic rare earth element such as Gd; the traditionally non-magnetic elements such as Sc are not viewed favourably with respect to potential magnetic applications.

EXPERIMENTAL METHODS (if relevant)

The experimental investigation was performed on polycrystalline arc-melted alloys. Magnetic measurements and temperature-dependent X-ray powder diffraction were main methods of the study. The local spin density approximation including onsite 4f electron correlation and spin orbit coupling (LSDA+U+SOC) approach has been employed to understand magnetostructural behaviours of $(Gd_{0.8}Sc_{0.2})_5Ge_4$.

RESULTS AND DISCUSSION

In the Gd_5Ge_4 – Sc_5Ge_4 pseudobinary system the end members are not ferromagnetic: Gd_5Ge_4 orders antiferromagnetically at 130 K and Sc_5Ge_4 is a Curie-Weiss paramagnet with no magnetic ordering. However, the alloys in the middle part of the system (leaning towards the Gd-rich side of the diagram) show ferromagnetic behaviour (see Figure 1).
A combination of many factors such as chemical pressure effect, increased strength of the *3d-5d* hybridization, and the introduction of Sc at the magnetically important intraslab 4(c) site occupied by rare earth element, is responsible for this interesting phenomenon.

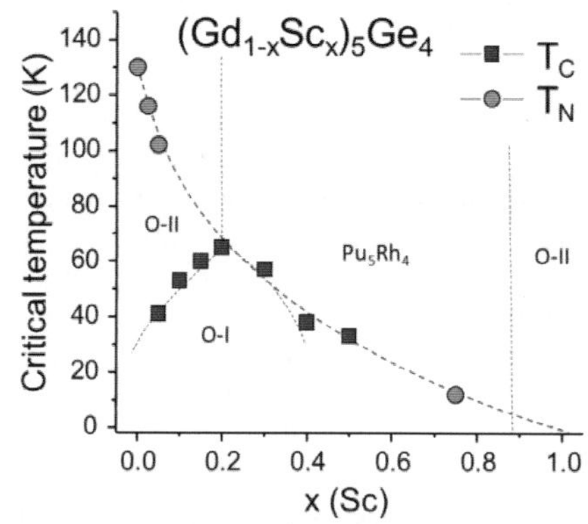

Figure 1. Tentative composition (x)-temperature (T) magnetic phase diagram of the $(Gd_{1-x}Sc_x)_5Ge_4$ pseudobinary system.

Further increase in Sc concentration results in dilution effect and loss of magnetism. The low-temperature crystallographic changes and their correlation with the measured magnetocaloric effect in both first- and second-order alloys are determined.

CONCLUSIONS

Our study proves that it is not wise to overlook compound's magnetism based on the assumed non-magnetic behaviour of the substituting element: a replacement of strongly magnetic gadolinium with weakly magnetic scandium unexpectedly enhances ferromagnetic interactions in $(Gd_{1-x}Sc_x)_5Ge_4$.

REFERENCES

1. Mudryk, Y.; Pecharsky, V.K.; Gschneidner, Jr., K.A., R_5T_4 compounds. An extraordinary versatile model system for the solid state science, in: Bunzli, J.-C. and Pecharsky, V.K. Eds., Handbook on the Physics and Chemistry of Rare Earths, Elsevier, 2014, **44** pp. 283-449.

ACKNOWLEDGMENTS

The Ames Laboratory is operated for the U. S. Department of Energy by Iowa State University of Science and Technology under contract No. DE-AC02-07CH11358. This work was supported by the Department of Energy, Office of Basic Energy Sciences, Materials Sciences Division.

Evaluating Bean-Rodbell Assumption on the Linear Relation between the Exchange Constant and the Volume Change in La(Fe,Mn,Si)$_{13}$

Henrique Neves Bez[1], Yaroslav Mudryk[1], Kaspar K. Nielsen[2], Christian Bahl[2], Vitalij K. Pecharsky[1,3]

[1]The Ames Laboratory of the US DOE, Iowa State University, Ames, Iowa 50011-3020, USA, hnbez@ameslab.gov
[2]Department of Energy Conversion and Storage, Technical University of Denmark - Frederiksborgvej 399, DK-4000 Roskilde – Denmark
[3]Department of Materials Science and Engineering, Iowa State University, Ames, Iowa, USA

INTRODUCTION

Many representatives of the La(Fe,Si)$_{13}$–based family exhibit first order magneto-elastic transitions, during which the ferromagnetic ordering/disordering is accompanied by a phase volume change, without changing symmetry. Strong magneto-elastic coupling leads to strong magnetocaloric effects, which are attractive for magnetic cooling. The model commonly used to describe the coupling relevant to magnetocalorics is the Bean-Rodbell model [1]. This model assumes a linear relationship between the exchange constant and the phase volume change in order to describe the transition. In this work, we experimentally evaluate whether the assumption holds for La(Fe,Mn,Si)$_{13}$ through measurements of the unit-cell volume as a function of temperature and magnetic field using a custom-built X-ray diffractometer.

EXPERIMENTAL METHODS

La(Fe,Mn,Si)$_{13}$H$_y$ has been provided by Vacuumschmelze GmbH. The sample was dehydrogenated in a vacuum furnace at 773 K during one hour. The structure of the material was evaluated in a custom-built X-ray diffractometer under different fields and temperatures while collecting the powder diffraction data during heating. Rietveld refinement was used to calculate the unit cell parameters and phase fractions. Magnetic measurements were performed in a vibrating sample magnetometer LakeShore 7407.

RESULTS AND DISCUSSION

The magnetic measurements, Figure 1, show a discontinuous magnetic transition around 180 K with a hysteresis of about 3 K at 1 kOe.

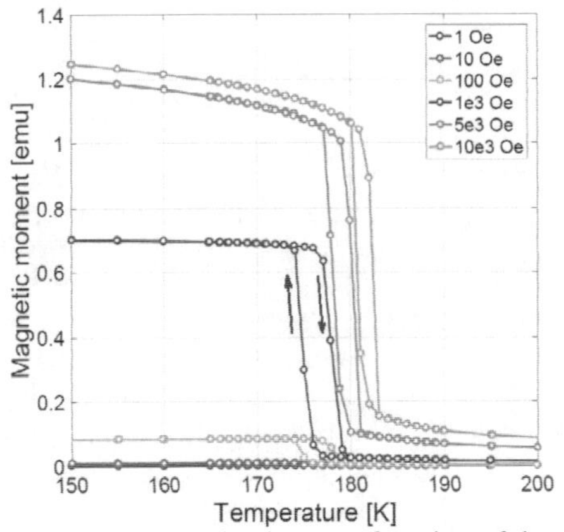

Figure 1: Magnetization as a function of the temperature for different applied fields.

Both the discontinuity and hysteresis signify a first order phase transition, which is further verified with the XRD measurements. Figure 2 depicts the unit cell volume as a function of temperature and field. There is a region of about 6 K where there is the co-existence of the paramagnetic and ferromagnetic phases. This has previously been observed on its hydrogenated twin [2]. The phase volume behaviour closely mimics the magnetization, which indeed has been predicted in general by Bean and Rodbell [1], when the FM phase adopts the larger volume, which is the case here.

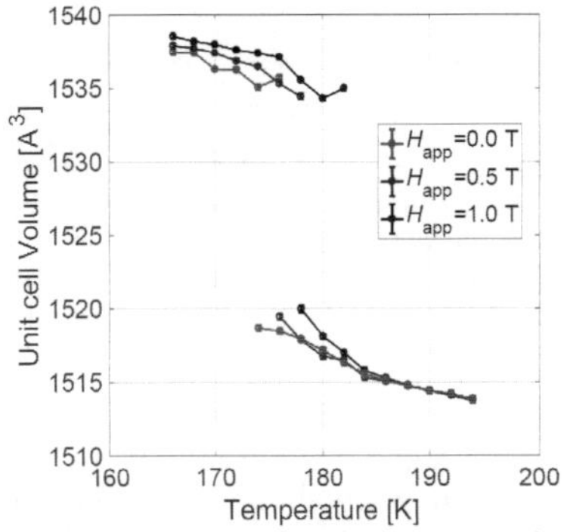

Figure 2: Unit cell volume of La(Fe,Mn,Si)$_{13}$ as a function of temperature and field.

CONCLUSIONS

The similarity of behaviours of the magnetization and phase volume indicates that at least the assumption that the exchange constant is related to the volume is correct. Our presentation will provide a detailed analysis of XRD and magnetization both measured as functions of temperature in various magnetic field using identical materials.

REFERENCES

1. Bean C.P. *et al.*, Phys. Rev. Let. 126:104-115, 1962.
2. Bez H.N. *et al.*, AIP Adv. 6:056217, 2016.

ACKNOWLEDGMENTS

The authors would like to thank Vacuumschmelze GmbH for providing the sample. CaloriCool™ is supported by the United States Department of Energy, Office of Energy Efficiency and Renewable Energy though its Advanced Manufacturing and Building Technologies Offices. Ames Laboratory is supported by the Office of Science, Basic Energy Sciences Programs and operated by Iowa State University under contract No. DE-AC02-07CH11358 with the United States Department of Energy. This work was in part financed by the ENOVHEAT project which is funded by Innovation Fund Denmark (contract no 12-132673).

Evidence of REE Mobility and Fossil Ion Adsorbed REE Deposits in Pennsylvania

Tracy Bank[1,2*], Bret Howard[1], Elliot Roth[1,2], and Evan Granite[1]

[1]National Energy Technology Laboratory, Pittsburgh, PA, USA
[2]AECOM, National Energy Technology Laboratory, Pittsburgh PA tracy.bank@netl.doe.gov

INTRODUCTION

Evidence for fossil ion adsorbed REE deposits in coal and sedimentary rocks surrounding coal seams in Pennsylvania is presented. These rocks formed under near tropical conditions with clastic input from REE-enriched granitic rocks of Appalachia[1], possibly similar to the conditions that formed the ion adsorbed REE deposits in China[2]. In this study, samples are investigated to determine their total REE concentration, the mineralogy of the REE phases, and for geochemical evidence of REE mobilization.

EXPERIMENTAL METHODS

For this study, 61 samples were collected from two geologically correlated sampling locations, the Lower Kittanning coal seam in the Allegheny Formation and the Buck Mountain coal seam in the Llewellyn Formation. Both deposits formed contemporaneously ~ 300Ma but experienced different geologic conditions post peat deposition. Where possible, samples of 2" drill core were collected at regular intervals through overburden, coal, and underclay. Samples were dried and ashed and REE and other element concentrations were determined using ICP-MS after molten salt fusion and acid digestion[3]. Scanning electron microscopy (SEM) with energy dispersive spectroscopy (EDS) was used to identify REE-enriched regions of interest and laser ablation (LA)-ICP-MS was used to measure the concentrations of REE in discrete mineral grains or regions of interest.

RESULTS AND DISCUSSION

The mean and range of concentrations of REE are similar in each sampling location and enriched approximately 50% compared to the North American Shale Composite (NASC). Samples collected in the Buck Mountain region display a positive correlation between REE and Ti/Al, indicating high weathering rates and a detrital source of REE. Samples collected in Lower Kittanning display no correlation between REE and Ti/Al and instead high REE samples occur with spikes in Ni, V, and Th/K. These data indicate an additional source of REE in the Lower Kittanning region.

SEM/EDS imaging identified diffuse regions of interest containing finely crystalline REE-enriched minerals intergrown with clay in the Lower Kittanning samples. LA-ICP-MS data indicate that these crystals are highly enriched in phosphorous and light REE (Fig.1). In some instances, the crystals are depleted in heavy REE compared with the surrounding clay material. These regions of interest are believed to be diagenetic monazite. Other occurrences of diagenetic monazite forming in mudrocks and shales are documented[4,5].

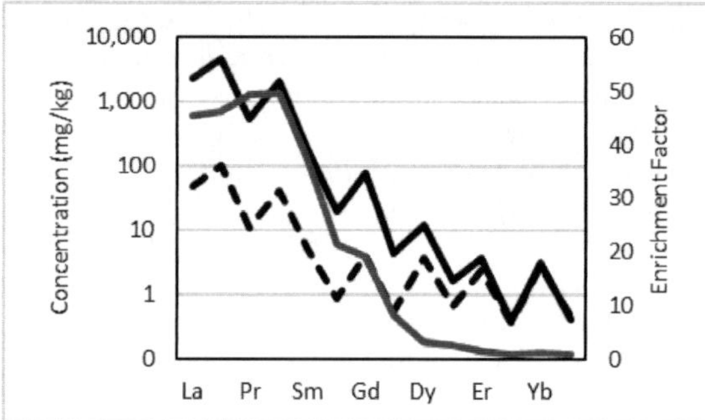

Fig. 1. Concentration of REE (solid line) and enrichment factor in diagenetic monazite (red line) compared to clay matrix (dashed line) in Lower Kittanning samples.

CONCLUSIONS

Diffuse regions of fine crystals of diagenetic monazite intergrown with clay were identified in sedimentary rocks collected near the Lower Kittanning bituminous coal beds. While the main component of REE in these rocks is likely detrital, the source of REE forming the diagenetic monazite may have come from seawater, as this coal bed is surrounded by units of marine limestone and shale. REE may have adsorbed onto the clay minerals from seawater forming temporary ion adsorbed deposits and later crystallized during changing conditions associated with diagenesis.

REFERENCES

1. Schatzel, S.J. and Stewart, B., Int.J. Coal.Geol.. 94:123-136, 2012.
2. Kanazawa, Y and Kamitani, M., J. Alloys Compd. 408-412: 339-1343, 2006.
3. Bank, T. et al., US DOE Topical Report DOE/NETL-2016/17942016.
4. Kryza, R., et a., Geol.Sudetica 36:39-51, 2004.
5. Alipour-AsII, M. et al., Miner Petrol 104:197-210, 2012.

ACKNOWLEDGMENTS

This technical effort was performed in support of the National Energy Technology Laboratory's ongoing research under the RES contract DE-FE0004000. The authors wish to thank Mary Anne Alvin for support and project management. Luca Fedele and Robert Bodnar contributed to LA-ICP-MS data collection.

Expanded Macrocyclic Ligands for Actinium Chelation

Justin J. Wilson[1*], Nikki A. Thiele[1], Valery Radchenko[2], Caterina F. Ramogida[2], Victoria Brown[2], Una Hermilova[2], and John W. Babich[3]

[1]Chemistry and Chemical Biology, Cornell University, Ithaca, NY 14853 USA
[2]Nuclear Medicine, TRIUMF, Vancouver, BC V6T 2A3 Canada
[3]Radiology, Cornell Weill Medicine, New York, NY 10065 USA
*jjw275@cornell.edu

INTRODUCTION

Targeted radionuclide therapy uses the cell-killing properties of radionuclides in conjunction with cancer cell-specific targeting vectors to selectively eradicate malignant tissue. The use of alpha-emitting radionuclides, in particular, has led to the development of potent therapeutic constructs with high efficacy against a range of different cancer types.[1] Among the alpha-emitters, actinium-225 ($t_{1/2}$ = 9.9 days) has found significant promise for therapeutic use due to its long half-life and extensive decay chain that yields four alpha particles.[2] A fundamental challenge in the widespread use of this radionuclide, however, is the lack of a suitable chelating agent that can stably retain it in vivo. In this talk, we will discuss our efforts to develop new effective chelating agents for actinium for use in targeted alpha therapy.

RESULTS AND DISCUSSION

We have explored novel expanded macrocyclic ligands for chelation of actinium and related lanthanide ions Specifically, the 18-membered macrocycle 1,7,10,16-tetraoxa-4,13-diazacyclooctadecane was functionalized with different pendant donor groups. The ligands were evaluated for their abilities to stably bind and retain La^{3+}, a non-radioactive surrogate for Ac^{3+}. Additionally, complexes of La^{3+} have been synthesized and fully characterized. Radiolabeling studies employing $^{225}Ac^{3+}$ demonstrate the ability of certain members of this class of ligands to achieve >95% radiochemical yield at room temperature. Additionally, the resulting Ac-complexes are stable with respect to challenges with human serum and excess cold metal ions. A bifunctional version of our best actinium ligand candidate was synthesized. Conjugation of this bifunctional ligand to the antibody trastuzamab was demonstrated, enabling us to achieve room temperature radiolabelling of an antibody with ^{225}Ac in a single-step process

CONCLUSIONS

This work had led to the discovery of a new, effective chelating agent for the actinium ion. The novel bifunctional version of the ligand enables facile room temperature radiolabelling of macromolecules, a feature that will be important for the design of new ^{225}Ac-based radiopharmaceuticals.

REFERENCES

1. Kim, Y. S.; Brechbiel, M. W. Tumor Biol. 33: 573–590, 2012..
2. Miederer, M.; Scheinberg, D. A.; McDevitt, M. R.. Adv. Drug Deliv. Rev. 60: 1371–1382, 2008.

ACKNOWLEDGMENTS

Cornell University is thanked for start-up funding. Additional funding was provided by a Pilot Award from the Cornell Weill Medical College Clinical and Translational Science Center, funded by NIH/NCATS UL1TR00457.

Extraordinary Magnetic and Structural Transformations in $HoCo_2$ via Competing 4f and Itinerant Interactions

Durga Paudyal,[1,*] Yaroslav Mudryk,[1] and Vitalij K. Pecharsky[1,2]

[1]Ames Laboratory of US DOE, Iowa State University, Ames, Iowa, 50011-3020, USA, *durga@ameslab.gov
[2]Department of Materials Science and Engineering, Iowa State University, Ames, Iowa, 50011-2300, USA,

INTRODUCTION

Often mysterious magnetism and magnetostructural transformations in lanthanide - transition element compounds arise from a complex interplay between 4f and 5d electrons of lanthanides and 3d electrons of transition metal counterparts. An unusual interplay between structure and magnetism of the low temperature orthorhombic phase in the simple Laves phase compound – $HoCo_2$ – was observed but its origin has not been fully understood. Here, we report an in-depth investigation and analysis of these extraordinary magnetic and structural behaviors from experiments and first-principles calculations.

THEORETICAL METHODS

The local spin density approximation including the onsite electron correlation and spin orbit coupling (LSDA+U+SOC) approach has been employed to investigate the electronic structure, magnetic properties, and magnetostructural transformations of $HoCo_2$. Calculations have been performed using all electron and fully relativistic full-potential linear augmented plane wave (FP-LAPW) method.

RESULTS AND DISCUSSION

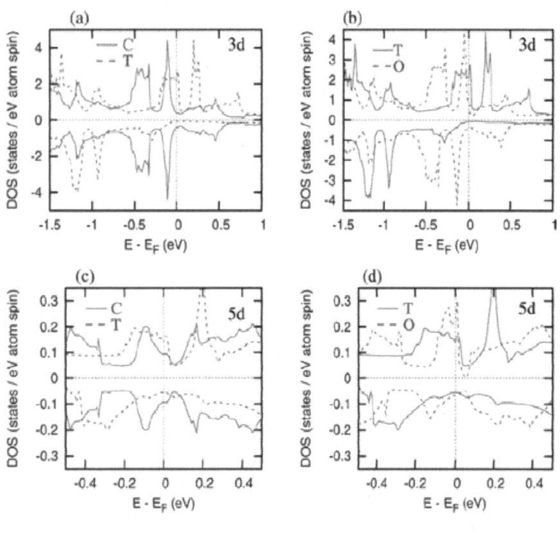

Figure 1. 3d and 5d density of states for the cubic (C), tetragonal (T), and orthorhombic (O – Fddd) structures of $HoCo_2$.

Detailed crystallographic and magnetic measurements complemented by first-principles calculations explain the unusual low-temperature polymorphic transformation in terms of competition between the itinerant ferrimagnetic and the low temperature state, where the Co magnetism is diminished, and both the magnetic and crystal structures are mainly defined by the magnetism of Ho in $HoCo_2$. Figure 1 shows Co-3d and Ho-5d density of states. There is a large exchange split Co-3d states in the tetragonal phase. This splitting is diminished in the orthorhombic structure allowing Ho atoms to be aligned ferromagnetically. Three crystal structures, each associated with a distinctly different magnetic state of the compound, exist in three different temperature regions: (1) the high-temperature cubic structure, associated with the paramagnetic state, (2) the intermediate temperature tetragonal phase with the ferrimagnetic order due to the itinerant electron metamagnetism, and (3) the low-temperature orthorhombic phase with the dominant ferromagnetism of Ho sub-lattice and the much reduced Co magnetism. Calculations show that the ground state orthorhombic structure is expected to adopt the Fddd space group. $HoCo_2$ responds strongly to the varying external magnetic field, including suppression of the structural distortions both at Curie (T_C) and spin reorientation (T_{SR}) temperatures, limiting the stability region of the tetragonal $HoCo_2$, changing the nature of the transition at T_C from the first order to the second order.

CONCLUSIONS

Unexpected increase of magnetization observed below the spin-reorientation temperature (T_{SR}) suggests that the low-temperature transition involves a reduction of Co moment. Calculations confirm that the paramagnetic cubic to ferrimagnetic tetragonal transformation at T_C is assisted by itinerant electron metamagnetism, and that the reduction of Co moment in $HoCo_2$ occurs in parallel with the ferrimagnetic tetragonal to the nearly ferromagnetic orthorhombic transformation at T_{SR} via the rearrangement of both 3d and 5d states.

REFERENCE

[1] Y. Mudryk, D. Paudyal, A. K. Pathak, V. K. Pecharsky, and K. A. Gschneidner, Jr., J. Mater. Chem. C **4**, 4521 (2016).

ACKNOWLEDGMENTS

The Ames Laboratory is operated for the U. S. Department of Energy by Iowa State University of Science and Technology under contract No. DE-AC02-07CH11358. This work was supported by the Department of Energy, Office of Basic Energy Sciences, Materials Sciences Division.

Fabrication of Yttrium-based Fluorides Through Topotatic Transformation of LREHs

Zhan Xu, He Feng, Zhijun Zhang, Jingtai Zhao*

College of Materials Science and Technology, Shanghai University, China, jtzhao@shu.edu.cn

INTRODUCTION

Rare-earth elements are often used for synthesizing fluorescent materials either as bases of host lattice or as dopants because of their abundant 4f electron energy levels allowing researchers to tune the emissions ranging from UV to NIR[1]. Their fluoride nanocrystals have low lattice phonon energy, which lead to low nonradiative relaxations and to the enhancement of the photoluminescence efficiency[2]. High temperature and long reaction time are normally required in synthesizing rare earth fluoride nanocrystals. Recently layered rare earth hydroxides (LREHs) can undergo topotactic transformation to other phases with structure similarity which owns to the anions exchange ability of LREHs[3]. By this route, we have successfully fabricated $Y_x(OH)_{3-x}F$, $NaYF_4$, and YF_3 nanocrystals by using F^- ions to exchange with NO_3^- and OH^- ions in the structure of $Y_2(OH)_5NO_3^- \cdot xH_2O$.

RESULTS AND DISCUSSION

The precursor LREH $Y_2(OH)_5NO_3^- \cdot xH_2O$ was fabricated by adopting the method reported by Takayoshi Sasaki research group[4]. In our work, NaF was employed to fluoridize the LREH to achieve diverse fluoride phases as shown in Fig. 1. It was found that PH values of reaction solutions play an important role to the fluorination process. The analysis of zeta potential of the product aqueous solutions indicates that moderate low PH value of solution benefits to the entrance of fluorine into the structure of LREH, which lead to remarkable change of the charged state of the remaining layers of LREH.

A model was proposed to interpret the process of the fluorination of LREH. The process can be divided into two steps of fluorine ion-exchange and topotatic transformation. Fluorine can exchange with -OH in the primary layers of LREH and NO_3^- in interlayers with the assistance of H^+. The topotactic transformation into a similar new structure can immediately occurs to reduce the system energy which results in a fluoride phase. Interestingly, when the PH values of solutions were lower than 7, the Y-F coordination polyhedra were changed into negatively charged and cations like Na^+ can occupy the interstitial positions of hexagon units in Fig. 4(a) to maintain charge balance, which results in the structure of $NaYF_4$.

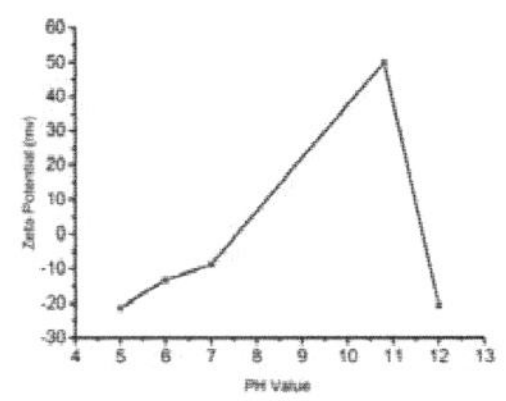

Fig.1 (a)LREHs; (b) $Y_x(OH)_{3-x}F$; (c) $NaYF_4$; (d) YF_3 Fig. 2 Zeta Potential

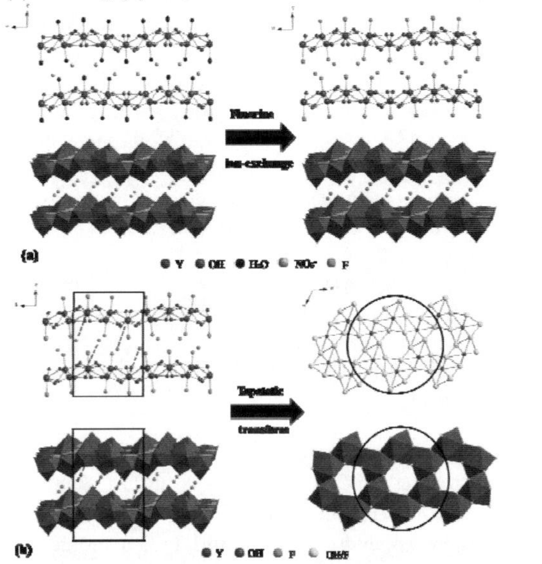

Fig. 3 Model of transformation of LREH to $Y_x(OH)_{3-x}F$

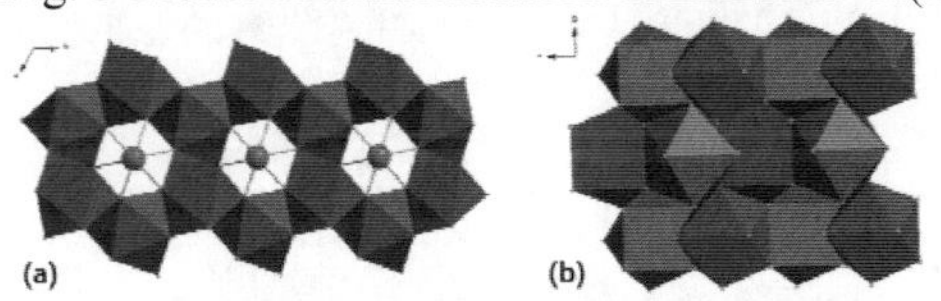

Fig. 4 (a)$NaYF_4$; (b) YF_3

CONCLUSIONS

Diverse Yttrium-based fluorides were successfully fabricated by fluorination of LREHs, in which the PH value plays an important role in the process. The lower the concentration of H^+ in the solution is, the higher the fluorination of LREHs. Two steps process model including ion-exchange and topotatic transformation was proposed for the fabrication of such fluorides.

REFERENCES

1. J.-C. G. Bünzli, et al., Chem. Sci. 2013, 4, 1939.
2. Liu, Y., et al., Adv. Mater. 2010, 22, 3266.
3. Shao, B., et al., Adv. Opt. Mater. 2015, 3, 583.
4. Geng, F., et al., Acc. Chem. Res. 2010, 43, 1177.

Facile Linking of Lanthanide Cations with Soft Metals *via* Thiosulfate

Eric M. Villa[1*] and Evan Z. Dalton[1]

[1]Department of Chemistry, Creighton University, 2500 California Plaza, USA, EricVilla@Creighton.edu

INTRODUCTION

The lanthanides have been applied to profitable innovations in medicine and technology,[1,2] yet a better understanding of the coordination chemistry of hard lanthanide cations with soft ligands could lead to innovative discoveries in medicine, industry and geochemistry. Here, we are using the soft ligand thiosulfate, which an effective metal complexing agent[3,4] that has industrial and medical applications.[5,6] As lanthanide mixed-metal compounds have been shown to be useful[7-9] and thiosulfate-based inorganic-organic frameworks have been shown to be effective photocatalysts,[10] we have also include 'soft' metal ions and organic ligands to the synthetic strategies. Herein, we discuss the syntheses and structures of these five new lanthanide thiosulfate structure types.

EXPERIMENTAL METHODS

All five of the lanthanide thiosulfate structure types presented herein were reacted at either room or elevated temperatures and crystallized *via* slow evaporation or solvent layering.

RESULTS AND DISCUSSION

Thus far, we have been able to successfully crystallize and obtain single crystal structures for five lanthanide thiosulfate compounds. **Compound 1** is a molecular species containing the complex anion $[Ln^{III}(S_2O_3)_6]^{9-}$, which is charge balanced by nine potassium cations in the crystal structure.

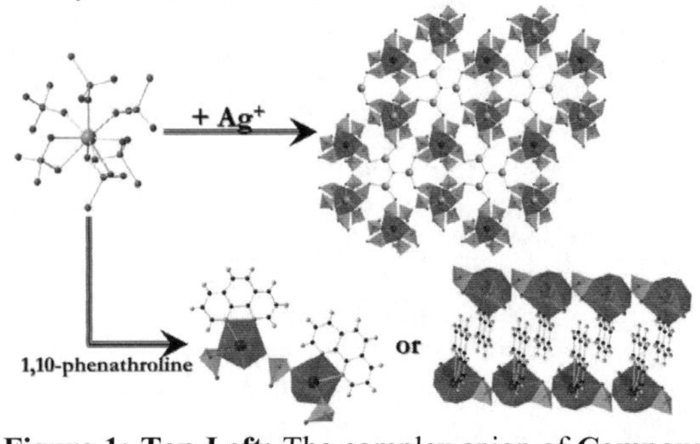

Figure 1: Top Left: The complex anion of **Compound 1** can be used in two different reactions. **Top Right** – When reacted with silver acetate **Compound 3** is formed. **Bottom** – When reacted with 1,10-phenathroline, two products form: the molecular **Compound 4** (left) and an extended network **Compound 5** (right). Here the lanthanide is in aqua or dark green, thiosulfate are mustard tetrahedra, silver are in silver, carbon are in black, hydrogen are light pink and oxygen are shown in red.

Compound 1 is formed by slow evaporation, but **Compound 2** (a mixed thiosulfate-sulfite species) is obtained *via* solvent layering. **Compound 1** was utilized in two subsequent reactions: one that employed silver as a soft cation and the other that used 1,10-phenathroline as a templating agent. **Compound 3** was crystallized from the former and **Compounds 4 & 5** were crystallized from the latter.

These five compounds presented here are the first steps into our investigations of the hard lanthanide cations with soft ligands, such as thiosulfate. We believe **Compound 1** can be employed as a flexible starting material in several different reaction types.

CONCLUSIONS

The coordination complexes and extended structures derived from hard lanthanide cations and soft anions is currently underexplored and represents a promising area for applications. We have shown that the lanthanide thiosulfate complex anion can be used as a starting material for mixed-metal complexes and that templating the structure with an organic ligand may lead to open and interesting structure types. We continue to explore this area and are excited for the results to follow.

REFERENCES

1. Aime S, Botta M, Fasano M & Terreno E. *Chem. Soc. Rev.* 27: 19-29, 1998.
2. Bünzli JG & Piguet C. *Chem. Soc. Rev.* 34: 1048-1077, 2005.
3. Chan EJ, Skelton BW & White AH. *Z. Anorg. Allg. Chem.* 2825-2844, 2008.
4. Freire E, Baggio S, Goeta A & Baggio R. *Aust. J. Chem.* 53: 785-790, 2000.
5. Tykodi RJ. *J. Chem. Ed.* 67: 146-149, 1990.
6. Dhawale SW. *J. Chem. Ed.* 70: 12-14, 1993.
7. Winpenny REP. *Chem. Soc. Rev.* 27: 447-452, 1998.
8. Colis JCF, Larochelle C, Staples R, Herbst-Irmer R, & Patterson H. *Dalton Trans.* 4: 675-679, 2005.
9. Sánchez-Barba LF, Hughes DL, Humphrey SM, & Bochmann M. *Organometallics.* 25: 1012-1020, 2006.
10. Paul AK, Madras G, & Natarajan S. *Phys. Chem. Chem. Phys.* 11: 11285-11296, 2009.

ACKNOWLEDGMENTS

Funding for this work was graciously provided by a Dr. Randolph M. and Mrs. Teresa Kolar Ferlic Research Scholar Award and the Creighton Center for Undergraduate Research and Scholarship. The authors would also like to thank Creighton University and the Creighton University Chemistry Department for continued support of undergraduate research.

f-Element Silylamide Complexes with Remarkable Bonding Motifs

David P. Mills[1*] and Conrad A. P. Goodwin[1]

[1*]School of Chemistry, The University of Manchester, UK, david.mills@manchester.ac.uk

INTRODUCTION

Investigations into low coordination number (CN) metal complexes are legion as they can exhibit interesting properties, including small molecule activation chemistry[1] and single molecule magnet (SMM) behaviour.[2] f-Element metal centers prefer high CNs, therefore judiciously selected ligands are required to prevent oligomerization and common decomposition pathways.[3]

Bulky monodentate amides (NR_2) are ubiquitous in f-element chemistry as the sterics and electronics of these hard donor ligands can be readily tuned.[4] At the vanguard of this class, bis(silylamides) have been most widely utilized since $\{N(SiMe_3)_2\}$ (N″) stabilized the first structurally characterized 3-coordinate f-element complexes [Ln(N″)$_3$] (Ln = lanthanide) nearly fifty years ago.[5]

Over the past four years our group has studied the physicochemical properties of f-block complexes with unusual bonding modes, geometries and/or oxidation states. We have recently developed a series of bulky silylamides that are suitable for supporting low CN f-element complexes.[6] Selected highlights of this research are presented herein.

RESULTS AND DISCUSSION

We disclosed synthetic routes to a wide range of bis(silylamides) in 2015, including $\{N(SiMe_2{}^tBu)_2\}$ (N**) and $\{N(Si^iPr_3)_2\}$ (N††).[6] A potassium transfer agent of N** was used to prepare the first trigonal planar actinide complex, [U(N**)$_3$] (**1**).[7] We have developed this work to investigate how changes in geometry affect the magnetic properties and reactivity profiles of U(III) silylamide complexes.

The bulkier bis(silylamide) N†† was then used to prepare the first near-linear f-element complex, [Sm(N††)$_2$] (**2**).[8] Calculations performed on a theoretical structurally analogous Dy(III) complex, [Dy(N††)$_2$]$^+$, indicated that such a combination of ligand and complex geometry could give a SMM with liquid nitrogen blocking temperatures. Synthetic efforts towards this target complex are currently being undertaken.

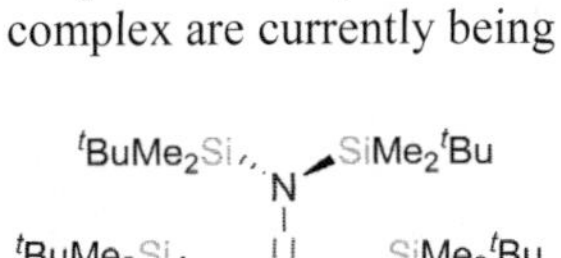

(**1**)

Ln = Sm (**2**), Eu (**3**), Tm (**4**), Yb (**5**)

More recently, we extended the series of near-linear f-element complexes to [Ln(N††)$_2$] (Ln = Eu, **3**; Tm, **4**; Yb, **5**).[9] We have discovered that this geometry engenders unique optical and magnetic properties and we have embarked on full reactivity studies of this series. Initial reactivity studies of **2** have shown that the sterically demanding ligand environment does not preclude inner-sphere single electron transfer processes.[9]

CONCLUSIONS

The introduction of the bulky bis(silylamides) N** and N†† to f-element chemistry has already provided new molecular geometries and interesting physicochemical properties. Given the ability of these ligands to stabilize unusual oxidation states [e.g. Tm(II)], current work in the group is focusing on extending these studies to include other low CN and oxidation state f-element bis(silylamide) complexes for future exploitation.[10]

REFERENCES

1. Lu, C. C.; Meyer, K. Eur. J. Inorg. Chem. 3731-3732, 2013, and references cited therein
2. *Molecular Nanomagnets*, Gatteschi, D.; Sessoli, R.; Villain, J. OUP, Oxford, 2006
3. *Ligand Design in Metal Chemistry: Reactivity and Catalysis*, Ed. Stradiotto, M.; Lundgren, R. L. John Wiley and Sons Ltd., Chichester, 2016
4. *Metal Amide Chemistry*, Lappert, M. F.; Protchenko, A.; Power, P. P.; Seeber, A. John Wiley and Sons Ltd., Chichester, 2009
5. Bradley, D. C.; Ghotra, J. S.; Hart, F. A. J. Chem. Soc., Chem. Commun., 1972, 349–350
6. Goodwin, C. A. P. *et al.* Organometallics, 2015, 34, 2314–2325
7. Goodwin, C. A. P. *et al.*, Chem. Eur. J., 2014, 20, 14579–14583
8. Chilton, N. F. *et al.*, Chem. Commun., 2015, 51, 101–103
9. Goodwin, C. A. P. *et al.*, Inorg. Chem., 2016, 55, 10057–10067
10. Goodwin, C. A. P.; Mills, D. P. Unpublished results

ACKNOWLEDGMENTS

The authors would like to thank the EPSRC (grant number EP/K039547/1), the Royal Society of Chemistry (Researcher Mobility Grant for C.A.P.G.), the University of Manchester, the Royal Society, the EPSRC UK National Electron Paramagnetic Resonance Service and the US National Science Foundation (CHE-1265396 to Prof. William Evans) for supporting this work financially. We also thank our many collaborators, who will be acknowledged in full during our presentations at this conference.

F-Element-Phosphinimide Chemistry: Probing metal-ligand covalency using ^{31}P NMR spectroscopy

Over the past several decades, theoretical, spectroscopic, and magnetic studies of actinide compounds have provided quantitative and qualitative evidence for metal-ligand covalency. However, the nature of actinide-ligand covalency still remains unclear. For instance, chemical bonding in actinide complexes is generally assumed to be more covalent in character, while lanthanide complexes are generally considered to be more ionic. However, recent spectroscopic and computational evidence suggests that isostructural tetravalent uranium and cerium complexes have a comparable degree of covalency, while isostructural thorium complexes seem to be much more ionic, more so than their cerium counterparts. These observations demonstrate that the ionic/covalent description of the chemical bonding for lanthanides versus actinides is not that simple, and there appears to be more to the story waiting to be uncovered. Herein, the synthesis and characterization of isostructural uranium, thorium, and cerium complexes bearing phosphinimide ligands is described. Phosphinimide ligands provide a unique opportunity to probe metal-ligand covalency using ^{31}P NMR spectroscopy. Interestingly, the phosphorus NMR chemical shifts for these isostructural complexes display the following trend: Th4+ < Ce4+ < U4+. Though the exact nature of this trend is unclear at this time, this observation nonetheless appears to be in line with recent spectroscopic and computational results from other groups demonstrating that tetravalent cerium engages in a degree of covalent bonding intermediate between that of uranium and thorium. The synthesis and characterization of additional homoleptic actinide-phosphinimide complexes, as well as mixed-ligand actinide-phosphinimide systems, are presented as well, with particular attention to their ^{31}P NMR chemical shifts. These complexes are ideal for further electronic structure studies, both experimental and computational, and could provide a unique opportunity to push our knowledge of actinide-ligand covalency forward.

Fluorescence Spectroscopy and Microscopy as Tools for Monitoring the Redox Chemistry of Uranium

Louise Natrajan,[1] Debbie Jones,[1] Michael Andrews,[1] Adam Swinburne,[1] Stanley Botchway,[2] Andrew Ward,[2] Jonathan Lloyd,[3]

[1*]Centre for Radiochemistry Research, School of Chemistry, The University of Manchester, Manchester, UK, louise.natrajan@manchester.ac.uk

[2]Central Laser Facility, Research Complex at Harwell, Rutherford Appleton Laboratory, Oxford, UK

[3]School of Earth and Environmental Sciences, Manchester, UK

INTRODUCTION

The contamination of land with radionuclides is an on going problem for the nuclear industry impacting on many sites due to accidental or intentional release and waste storage in addition to other activities. Of the contaminants, uranium dominates the radionuclide inventory causing concern due to its long half-life, chemical and environmental toxicity.[2] Under oxic conditions, uranium is most commonly found in the +VI oxidation state as the mobile uranyl cation UO_2^{2+}.[1] Alternatively the +IV oxidation state dominates in anoxic environments, forming, for example, an insoluble, cystalline uraninite phase (UO_2).[2] Certain bacteria have been known to respire under anaerobic conditions, by coupling the reduction of U(VI) to U(IV) to the oxidation of organic matter. Luminescence spectroscopy can be used to distinguish between these two oxidation states of uranium, since both the spectral fingerprint/features and luminescence lifetime are readily distinguished in solution.[3]

RESULTS AND DISCUSSION

We have been developing the use of luminescence spectroscopy to study redox reactions and the environmental fate of uranium in model conditions, and here demonstrate the first example of two photon absorption and emission studies of simple uranyl(VI) salts to image biosorption of uranyl onto bacteria (yeast, E-coli, *Geobacter sulfurreducens* and *Shewanella oneidensis*). Further insight into the mechanisms of biosorption coupled with microbially-catalysed redox reactions have been gained through the use of confocal microscopy techniques, including one and two photon FLIM (fluorescence lifetime image mapping) and PLIM (phosphorescence lifetime image mapping). These powerful techniques offer the potential to image redox reactions occurring *in situ*, at sub micron spatial resolution proving an excellent and sensitive probe to assess the chemistry of uranium during the bioreduction process (Figure 1).[4]

Geobacter sulfurreducens and *Shewanella oneidensis* are the focus of this study; both are anaerobic, Gram-negative bacteria known for their metal reduction capabilities. For the first time, one and two photon luminescence spectroscopy has been used to study the uranium reduction *in situ,* following the fate of uranyl(VI) during the bioreduction process over 24 hours. Furthermore, we have shown that luminescence microscopy can be used to study the mechanism of electron flow in *Shewanella oneidensis* when combined with mutants lacking key electron transport proteins.

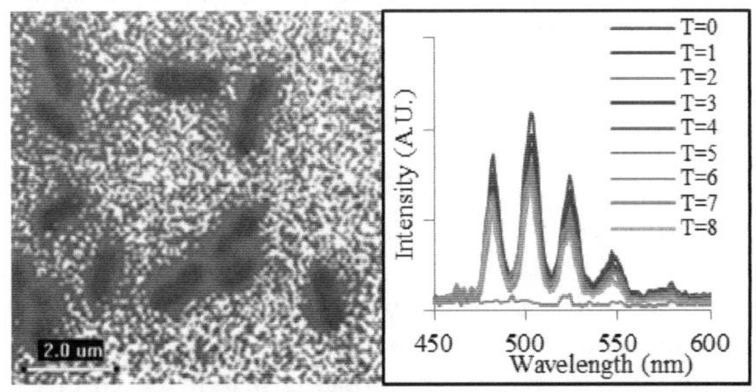

Figure 1. Confocal fluorescence microscope image of *Geobacter sulfurreducens* and UO_2^{2+} at 2 hours during the bioreduction (left) and total emission spectra over the course of 8 hours following 405 nm excitation.[4]

Time resolved fluorescence spectroscopy and microscopy data thus aid in the interpretation of the bioreduction mechanisms in two contrasting model metal-reducing bacteria.

CONCLUSIONS

We have demonstrated the utility of readily available, facile luminescence spectroscopy in studying the bioreduction of UO_2^{2+} by *Geobacter sulfurreducens* over a 24 hour time period. The speed at which a spectrum can be obtained, and the minimal sample preparation involved has allowed the progress of the reaction to be studied *in situ* providing supporting evidence towards the role of a $U^VO_2^+$ intermediate. We have also demonstrated for the first time the viability of using the inherent fluorescent properties of UO_2^{2+} to report directly on the localisation of uranium by fluorescence and phosphorescence lifetime image mapping

REFERENCES

1. George S.A. *et al.*, Res. Toxicol., 24:1830-2, 2011.
2. Lovley D.R. Annu. Rev. Microbiol. 47:263-90, 1993.
3. Hashem E. *et al.*, RSC Adv., 3:4350-61, 2013.
4. Jones D. *et al.*, Chem. Sci. 6:5133-8, 2015.

ACKNOWLEDGMENTS

"The authors would like to thank the EPSRC (Grant no: EP/G004846/1), the Leverhulme Trust (Grant no: RL-2012-072) and the STFC for (Reference no: 13230018) for providing financial support to this project".

Follow the Thread: Unravelling the Mechanism of Substrate Reduction by Sm(II)-Water Complexes

Robert A. Flowers II, Tesia V. Chciuk, and Caroline O. Bartulovich

Department of Chemistry, Lehigh University, Bethlehem PA 18015 USA
rof2@lehigh.edu

INTRODUCTION

The addition of water to samarium diiodide (SmI_2) in THF has a significant impact on the reactivity and selectivity of the reagent.[1] The reactivity of the reagent combination and the relationship to the elementary reaction steps are poorly understood. During the last several years, we have carried out a wide range of thermochemical, kinetic, and computational studies designed to determine whether reduction occurs through a sequential electron-proton transfer (ET-PT), or a concerted proton-coupled electron-transfer (PCET) process. Recent work demonstrates that most unactivated functional groups containing carbonyls proceed through a PCET. Substrates containing unactivated aldehydes or ketones are reduced through an asynchronous PCET whereas lactones are reduced through a concerted PCET. In the case of lactones, a concerted PCET is likely a consequence of a significant endergonic initial ET.

RESULTS AND DISCUSSION

To evaluate the mechanism of the reduction of carbonyls by SmI_2-water, kinetic experiments were initiated to determine the rate constants and rate orders for each component. Figure 1 shows the impact of

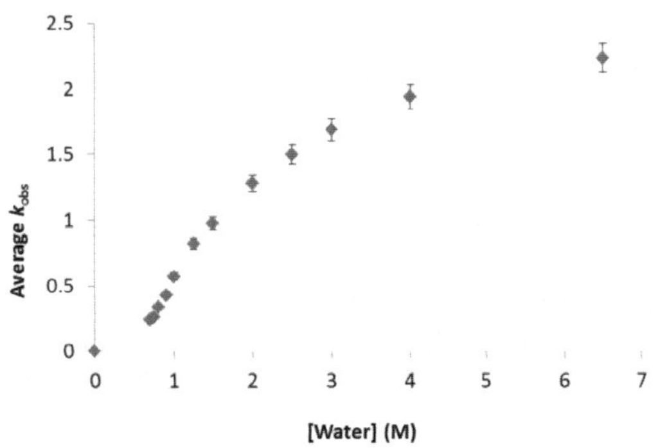

Figure 1. Impact of water concentration on the rate of reduction of cyclohexanone by SmI_2. Conditions: 10 mM SmI_2 , 100 mM cyclohexanone, 0.7-6.5 M H_2O measured at 560 nm and 25 °C.

[water] on the rate of reduction of cyclohexanone by SmI_2-water. The rate plot demonstrates a curvature consistent with a rate order of water greater than unity at water concentrations up to 1.5 M. This was consistent for all substrates examined. Rate studies to determine the role of all components demonstrated that in each case, water was second order, and both substrate and SmI_2 were first-order. Further studies with D_2O showed primary deuterium isotope effects in the range of 1.7-2.1 ± 0.1.

Further mechanistic studies provided activation parameters with very low enthalpies ($\Delta H^\ddagger$) and entropies ($\Delta S^\ddagger$) of activation. In addition, computational studies were employed to determine the charge on the carbonyl oxygen for each substrate and the associated radical anion. This analysis demonstrated a strong correlation between the $\Delta H^\ddagger$ values and change in charge between the parent substrate and its associated radical anion.

Scheme 1

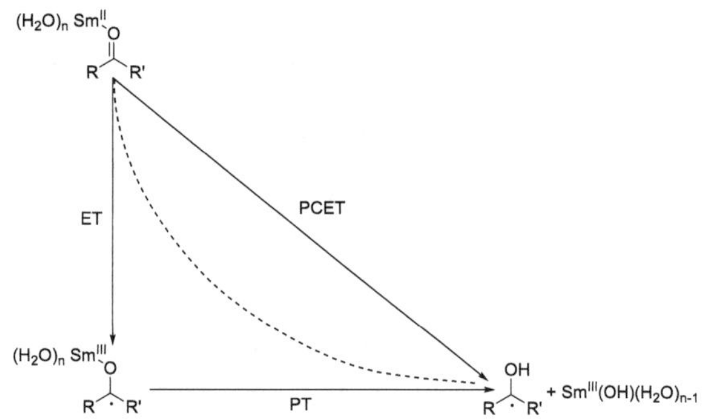

CONCLUSIONS

Overall, our studies are consistent with a continuum where substrates are reduced by SmI_2-water through either a stepwise or concerted PCET that is dependent on the degree of endergonicity of ET from Sm(II) as demonstrated in Scheme 1 above. Ongoing work on this project will be presented.[2]

REFERENCES

References must be numbered. Keep the same style.
1. a) Szostak M. *et al., Chem. Commun. 48*:330-346, 2012; b) Flowers R. A. II *Synlett* 1427-1439, 2008.
2. a) Chciuk T. V. *et al., J. Am. Chem. Soc. 137*:11526-11531, 2015; b) Chciuk T. V. *et al., Angew. Chem., Int. Ed. 55*:6033-6036, 2016; c) Chciuk T. V. *et al., J. Am. Chem. Soc. 138*:8738-8741, 2016; d) Chciuk T. V. *et al., Org. Lett. 19*:290-293, 2017.

ACKNOWLEDGMENTS

The authors would like to thank the US National Science Foundation (CHE 1565741) for support of this project. T. V. C. thanks the Department of Chemistry at Lehigh University for a graduate research fellowship.

Formation and Characterization of Praseodymium Oxides and Nitride-Oxides with Pr(V) Oxidation State

Qingnan Zhang[1], Shuxian Hu[2], Hui Qu[1], Jun Li[2*] and Mingfei Zhou[1*]

[1]Department of Chemistry, Fudan University, Shanghai 200433, China, mfzhou@fudan.edu.cn
[2]Department of Chemistry, Tsinghua University, Beijing 100084, China

INTRODUCTION

Oxidation state is a central concept in chemistry used for elucidating structure and bonding of molecules and for rationalizing redox reactions.[1] Although exotically high oxidation state of +VIII and even +IX are known for some main-group and transition-metal elements in the Periodic Table,[2,3] the chemistry of lanthanides is dominated by the low-valent +III or +II oxidation state due to the chemical inertness of the valence 4f electrons.[4,5] The highest known oxidation state of the whole lanthanide series is +IV, as found commonly in Ce as well as in Pr, Nd, Tb, and Dy. No pentavalent lanthanide compounds with +V oxidation state had been experimentally confirmed. Here, we report a combined experimental and theoretical study on the lanthanide oxide species PrO_4 and NPrO in solid noble-gas matrixes as well as the PrO_2^+ ion complexes in the gas phase, which are characterized to have the Pr(V) oxidation state.[6]

EXPERIMENTAL AND THEORETICAL METHODS

The infrared photodissociation spectra were measured by a collinear tandem time-of-flight mass spectrometer. The cation complexes were produced in a laser vaporization supersonic ion source. The cations of interest were mass separated and selected for photodissociation by a tunable IR laser. The photodissociation spectrum is obtained by monitoring the yield of the fragment ion as a function of the dissociation IR laser wavelength. The praseodymium oxide and nitride-oxide neutral species were prepared by the reactions laser-evaporated praseodymium atoms and dioxygen or nitric oxide in solid argon or neon and were studied by infrared absorption spectroscopy in the mid-infrared region using a Bruker Vertex 80V spectrometer.

Quantum chemical calculations were performed at the levels of density functional theory (DFT) and wavefunction theory (WFT) to gain insight into the geometries, electronic structure, bonding, and oxidation states of the observed species. The hybrid B3LYP functional and the *ab initio* CCSD(T) method were used to optimize the geometry structures and to calculate the harmonic vibrational frequencies. The ECP28MWB pseudopotential and ECP28MWB_ANO basis set for Pr atom and the Dunning's correlation consistent basis set with polarized triple-zeta plus diffuse functions for N and O atoms were employed.

RESULTS AND DISCUSSION

The infrared photodissociation spectrum of $[(PrO_2)^+(O_2)_6]$ in the gas phase confirms the involvement of a linear PrO_2^+ core ion, which has a linear $^1\Sigma_g^-$ ground state with a (core) $(1\pi_g)^4(1\sigma_g)^2(2\pi_u)^4(2\sigma_u)^2(1\delta_u\text{-}4f)^0$ configuration. The $1\pi_g$ and $1\sigma_g$ MOs are the bonding orbitals between Pr 5d and O 2p atomic orbitals, while the $2\pi_u$ and $2\sigma_u$ MOs are weak bonding orbitals formed by Pr 4f and O 2p atomic orbitals, indicating that the Pr center has a (f^0d^0) configuration. Following the general rules for the determination of formal oxidation states, the linear PrO_2^+ cation can be classified as a Pr(+V) species. The significant covalency of Pr 5d/4f orbitals in PrO_2^+ underscores the importance of the less contracted early lanthanide 5d/4f orbitals in stabilizing the high-valent oxides.

In addition to the gas phase study, the praseodymium oxide $(\eta^2\text{-}O_2)PrO_2$ and nitride-oxide NPrO neutal molecules are also prepared in solid neon and argon matrices and are identified based on isotopic substitution studies. The NPrO neutral molecule that is isoelectronic with PrO_2^+ is characterized to have a linear structure with both Pr≡N and Pr=O multiple bonds. Thus, the neutral NPrO molecule is another pentavalent species with Pr(V) oxidation state. The infrared spectra and theoretical calculations confirm that the O_2 unit in $(\eta^2\text{-}O_2)PrO_2$ is a superoxide ligand and the two oxygen atoms in the PrO_2 fragment are both $(O)^{2-}$. Accordingly, a Pr(V) oxidation state can also be assigned in PrO_4, which is formed by the electron-transfer from PrO_2 to O_2.

CONCLUSIONS

Matrix-isolation infrared absorption spectroscopic experiments and infrared photodissociation spectroscopic experiments with mass-selected $[(PrO_2)^+(O_2)_n]$ cations in the gas phase have shown that the cation complexes with a linear PrO_2^+ core ion and the PrO_4 and NPrO neutral molecules are formed. The experimental observations are consistent with high-level quantum-chemical calculations showing that these species all have an f^0d^0 electron configuration for Pr. Thus, the formal oxidation state +V for praseodymium can be assigned, which for the first time reveals an oxidation state higher than +IV for lanthanide elements.

REFERENCES

1. Karen P. *et al.*, Pure Appl. Chem. 86:1017, 2014
2. Riedel S. *et al.*, Coord. Chem. Rev. 253:606, 2009.
3. Wang G. J. *et al.*, Nature 514:475, 2014.
4. Schulz A. *et al.* Struct. Chem. 19:633, 2008.
5. Hitchcock P. B. *et al.*, Angew. Chem. Int. Ed. 47:1488, 2008.
6. Zhang Q. *et al.* Angew. Chem. Int. Ed. 55:6896, 2016.

Formation of 1:13 Phase in La(Fe,Si) -Based Alloys

S. T. Zong, Y. Long*, H.Zhang

[1]School of Material Science and Engineering, University of Science and Technology Beijing, Beijing, China, *longy@mater.ustb.edu.cn

INTRODUCTION

$La(Fe,Si)_{13}$ alloys with $NaZn_{13}$ structure (1:13 phase) are paid attention because the alloys have giant magnetocaloric effect (MCE)[1]. A lot of papers investigated the effect of element substitution in $La(Fe,Si)_{13}$ alloys on the magnetic properties[2-4].The almost substitution elements for Fe decreased MCE of $La(Fe,Si)_{13}$ alloys, while the magnetocaloric effect of the alloys can be enhanced by replacing La with Ce and Nd elements.

In this paper, we investigated the effect of the substitution elements for Fe on the 1:13 phase formation to understand the decrease of MCE. The link between the change in microstructure and the magnetic properties was discussed.

EXPERIMENTAL METHODS

The ingots of $LaFe_{11}M_{0.5}Si_{1.5}$ (M=Cr, Ni, Nb) and $LaFe_{13-y}Si_yC_{0.2}$(y=0.8-1.4) were prepared by arc-melting in an argon atmosphere. The as-cast ingots were then annealed. XRD, SEM with EDS and VSM were used to characterize the phase microstructures, compositions, and magnetic properties, respectively.

RESULTS AND DISCUSSION

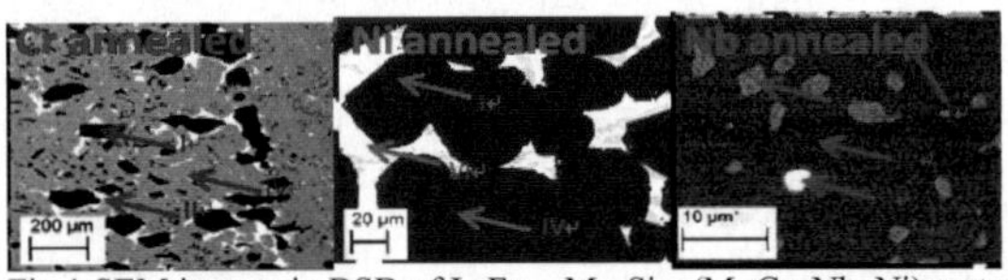

Fig.1 SEM images in BSD of $LaFe_{11.0}M_{0.5}Si_{1.5}$ (M=Cr, Nb, Ni) casted and annealedalloys.(I:α-Fe, II:La_5Si_3, III:LaFeSi, IV:1:13phase, V:Fe_2Nb, VI:La-rich)

As shown in Fig.1, 1:13 phase became the the matrix phase after annealing $LaFe_{11.0}M_{0.5}Si_{1.5}$ (M=Ni,Cr) alloys for 14 days. But large amount of residual α-Fe(black phase) and LaFeSi phases(white phase) remained after annealed still. The EDS analysis revealed Ni mainly exists in LaFeSi phase and La-rich phase for as-cast ingot, while Cr element mainly exists in α-Fe phase in the as-cast ingot. After annealed, Ni and Cr elements were detected in α-Fe phase rather than in 1:13 phase. Obviously, the solubility of Ni and Cr in 1:13 phase is very small. For the case of the substitution Nb for Fe, Fe_2Nb phase(V in Fig.1) was found in the as-cast $LaFe_{11}Nb_{0.5}Si_{1.5}$ sample. EDS results showed that almost all of Nb element was in $(Fe,Si)_2Nb$ phase and no Nb element was found in other phases.

Fig.2 shows the magnetic properties of the $LaFe_{11.0}M_{0.5}Si_{1.5}$ (M=Ni, Cr, Nb). $LaFe_{11.0}M_{0.5}Si_{1.5}$ (M=Ni,Cr) samples showed a second order magnetic phase transition character and very small magnetic entropy change value, while $LaFe_{11.0}Nb_{0.5}Si_{1.5}$ showed the first order magnetic phase transition character and larger magnetic entropy change value.

Fig.3 shows SEM for $LaFe_{13-y}Si_yC_{0.2}$ alloys. Obviously, Carbon doping could accelerate the formation of 1:13 phase. But C can not inhibit the growth of α-Fe.

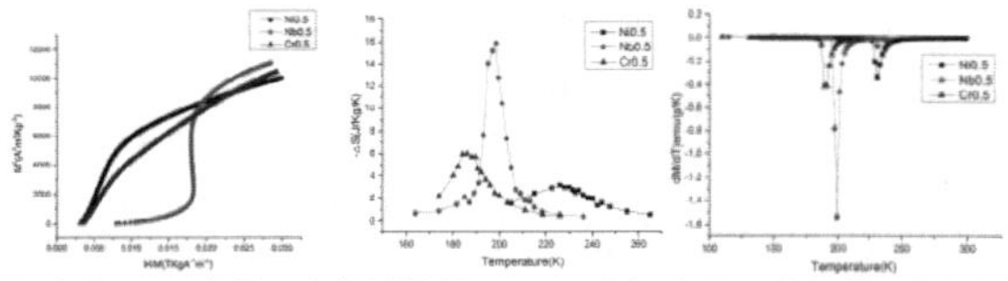

Fig.2 Arrot, ΔS-T and dM/dT-T curves of the $LaFe_{11.0}M_{0.5}Si_{1.5}$ (M=Ni, Cr, Nb)

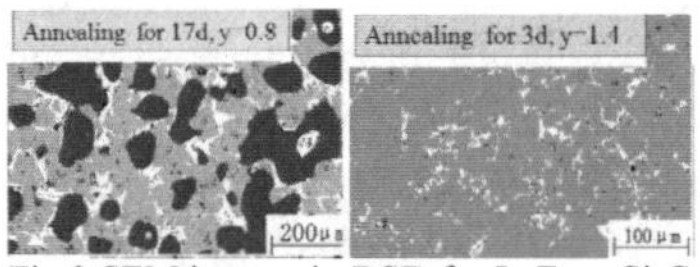

Fig.3 SEM images in BSD for $LaFe_{13-y}Si_yC_{0.2}$ alloys

CONCLUSIONS

Substitution elements Ni and Cr have limited solubility in 1:13 phase. It leads to a reduction of 1:13 phase after a peritectoid reaction. The solid solubility of Ni and Cr in 1:13 phase may be caused the disappearance of first order magnetic phase transition. A new phase $(Fe, Si)_2Nb$ formed due to the larger Nb atom radius when Nb substitutes Fe in $La(Fe,Si)_{13}$. The alloy with Nb substitution maintains first order magnetic phase transition character. Enough Si content is required to obtain $LaFe_{13-y}Si_yC_x$ alloys of almost single 1:13 phase.

REFERENCES

1. S. Jacobs et al., International journal of refrigeration. 37, 84(2014)
2. K. Niitsu a et al., Journal of Alloys and Compounds. 578, 220 (2013)
3. Chaolun Wang, Yi Long, Journal of Applied Physics. 113, 143902 (2013)
4. Maria Krautz, Konstantin Skokov, et al., Journal of Alloys and Compounds. 598, 27 (2014)

ACKNOWLEDGMENTS

This work is supported by the National Science Foundation of China (51571018, 51371026)

Fundamentals of URu_2Si_2 Probed through Electron Doping and High Magnetic Fields

Ryan E. Baumbach[1]

[1]National High Magnetic Field Laboratory, Florida State University, USA
baumbach@magnet.fsu.edu

INTRODUCTION

URu_2Si_2 continues to attract interest because it hosts an unidentified ordered state ("hidden order") below $T_0 \approx$ 17.6 K and unconventional superconductivity below $T_c \approx$ 1.5 K.[1] These phenomena occur within a strongly hybridized *f*-electron lattice that is reminiscent of that seen for related systems with magnetically ordered ground states. Despite this parallel, various measurements show that the ordered state does not have an intrinsic magnetic moment.[2] A multitude of theories have been proposed to describe this behavior, but no consensus has been reached regarding their applicability.[1] This is at least partly because there are several important energy scales in 5*f*-electron materials including the Kondo, RKKY, Coulomb, and spin orbit interactions and their relative importance is not obvious.

In order to elucidate this situation, we have tuned URu_2Si_2 using electron doping (Si → P) and probed the resulting electronic states using large magnetic fields.[3,4,5] This work reveals rich phenomena, where fine changes in the details of the Fermi surface, induced by electronic shell filling or magnetic fields, profoundly influence the ground state behavior while leaving the overarching strongly hybridized correlated electron lattice intact.

EXPERIMENTAL METHODS

Single crystal specimens of $URu_2Si_{2-x}P_x$ were produced using the molten metal flux growth technique described in Refs. [3,4]. Bulk electrical transport and thermodynamic properties were determined for magnetic field H = 0 and temperatures 400 mK < T < 300 K using a Quantum Design PPMS and MPMS. Electrical transport data was additionally collected using a 4-point AC lock-in method in pulsed magnetic fields H < 65 T applied parallel to the c-axis. Magnetization measurements were performed using an extraction magnetometer technique for H < 65 T applied parallel to the c-axis. These measurements took place at the National High Magnetic Field Laboratory.

RESULTS AND DISCUSSION

We find that the parent compound hidden order and superconductivity are rapidly suppressed for x < 0.03 and are replaced by a region with persistent Kondo lattice behavior but no low temperature ordered state. This no-order region persists for 0.03 < x < 0.26, after which antiferromagnetism emerges from the *f*-electron lattice for 0.26 < x < 0.5. Recent NMR measurements elucidate the phenomena in these regimes, where the Kondo lattice behavior of the no-order region is similar to that seen above T_0 at low x.[6] These measurements also reveal that the antiferromagnetism occurs in the bulk and has a commensurate wave vector. Application of a large magnetic field produces a complex phase diagram.[5] Within the hidden order x-region, the previously reported field induced state[7] is robust, even as T_0 is suppressed. Within the no-order x-region, the high field ordering is persistent, and eventually is even enhanced in the AFM region. We remark that this behaviour is notably similar to that seen for Ru → Rh substitution,[8] establishing electronic band filling (regardless of whether it is through the *s/p* or *d*-shell) as a fundamental axis for controlling the ground state behaviour of URu_2Si_2.

CONCLUSIONS

We find that Si → P substitution destroys hidden order, leaves behind an unmodified Kondo lattice without ordering, and eventually produces an atiferromagnetically ordered ground state. Application of a large magnetic field (H = 30-40 T) induces high field ordering across the entire x-range, revealing a close similarity to the $URu_{2-x}Rh_xSi_2$ substitution series. These intriguing results (1) give insight into what factors are essential to hidden order and superconductivity in URu_2Si_2, and (2) amplify investigations of complex electronic ordering in the Kondo lattice.

REFERENCES

1. Mydosh J. A. *et al.*, Phil. Mag. 94:3642-3662, 2014
2. Das P. *et al.*, New J. Phys. 15:053031, 2013
3. Gallagher A. *et al.*, Nat. Comm. 7:10712 2016
4. Gallagher A. *et al.*, J. Phys. C. Mat. 29:024004 2016
5. Wartenbe M. R. *et al.*, Manuscript in prep.
6. Shirer K. R. *et al.*, Phys. Rev. B 95:041107(R) 2016
7. Scheerer G. W. *et al.*, Phys. Rev. B 85:094402 2012
8. Burlet P. *et al.*, J. Mag. Magn. Mat. 109:202 1992

ACKNOWLEDGMENTS

This work was performed at the National High Magnetic Field Laboratory, which is supported by National Science Foundation Cooperative Agreement No. DMR-1157490, the State of Florida and the DOE. A portion of this work was supported by the NHMFL User Collaboration Grant Program (UCGP).

Further Investigations of Curium Photoluminescence Efficiency with Hydroxypyridinone Ligands

Christopher M. Andolina,[1*] Ga-Lai Law,[1] Jide Xu,[1] David K. Shuh,[2] and Kenneth N. Raymond[1,2]

[1]Department of Chemistry, University of California, Berkeley, United States
[2]Chemical Sciences Division, Lawrence Berkeley National Laboratory, United States
*christopher.andolina@gmail.com

INTRODUCTION

Few studies exist investigating the photoluminescence sensitization efficiency of actinide(III) centers due to the inherent cost and potential health hazards of examining these materials compared to lanthanide complexes.[1,2,3] In this study, we make a more detailed examination the photoluminescence (PL) figures of merit for the sensitization of Cm(III) in the presence of three types of all oxyanion ligands 2-hydroxyisophthalamide (**IAM**), 1-hydroxy-2-pyridinone[1] (**1,2-HOPO**) and 3,2-hydroxypyridinone (**3,2-HOPO**). This work is a fundamental study using well-described actinide and lanthanide chelator-chromophores (ML_2) in order to assess which of these ligands are best suited of the efficient sensitization of Cm(III) in solution. Evaluation of the quantum yield (Φ), molar absorptivity (ε), brightness (Φ x ε), and "Q" number was used to compare complexes. This work illuminates which ligands are best suited for the development of highly sensitive An(III) optical probes and furthers fundamental studies of Cm(III) photophysics *in situ*.

EXPERIMENTAL METHODS

Complexes were prepared *in situ* by mixing solutions of ligand and Cm(III) solutions in ratios (~2:1) to ensure octadentate complexation in dry methanol to reduce excited state quenching. Cm(III) complexation was characterized by electrospray ionization-mass spectroscopy (ESI-MS, Gibson Group).[1] Photoluminescence (PL) measurements were collected for using a Fluorolog-3 and Cary 6000i instruments (methodology described in detail elsewhere).[1]

RESULTS AND DISCUSSION

When a simple superposition of the ligand (**IAM**, **1,2-HOPO**, and **3,2-HOPO**) triplet states are overlaid onto the Cm(III) transitions (Figure 1) we would expect that the 3,2-HOPO chelator would provide the most efficient sensitization[4] of the Cm(III) center assuming all

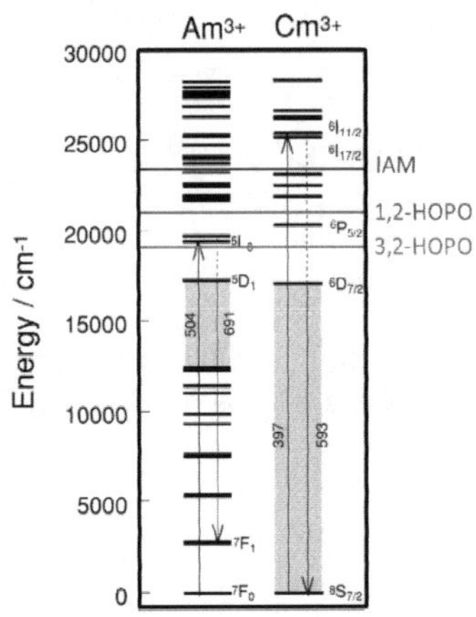

chelators provide octadentate coordination sphere. However, the PL figures of merit shown in Table 1 indicate that coordination of Cm(III) with a 1,2-HOPO and IAM produce most favorable Φ.

Ligand	Φ (λ_{EX} at ~345 nm ±15%)	ε (M^{-1} cm^{-1} ±15%)	(Φ x ε) (M^{-1} cm^{-1})
Et(H2,2)**IAM**[1]	0.50	19,300	8,600
m-Phen-**1,2-HOPO**	0.67	20,700	13,900
5LI-**3,2-HOPO**	0.11	26,300	2,800

These results suggest that **1,2-HOPO** ligands are the most useful sensitizing chromophores in this class ligands when relative brightness (Ω x ε) is considered. The ineffectiveness, of the **3,2-HOPO** sensitization, is unexpected and may arise from complex instability or an efficient back-energy transfer pathway that deactivates the excited Cm(III) center. The use of chelators with similar coordination geometries (ML_2) provides a more detailed comparison/contrasting of these HOPO systems.

CONCLUSIONS

This class of ligands (**1,2-HOPO**) are ideal to use to further develop highly sensitive ligand systems for the optical PL detection of Cm(III) cations (< 1 μM) in either organic, aqueous, and biological media.

REFERENCES

1. Law G-L. *et al.*, J. Am. Soc. Chem. 134:15545-1554911, 2013
2. Strurzbecher-Hoehne, M. *et al.*, Chem. Euro. J. 20:9962-9968, 2014
3. Strurzbecher-Hoehne, M. *et al.*, Dalton Trans. 45:9912-9919, 2016
4. Moore, E. G. *et al.*, Acc. Chem. Res. 42:542–552, 2009

ACKNOWLEDGMENTS

The authors thank Drs. D. S. Tatum for synthesizing the *m*-Phen-1,2-HOPO ligand, J. K. Gibson, P. Rutkowski, and D. Rios for ESI-MS of the Cm complexes. The authors thank the NIH (Grant No. HL069832) and the Director, Office of Science, Office of Basic Energy Sciences, Division of Chemical Sciences, Geosciences, and Biosciences Heavy Element Chemistry program of the U.S. Depart. of Energy at LBNL under Contract DE-AC02-05CH11 for support of this work.

Green processing of Canadian REE ores by combining attrition and lixiviation

Nicolas Reynier[1*], Cheryl Laviolette[1], Magella Bilodeau[1], André Demers[1] Jean-François Fiset[1] and Janice Zinck[1]
[1*]CanmetMINING, Natural Resources Canada, Canada, nicolas.reynier@canada.ca

INTRODUCTION

Important needs for the Canadian mining industry in relation to the actinide elements (U and Th) are the management of radioactive mine wastes and the separation of actinide elements during rare earth elements (REE) production. In Canada, the wastes inventory at the end of 2010 reached 214 million tons of uranium tailings and 174 million tonnes of U mining waste rock[1]. The separation of U and Th from rare earths is often a big concern in the rare earth industry in order to manage the radioactive nuclides[2]. Combining attrition and lixiviation in one step is an important optimization for the green processing of ore. This will increase leaching and grinding efficiencies, reduce process steps and energy consumption, and simplify equipment. An innovative process, using sulphuric acid and potassium iodide, was developed for actinides and lanthanides recovery from nuclear wastes, mine wastes and REE ore[3]. The objectives of this research were 1) to optimize this process for the leaching of REE and separation of radioactive impurities from ore, and 2) to evaluate the efficiency of combining attrition and lixiviation in leaching and recovering actinide and lanthanide from 6 different Canadian ores.

EXPERIMENTAL METHODS

The leaching tests were performed in a 2 L beaker by mixing 50 g of ore with 500 mL of solution (ratio of solids to solution of 10%). Agitation was performed using an immersed axial impeller. Three attrition cells (1.2 L, 12 L and 120 L) were designed to conduct experiments on 6 REE ores at different pulp density and using different particle size materials. Liquid samples were analyzed by ICP-MS. Solid samples acid digestion was done by microwave and analyzed by ICP-MS and compared with certified reference materials of REE ore.

RESULTS AND DISCUSSION

Ore leaching with dilute aqua regia, HNO_3 with Hg as catalyst, and H_2SO_4 with KI were conducted according to literature[3,4,5]. A leaching process, using sulphuric acid and potassium iodide, was optimized for actinides and lanthanides recovery from a REE ore. A Box–Behnken design (surface response methodology) was used to optimize the operating parameters and ensure statistical significance. Figure 1 shows the solubilisation of 75% of U, 90% of Th, 55% of LREE, and 75% of HREE in the optimized conditions ([H_2SO_4]: 4M, [KI]: 0.2M, T°C: 75°C). Leaching tests were done in the small attrition cell (1.2 L) at different pulp density and using different particle size materials. NaOH pellets addition follow by HNO_3 or H_2SO_4 lixiviation were investigated to increase leaching efficiencies.

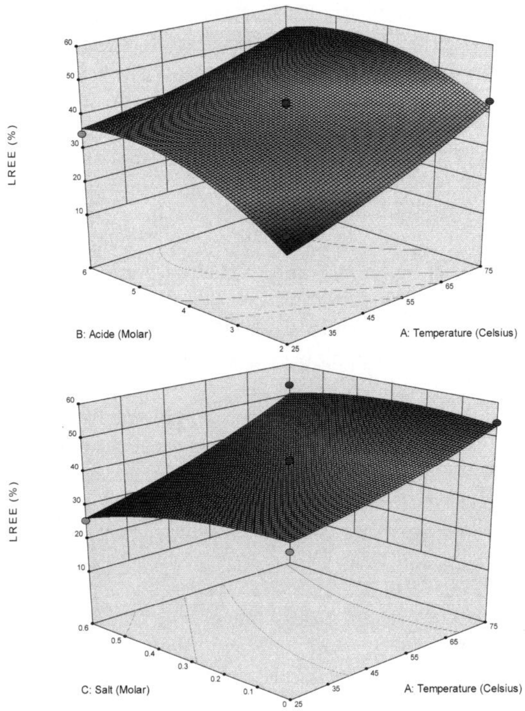

Figure 1: Surface response for Light REE solubilisation from 50 g of ore at 10% solids ([H_2SO_4]: 2-6M, [KI]: 0-0.6M, T°C: 25-75°C)

Results show that attrition and lixiviation were performed in 1 step and in 1 hour only. Processing of coarse ore at 20-30% pulp was efficiently achieved and key factors are the pH & the acid/ore ratio, depending of particle size and pulp density. The NaOH addition increases the temperature and the efficiency of metals solubilisation. Salt addition may increase significantly metals solubilisation as observed during optimization.

CONCLUSIONS

Sulphuric leaching using KI improves the solubilisation by a combination of acidification, complexation, and oxidation and form tetraiodide complex amenable to IX technology for actinide separation[6]. Combining attrition and lixiviation in one step increases leaching and grinding efficiencies and reduces both CAPEX/OPEX.

REFERENCES

1. LLRWMO-01613-041-10003: Ottawa, ON, 2012
2. Zhu Z. *et al.*, J. Miner. Eng. 77:185-196, 2015
3. Reynier N. *et al.*, Minerals, 5(4):744-757, 2015
4. Kitts F.G. *et al.*, Oak Ridge National Lab., 1957
5. Mellor J.W. Comprehensive Treatise on Inorganic and Theoretical Chemistry, 31:432-444, UK, 1930.
6. Reynier N. *et al.*, Solv. Extr. Ion Exch. 34(2):188-200, 2016

Growth Mechanism and Rational Synthesis of Monodisperse Upconversion Nanocrystals based on Sodium Rare-Earth Fluorides

Christian Homann, Thorben Rinkel, Athira N. Raj, Rajesh Komban, Benjamin Voss, Simon Dühnen, Markus Haase*

Institute of Chemistry of New Materials, University of Osnabrueck, Germany, markus.haase@uos.de

INTRODUCTION

Luminescent nanocrystals based on sodium rare-earth fluorides (β-$NaREF_4$) are currently are currently investigated for a variety of application such as biological labelling, sensing and imaging,[1–3] photo-dynamic therapy,[3] spectral conversion in solar cells,[1,4] 3D optical displays and optical storage,[1,5] laser,[6] and security printing.[1,7] Well-known examples are Yb^{3+}/Er^{3+} or Yb^{3+}/Tm^{3+} doped β-$NaYF_4$ particles showing energy transfer upconversion emission (ETU). In recent years, solution-based synthesis procedures have been developed yielding well-separated, re-dispersible β-$NaREF_4$ nanocrystals with amazingly narrow size distributions. Moreover, the mean particle size can be varied over a wide range and core/shell particles can be prepared displaying improved optical properties due to decreased energy losses on the particle surface.

RESULTS AND DISCUSSION

We have investigated in detail the growth mechanism of such monodisperse core and core/shell particles by making use of the distinct optical and magnetic properties of the rare-earth ions. EPR and luminescence spectroscopy show that the particles grow by Ostwald ripening, that is, by the exchange of molecular compounds ("monomers") via the solution.[8,9] X-ray powder diffraction shows that $NaREF_4$ particles of two different crystal phases are formed at early stages of the synthesis. The simultaneous presence of particles of the cubic α-phase and the hexagonal β-phase of $NaREF_4$ is the key feature of the synthesis method, since only Ostwald ripening of such mixtures results in focusing of the particle size distribution.[9-12] The resulting formation of β-$NaREF_4$ particles with very narrow size distribution has been confirmed experimentally as well as by numerical simulations of the growth process.[11] The final size of the resulting monodisperse particles depends on the number of β-phase seeds nucleating during the synthesis. One key parameter for controlling the number of these seeds is the ratio of sodium to rare-earth ion allowing to vary the mean particle size over a wide range.[12,13] Control of the nucleation is also important in the synthesis of core/shell particles where the formation of additional particles of pure shell material must be avoided as well as the intermixing of core and shell material during deposition of the shell.[14] The degree of intermixing depends on the synthesis method as can be shown spectroscopically for europium-doped systems.[15]

Fig.: TEM image of $NaREF_4$ nanocrystals with hexagonal prismatic shape (inverted greyscale)

CONCLUSIONS

$NaREF_4$ nanocrystals represent a textbook example for monodisperse nanoparticles. The optical and magnetic properties of the rare-earth ions can be used to unravel the growth mechanism of the particles and the origin of their narrow size distributions.

REFERENCES

1. Vennerberg D. *et al.*, Sci. Adv. Mater. 3:26–40, 2011
2. Wang F. *et al.*, Chem. Soc. Rev. 38:976–989, 2009
3. Chatterjee D. K. *et al.*, Small 6:2781–2795, 2010
4. van der Ende B. M. *et al.*, Phys. Chem. Chem. Phys. 11:11081–11095, 2009
5. Zhang C. *et al.*, Adv. Mater. 22: 633–637, 2010
6. Baldelli S., Nat. Photon. 5, 75–76, 2011
7. Meruga J. M. *et al.*, J. Mater. Chem. C 2:2221–2227, 2014
2. Komban R. *et al.*,Angew. Chem. 51:6506-6510, 2012
9. Voss B. *et al.*, Nanoscale 5: 806-812, 2013
10. Raj A.N. *et al.*, Chem. Mater. 26:5689–5694, 2014
11. Voss B. ACS Nano, 7:11242–11254, 2013
12. Rinkel T. *et al.*, Nanoscale, 6:14523 – 14530, 2014
13. Dühnen S. et al., Chem. Mater. 27:4033–4039, 2015
14. Rinkel T. *et al.*, Angew. Chem., 55:1164-1167, 2016
15. Dühnen S. *et al.*, Chem. Mater. 27:8375–8386, 2015

ACKNOWLEDGMENTS

The authors would like to thank the German Science Foundation DFG (Grant no: HA 1649/7-1) for providing financial support to this project.

High-pressure synthesis and thermoelectric properties of partially filled skutterudite compounds $Ce_xCo_4Sb_{12}$

Yuttana Mona[1], Jirattagan Sirimart[1], Yukihiro Kawamura[1], Junichi Hayashi[1], and Chihiro Sekine[1*]

[1]Graduate School of Engineering, Muroran Institute of Technology, Muroran, Hokkaido 050-8585, Japan
*sekine@mmm.muroran-it.ac.jp

INTRODUCTION

Thermoelectric (TE) materials have attracted worldwide attentions from the perspective of environmental protection and clean energy. TE materials can directly convert heat energy to electric energy or vice versa. The performance of TE materials is estimated by the dimensionless figure of merit $ZT = S^2\sigma T/\kappa$, where S is the Seebeck coefficient, T is the absolute temperature, σ is the electrical conductivity, κ is total thermal conductivity, the total thermal conductivity can be separated as $\kappa = \kappa_e + \kappa_l$, where κ_e is the electronic contribution and κ_l is the lattice contribution[1]. Following Slack's ideal TE material Phonon Glass and Electron Crystal (PGEC) concept[2], $CoSb_3$ based skutterudite compounds have been intensely researched as one of the most promising TE materials.

Unfilled skutterudite compound $CoSb_3$ has high Seebeck coefficient S and high hole mobility, while the lattice thermal conductivity κ_l is too high. However, κ_l can be reduced by partially filling of rare-earth ions, such as Yb and Eu in the two voids of the structure ($Yb_xCo_4Sb_{12}$, $Eu_xCo_4Sb_{12}$). The filled rare earth ions are located inside the oversized cages and show a random motion inside the void (rattling effect). The effect increases phonon scattering and reduces κ_l[3]. By using a high-pressure synthesis method, the maximum actual filling rate x can be increased. In fact, the maximum actual Yb filling rate x was increased to 0.29, which is close to a theoretical limit and the highest for any $Yb_xCo_4Sb_{12}$ reported so far[4]. In this study, we focus on Ce as a guest ion to fill in the voids of $CoSb_3$ and synthesized partially filled skutterudite compounds $Ce_xCo_4Sb_{12}$ with high filling rate by using a high-pressure synthesis technique.

EXPERIMENTAL METHODS

The samples of $Ce_xCo_4Sb_{12}$ (nominal filling rate $x = 0.2$, 0.4, 0.6, and 0.8) were prepared at 4GPa by using a cubic anvil press. Mixture of stoichiometric amounts of each metal and antimony powder was used as starting materials. The reaction temperature and time were 600°C and 120 min, respectively. The samples were characterized by powder X-ray diffraction using CoK α_1 radiation and silicon as a standard. The actual composition was estimated by a scanning electron microscope (SEM) with an energy dispersive X-ray spectrometer (EDX). The resistivity was measured with a standard dc four-probe method. Thermoelectric properties were measured by using a Quantum Design Physical Property Measurement system (PPMS) with a thermal transport option (TTO).

RESULTS AND DISCUSSION

Fig. 1 shows the lattice constant and actual filling fraction of $Ce_xCo_4Sb_{12}$ vs the nominal composition x. The lattice constant increases with increasing x and shows a maximum value of 9.120 Å. The change of the lattice constant saturates for more than $x = 0.4$. The maximum actual filling fraction of Ce was found to be 0.24.

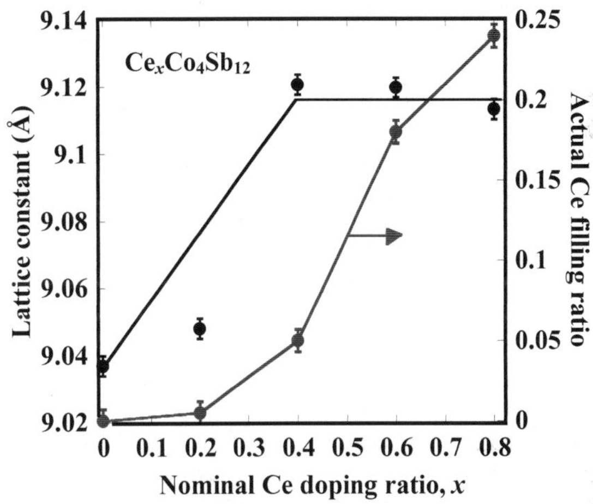

Fig. 1. Lattice constant and actual composition of $Ce_xCo_4Sb_{12}$.

CONCLUSIONS

We synthesized the partially filled skutterudite compounds $Ce_xCo_4Sb_{12}$ at 4 GPa. The actual filling fraction was improved to 0.24 for nominal composition of $Ce_{0.8}Co_4Sb_{12}$. A remarkable rattling effect is expected due to the high filling fraction of Ce. The thermoelectric properties of $Ce_xCo_4Sb_{12}$ prepared under high pressure will be reported.

REFERENCES

1. D. M. Rowe, *CRC handbook of thermoelectrics* (CRC press, 1995)
2. Nolas G. S. *et al.*, J. Phys. Rev. B 58:164, 1998
3. Chen Y. *et al.*, J. Applied Physics 54, 055501, 2015

ACKNOWLEDGMENTS

This work was supported by a Grant-in-Aid for Scientific Research (B) (No. 23340092) from the Japan Society for the Promotion of Science.

Homoleptic Organolanthanide Compounds Supported by β-SiH-containing Ligands

Aaron D. Sadow, KaKing Yan, Aradhana Pindwal, and Naresh Eedugurala
Iowa State University, Ames IA, 50011, United States, sadow@iastate.edu

INTRODUCTION

Homoleptic rare earth alkyls or amides can be starting materials for heteroleptic compounds, homogeneous or single-site grafted catalysts, or as precursors for new materials.[1] Some of the classic ligands, such as benzyl, trimethylsilylmethyl, bis(trimethylsilyl)methyl, hexamethyl- or tetramethyldisilazide, or their derivatives have proven useful, yet all have limitations that include thermal instability, challenging preparations, solvent or salt adduct formation, or .ease of substitution. We have targeted bulky alkyl and amide ligands containing β -SiH groups as a strategy to stabilize homoleptic compounds through secondary interactions, provide a spectroscopic handle for structural characterization, and allow derivatization reactivity directed away from the rare earth center.

RESULTS AND DISCUSSION

We have synthesized homoleptic trivalent and divalent organometallic compounds supported by tris(dimethylsilyl)methyl[2] $-C(SiHMe_2)_3$ and pseudo-organometallic supported by t-butyl(dimethylsilyl)amide $-N(SiHMe_2)_2t$Bu.

The series of early lanthanides $Ln\{C(SiHMe_2)_3\}_3$ (Ln = La, Ce, Pr, Nd) are isostructural, as determined by single crystal X-ray diffraction, NMR spectroscopy, and IR spectroscopy. These and related Y and Lu compounds[2] contain six secondary interactions involving β-SiH groups, and the dynamics of exchange are revealed by variable temperature NMR spectroscopy. The nature of the secondary interactions are further explored by NMR and infrared spectroscopy. The reaction of $Ln\{C(SiHMe_2)_3\}_3$ and $B(C_6F_5)_3$ results in hydride abstraction to give $Ln\{C(SiHMe_2)_3\}_2HB(C_6F_5)_3$.

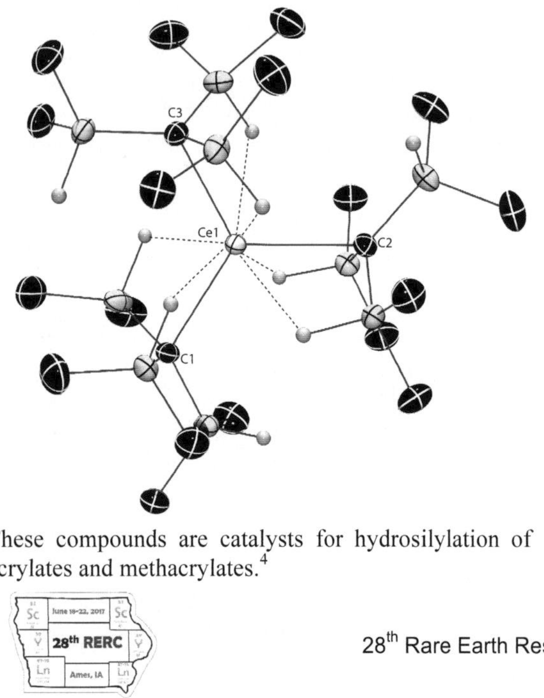

These compounds are catalysts for hydrosilylation of acrylates and methacrylates.[4]

Homoleptic silazido scandium, yttrium, and lutetium compounds $Ln\{N(SiHMe_2)t\text{Bu}\}_3$ and their THF adducts $Ln\{N(SiHMe_2)t\text{Bu}\}_3$THF contain three secondary interactions of the β-SiH groups.[3] Previously, Schumann reported that $Er\{N(SiHMe_2)t\text{Bu}\}_3$ contains three bridging Er←H–Si groups, but the NMR properties of the compound were not analyzed presumably due to the paramagnetic erbium center.[5] The secondary interactions in scandium, yttrium, and lutetium analogues are maintained from low temperature to room temperature, as determined by variable temperature NMR spectroscopy. Interestingly, the $^1J_{SiH}$ values for $Ln\{N(SiHMe_2)t\text{Bu}\}_3$ are reduced from $Ln\{N(SiHMe_2)t\text{Bu}\}_3$THF. The higher $^1J_{SiH}$ feature is reproduced when $Ln\{N(SiHMe_2)t\text{Bu}\}_3$ are grafted onto mesoporous silica, suggesting that surface siloxanes replace THF as an additional donor. Thus, the secondary interactions help characterize the coordination environment in surface sites. These surface grafted sites were studied in hydroamination/bicyclization of aminodialkenes. Remarkably, the diastereomeric ratios (cis/trans) of the products are equivalent from homogeneous and surface grafted species.

CONCLUSIONS

Rare earth organometallics and pseudo-organometallics containing β-SiH groups provide advantageous precursors for catalytic chemistry, both under homogeneous and interfacial conditions.

REFERENCES

1. Zimmermann, M.; Anwander, R., *Chem. Rev.* **2010,** *110*, 6194-6259.
2. (a) Yan, K.; et. al. *Chem. Commun.* **2009**, 656-658. (b) Yan, K.; et. al. *J. Am. Chem. Soc.* **2009**, *131*, 15110-15111.
3. Eedugurala, N.; et. al. Organometallics, 2017. doi: 10.1021/acs.organomet.6b00956
4. Pindwal, A.; et. al *Angew. Chem. Int. Ed.* **2017**, *56*, 628-631.
5. Rees, W.S. J.; Just, O.; Schumann, H.; Weimann, R., *Angew. Chem. Int. Ed. Engl.* **1996**, *35*, 419-422.

ACKNOWLEDGMENTS

$Ln\{C(SiHMe_2)_3\}_3$ were prepared with support from Critical Materials Institute, an Energy Innovation Hub funded by the U.S. DOE Advanced Manufacturing Office. Part of this work was supported by NSF (CHE-1464774) for financial support. Interfacial catalytic chemistry is supported by U.S. Department of Energy Office of Basic Energy Sciences, Division of Chemical Sciences, Geosciences, and Biosciences through the Ames Laboratory (Contract No. DE-AC02-07CH11358).

In Search of New Rare-Earth Rich Intermetallics

Melissa Rhodehouse[1,2*]; Volodymyr Smetana[2]; Gordon Miller[1,2]; Anja-Verena Mudring[2,3]; Gerd Meyer[1,2]

[1] Department of Chemistry, Iowa State University, Ames, IA, 50011, USA, mlrhode@iastate.edu
[2] Ames Laboratory, USDOE, Iowa State University, Ames, IA 50011, USA
[3] Department of Materials Science and Engineering, Iowa State University, Ames, IA 50011, USA

INTRODUCTION

The Pt/Pr phase diagram has been well explored.[1] Previous work has shown the existence of a new binary compound within the Pt/Pr system that is not present within the known phase diagram. This compound, Pt_3Pr_4, crystallizes in a new structure type [a = 12.353(2) Å, b = 7.4837(9) Å, c = 17.279(2) Å, β = 118.003(7)°, $P2_1/c$, Z = 8].[1] Here, we report on further investigation of the binary Pt/Pr system, including $Pt_{1.99}Pr_3$ [a = 11.931(9) Å, c = 14.45(1) Å, $I4/mcm$, Z = 16] crystallizing with the Gd_3Ga_2 structure type, unlike that reported for Pt_2Pr_3 in the phase diagram, as well as on the ternary Pt/Pr/Sn system.

EXPERIMENTAL METHODS

Starting materials were Pt beads (99.9% purity), Pr and Sn pieces (99.9% purity), and NaCl (dried). All samples were weighed and loaded into tantalum ampules inside an argon-filled glovebox. Pt/Pr binary samples were loaded with either NaCl or Sn (approx. 250 mg) as a flux. Ampules were sealed under argon followed by sealing in evacuated silica tubes. Samples were placed in a furnace at 1000°C for 24 hours followed by slow cooling (-20°C/hr) to 700°C and held there for 72 hours. Powder and single crystal X-ray diffraction were used to characterize the reaction products. Single crystal X-ray diffraction was performed on a Bruker APEX CCD diffractometer with refinements completed with SHELXTL.

RESULTS AND DISCUSSION

During the exploration of the binary Pt/Pr system using NaCl flux, single crystals of $Pt_{1.99}Pr_3$ [a = 11.931(9) Å, c = 14.45(1) Å, $I4/mcm$, Z = 16] were isolated. $Pt_{1.99}Pr_3$ adopts the Gd_3Ga_2 structure type where Pt atoms are coordinated by Pr atoms in chains of alternating cubes and square antiprisms along the c axis. The remaining Pr positions form trigonal prisms and distorted tetrahedra around Pt atoms in 1:2 chains. The previously reported crystal structure of Pt_2Pr_3 is that of hexagonal Ho_3Ni_2.[2]

Subsequent loading of "Pt_2Pr_3" in a Sn flux yielded two new ternary intermetallics within the Pt/Pr/Sn system. Single crystal analysis identified $Pr_4Pt_{12}Sn_{25}$ [a = 12.274(1) Å, Im-3, Z = 2] and $Pt_3Pt_4Sn_6$ [a = 27.62(1) Å, b = 4.5958(2) Å, c = 9.3499(5) Å, $Pnma$, Z = 4]. The first is isostructural with $Gd_3Ni_8Sn_{16}$ while the latter is a new ternary compound in the $R_3T_4E_6$ family (R = rare earth metal; T = Pt, Pd; E = Ge, Al, Sn, Si) where Pt and Sn form hexagonal and pentagonal nets in the ac plane.[3]

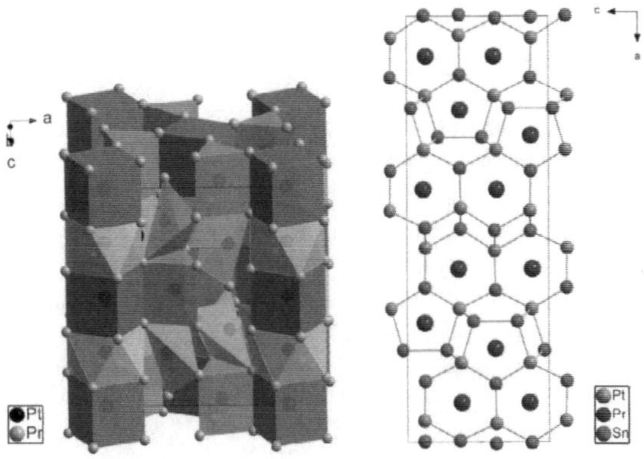

Figure 1. Structural features of $Pt_{1.99}Pr_3$ (left) and $Pt_3Pt_4Sn_6$ (right).

CONCLUSIONS

Further investigation in the Pt/Pr system utilizing flux syntheses reveals previously unidentified binaries not reported in the phase diagram. Using a Sn flux, several Pt/Pr/Sn ternary intermetallics with known structure types have been identified.

REFERENCES

1. T. Bell, *Dr. rer. nat. Thesis*, University of Cologne, Germany, 2014.
2. le Roy, J.; Moreau, J. M. ; Paccard, D. ; Parthé, E., Acta Cryst. B, 1977, 33, 2414-2417.
3. Gribanov, A.; Rogl, P.; Grytsiv, A.; Seropegin, Y.; Giester, GJ. Alloys and Compounds., 2013, 571, 93-97.

ACKNOWLEDGMENTS

Research was supported by the Office of the Basic Energy Sciences, Materials Sciences Division, U. S. Department of Energy (DOE), and the Department of Chemistry at Iowa State University (ISU). Ames Laboratory is operated for DOE by ISU under contract No. DE-AC02-07CH11358.

In vitro and *in vivo* detection of bacteria with lanthanide complexes

Sylvie L. Pailloux,*[1] Andrey Joaqui Joaqui,[1] Fiona Armstrong-Pavlik,[1] and Valerie C. Pierre[1]

[1]Department of Chemistry, University of Minnesota, USA
* spaillou@umn.edu

INTRODUCTION

The rise in resistance to antibiotics is increasingly concerning. Statistics from the CDC indicate that each year, 2 million people in the US become infected with bacteria that are resistant to at least one antibiotic. Each year, these directly result in over 23,000 deaths; many more die from conditions that were complicated by these infections. The need to respond to the increasing threat of bacterial infections and antibiotic resistance is now urgent. This need has many facets. One involves the development of novel antibiotics, including antibiotics with novel mechanisms of action and targets. Another one involves the development of new diagnostic tools for rapid in vitro and in vivo detection of bacterial infections. We will present a coordination approach to solve both of these needs with rationally-designed siderophore-inspired lanthanide complexes that function either as *in vitro* or *in vivo* diagnostic probes. These metal complexes, which function by targeting the FepA receptors of Enterobacteriaceae, open a new alley in the fight against resistant Gram-negative pathogens.

RESULTS AND DISCUSSION

All the Enterobacteriaceae rely on two siderophores for their iron needs: the catechol-based enterobactin (also called enterochelin) and the hydroxamate-based aerobactin. Of these two siderophores, enterobactin has a substantially higher affinity for iron than aerobactin ($K_a = 10^{49}$ M^{-1} for Fe^{III}-enterobactin *vs* $K_a = 10^{23}$ M^{-1} for Fe^{III}-aerobactin). Consequently, enterobactin was repeatedly demonstrated to be one of the primary virulence factors for *E. coli* and other Enterobacteriaceae: knocking-down the synthesis and excretion of enterobactin or the uptake of ferric enterobactin result in strains that are no longer virulent. These observations strongly suggest that molecules that can exploit the enterobactin uptake systems could accumulate rapidly and selectively in bacteria in sufficient concentration to enable in vivo imaging. Importantly, enterobactin is secreted and taken up by all the Enterobacteriaceae, including *Klebsiella* spp., *Escherichia coli*, *Salmonella*, *Shigella*, *Yersinia* and *Enterobacter*.

Interestingly, we have demonstrated that lanthanide complexes of enterobactin are also recognized and actively taken-up by bacteria. This ability makes them uniquely powerful for both in vitro and in vivo detection of bacteria. For in vivo applications, since siderophores are primarily hexadentate, gadolinium complexes of siderophores have two or more open coordination sites that can be filled by water. As such, siderophores, and siderophore mimics all behave as powerful MRI contrast agents. For in vitro application, complexes of terbium and holmium, which have primarily one isotope, enable rapid, sensitive and selective detection of bacteria by mass cytometry – a technique for which lanthanide complexes are uniquely suited.

Herein we will discuss how the coordination chemistry of the lanthanide complex affects their recognition by bacteria and their turn-on response, both by MRI and by mass cytometry.

CONCLUSIONS

Our results demonstrate that lanthanide complexes of siderophores and siderophore analogues are rapidly and efficiently taken up by bacteria. This process is highly selective: whereas a wide range of bacterial pathogens can recognize the lanthanide complexes, mammalian cells do not. Moreover, the coordination chemistry of the lanthanide complex, and the geometry inferred by the ligand, does affect uptake by the bacteria.

ACKNOWLEDGMENTS

The authors acknowledge the support of the National Institutes of Health and the National Institute of Allergy and Infectious Diseases.

Inert-Shell Effects on the Upconversion Dynamics of β-NaYF$_4$:Yb,Er Nanomaterials

Md Yeathad Hossan, Mary T. Berry, P. Stanley May

Chemistry Department, University of South Dakota, USA, md.hossan@coyotes.usd.edu

INTRODUCTION

High-quality micron-size crystals of β-NaYF$_4$:Yb^{3+}, Er^{3+} are known to have NIR-to-visible upconversion (UC) quantum efficiency (QE) as high as 7% at moderate pump power densities.[1] However, β-NaYF$_4$:Yb^{3+}, Er^{3+} nanocrystals exhibit more rapid UC dynamics with lower quantum efficiency.[2] This is presumed to be due primarily to increased surface quenching effects for nanomaterials. Adding a passive shell of optically-inert un-doped material onto the core UC nanocrystals can effectively reduce the surface quenching. We have studied the optical properties of core and core-shell β-NaYF$_4$:Yb^{3+}, Er^{3+} UC nanocrystals and have shown that with sufficient shell thickness, the UC dynamics, and therefore presumably the QE, are similar to those in the bulk.

EXPERIMENTAL METHODS

Core β-NaYF$_4$:17%Yb^{3+}, 3%Er^{3+} UC nanocrystals were coated with NaYF$_4$ inert-shells of varying thickness. The shell addition was confirmed by TEM images. Optical characterization in this study involves steady-state and time-resolved emission spectroscopy on core, core-shell nanocrystals, and standard micron-size β-NaYF$_4$:Yb^{3+}, Er^{3+} UC crystals.

RESULTS AND DISCUSSION

Addition of an inert-shell changes the time dependence of UC emission following pulsed excitation at 1 μm and results in a significant increase in emission intensity for both pulsed and cw excitation. Our models demonstrate that the enhancement of UC emission results from minimizing surface quenching of Yb^{3+} ($^2F_{5/2}$) and Er^{3+} ($^4I_{11/2}$), energy levels which participate in rapid energy migration across tens of nanometers.[3] Shell thicknesses greater than ~5 nm appear to block the surface quenching and restore bulk UC behavior to the nanomaterials.

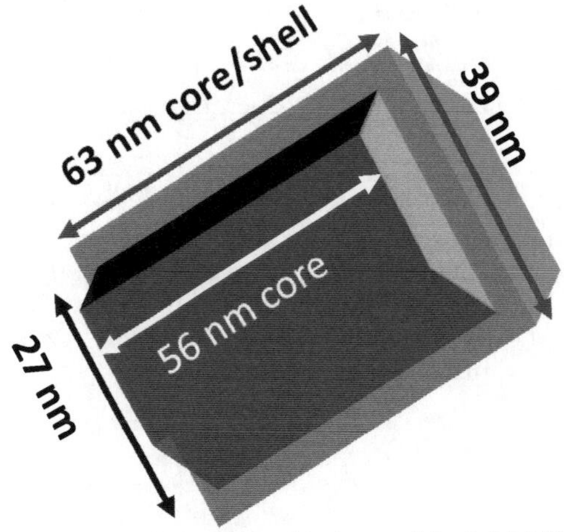

Fig. Inert-shell (green) added core (blue) β-NaYF$_4$:Yb^{3+}, Er^{3+} UC nanocrystals.

CONCLUSIONS

In β-NaYF$_4$:Yb^{3+}, Er^{3+} nanocrystals, energy migration and surface quenching of the Yb^{3+} ($^2F_{5/2}$) and Er^{3+} ($^4I_{11/2}$) reservoir states is the dominant cause of reduced QE in nanomaterials. This effect can be entirely removed by addition of an inert-shell.

REFERENCES

1. Anderson, R. B.; Smith, S. J.; May, P. S.; Berry, M. T. *J. Phys. Chem. Lett.* **2014,** *5* (1), 36-42.
2. Boyer, J.-C.; van Veggel, F. C. J. M. *Nanoscale* **2010,** *2* (8), 1417-1419.
3. Fischer, S.; Bronstein, N. D.; Swabeck, J. K.; Chan, E. M.; Alivisatos, A. P. *Nano Letters* **2016.**

ACKNOWLEDGEMENTS

The authors acknowledge financial support from NSF-PFI, SDBOR-SPACT, and NASA EPSCOR.

Influence of Monovalent Counter Cations on Neptunium(V/VI) Chloride Chemistry in Aqueous Solution

Geng Bang Jin and Shanna L. Estes

Chemical Sciences and Engineering Division, Argonne National Laboratory, USA, gjin@anl.gov

INTRODUCTION

Understanding actinide speciation in solution is critical for developing actinide separation schemes and rational syntheses for actinide materials. Despite decades of research, considerable discrepancies exist between the existing knowledge of neptunyl chloride solution complexes and solid-state products. To improve our understanding of neptunyl-chloride interactions and their formation mechanisms, we conducted systematic studies on the electrochemical behavior of Np(V/VI) chloride solution species and the structural characteristics of solids resulting from evaporation of Np(V) chloride solutions.

RESULTS AND DISCUSSION

Evaporation reactions of Np(V) chloride solutions with small counter cations (e.g. Na^+) or without any other cations have resulted in several new compounds that consist of aquo and chloro complexes of Np(V).[1,2] NpO_2^+ units further participate in Cation-Cation Interactions with each other to form extended Np(V) chloride networks. In contrast, similar reactions in the presence of larger counter cations (e.g. Cs^+ and PPh_4^+) have resulted in solid products that consist of discrete tetrachloro complexes of Np(VI). These $[NpO_2Cl_4]^{2-}$ units are separated by counter cations via extensive bonding interactions.[3] Interconversion of these Np(V) and Np(VI) compounds were observed during evaporation when the solution is close to dryness (Figure 1).

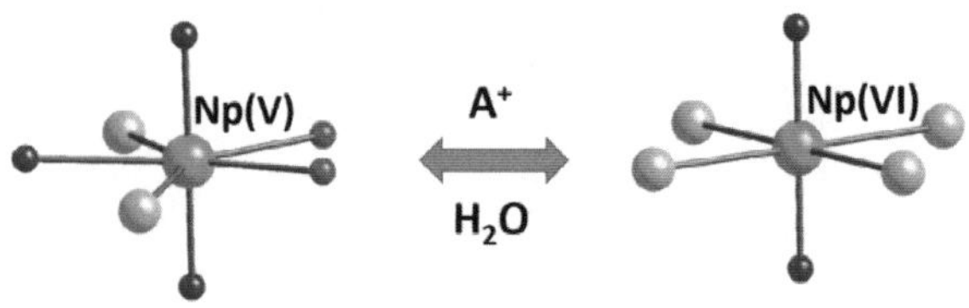

Figure 1. Interconversion between Aquo and Chloro complexes of Np(V) and tetrachloro complexes of Np(VI) during evaporation of Np(V) aqueous solutions in the presence of large counter cations. (Orange ball: Chloride; Red ball: Neptunyl Oxygen; Pink ball: water)

Preliminary electrochemical measurements indicate substantial influence of large counter cations on the redox behaviour of Np(V/VI) species in solution (Figure 2).[3] Together, these results suggest that large counter cations tend to stabilize higher orders of neptunyl chloride complexes, presumably by forming ion-pairs.

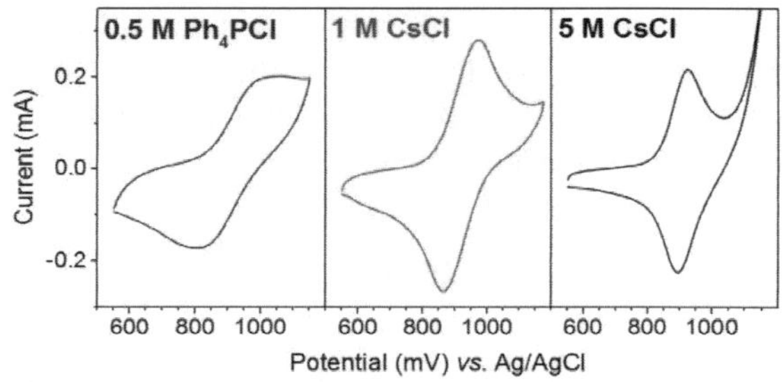

Figure 2. Cyclic voltammograms (4th cycle) of 5 mM Np(V) dissolved in 0.5 M Ph_4PCl (left), 1 M CsCl (middle), or 5 M CsCl (right). Initial and cathodic switching potentials = 550 mV, anodic switching potential = 1150 - 1175 mV, v = 16 mV s^{-1}.

CONCLUSIONS

Counter cations have displayed significant influence on the formation of neptunyl chloride solid products from evaporating Np(V) aqueous solutions. Other than traditionally recognized structure-directing effects, those larger cations also exhibit strong impact on the speciation of neptunyl chloride complexes in solution. These findings open a new avenue for understanding and ultimately controlling the formation mechanism of actinide complexes.

REFERENCES

1. G. B. Jin, *Inorg. Chem.*, 2013, **52**, 12317-12319.
2. G. B. Jin, *Inorg. Chem.*, 2016, **55**, 2612-2619.
3. G. B. Jin and S. L. Estes, *Unpublished results*.

ACKNOWLEDGMENTS

This work was performed at Argonne National Laboratory, operated by UChicago Argonne LLC for the United States Department of Energy under contract number DE-AC02-06CH11357 and was supported by the DOE Office of Basic Energy Sciences, Chemical Sciences, Heavy Elements Chemistry.

Innovative Use of Light Rare-earth Fibers as Antibacterial Agents

Shinji Hirai[1*], Naohisa Gandou[2], Oomori Hideki[3] and Eiji Nakamura[4]

[1]Research Center for Environmentally Friendly Materials Engineering, Muroran Institute of Technology, Japan
[2]Graduate School of Engineering, Muroran Institute of Technology, Japan
[3]The Japan Wool Textile Co., Ltd., Japan
[4]Santoku Corporation, Japan
[*]hirai@mmm.muroran-it.ac.jp

We discovered that wool, silk, and cotton with Ce ions adsorbed on to the fibers demonstrate outstanding antibacterial properties against *Staphylococcus aureus*.

INTRODUCTION

Hard bases form bonds with hard acids in rare earth ions, with the -COOH of keratin protein in wool, with the -OH of fibroin protein in silk, and with the -OH of cellulose in cotton. According to the hard and soft acid and bases (HSAB) concept, these are strongly attracted to each other, allowing adsorption of Ce on the fiber surface. Here, we facilitated the adsorption of Ce ions on to wool, silk, cotton, and polyester fabrics to evaluate their antibacterial properties.

EXPERIMENTAL METHODS

Wool, silk, cotton, and polyester fabrics with a mass ratio of 1:100 were soaked in an aqueous solution of cerium(III) nitrate with Ce ion concentration of 400 ppm, and were left disturbed at 80 °C for 2 h. Wool fabrics were soaked in aqueous solutions of lanthanum(III) nitrate and gadolinium(III) nitrate of the same concentration. As a control, fabrics sterilized in an autoclave were inoculated with test bacterial cultures of *Staphylococcus aureus* containing 0.05% non-ionic surfactant, and were then cultured for 18 h in a dark room. Thereafter, the bacteria were washed from the fabrics using normal saline, and the number of bacteria present in the post-wash saline solution was measured. The antibacterial activity value was obtained by dividing the number of bacteria present after 18 h of cultivation on the Ce ion-adsorbed fabric by that present after 18 h of cultivation on the Ce ion-free fabric. The compounds with an antibacterial activity value of 2.0 or more ($\geq$99% killing ratio) were considered to have an antibacterial effect. The antibacterial properties were evaluated after the Ce ion-adsorbed fabric samples were washed for 50 cycles in an electric washing machine at 40 °C using a detergent, with each cycle being: 5-minute wash, dewatering, 2-min rinse, dewatering, 2-min rinse, dewatering, and line drying.

RESULTS AND DISCUSSION

The antibacterial activity value of wool prior to the adsorption of Ce ions was -0.6, indicating poor antibacterial properties. The antibacterial activity values of wool, silk, cotton, and polyester fabrics after adsorption of Ce ions were 5.1, 5.8, 5.2, and 2.2 respectively, with all fabrics demonstrating outstanding antibacterial effect except for polyester, on to which the Ce ions do not get adsorbed. In particular, the antibacterial effect was maintained after 50 wash cycles in wool, silk, and cotton, with their respective antibacterial activity values being 3.8, 4.8, and 2.9. Even the adsorption of similar rare earth elements La and Gd on to wool fibers demonstrated outstanding antibacterial properties, with an antibacterial activity value of 5.9 for both elements.

An aqueous solution of cerium nitrate easily converts from (III) to (IV) in the presence of oxygen. The energy metabolism process in the mitochondria of cells produces hydrogen peroxide; however, in the presence of tetravalent ions, cerium(IV), which has a positive and high standard electrode potential, acts as a powerful oxidant and the Fenton/Haber Weiss reactions successively generate reactive oxygen such as hydroxyl radical or superoxide anion. Even among the reactive oxygen species, the standard electrode potential of hydroxyl radical is higher and more positive than that of hydrogen peroxide, ozone, and hypochlorous acid, making it a powerful oxidant, which may damage the bacteria and the cell walls. In contrast, in the case of La and Gd, which have superior trivalent ions, there are examples of good antimicrobial activity of Al[1]. The reaction between trivalent Al ions and superoxide anion produces Al superoxide semi-reduced radical ions. These ions then react inside the cells with trivalent ferric ions to produce divalent ferric ions. Next, these divalent ferric ions react with hydrogen peroxide to produce reactive oxygen by the Fenton reaction. Even in the case of trivalent La and Gd, generation of their superoxide semi-reduced radical ions can be expected.

CONCLUSIONS

Fabrics with La, Ce, and Gd ions adsorbed on to fibers demonstrated outstanding antibacterial properties against *Staphylococcus aureus*. The antibacterial properties were maintained in the fabrics even after 30 wash cycles. The antibacterial effect of light rare-earth ions is thought to be caused by the easy generation of reactive oxygen during the energy metabolism process, which damages the bacteria and cell walls.

REFERENCE

1. Yuan Chia-Yi *et al.*, Journal of Biomedical Science, 19(1):51, 2012.

Inspiration from Karl A. Gschneidner, Jr.: Rare Earth containing Zintl Phases and their Thermoelectric and Magnetic Properties

Kasey P. Devlin,[1] Nasrin Kazem,[1] Julia Zaikina,[1] and Susan M. Kauzlarich[1]

[1]Department of Chemistry, One Shields Ave, University of California, Davis, California 95616 USA, smkauzlarich@ucdavis.edu

INTRODUCTION

In the past decade, Zintl phases have seen a significant increase in research because they provide a class of compounds where one can easily correlate the structure and bonding to physical properties.[1] Zintl phases are a subset of the intermetallics where the structure can be understood using simple electron counting rules and in general, the stoichiometry is consistent with an electron precise semiconductor.[2] Zintl phases are made up of electropositive elements that donate their electrons to the more electronegative metalloids to form isolated anions, polyhedra, or extended structures. The structures can be described via a combination of ionic bonding between the cations and anions and covalent bonding for polyhedra or extended structures. In the case of isolated anions, the element can be considered to complete its octet.

The addition of both rare earth metals and transition metals to Zintl phases has been met with remarkable success in discovering new compounds with good thermoelectric and novel magnetic properties. The addition of rare earth metals to these phases to form solid solutions has expanded the crystal chemistry and properties further. There are two different compounds that will be presented: $Yb_{14-x}Ce_xMnSb_{11}$ and $Eu_{11}Zn_4Sn_2As_{12}$.

EXPERIMENTAL METHODS

$Yb_{14-x}Ce_xMnSb_{11}$ has been prepared as single crystals from Sn flux and via stoichiometry mixing of the elements. Fully dense pellets were prepared via spark plasma sintering for thermoelectric measurements. Thermoelectric measurements were obtained from 300-1000 K. $Eu_{11}Zn_4Sn_2As_{12}$ is prepared as single crystals via Sn flux. Magnetization and electrical resistivity as a function of temperature where obtained from 300 – 2 K. The electronic structure and chemical bonding were elucidated using first-principles quantum chemical calculations (TB-LMTO-ASA).

RESULTS AND DISCUSSION

The synthesis and structure of $Yb_{14-x}Ce_xMnSb_{11}$ will be presented, along with thermoelectric properties. $Yb_{14-x}Ce_xMnSb_{11}$ crystallizes in the $Ca_{14}AlSb_{11}$ structure type, which can be described as a Zintl phase. The structure can be described as consisting of Ca^{2+} cations, $AlSb_4^{9-}$ tetrahedra, Sb_3^{7-} linear anions, and Sb^{3-} anions. In the case of $Yb_{14-x}Ce_xMnSb_{11}$, Yb is Yb^{2+}, Ce is Ce^{3+} and $MnSb_4^{9-}$ is considered $[Mn^{2+}(Sb_4)^{12-}]$, one electron less than $[Al^{3+}(Sb_4)^{12-}]$. The addition of Ce to $Yb_{14}MnSb_{11}$ results in reduction of the carrier concentration. The Seebeck is enhanced compared with $Yb_{14}MnSb_{11}$ and thermal conductivity is comparable. However, the electrical resistivity is slightly larger providing an overall similar figure of merit, zT.

$Eu_{11}Zn_4Sn_2As_{12}$ has an anisotropic structure that features staggered ethane-like $[Sn_2As_6]^{-12}$ units and infinite 2D $[Zn_2As_3]^{-5}$ sheets extended in *ab* plane. Eu cations fill the space between these anionic motifs. Temperature-dependent transport properties of the new $Eu_{11}Zn_4Sn_2As_{12}$ Zintl phase between 2 and 300 K have been studied, and the electronic structure and chemical bonding were elucidated using first-principles quantum chemical calculations (TB-LMTO-ASA). The structure, quantum chemical calculations, and properties will be presented and discussed.

CONCLUSIONS

Two compounds will be presented and their structure, thermoelectric and magnetic properties described.

REFERENCES

1. Hu, Y.; Chen, C.-W.; Cao, H.; Makhmudov, F.; Grebenkemper, J. H.; Abdusalyamova, M. N.; Morosan, E.; Kauzlarich, S. M. Tuning Magnetism of $[MnSb_4]^{9-}$ Cluster in $Yb_{14}MnSb_{11}$ through Chemical Substitutions on Yb Sites: Appearance and Disappearance of Spin Reorientation. *J. Am. Chem. Soc.* 138: 12422−12431, 2016.
2. Kauzlarich, S. M. *Chemistry, Structure, and Bonding of Zintl Phases and Ions*. VCH Publishers, Inc: New York, 1996.

ACKNOWLEDGMENTS

The authors would like to thank the National Science Foundation (DMR-1405973) for providing financial support for this project.

Interaction of neptunium with crown ether and cucurbituril macrocycles

Tori Z. Forbes[1*] and Madeline Basile[1]

[1*]Department of Chemistry, University of Iowa, USA, tori-forbes@uiowa.edu

INTRODUCTION

Actinide interactions with macrocyclic ligands have been previously investigated due to their potential use in separations technologies.[1] Many studies present in the literature focused on uranyl complexes and much less is known regarding the binding of macrocycles with neptunium moieties. Herein we explore the interaction of both crown ether and cucurbituril macrocycles with Np(IV, V, and VI) and note significant differences in the formation of inclusion and exclusion complexes depending on the oxidation state of the metal.

EXPERIMENTAL METHODS

Neptunium compounds were synthesized from aqueous solutions under room-temperature conditions. The structural features of the solid state compounds were determined using single crystal X-ray diffraction. Both solution and solid phase vibrational spectroscopy was utilized for additional characterization of the crown ether and cucurbituril systems.

RESULTS AND DISCUSSION

Our initial work built upon previous investigations by Danis et al.,[2] which indicated limited interaction between the oxo group in uranyl tetrachloro complexes and nearby alkali metals. Np(VI) shows similar interactions, but the Np(V) tetrachloro species were not readily crystalized (Figure 1). Instead, crown ether inclusion complexes were formed with significant interactions to neighbouring cations. These results are similar to the report by Clark et al.[3] that described the formation of an inclusion compound for the Np(V) 18-crown-6 system. Clark et al.[3] indicated in this previous report that the Np(V) complex formed from the reduction of Np(VI). We also observed redox activity in these systems where the Np(V) stock solution would reduce to Np(IV) to form novel sandwich complexes with 12-crown-4 and 15-crown-5.

The formation of the Np(V) crown ether inclusion complexes inspired additional work investigating the U(VI) and Np(V) cucurbit[5]uril system. The identity of the metals in the cucurbituril inclusion complexes can be tuned to investigate specific interactions to the neptunyl oxo. In addition, encapsulation of metals and molecules inside these macrocycles can lead unique redox properties and greater control over chemical reactivity.

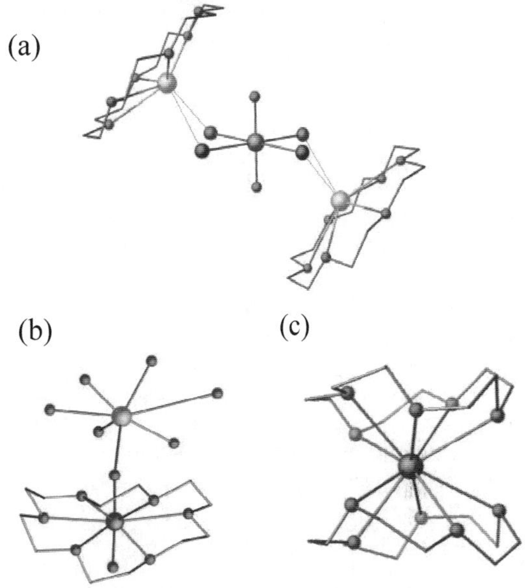

Figure 1. **(a)** the $[Np(VI)O_2Cl_4]^{3-}$ anion interacting with K^+ 18-crown-6. **(b)** Np(V) 18-crown-6 complex displays oxo interactions with sodium cation (pink sphere). **(c)** Np(IV) sandwich complex with 15-crown-5.

CONCLUSIONS

Formation of exclusion and inclusion crown ether complexes is dependent on the oxidation state of the Np cation. Additional studies are ongoing to explore the chemical properties of the coordination complexes, particularly for the cucurbituril system where the close interatomic distances between the two metal centres may lead to enhanced chemical reactivity.

REFERENCES

1. Morss *et al.*, "The Chemistry of the Actinide and Transactinide Elements." 2448-2456, 2010
2. Danis J. *et al.*, Inorg. Chem. 40:3389-3394, 2001
3. Clark D. *et al.*, Angew. Chem. 37:164-166, 1998

ACKNOWLEDGMENTS

The authors would like to thank the Department of Energy Heavy Elements Chemistry program (DE-SC0013980) for providing financial support to this project.

Interplay Between Localized and Itinerant Magnetism in Eu-Containing Pnictides of the $ThCr_2Si_2$ Structure Type

Xiaoyan Tan,[1] Alexander A. Yaroslavtsev,[2] Daniel Haskel,[3] Francois Guillou,[4] Andrei Rogalev,[4] Michael Shatruk[1*]

[1]Department of Chemistry and Biochemistry, Florida State University, Tallahassee, FL 32306, USA
[2]European XFEL GmbH, 22869 Schenefeld, Germany
[3]Advanced Photon Source, Argonne National Laboratory, Argonne, Illinois 60439, USA
[4]European Synchrotron Radiation Facility, 71, Avenue des Martyrs, CS40220, 38000 Grenoble, France
shatruk@chem.fsu.edu

INTRODUCTION

Itinerant magnets exhibit strong correlations between their electronic structure and magnetic properties. As a result, their magnetic behavior is highly sensitive to external perturbations, e.g., to applied pressure or magnetic field. In particular, intermetallics of the $ThCr_2Si_2$ structure type have been shown to exhibit pressure-induced structural, magnetic, and electronic phase transitions. One of the most impressive recent manifestations of such behavior For example, pnictides AFe_2As_2 (A = Ca, Sr, Eu) exhibit antiferromagnetic (AFM) ordering under ambient conditions but become superconducting at high pressure.[1-3]

RESULTS AND DISCUSSION

We report a remarkable emergence of itinerant 3*d* ferromagnetism (FM) caused by pressure-induced 4*f* valence fluctuations in $EuCo_2As_2$. In striking contrast to high-pressure $EuCo_2P_2$, which contains mainly Eu^{3+} ions [4], high-pressure $EuCo_2As_2$ exhibits the stabilization of a mixed-valent $Eu^{2.25+}$ state that persists up to 12.6 GPa, the maximum pressure achieved in our experiments. Furthermore, we demonstrate the generality of this phenomenon in the present system by inducing valence changes via chemical compression in $Ca_{0.9}Eu_{0.1}Co_2As_2$ and via direct electron doping in $Ca_{0.9}La_{0.1}Co_2As_2$.

We show the emergence of itinerant magnetism in $EuCo_2As_2$ under pressure by using X-ray absorption near-edge structure (XANES) and X-ray magnetic circular dichroism (XMCD) spectroscopy supported by the band structure calculations. The second-order structural phase transition around 4.7 GPa induces mixed valence in the Eu sublattice and breaks down the AFM ordering of the Eu^{2+} magnetic moments, as both Eu (4*f*) and Co (3*d*) moments become FM ordered with T_C = 125 K. While the FM ordering of Eu 4*f* moments was confirmed directly by XMCD experiment, the FM ordering of Co 3*d* moments was established indirectly by electronic band structure calculations and by investigation of model compounds, $Ca_{0.9}R'_{0.1}Co_2As_2$ (R′ = Eu, La). In the latter, the itinerant 3*d* ferromagnetism in the Co sublattice is triggered by mixed valence of Eu induced by chemical compression and by direct electronic doping into the Co layer due to aliovalent substitution, respectively.

Finally, we discuss the problem of the electron density delocalization upon the electron transfer from the Eu sites to the $[Co_2Pn_2]$ layer. Our XANES measurements on the phosphorus *K*-edge clearly prove that the present systems cannot be treated in terms of the simplistic electron-counting formalism and must be approached as truly itinerant-electron materials.

REFERENCES

1. Baek, S.-H. *et al.*, Phys. Rev. Lett. 102:227601, 2009.
2. Lee, H. *et al.*, Phys. Rev. B 80:024519, 2009.
3. Kotegawa, H. *et al.*, J. Phys. Soc. Jpn. 78:013709, 2009.
4. Chefki, M. *et al.*, Phys. Rev. Lett. 80:802, 1998.

ACKNOWLEDGMENTS

We would like to thank the National Science Foundation (grant DMR-1507233) for providing financial support to this project. The work at the Oak Ridge National Laboratory was sponsored by the Scientific User Facilities Division, Office of Basic Energy Sciences, U.S. Department of Energy (DOE). The use of the Advanced Photon Source at Argonne National Laboratory was supported by U.S. DOE under Contract No. DEAC02−06CH11357.

Interplay between the Sb/Bi Bonding and Charge Transport Properties in Rare-Earth Antimonide and Bismuthide Oxides

Yurij Mozharivskyj

Department of Chemistry and Chemical Biology, McMaster University, Hamilton, Ontario, Canada, L8S 4M1, mozhar@mcmaster.ca

Rare-earth antimonide and bismuthides oxides with the $RE_n(Sb,Bi)_mO_n$ composition were investigated as potential thermoelectric materials. They consist of the Sb/Bi and *RE*-O building blocks with different transport properties. The Sb/Bi layers define the density of states around the Fermi level and controls the charge transport properties. The electronic properties of the Sb/Bi substructure can be tuned through the size of the *RE* atoms in the *RE*-O blocks. E.g. in the *anti*-$ThCr_2Si_2$ type $RE_2(Sb,Bi)O_2$ phases, substitution of smaller *RE* atoms into the *RE*O slabs or larger Bi atoms into the Sb layer yields different charge carrier activation mechanisms.[1,2] As a result, the electrical conductivity and Seebeck coefficient are improved, while the number of charge carriers in the series remains constant. The charge transport properties can be explained within the framework of the Anderson localization.

References:

1. Wang, P.L., T. Kolodiazhnyi, J. Yao, and Y. Mozharivskyj, *JACS*, 2012. **134**(3): p. 1426-1429.

2. Wang, P.L., T. Kolodiazhnyi, J. Yao, and Y. Mozharivskyj, *Chem. Mater.*, 2013. **25**(5): p. 699-703.

Investigating Sm-Co Permanent Magnets as an Alternative to Dy-containing Nd-Fe-B in Direct-Drive Wind Turbines

H. A. Khazdozian[1] A. Vedantam[2], D. D. Imholte[3], A. Iyer[4], and I. C. Nlebedim[1]

[1]Ames Laboratory, USA, helenak@ameslab.gov
[2]Operations Management and Strategy Dept., University of Buffalo, USA
[3]Lawrence Livermore Laboratory, USA
[4]Operations Management, Purdue University, USA

INTRODUCTION

Rare earth elements (REEs) provide the high energy product necessary for permanent magnets in many applications including direct-drive wind turbines. Nd-Fe-B permanent magnets in direct-drive wind turbine generators must contain Dy to prevent demagnetization in the high temperature operating environment. However, the criticality of materials like Dy necessitate the investigation of alternatives. The high curie temperature of Sm-Co makes it attractive for high temperature applications, but its relatively low energy product makes it challenging to incorporate into high energy density applications. Varying grades of commercial Sm-Co permanent magnets were investigated as a substitute for Dy-containing N42 SH grade Nd-Fe-B permanent magnets in a 3 MW direct-drive permanent magnet generator (DDPMG).

METHODS

A 3 MW DDPMG with an external rotor and surface permanent magnet topology was designed based on an existing system[1]. Design specifications are shown in Table 1. Table 2 shows the properties of N42 SH magnets used as baseline for evaluating the potential of substituting Nd-Fe-B with Sm_2Co_{17} magnets. 3D transient finite element methods were used to calculate the performance of each material in the 3 MW DDPMG under steady-state conditions. Demagnetization was considered for operating temperatures ranging from -25 to 75 °C. The magnet volume was increased until rated performance was achieved where possible. Rotor dimensions and stator design were kept constant. Finally, the weight and cost of the permanent magnets were compared for each design.

Table 1: Specifications of a 3 MW DDPMG.

Specification	Value
Rated Power (MW)	3
Rated Speed (rpm)	15
Number of Poles	160
Number of Slots	480
Stator Outer Diameter (mm)	5,000
Stator Inner Diameter (mm)	4,760
Rotor Outer Diameter (mm)	5,120
Stack Length (mm)	1,200
Magnet Length (mm)	25

RESULTS AND DISCUSSION

For direct-substitution (constant magnet volume) of the Sm-Co permanent magnets for N42 SH in the 3MW DDPMG, the energy product of the Sm-Co permanent magnets was not sufficient to achieve the rated power (Fig. 1). Although Sm-Co offers better temperature dependence of magnetic properties than Nd-Fe-B, increased magnet volume was necessary to allow for achievement of rated power with high-grade Sm-Co magnets. This increases the weight of the generator, but may offer a cost savings depending on the market price of Dy.

Table 2: Magnetic Properties of the N42 SH and Sm_2Co_{17} Magnets

Material	Coercivity (kA/m)	Remanence (T)
N42 SH	767	1.30
SmCo 26HE	800	1.07
SmCo 30 S	845	1.12
SmCo 32 S	850	1.15
SmCo 35E	860	1.17

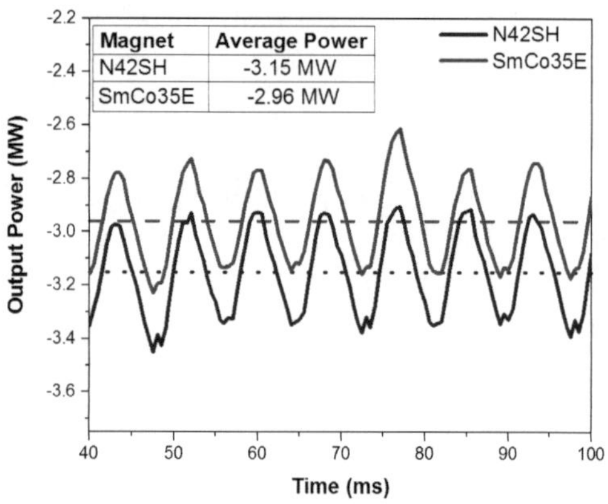

Fig. 1. Output power of a 3 MW DDPMG with varying magnet material and constant magnet volume at 273 K.

CONCLUSIONS

High grade Sm-Co permanent magnets may be a technically feasible substitute for Dy-containing Nd-Fe-B permanent magnets in direct-drive wind turbines. The magnetic and thermal performance will be determined for each Sm-Co magnet in the 3 MW DDPMG, and magnet volume scaled as needed. The magnet weight and cost will be determined for each Sm-Co magnet and compared to the baseline case (N42 SH).

REFERENCES

1. Anvari B. *et al.*, IEMDC. 181-186, 2015.

ACKNOWLEDGMENTS

This research was supported by the Critical Materials Institute, an Energy Innovation Hub funded by the U.S. Department of Energy, Office of Energy Efficiency and Renewable Energy, Advanced Manufacturing Office.

Investigation of Luminescence Properties of Ce^{3+}-Doped $Tb_3(Al,Ga)_5O_{12}$ Phosphor

Stefan Fischer and Thomas Jüstel

Department of Chemical Engineering, Münster University of Applied Sciences, Germany
corresponding authors: stefan-fischer@fh-muenster.de, tj@fh-muenster.de

INTRODUCTION

A general restriction for phosphor converted light emitting diodes (pcLEDs) is the small amount of suitable red emitting phosphors to enhance lumen equivalent and color rendering at the same time.[1] Theoretically, sensitizing Eu^{3+} with Ce^{3+} would be an appropriate combination, since the $^2D_J \rightarrow {}^2F_{5/2}$ interconfigurational transition of Ce^{3+} emission is energetically in the range of $^5D_0 \rightarrow {}^7F_J$ interconfigurational transitions of Eu^{3+} excitation.[2] You also have to keep in mind the ideal excitation of Ce^{3+}, which could be shifted to higher energies with substitution of Al^{3+} with Ga^{3+}, if excitation with higher energies in the blue range is preferred. This is already known for other garnets like $Gd_3(Al,Ga)_5O_{12}$:Ce^{3+}.[3] In this study, we investigated the luminescence properties of Ce^{3+}-doped $Tb_3(Al,Ga)_5O_{12}$ (TAG) with increasing Ga^{3+} concentration.

EXPERIMENTAL METHODS

Preparation

The samples were prepared by citric acid sol-gel synthesis. Starting materials were dissolved in deionized water or nitric acid, depending if nitrates or oxides were used. The solutions were mixed and $C_6H_8O_7{\cdot}H_2O$ was added under vigorous stirring for several hours at 65 °C to obtain the sol. This sol was placed in a drier overnight at 130 °C to turn it into an orange colored gel. The gel was annealed at 1000 °C in air for several hours to remove organic components and subsequently calcined at 1450 °C in reducing CO atmosphere.

Measurements

Phase formation was determined by powder X-ray diffraction. Diffuse reflectance spectra, excitation spectra and emission spectra were recorded by different spectrometers. Quantum efficiencies were calculated by measuring the emission spectrum of $BaSO_4$ in an integration sphere out of PTFE. Also temperature dependent emission spectra were measured to investigate thermal quenching behavior.

RESULTS AND DISCUSSION

A series with different Ga^{3+} concentrations in TAG:Ce^{3+} has been synthesized, viz $(Tb_{0,97}Ce_{0,03})_3(Al_{1-x}Ga_x)_5O_{12}$ with x = 0, 0,2, 0,4, 0,6, 0,8 and 1. The samples have been prepared nearly phase pure with a small amount of a perovskite impurity phase, which do not disturb the optical characterization. Excitation and emission intensities and thermal quenching behavior of $(Tb_{0,97}Ce_{0,03})_3(Al_{1-x}Ga_x)_5O_{12}$ vary with increasing Ga^{3+} concentration due to different effects. While the conduction band nearly remains at the same energy level for up to 40 % Ga^{3+} substitution, the Ce^{3+} 5d band is already contracting just from 20 % Ga^{3+} substitution upwards. This is why the highest $TQ_{1/2}$ value and emission intensity is observed for $(Tb_{0,97}Ce_{0,03})_3(Al_{0,6}Ga_{0,4})_5O_{12}$. Excitation and emission spectra for $(Tb_{0,97}Ce_{0,03})_3Al_5O_{12}$ and $(Tb_{0,97}Ce_{0,03})_3(Al_{0,6}Ga_{0,4})_5O_{12}$ are depicted in Fig. 1.

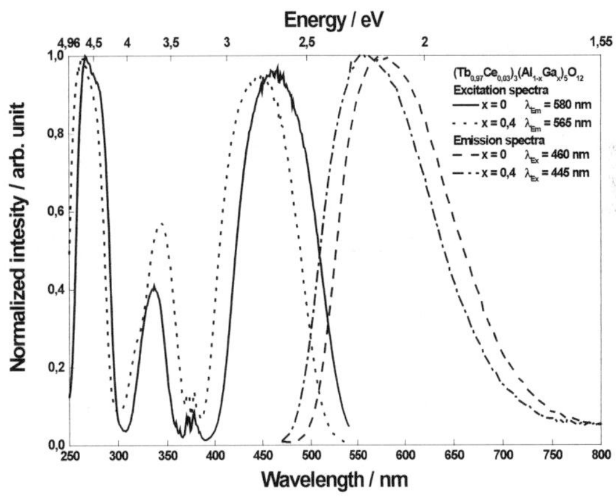

Fig. 1: Excitation and emission spectra of $(Tb_{0,97}Ce_{0,03})_3Al_5O_{12}$ and $(Tb_{0,97}Ce_{0,03})_3(Al_{0,6}Ga_{0,4})_5O_{12}$

CONCLUSIONS

By substituting Al^{3+} with Ga^{3+} in Ce^{3+}-doped $Tb_3Al_5O_{12}$ garnet, luminescence properties were modified. The sample with 40 % Ga^{3+} substitution shows the highest emission and excitation intensity as well as the highest $TQ_{1/2}$ value, since the Ce^{3+} 5d bands are just contracted, but the conduction band is nearly on the same energy level as in Ga^{3+}-undoped TAG:Ce^{3+}. Therefore, the activation energy in $(Tb_{0,97}Ce_{0,03})_3(Al_{0,6}Ga_{0,4})_5O_{12}$ for the process involved in thermal quenching is higher than those of the other samples, which leads to the highest observed $TQ_{1/2}$ value. Furthermore, emission band maximum and excitation band maximum of the higher 5d excitation band are shifted to higher energies due to decreased crystal field splitting.

REFERENCES

1. Pust P. *et al.*, Nat. Mater. 9:891-896, 2014
2. Blasse G., Phys. Status Solidi A 1:205-208, 1982
3. Ogiegło J. *et al.*, J. Phys. Chem. A 117:2479-2484, 2013

ACKNOWLEDGMENTS

The authors are grateful to Merck KGaA Darmstadt, Germany for generous financial support.

Lanthanide Complexes for Luminescence and Singlet Oxygen Generation

Jorge H.S.K. Monteiro, Alejandro Gracia-Nava and Ana de Bettencourt-Dias[1*]

[1*]Department of Chemistry, University of Nevada, Reno, USA, abd@unr.edu

INTRODUCTION

The luminescence of lanthanide ions, which can be sensitized through organic ligands, is useful in imaging, sensing and lighting applications.[1] For bioimaging applications, low energy excitation enables better penetration of the radiation and helps protect tissues from damage. Thus, ligands which enable two-photon excitation are desirable.[1] Finally, due to the location of the energy levels of the ligands, these are also photoactive on their own,[2] and it should be possible to use them as generators of singlet oxygen, especially if the intersystem crossing efficiency is increased through the presence of the lanthanide heavy atom.

RESULTS AND DISCUSSION

We have isolated several new ligands based on dipicolinic acid and a variety of functional groups.

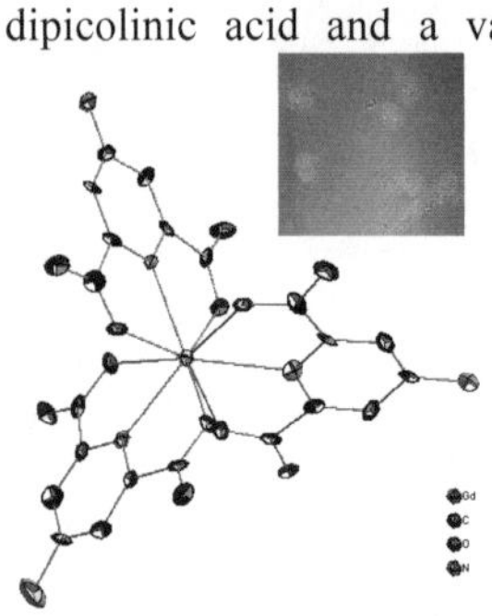

When the para-position of the dipicolinate is $-NH_2$, the resulting complexes cross the cell membrane of cancer cells. They thus enable fluorescence imaging, which is combined with moderate cytotoxicity. This effect is not observed in healthy cells. If the amine functional group is exchanged for a group capable of two-photon absorption, red EuIII-centered emission is observed upon excitation at 800 nm in a two-photon process. The third type of functional group is known for its singlet oxygen generation ability and the resulting lanthanide ion complexes show concurrent Ln^{III}-centered emission as well as singlet oxygen generation.

CONCLUSIONS

New lanthanide complexes show the ability to penetrate cell membrane to be used in fluorescence imaging. Other complexes are capable of red Eu^{III} emission upon low energy two-photon excitation. Finally, a new family of ligands leads to complexes that display Ln^{III}-centered emission and simultaneous singlet oxygen generation.

REFERENCES

1.Luminescence of Lanthanide Ions in Coordination Compounds and Nanomaterials. de Bettencourt-Dias, A., Ed. John Wiley and Sons: 2014.

2.de Bettencourt-Dias, A.; Poloukhtine, A., Phenylthiophene-dipicolinic acid-based emitters with strong solution blue and solid state green emission. *J. Phys. Chem. B* **2006,** *110*, 25638-25645.

ACKNOWLEDGMENTS

"The authors would like to thank the National Science Foundation (Grant CHE-1363325) for providing financial support to this project".

Lanthanide Extraction by TBP into bis(trifluoromethylsulfonyl)imide Ionic Liquids

Christopher Lee Klug[1], Jessica Bass[2], Charity Buggs[2], David Dang[2], Quyen Dang[2], David Shannon Meeker[2], and Yen Yim[2]

[1]Assistant Professor, Department of Chemistry and Physics, Augusta University, GA, USA cklug@augusta.edu
[2]Undergraduate Chemistry major, Department of Chemistry and Physics, Augusta University, GA, USA

INTRODUCTION

The recovery of reusable fuel from nuclear waste is a crucial step in is closing the nuclear fuel cycle, while reductions in waste volume are necessary to ensure efficient use of geologic repositories. Currently, most methods for actinide extraction are based on the PUREX process. A growing area of interest is the search for significantly safer and more eco-friendly improvements to nuclear fuel reprocessing. The use of room temperature ionic liquids (RTIL) in place of traditional organic solvents could show such improvements. Unlike the currently used solvents, most ionic liquids have low vapor pressures and low flammability, which greatly reduces the risk of industrial accidents and environmental contamination.[1] The ionic nature of RTILs also allow for the use of electrochemistry, either for more selective redox based separations or for non-proliferation monitoring. Most of the prior work reported in literature has been focused on the methylimadazolium cations, or on anions other than bis(trifluoromethylsulfonyl)imide.[2]

In order to explore the possible use of RTILs for fuel cycle separations, we have determined the distribution ratios and extraction stoichiometries for neodymium and holmium with tributyl phosphate loaded RTILs from bistriflimic acid solutions. Extraction times were determined based on a kinetics study of Ho extraction in the most viscous of the tested RTILs, and four RTILs representing three classes of cations were tested. The distribution ratios were found using the peak absorbance at a non-hypersensitive peak in each phase, and we are also currently reanalysing the same data using a mass balance approach. The log D-log [TBP] plots were analysed, and generally showed that one TBP was in the lanthanide complexes extracted by each of the RTILs.

EXPERIMENTAL METHODS

Hydrates of neodymium (III) carbonate and holmium (III) chloride were dissolved in excess bistriflimic acid, followed by dilution to make 0.2 M solutions. TBP-RTIL solutions were made to contain 0.02 M, 0.04 M, 0.1 M, 0.2 M, or 0.4 M TBP. The ionic liquids used were C_4mpiper Tf_2N, C_3mpiper Tf_2N, C_4mpyrr Tf_2N, and TMBA Tf_2N. The TBP-RTIL solutions were pre-equilibrated with 0.6 M solutions of bistriflimic acid, prior to contacting replicated aliquots with the appropriate 0.2 M lanthanide solution. The phases were separated, and spectra of the stock lanthanide solution and each of the contacted phases were collected with an Ocean Optics HR2000+ spectrometer, for analysis using peak fitting functions in OriginPro 9.1. Extractions were also performed with TBP in n-dodecane for comparison.

RESULTS AND DISCUSSION

Holmium was minimally extracted into the ionic liquids, with D values << 1. Unlike extractions into n-dodecane, there was a slope of ~1 for each of the systems. Ho^{3+} was minimally extracted into n-dodecane, with no correlation to [TBP]. The extraction of Ho^{3+} into the bistriflimide ionic liquids however was caused by a single TBP complexing each Ho^{3+} atom, as supported by the slope analysis. Future work should examine whether this trend continues at higher [TBP], and with counter-ions like nitrate. Analysis of the Nd extractions is on-going, and will be discussed in the oral presentation.

The different extraction behavior of Ho^{3+} in n-dodecane compared to RTIL further supports the previously reported cation exchange mechanism for ionic liquid extraction of lanthanides.[3] Our results show a difference based on the cation type, with D values for the TMBA 1:2 trials being nearly 1.5 times greater than the C_4mpiper ones. This is to be expected as the bulkier, more hydrophobic C_4mpiper cation would seem to have more difficulty passing into an aqueous solution than the TMBA cation.

CONCLUSIONS

Equilibrium with the most viscous RTIL was reached within one hour of mixing. All RTIL trials show the presence of one TBP ligand in the extracted cationic holmium complex, in contrast to n-dodecane extractions. As the TMBA Tf_2N has significantly higher D values, it appears that it is better suited for this exchange than the bulkier C_4mpiper Tf_2N, but loss of RTIL cations to the aqueous phase would require recovery steps to conserve the extraction solvent in an industrial setting. Future work should examine the extraction of actinides.

REFERENCES

Chiappe, C; Pieraccini, D. (2005) *J. Phys. Org. Chem.* 18(4), 275–297.
Binnemans. K. (2007) *Chem. Rev.* 107(6), 2592–2614.
Hawkins, C.; Garvey, S.; Dietz, M. (2012) *Sep. and Pur. Tech.* 89, 31-38.

ACKNOWLEDGMENTS

The authors wish to thank the Department of Chemistry and Physics, and the Augusta University Center for Undergraduate Research and Scholarship for funding and support.

Lanthanide helicates: from self-assembly to functionality

Jean-Claude G. Bünzli

Swiss Federal Institute of Technology, Lausanne (EPFL), Switzerland
Hong Kong Baptist University, Hong Kong, S.A.R., P.R. China
jean-claude.bunzli@epfl.ch

INTRODUCTION

Helical molecular structures have fascinated scientists since the secondary structure of proteins produced by 3-D helical patterns was recognized by L. Pauling in the 1950's. However, coordination chemists had to wait until 1987 when J.-M. Lehn connected two covalent helical strands around an axis defined by three Cu^{I} ions for producing a molecular helical architecture embedding metal ions. The recognition of specific metal ions through self-assembly opened a new playground for coordination chemists and the underlying principles were soon successfully applied to lanthanides.[1]

RESULTS AND DISCUSSION

We show here how ditopic hexadentate ligands with bis(benzimidazole)pyridine core self-assemble with Ln^{III} ions in stoichiometric ratio to yield $[Ln_2(L)_3]$ helicates in which the coordination sphere of the metal ion is saturated. Faint secondary electrostatic or non-covalent interactions between tridentate ligand strands generate size discriminating effects. Careful design of the two ligand compartments in unsymmetrical hexadentate receptors results in the supramolecular recognition of heteropairs of Ln^{III} ions, thus opening exciting perspectives for the elaboration of bi-color luminescent probes or of bifunctional, luminescent and magnetic stains. Extension of the concept leads to heterometallic 4f-*n*d helicates, as well as edifices with higher nuclearity.

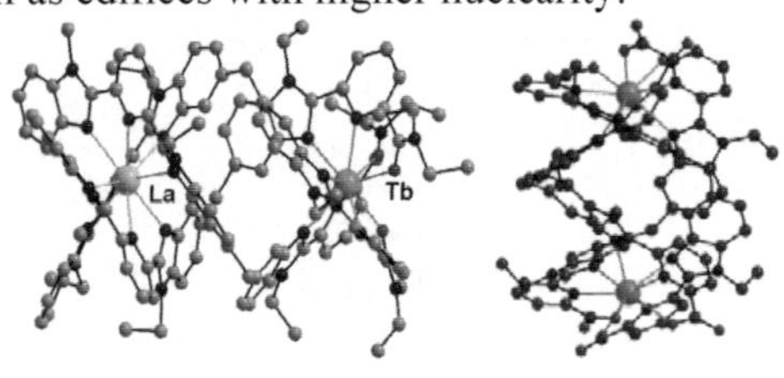

Thermodynamics parameters controlling the formation of these edifices, kinetics of formation, solid-state and solution structure are presented. Metal-metal communication along the nuclear axis is used to control the properties of one ion by the other.

Another series of ligand self-assemble in water with Ln^{III} ions to form highly luminescent bioprobes that are non-cytotoxic for a number of cancerous and non-cancerous cells (IC_{50}>500 μM). The helicates can be conjugated to various monoclonal antibodies. When combined with microfluidic devices, the dinuclear probes are highly efficient in simultaneous imaging of estrogen and human epidermal growth factor receptors of breast cancer MCF-7 cells.[2]

CONCLUSIONS

The new family of self-assembled helicates represents a promising tool for controlling intermetallic interactions and for developing highly sensitive bioprobes.

REFERENCES

(1) Piguet, C. and Bünzli, J.-C. G. Self-assembled Lanthanide Helicates: from Basic Thermodynamics to Applications. In *Handbook on the Physics and Chemistry of Rare Earths*; Gschneidner Jr., K. A., Bünzli, J.-C. G., and Pecharsky, V. K., Eds. Elsevier Science, B.V.: Amsterdam; 2010; Vol. 40, Ch. 247, 301-553.

(2) Chauvin, A.-S.; Thomas, F.; Song, B.; Vandevyver, C. D. B.; Bünzli, J.-C. G. Synthesis and Cell Localization of Self-assembled Dinuclear Lanthanide Bioprobes. *Phil. Trans. Roy. Soc. A* 2013, *371*, Art. Nr. 20120295.

Lanthanide Texaphyrin-based Theranostics

Jonathan L. Sessler[1,2]

[1]Department of Chemistry, The University of Texas at Austin
105 E. 24th Street – A5300, Austin, TX 78712-1224 USA
and
[2]Center for Supramolecular Chemistry and Catalysis, Shanghai University
Shanghai 200444, P. R. China, sessler@cm.utexas.edu

Please download and read the instruction sheet to ensure all necessary formatting and abstract criteria are adhered to.

INTRODUCTION

Expanded porphyrin is a term we introduced into the literature in 1988 to describe larger homologues of natural and synthetic tetrapyrrolic macrocycles. Expanded porphyrins, along with many other contracted, isomeric, and core-modified porphyrin analogues, are now known. Expanded porphyrins, in particular, have seen application in areas as diverse as anion recognition and transport, self-assembly, liquid-liquid ion extraction, photodynamic therapy, and anticancer drug development.[1-6] In recent years expanded porphyrins have helped increase our understanding of aromaticity and antiaromaticity.

RESULTS AND DISCUSSION

In this lecture, an update on recent advances in expanded porphyrin chemistry will be presented. Systems that have been synthesized as possible drug leads will be discussed in detail. Particular emphasis will be placed on texaphyrin conjugates that can function as so-called theranostics and permit both imaging and therapy using the same molecular construct. In these systems, the texaphyrin acts as a complexant for a rare earth cation, such as gadolinium or lutetium, that allows for sensing. It also serves as a cancer targeting moiety. Several cytotoxic payloads have been tested and a summary of this progress will be given, including an update on new entrepreneurial efforts. Work involving targeting agents other than texaphyrins or that involves fluorescence-based sensing will be presented as time permits.

CONCLUSIONS

Expanded porphyrins and texaphyrins in particular may have an important role to play as cancer-localizing entities that facilitate sensing at the cellular and sub-cellular level.

REFERENCES

References must be numbered. Keep the same style.

1. Sessler, J. L.; Murai, T.; Lynch, V.; Cyr, M. *J. Am. Chem. Soc.* **1988**, *110*, 5586-5588.
2. Sessler, J. L.; Miller, R. A. *Biochemical Pharmacology* **2000**, *59*, 733-739.
3. Ishida, ; Kim, S.-J.; Preihs, C.; Ohkubo, K.; Lim, J. M.; Lee, B. S.; Park, J. S.; Lynch, V. M.; Roznyatovskiy, V. V.; Sarma, T.; Panda, P. K.; Lee, C. H.; Fukuzumi, S.; Kim, D.; Sessler, J. L. *Nature Chem.* **2013**, *5*, 15-20.
4. Thiabaud, G.; Arambula, J. F.; Siddik, Z. H.; Sessler, J. L. *Chem. Eur. J.* **2014**, *20*, 8942-8947.
5. Zhang, Z.; Kim, D. S.; Lin, C.-Y.; Zhang, H.; Lammer, A. D.; Lynch, V. M.; Popov, I.; Miljanić, O. S.; Anslyn, E. V.; Sessler, J. L. *J. Am. Chem. Soc.* **2015**, *137*, 7769-7774.
6. Lee, M. H. Kim, E.-J.; Lee, H.; Kim, H. M.; Chang, M. J.; Park, S. Y.; Hong, K. S.; Kim, J. S.; Sessler, J. L. *J. Am. Chem. Soc.* **2016**, *138*, 16380–16387.

ACKNOWLEDGMENTS

This work has benefited from support from the U.S. National Science Foundation, The National Institutes of Health, the Cancer Research and Prevention Institute of Texas, as well as the R. A. Welch Foundation and the U.S. Department of Energy. Productive collaborations with a number of groups, including those of Profs. Alan Watts, Rick Finch, Zahid Siddik, Jonathan F. Arambula, Dongho Kim, Jong Seung Kim, Shunichi Fukuzumi, T.K. Chandrashekar, Christophe Bucher, Dirk Guldi, Pradeepta Panda, Changhee Lee, Jan Jeppesen, Masatoshi Ishida, and Tomas Torres, are also gratefully acknowledged.

Lanthanide-activated ternary sulfides: design of materials and light-emitting devices

Jonas J. Joos*, Lisa I. D. J. Martin, Sofie Abe, Dirk Poelman, Philippe F. Smet

LumiLab, Department of Solid State Sciences, Ghent University, Belgium
Center for Nano and Biophotonics (NB Photonics), Ghent University, Belgium
* jonas.joos@ugent.be

INTRODUCTION

Bright phosphors are obtained upon doping ternary alkaline earth thioaluminates and thiogallates with lanthanide ions. Especially when divalent europium (Eu^{2+}) is selected as optical dopant, technologically interesting phosphors are obtained, showing an emission band with a saturated color that is tunable from blue to yellow by interchanging the alkaline earth ion (Ca, Sr, Ba) or the trivalent metal (Al, Ga) (see Fig. 1).

EXPERIMENTAL METHODS

Lanthanide doped sulfide powders were prepared by solid state synthesis under a H_2S flow. Luminescence spectroscopy as function of temperature was performed[1]. Spectroscopic properties are empirically modelled to study the systematics of the luminescent properties of the lanthanide doped sulfides[2,3].

A detailed microscopic investigation is performed, combined with the simultaneous detection of element-specific x-ray emission and cathodoluminescence. This yields information on the dispersion of the dopant in the samples, which can be connected to the luminescence spectra and phosphor performance at elevated temperature (thermal quenching).

RESULTS AND DISCUSSION

In a first part, the luminescence properties of the solid solution series $SrGa_{2(1-x)}Al_{2x}S_4$ are investigated in detail. Next to the emission color that can be continuously tuned from green ($x = 0$) to blue ($x = 1$), other properties such as the decay dynamics of the excited state and the thermal quenching behaviour vary as a function of x. It is reviewed how the color point of the green component in white light-emitting diodes (LEDs) can be optimized by finetuning x. The suitability of these sulfide phosphors is critically evaluated for the use in white light-emitting diodes for both indoor lighting as well as display applications, using well-defined requirements[4].

Subsequently, the phenomenologically determined luminescent properties are further interpreted in terms of band diagrams displaying the charge-state transition levels of the lanthanide ions in the thiogallates. The locations of these levels are determined by exploiting systematics in lanthanide photophysics using the spectra of various lanthanides in the sulfide hosts. The emission color is explained by a simple crystal field model and the thermal quenching behaviour is interpreted based on the locations of the charge-state transitions levels.

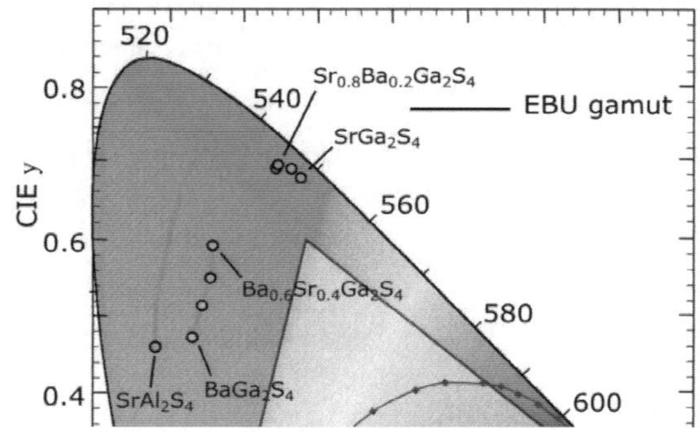

Figure 1. CIE (x,y) chromaticity diagram showing the tunability of the green color point of $SrGa_2S_4:Eu^{2+}$ by replacing Sr by Ba or Ga by Al.

Finally, state of the art remote phosphor LEDs for displays are constructed from a blue high-power LED, the optimized green-emitting sulfide and red-emitting CdSe/CdS quantum dots. It is demonstrated how the cost-effectiveness and performance can be optimized by tweaking the design of the remote phosphor layers[5].

CONCLUSIONS

This contribution demonstrates the color tunability and potential of sulfide phosphors in LED applications. It is shown how energy level diagrams can be easily constructed and used to explain physical properties or design new phosphors. The success of the considered materials is proven by combining them with red-emitting quantum dots in white light-emitting diodes, achieving a state-of-the-art performance while minimizing the materials cost.

REFERENCES

1. Joos J. J. *et al.*, Opt. Mater. 34:1902-1907, 2012
2. Dorenbos P., ECS J. Solid State Sci. Technol. 2:R3001-R3011, 2013
3. Joos J. J. *et al.*, Phys. Chem. Chem. Phys. 17:19058-19078, 2015
4. Smet P. F. *et al.*, J. Electrochem. Soc. 158:R37-R54, 2011
5. Abe S. et al., Light Sci Appl. 6:e16271, 2017

Lanthanide-based Nanoparticles: a versatile platform for optical probes

Fernando A. Sigoli
Laboratory of Functional Materials, Institute of Chemistry, University of Campinas-SP, Brazil
fsigoli@iqm.unicamp.br

INTRODUCTION

The lanthanide ions (Ln) exhibit interesting luminescent properties due the 4f-4f intra-configurational transitions. The degeneration of $4f^n$ orbitals is broken by (i) the high interelectronic repulsion, (ii) by the spin-orbital coupling and (iii) by the ligand field effect that promotes energy difference about 100 cm^{-1} between the electronic levels. Therefore, the choice of the host matrices is important and among the possibilities are inorganic matrixes such as oxide, fluoride and chalcogenides, complexes, organic polymers and hybrid materials. One of the major scope of our research group concern about the investigation of different lanthanide-doped matrices suitable for several applications such as optical probes. The upconversion and the NIR emission characteristic of some lanthanide ions are important for bioimaging, temperature sensing and photodynamic therapy (PDT). It is known that the temperature plays an important role in many biological processes and deregulations of this physical property at bio-micro-environment are normally correlated to cellular dysfunctions or pathologies. In this context, the design of temperature optical probes with high sensitivity, biocompatibility and accuracy at cellular level is of great interest for the understanding biological events and development of diagnostic and therapeutic techniques for some diseases.[1] Upconverting nanoparticles (UCNP) and luminescent molecular thermometers have received considerable attention, since they may be used for mapping of temperature variations in small or large areas such as cells, electronic microdevices or spacecraft in wind tunnel tests. Also, luminescent nanoparticles based on poly(N-isopropylacrylamide) (pNIPAM) have been considered as promising candidates for molecular thermometry. This organic polymer is water-dispersible, biocompatible and undergoes a phase transition near to the physiological temperature. The induced structural changes of pNIPAM may provide variations of emission intensities of incorporated fluorophores. Among these fluorophores, lanthanide(III) complexes have several advantages over organic molecules, due to their intrinsic photophysical properties.

RESULTS AND DISCUSSION

Our research group has used a specific methodology for preparing spheroidal and rod-like lanthanide-doped $NaLnF_4$ nanoparticles crystalized as pure α or β crystalline phases to evaluate the relative thermal sensitivities as a function of crystalline phases, morphology and particle sizes. The choice of $NaGdF_4$ matrix opens also the possibility to use these samples in Magnetic Resonance Imaging (MRI), resulting in a multiplex UCNP for simultaneous diagnosis by imaging and thermal sensing. Core@shell UCNP were also prepared and functionalized with a photosensitizer molecule for generation of oxygen reactive species aiming PDT applications. This work will also discuss about the possibilities of synthesis and the photophysical properties of luminescent molecular thermometers based on lanthanides(III) complexes grafted on pNIPAM nanoparticles as potential candidates for optical thermal sensing. As the pNIPAM phase transition results in different responses of the lanthanide emissions, Tb^{III} and Eu^{III} complexes grafted on pNIPAM were prepared and the luminescent intensity ratio (LIR) of the emissions at 612 and 543 nm was obtained resulting in a ratiometric and reversible temperature optical probe with a maximum relative thermal sensitivity (S_m = 8.4 % °C^{-1}, 40 °C) in the biological temperature range.

CONCLUSIONS

This work shows the possibilities of obtaining UCNP with different morphologies, sizes, crystalline phases as core-only or core@shell structures. The core nanoparticles in both α or β crystalline phases showed high S_m for lanthanide containing systems and the $^1O_2$ was detected by exciting the photosensitizer-core@shell UCNP at 980 nm. The new ratiometric luminescent nanothermometer based on lanthanides(III) complexes grafted on pNIPAM nanoparticles is reversible and has one of the highest values of the relative thermal sensitivity reported in the literature for lanthanide complexes in the biological temperature range.

REFERENCES

1. Bai T. *et al.*, Small. 12:4590-4610, 2016

ACKNOWLEDGMENTS

The author would like to thank CNPq, CAPES, INOMAT and FAPESP for financial support

Lanthanides in Bondage: From Basics to Business

Raymond, Kenneth N.

Department of Chemistry, University of California, Berkeley 94720-1460, United States
Ken Raymond knr4uc@gmail.com

Selective complexation of metal ions is a matter of fundamental science and of practical application. Nature provides many examples of selectivity in metal complexation and transport that can be used to design selective chelators. However, the lanthanides, because of their variable coordination number, highly ionic nature and lability, provide a particular challenge to the chemist. Throughout much of his career this chemist has sought to put the lanthanides in bondage, as illustrated to the right. The donor atoms in the figure are nitrogen, because some early chemistry involved using the Ln metal ion as a template to make nitrogen macrocycle ligands.[1] However, the lanthanides are oxophiles and so selective and strong chelators for them should contain oxygen donors. Before describing some of the ligand families that this has led to, a brief discussion of arguably the most fundamental lanthanide complexes, the aquo ions, will be presented as part of a discussion of the lanthanide ionic radii contraction.[2]

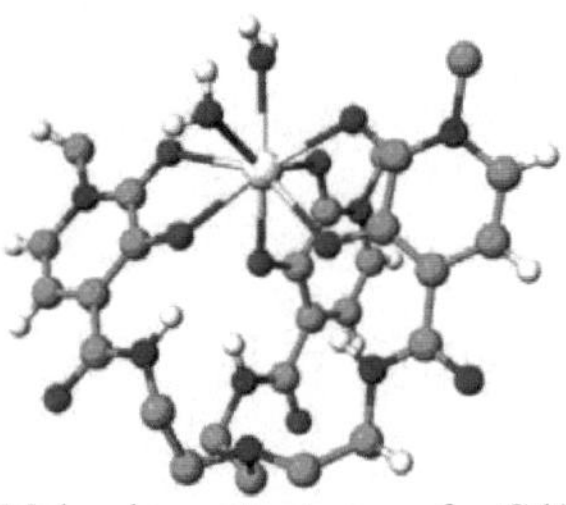

Molecular structure of Gd(III) (TREN-Me-3,2-HOPO), representative of potential MRI contrast agents with faster water exchange rates, high stability, and high selectivity over the physiologically important ions Ca(II), Zn(II), and Cu(II).

Magnetic Resonance Imaging (MRI) has fundamentally altered diagnostic medicine by providing non-invasive views of internal organs. Approximately 40% of all MRI's performed utilize paramagnetic shift reagents to enhance the image. Because of its $4f^7$ electronic configuration, gadolinium is used in the form of stable complexes. Currently, the relaxivity of commercial poly(amino-carboxylate) complexes is restricted by slow water exchange. There is also a growing concern about gadolinium deposited in the patient and consequent toxicity. A series of Gd^{III} complexes based on 3,2 hydroxypyridonate (3,2 HOPO) shown to the left were developed and investigated.[3] While there are no pending clinical applications for these agents, there are pending other clinical applications for octadentate varieties.[4-5]

Highly luminescent Ln(III) complexes (with Ln = Tb, Eu) for applications in biotechnology have been developed and will be presented in the award symposium. In particular, the 2-hydroxyisophthalamide (IAM) chelate for Tb(III) exhibits highly efficient emission ($\Phi_{total} > 50\%$), large extinction coefficients (ε_{max} >20,000 M^{-1} cm^{-1}), and long luminescence lifetimes ($\tau_{H2O} > 2.45$ ms) at dilute concentrations in standard biological buffers. Tri-macrocyclic Tb(III) complexes in this class (shown below) display long-term stability, with little if any change in their spectral properties (including lifetime, quantum yield, and emission spectrum) over time or in different chemical environments. Functionalized derivatives with terminal amine, carboxylate, and N-hydroxysuccinimide groups suitable for derivatization and protein bioconjugation have also been developed.[6]

Circularly polarized luminescence (CPL) is a rarely used spectroscopy that discriminates between luminescent chiral complexes. Using several different chiral ligands based on the IAM chelate, the first actinide circularly polarized luminescence spectra were reported.[7]

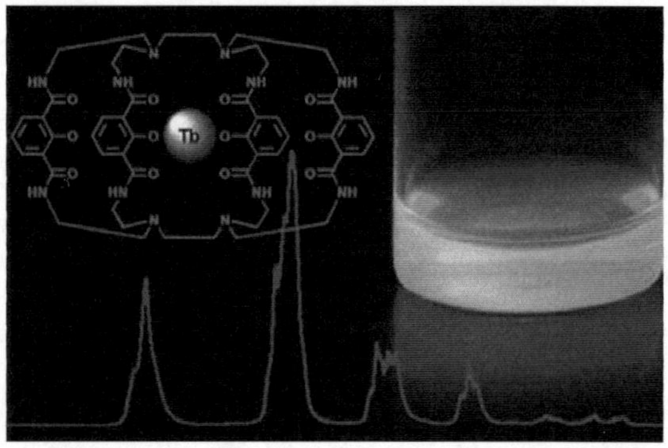

1. Smith P.H. *et al.,* Inorg. Chem. 24: 3469-3477, 1985.
2. Raymond, K.N. *et al.*, Comptes Rendus Chimie, 13: 849-852, 2010.
3. Datta, A. *et al.*, Accts. Chem. Res., 42: 938–947, 2009.
4. Gorden, A.E.V. *et al.*, Chem. Rev., 103: 4207-4282, 2003.
5. Liu, M. *et al.*, J. Pharm. Biomed. Anal. 102: 443–449, 2015.
6. Xu, J. *et al.*, J. Am. Chem. Soc., 133: 19900–19910, 2011.
7. Law, G-L. *et al.*, J. Am Chem Soc., 134: 15545–15549 2012.

Large spectral difference in triboluminescence and photoluminescence of Tb(III)/Eu(III) coordination polymers

Yuichi Hirai, Takayuki Nakanishi, Yuichi Kitagawa, Koji Fushimi, Tomohiro Seki, Hajime Ito, and Yasuchika Hasegawa

Faculty of Engineering, Hokkaido University, Japan, y-hirai@cse.hokudai.ac.jp

INTRODUCTION

Luminescence upon grinding solid materials (triboluminescence: TL) has long been a puzzling phenomenon. TL materials, unlike photoluminescence (PL) materials, have the advantage of emission without a light source. TL has been observed in inorganic, organic, and coordination compounds.[1] These TL-active compounds usually show PL with similar profiles, although they occur from distinct stimuli.[2] In order to provide new insights into TL, we here select lanthanide(III) (Ln(III)) complexes due to their high spectral resolution.[3] In this study, novel Ln(III) coordination polymers composed of Ln(III) ions, hexafluoroacetylacetonato (hfa), and furan-based bridging ligand (2,5-bis(diphenylphosphoryl)furan: dpf) are designed to induce high energy transfer efficiency for strong PL and mechanically unstable structure for strong TL. The reported Ln(III) coordination polymers with thiophene-based bridges (3,4-bis(diphenylphosphoryl)ethylenedioxythiophene: dpedot and 2,5-bis(diphenylphosphoryl)thiophene: dpt) are also prepared for comparison of TL and PL properties.[4]

EXPERIMENTAL METHODS

The bridging ligands were synthesized using α-substitution reaction of hetero(aromatic) compounds. The corresponding Eu(III) coordination polymers $[Eu(hfa)_3(X)]_n$ (X = dpf: **1**, dpedot: **2**, dpt: **3**) were prepared by refluxing $Eu(hfa)_3(H_2O)_2$ and bridging ligands in methanol.

RESULTS AND DISCUSSION

The ability for generation of TL was ascertained by observing Eu(III)-characteristic red emission upon grinding solid Eu(III) coordination polymers. We observed quite weak TL for **3**, which exhibited an intermolecular alternate arrangement of CF_3 groups to form a mechanically stable structure. The TL activity was clearly high in **1** and **2** with face-to-face CF_3 arrangement between single polymer chains. These observations indicate that the TL-activity strongly depends on the arrangement of bulky CF_3 groups in crystal packing structures.

When the same excitation process in TL and PL is assumed, the emission colours of Tb(III)/Eu(III) mixed coordination polymers should be the same under grinding and UV irradiation. We thus prepared $[Tb,Eu(hfa)_3(dpf)]_n$ (Tb/Eu = 1, 10), and the obtained compounds exhibited strong TL and PL. PL colours are dominated by Eu(III)-centred emission under UV irradiation, since the Tb(III)-centred emission is quenched by both back energy transfer from Tb(III) ions to hfa ligands and excitation energy transfer from Tb(III) to Eu(III) ions at room temperature. On the contrary to PL, the observed TL colours were dominated by Tb(III)-centred emission (Table 1). The large TL/PL spectral differences were independently recorded using CCD system. We here considered the TL of Ln(III) coordination compounds to be responsible for both the ligand-excitation and direct Ln(III)-excitation, unlike photo-excitation with specific wavelength in PL.

Table 1 TL and PL colours of Tb(III)/Eu(III) mixed coordination polymers $[Tb,Eu(hfa)_3(dpf)]_n$.

Sample	PL colour	TL colour
Tb/Eu = 1	Reddish-orange	Yellow
Tb/Eu = 10	Greenish-yellow	Green

CONCLUSIONS

Strong TL/PL Ln(III) coordination polymers were successfully prepared by introducing a furan-based bridging ligand. The relationships between crystal structures and TL activity were systematically demonstrated using Eu(III) coordination polymers with face-to-face/alternate intermolecular packing structures. The intermolecular arrangement of bulky substituents plays an important role in providing efficient TL property. The potential differences between TL and PL were clearly demonstrated by the TL and PL spectra of Tb(III)/Eu(III) mixed coordination polymers for the first time. Based on the results, the existence of discrete excitation mechanisms for TL and PL were suggested.

REFERENCES

1. Yamashita Y. *et al.*, Chem. Mater. 24:671-676, 2012
2. Chi Y. *et al.*, ACS Appl. Mater. Inter. 8:33888-33898, 2016
3. Bünzli J.-C. G. *et al.*, Chem. Soc. Rev. 34:1048-1077, 2005
4. Hirai Y. *et al.*, Angew. Chem. Int. Ed. 55:12059-12062, 2016

ACKNOWLEDGMENTS

We appreciate Grants-in-Aid for Scientific Research on Innovative Areas of "New Polymeric Materials Based on Element-Blocks (no. 2401)" (no. 24102012) of the Ministry of Education, Culture, Sports, Science and Technology (MEXT), Japan.

Low Percentage Doping in Multinuclear Lanthanide Polyoxometalates (POMs) and Their Use as Ratiometric Cryogenic Thermometers

Anna M. Kaczmarek[1], Jing Liu[1], Brecht Laforce[2], Laszlo Vincze[2], Kristof Van Hecke[1], Rik Van Deun[1]

[1]Department of Inorganic and Physical Chemistry, Ghent University, Belgium, anna.kaczmarek@ugent.be
[2] Department of Analytical Chemistry, Ghent University, Belgium

INTRODUCTION

Accurate, self-referenced temperature measurements are of great interest and importance in a wide range of fields.[1] Here, the thermal sensing capabilities of multinuclear lanthanide polyoxometalates (LnPOMs) in the cryogenic region (<100 K) are presented. Currently, most reported lanthanide based luminescent thermometers are focused on lanthanide metal organic frameworks.[2] Up till now, to the best of our knowledge, no LnPOMs have been studied for their use as ratiometric luminescent thermometers. In order to develop ratiometric LnLn' POM thermometers multinuclear LnPOMs co-doped at low ratios of the Ln' ions need to be designed.

RESULTS AND DISCUSSION

A series of LnPOMs built from the $[Mo_7O_{24}]^{6-}$ polyanion with La^{3+} , Eu^{3+}, Tb^{3+}, Sm^{3+}, Dy^{3+}, Nd^{3+}, Er^{3+} and Yb^{3+} were synthesized and a detailed analysis revealed that the tetranuclear clusters formed monomers or dimers linked through oxygen bridges. The smaller lanthanide ions - Er^{3+} and Yb^{3+} formed mononuclear sandwich type POMs. The $[La_4(MoO_4)(H_2O)_{16}(Mo_7O_{24})_4]^{14-}$ **(LaPOM)** polyanion has been chosen as the host material for the study on low percentage Ln^{3+} doping. The cluster consists of four La^{3+} ions, four $[Mo_7O_{24}]^{6-}$ and one $[MoO_4]^{2-}$ tetrahedron (Fig. 1). **5%Ln: LaPOM** compounds (Ln^{3+} = Eu^{3+}, Tb^{3+}, Sm^{3+}, Dy^{3+}, Nd^{3+}, Er^{3+}, Yb^{3+}) were prepared and their crystal structures and luminescence properties were studied to gain knowledge on how the low percentage doping affects the properties of the materials.

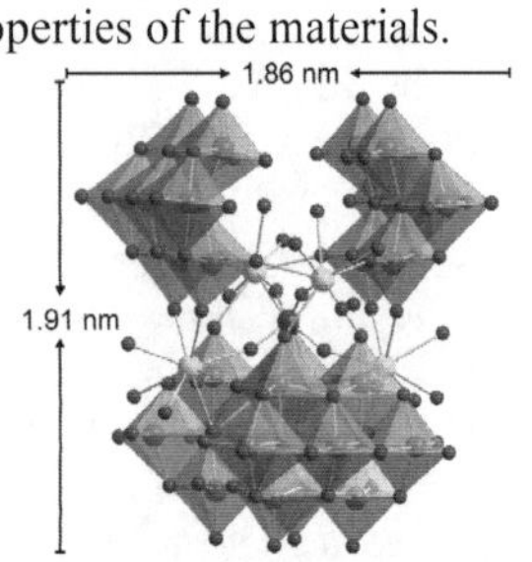

Fig. 1. Structure of $[La_4(MoO_4)(H_2O)_{16}(Mo_7O_{24})_4]^{14-}$.

The successful incorporation of a second Ln^{3+} ion (Ln') into the structure of the $[La_4(MoO_4)(H_2O)_{16}(Mo_7O_{24})_4]^{14-}$ has led us to further investigate these type of compounds. $[Tb_8(MoO_4)_2(H_2O)_{32}(Mo_7O_{24})_8]^{28-}$ **(TbPOM)** was employed for this investigation and doped with 5% or 10% of Eu^{3+}. Single crystal XRD revealed that the presence of Eu^{3+} ions influenced the structure formation of the compounds. Temperature dependent luminescence measurements for **5%Eu: TbPOM** over a 10 – 100 K temperature range were performed to evaluate the luminescence properties of the material (Fig. 2). With temperature increase from 10 K towards 100 K the emission of the Tb^{3+} transition peaks gradually decreased, whereas the intensity of the Eu^{3+} transition peaks stayed at a similar intensity level. As already around 100 K the Tb^{3+} transition peaks become very weak the material is considered for a thermometer in the cryogenic region. The maximum values of S_a and S_r are 0.13 K^{-1} (at 80 K) and 2.90 % K^{-1} (at 50 K), respectively, which indicate that this material has very good signal sensing capability and could be used as a cryogenic thermometer.

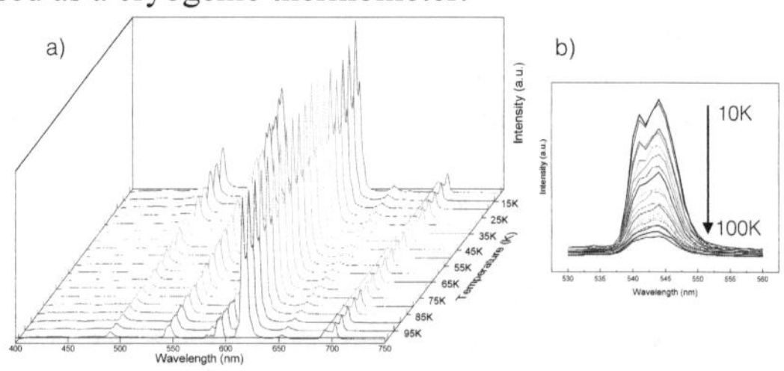

Fig. 2. a) Emission map of spectra recorded in the 10 – 100 K range, b) zoom in on the $^5D_4 \rightarrow {}^7F_5$ (Tb^{3+}) transition peak, which gradually decreases in intensity with temperature increase.

CONCLUSIONS

In this work we show a rare study in which multinuclear LnPOMs of a non-emissive lanthanide can be co-doped with emissive ions at a low doping percentage to obtain very good luminescence properties. We overview the influence of the Ln^{3+} doping on the crystal structures and compare the structures and luminescence properties to the 100% LnPOMs. With the same approach in mind we prepared Eu: TbPOM (where Eu is doped at 5% or 10%) and explored the use of these compounds as first examples of LnPOM ratiometric thermometers.

REFERENCES

1. Jaque D., Vetrone F., Nanoscale 4:4301-26, 2012
2. Ananias D., Brites C. D. S., Carlos L. D., Rocha J., Eur. J. Inorg. Chem. 13-14:1967-71, 2016

ACKNOWLEDGMENTS

AMK acknowledges Ghent University's Special Research Fund (BOF) for a Postdoctoral Mandate (project BOF15/PDO/091). KVH and RVD thank the Hercules Foundation (project AUGE/11/029 "3D-SPACE: 3D Structural Platform Aiming for Chemical Excellence") for funding. RVD acknowledges the Hercules Foundation (project AUGE/09/024 "Advanced Luminescence Setup") for funding.

Low-temperature Electrochemical Deposition of Neodymium and Dysprosium Using Ionic Liquids

Anne Willert[1], Denis Prodius[1], Anja Mudring[1]

[1]U.S. Department of Energy/Ames Laboratory, Iowa State University, USA
willert@ameslab.gov

INTRODUCTION

One way to produce rare earth elements as pure metals is electrodeposition. Although this method can be very selective for a certain element, in the current form, it is a highly energy consuming process. As the electrodeposition is carried out from a flux of molten alkali metal chlorides the operation temperatures range from 700K to over 1000K, rendering this an extremely energy intensive process.[1,2,3] In the seek for a more energy efficient and more economic process ionic liquids (ILs) have come into focus. These liquids are a new class of molten salts being liquid at temperatures below 100°C or even at room temperature (which applies to most of the commonly used ILs). The replacement of the high temperature molten salts by these ILs provides a high potential to make the deposition process more energy efficient and lower the deposition temperature drastically, possibly even to ambient temperature. ILs offer an additional advantage due to their high variability with regard to the constituting anions and cations. First investigations on the deposition of rare earth metals, e.g. Sm, from ionic liquids have already been reported.[4,5]

EXPERIMENTAL METHODS (if relevant)

However, ionic liquids were used as solvent all experiments were carried out with an open experimental setup. Therefore no special precautions had to be taken. We used a simple 3-electrode setup with a working, reference and counter electrode. As working electrode, also the substrate, we used gold on glass (Au(111)). Reference and counter electrode were made of Pt wire, respectively.

RESULTS AND DISCUSSION

After scanning different ionic liquids for their electrochemical stability window, 1-butyl-1-methylpyrrolidinium bis(trifluoromethylsulfonyl)imide ($[Py_{1,4}]$ TFSI) was the IL of choice for our electrochemical investigations. We will report about the successful deposition of Nd from a solution consisting of Nd $(TFSI)_3$ dissolved in $[Py_{1,4}]$ TFSI. This solution was then used to deposit the Nd by using different deposition techniques like potentiostatic or pulsed deposition. In the next step we investigated the electrochemical behaviour and deposition of Dy from an ionic liquid. The voltammograms of this solution showed two expressed reduction steps which suggested that Dy^{3+} is reduced to Dy^0 in 2 steps. Further we could show first attempts on selectively depositing Nd or Dy from a solution of a mixture of both metal salts. SEM and EDS as well as XPS measurements were used for the analysis of the deposited metal(oxides).

CONCLUSIONS

With our work we could show that it is possible to deposit Nd from an ionic liquid. This deposition seemed to occur in a fairly thick layer on the gold substrate. By changing the deposition technique the thickness of the deposited Nd can be even thicker. On the other hand it seemed more difficult to get a thick and even layer of Dy from the solution

REFERENCES

References must be numbered. Keep the same style.
1. Konishi H. et al., J. Electrochem. Soc. 148:C506-C511, 2001
2. Castrillejo Y. et al., J. Electroanal. Chem. 575:61-74, 2005
3. Castrillejo Y. et al., Electrochim. Acta 50:2047-2057, 2005
4. Ispas A. et al., ECS Trans. 13:113-119, 2008
5. Ispas A. et al., ECS Trans. 16:119-127, 2009

ACKNOWLEDGMENTS

I would like to thank Jim Anderegg for the XPS measurements as well as the CMI for the financial support.

Luminescence and Energy Transfer in Eu^{3+} Activated $Li_3Ba_2Tb_3(MoO_4)_8$

David Böhnisch*, Florian Baur, and Thomas Jüstel*

Department of Chemical Engineering, Münster University of Applied Sciences, Germany
*corresponding authors: boehnisch.david@fh-muenster.de, tj@fh-muenster.de

INTRODUCTION

Light emitting diodes (LEDs) widely replaced conventional light sources like incandescent and fluorescent lamps. This is due to their longer lifetime, higher wall plug efficiency, and higher colour rendering.[1-2] Commercially available LEDs comprising a blue LED chip coated with a yellow emitting phosphor such as YAG:Ce exhibit high luminous efficacies. However, due to the weak intensity of red emission, the overall spectrum results in a high correlated colour temperature (CCT) and a low colour rendering index (CRI).[3] Therefore, introducing a red emitting phosphor inevitable to achieve lower CCTs and higher CRIs.[4] Eu^{3+} activated phosphors show great potential use in LEDs, but suffer from low absorption in the blue spectral range.[5] Hence, a sensitizer such as Tb^{3+} has to be used.[1]

In this study we investigated the luminescence and energy transfer properties of Eu^{3+} activated $Li_3Ba_2Tb_3(MoO_4)_8$.

EXPERIMENTAL METHODS

Samples of $Li_3Ba_2(Tb_{1-x}Eu_x)_3(MoO_4)_8$ with varying Eu^{3+} concentrations ($0 \leq x \leq 1$) were prepared by a solid state reaction. To this end, the high purity educts Li_2CO_3 (99% AlfaAesar). $BaCO_3$ (99.95% AlfaAesar), Tb_4O_7 (99.99% Treibacher), MoO_3 (99.95% AlfaAesar), and Eu_2O_3 (99.99% Treibacher) were weighed in stoichiometric amounts. Then they were thoroughly grounded with acetone in an agate mortar. The dried mixtures were placed in a porcelain crucible and were sintered at 800 °C for 10 h in air.

XRD, fluorescence, reflection and decay measurements were performed in order to investigate phase purity and photoluminescence and energy transfer properties.

RESULTS AND DISCUSSION

Figure 1 shows that the monoclinic phase with the space group C2/c (15) could be synthesized.

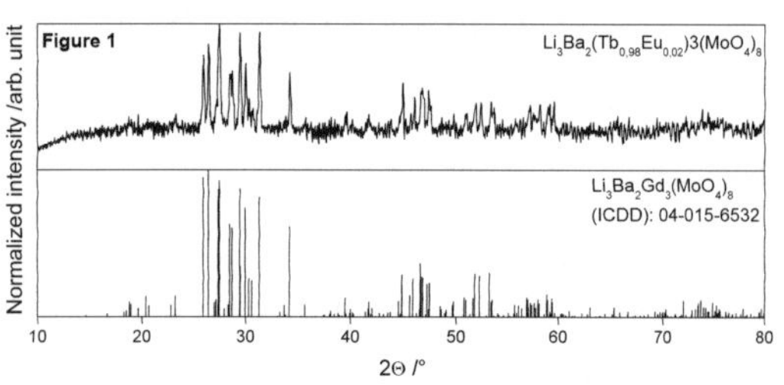

Figure 2 depicts the excitation and emission spectrum of $Li_3Ba_2(Tb_{0,98}Eu_{0,02})_3(MoO_4)_8$. It turns out that the emission originating from the $^5D_0$ to $^7F_2$ transition of Eu^{3+} is observable when excited at 486 nm. Hence, the emission of Eu^{3+} can be sensitized by Tb^{3+}. Thermal quenching curves show that the investigated phosphor is of interest for phosphor converted LEDs (pcLEDs).

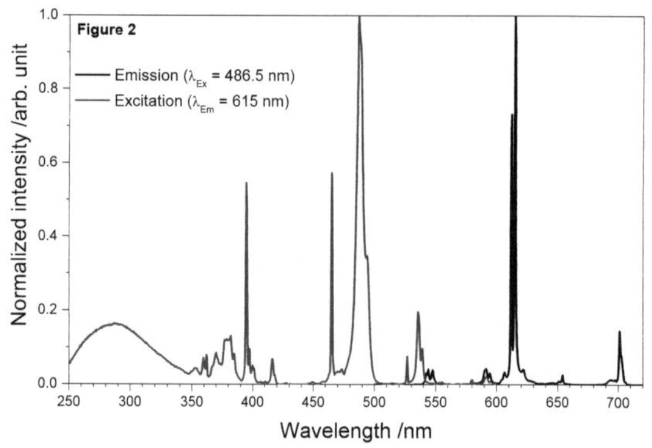

CONCLUSIONS

By using Tb^{3+} as a sensitizer in $Li_3Ba_2Tb(MoO_4)_8$ it is possible to achieve increased absorption in the blue spectral range and solely red emission originating from Eu^{3+} at high substitution concentrations. Moreover, it could be shown that the investigated phosphor has potential to be used in LED applications due to its high thermal stability.

REFERENCES

1. Feldmann C. *et al.*, Adv. Funct. Mater. 13: 511-516, 2005
2. Schubert E. F. *et al.*, Science 308: 1274-1278, 2005
3. Narukawa Y. *et al.*, J. Phys. D: Appl. Phys. 43: 354002, 2010
4. Debasu M. L. *et al.*, Phys. Chem. Chem. Phys. 15: 15565, 2013
5. Baur F. *et al.*, J. Mater. Chem. C 3: 2054-2064, 2015

Luminescent Properties of Lanthanide-Based Metal-Organic Frameworks with Rigid Tritopic Carboxylate Ligands

Hani Nasser Abdelhamid[1], Magdalena Wilk-Kozubek[2,3,4], Xiaodong Zou[1] and Anja-Verena Mudring[2,3]

[1]Inorganic and Structural Chemistry and Berzelii Centre EXSELENT on Porous Materials, Department of Materials and Environmental Chemistry, Stockholm University, Sweden
[2]Department of Materials Sciences and Engineering, Iowa State University, USA, mwilk@iastate.edu
[3]Ames Laboratory, US Department of Energy and Critical Materials Institute, USA
[4]Department of Nanotechnology, Wrocław Research Centre EIT+, Poland

INTRODUCTION

Lanthanide-based metal-organic frameworks (Ln-MOFs), constructed from Ln^{3+} ions and organic ligands, have attracted a great attention not only for their impressive structural diversity, but also for their potential application in the field of luminescence.[1] Most Ln^{3+} ions have excellent luminescent properties such as high color purity and relatively long luminescence lifetimes. Unfortunately, it is well-known that a direct excitation of Ln^{3+} ions is compromised because of their low absorption coefficients. However, the excitation process can be enhanced by introducing organic ligands into the local environment of Ln^{3+} ions. The effective excitation is then realized by so-called antenna effect. This phenomenon relies on the fact that the incident excitation is first absorbed by organic ligands. Then the energy is transferred to the Ln^{3+} ions, which become excited and subsequently emit a characteristic radiation. In this context, luminescent properties of Ln-MOFs with 4,4',4"-(benzene-1,3,5-triyl)tribenzoate (L1) and 4,4',4"-(pyridine-2,4,6-triyl)tribenzoate (L2) ligands will be presented.

EXPERIMENTAL METHODS

The Ln-MOFs with L1 and L2 ligands were synthesized and their luminescent properties were characterized. The corrected steady-state excitation and emission spectra were recorded on a HORIBA Jobin Yvon Fluorolog 3-222 spectrofluorometer, equipped with a 450 W xenon arc lamp and a R928P PMT detector for the UV-VIS range, and a Hamamatsu H10330B-75 NIR PMT detector for the NIR range. The decay curve measurements were performed on the same instrument operating in a time-correlated single-photon counting mode or in a phosphorescence mode. A Fianium WhiteLase supercontinuum laser or a xenon flash lamp were used as an excitation source, respectively.

RESULTS AND DISCUSSION

Luminescent properties of Ln-MOFs with L1 and L2 ligands have been investigated by measuring excitation and emission spectra, and luminescence lifetimes. In Figure 1, the emission spectrum of Sm-MOF with L2 ligand is presented. Upon excitation at 375 nm, an intense broad emission band at 545 nm is found which can be assigned to a $\pi^* \rightarrow \pi$ ligand-centred transition. The other two peaks at 597 and 643 nm are attributed to the f-f forbidden transitions of the Sm^{3+} ion.

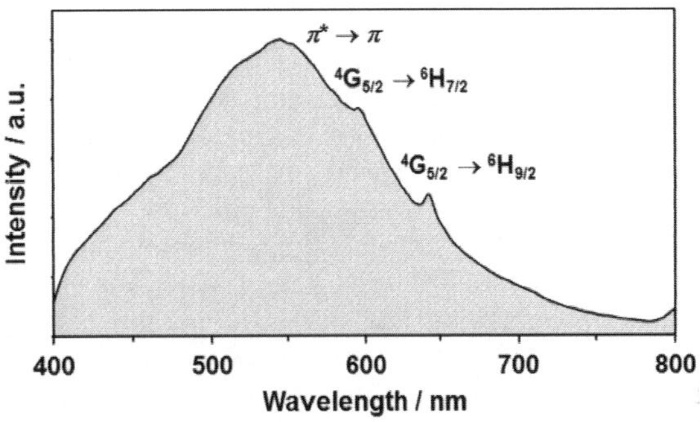

Figure 1. The emission spectrum of Sm-MOF with L2 ligand excited at 375 nm.

In Figure 2, the luminescence decay curve of Sm-MOF with L2 ligand is shown. The curve can be fitted with two exponential decay components of 1.71 and 0.08 ms.

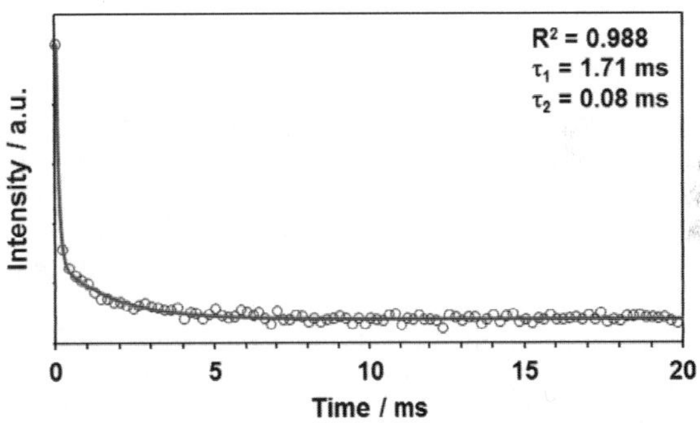

Figure 2. The luminescence decay curve of Sm-MOF with L2 excited at 375 nm. The luminescence was monitored at 597 nm.

CONCLUSIONS

The Ln-MOFs with L1 and L2 ligands display emission in the visible range from blue to red and in the near-infrared range. They possess long luminescence lifetimes (up to milliseconds), which allows for their potential use in light-emitting devices.

REFERENCES

1. Cui Y. *et al.*, Struct Bond 157:27-88, 2014

ACKNOWLEDGMENTS

The authors would like to thank the National Science Foundation (CHE-1465071) for providing financial support to this project.

Luminescent Properties of Mechanochemically Synthesized Rare-Earth Containing MIL-78 MOF

Tarek Alammar[1]*, Ihor Hlova[2], Shalabh Gupta[2], Viktor Balema[2], Anja-Verena Mudring[1,2], Vitalij Pecharsky[1,2]

[1*]Department of Materials Science and Engineering, Iowa State University,
Ames, Iowa, 50011-2300, USA, talammar@iastate.edu
[2]Division of Materials Science and Engineering, Ames Laboratory of US DOE, Iowa State University,
Ames, Iowa, 50011-3020, USA,

INTRODUCTION

Search for increasingly efficient and inexpensive phosphors is actively pursued to continually improve the quality of light and bring down the cost. Aside from efficient emission, hence a high photoluminescence quantum yield, high color purity emission in the desired color needs to be achieved. For this reason, rare earth-containing phosphors move into focus. Among a wide variety of materials, metal-organic frameworks (MOFs) containing rare-earth elements represent an attractive class of luminescent materials with a considerable potential in lighting applications [1,2]. The conventional solvothermal or other solvent based approaches used to obtain MOFs may not always be easy to scale up, and hence other synthetic routes must be developed. Solvent-free mechanochemical preparation of MOFs has proven to be simple and easily scalable [3]. However, this route may lead to material with high defect concentration and low emission quantum yields. For this reason, we have studied the luminescent properties of mechanochemically synthesised MOFs for phosphor applications.

EXPERIMENTAL METHODS

Solid-state ball-milling (SPEX) was successfully utilized to prepare rare-earth containing MOFs with MIL-78 structure [2]. The sample characterization was carried out by X-ray diffraction and scanning electron microscopy. The luminescence studies were performed on a HORIBA Jobin Yvon Fluorolog-3 spectrofluorometer.

RESULTS AND DISCUSSION

Tb-containing MIL-78 MOF was prepared according to the procedures described by Singh *et al.* [3] and its structural identity confirmed by PXRD [4]. Fig. 2 shows the emission spectrum of Tb-containing MIL-78 MOF at room temperature upon excitation at 290 nm. A set of narrow emission peaks is observed that can be attributed to the transitions of $^5D_4 \rightarrow {}^7F_J$ (J =6-2) levels of Tb^{3+}, with maximum emission in the green region of light ($^5D_4 \rightarrow {}^7F_5$), Fig. 1. Intensity decay measurements for the $^5D_4 \rightarrow {}^7F_5$ transition at 544 nm under 290 nm excitation yielded a lifetime of 2.15 ms (Fig. 2 (inset).

CONCLUSIONS

The mechanochemically synthesized Tb-MOF with the MIL-78 structure represents a good example of a highly luminescent MOF prepared by ball milling. Other MOFs or coordination polymers showing good luminescence properties are likely to be prepared via this simple, scalable and solvent-free route.

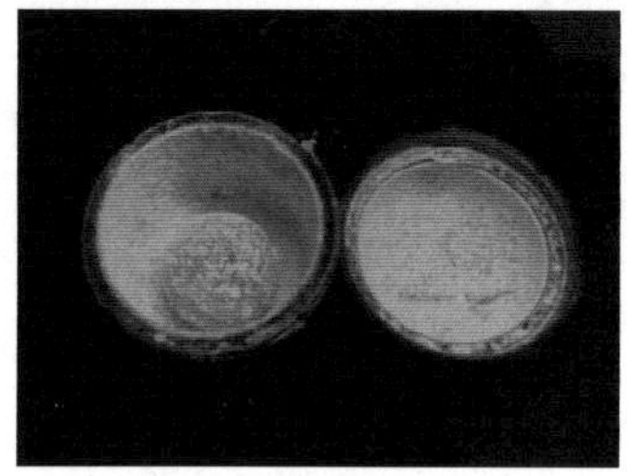

Fig. 1. Photoluminescence under UV light of Tb-doped MIL-78 MOF.

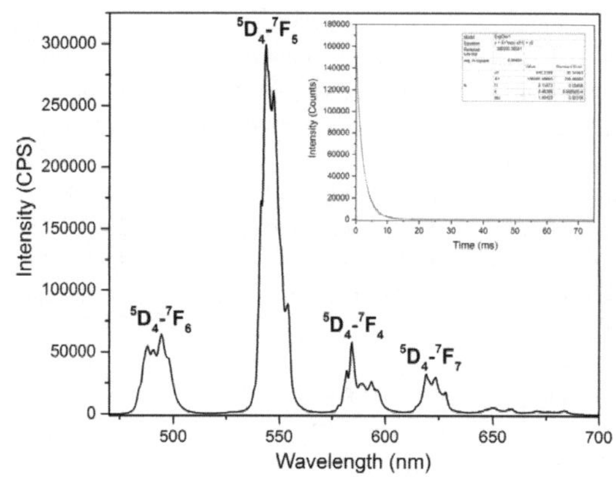

Fig. 2. Emission spectrum recorded with 290 nm excitation and decay curve (inset) of Tb-doped MOF.

REFERENCES

1. Cui, Y. *et al.*, Chem. Rev., 112(2):1126-1162, 2011.
2. Allendorf *et al.*, Chem. Soc. Rev., 38:1330-1352, 2009.
3. Singh, N.K. *et al.*, J. Alloys Compd., 696:118-122, 2017.
4. Serre, C. *et al.*, J. Mater. Chem., 14:1540-1543, 2004.

ACKNOWLEDGMENTS

The Ames Laboratory is operated for the U. S. Department of Energy by Iowa State University of Science and Technology under contract No. DE-AC02-07CH11358. This work was supported by the Department of Energy, Office of Basic Energy Sciences, Materials Sciences Division.

Magnetic and Structural Phase Transitions in $Pr_{1-x}Er_xAl_2$

Arjun K. Pathak,[1] Y. Mudryk,[1] and V. K. Pecharsky[1,2]

[1]*The Ames Laboratory, U.S. Department of Energy, Iowa State University, Ames, Iowa 50011, USA*
[2]*Department of Materials Science and Engineering, Iowa State University, Ames, Iowa 50011, USA*
pathak138@ameslab.gov

INTRODUCTION

Rare earth (R) intermetallic compounds often show interesting fundamental physics due to localized *4f* electrons, and are, therefore, intriguing subject for materials science and condensed matter physics communities. Among those, rare earth dialuminides reveal a plethora of interesting phenomena, such as strong crystal field (CF) effects due to anisotropic *4f* charge density as well as the competition between different magnetic and crystal structures. In particular, mixing one lanthanide with another in pseudobinary $R_{1-x}R'_xAl_2$ creates an interesting playground that results in low-temperature anomalies in heat capacity and multiple ordering phenomena.[1] For example, cubic lattice of $PrAl_2$ distorts tetragonally at the Curie temperature T_C, but the distortion is rhombohedral in $ErAl_2$, creating a potential for instability in the pseudobinary $PrAl_2$-$ErAl_2$ system. In this presentation, we will discuss the crystal structure, magnetic, specific heat and transport properties of $Pr_{1-x}Er_xAl_2$.[1, 2, 3]

RESULTS AND DISCUSSION

The room temperature XRD measurements confirm that all $Pr_{1-x}Er_xAl_2$ alloys crystallize in the $MgCu_2$ type cubic Laves-phase structure with no detectable secondary phases. The temperature-dependent XRD data collected in a zero magnetic field for $PrAl_2$ (T_C = 32K) and $ErAl_2$ (T_C = 13K) reflect a cubic to tetragonal, and cubic to rhombohedral transformation at their respective T_Cs. Moreover, cubic to tetragonal distortion has also been observed for x = 0.05, and 0.1, and rhombohedral distortion for x = 0.95. Interestingly no distortion is observed for x = 0.4. Comparison of $ErAl_2$ with $ErCo_2$ (both have the cubic C15 crystal structure at room temperature and distort rhombohedrally at T_C, but contrary to $ErAl_2$ the $ErCo_2$ has a discontinuous volume change and a first-order transition) indicates that in these intermetallic Laves phases the type of the distortion is defined by the rare-earth element, while the nature of the transition is defined by the presence of itinerant electron metamagnetism in $ErCo_2$ and its absence in $ErAl_2$.

Heat capacity, C_P, measurement at zero magnetic field for $Pr_{1-x}Er_xAl_2$ ($0 \leq x \leq 1$) showed a λ-type anomaly for all samples, suggesting a second order phase transition. Anomaly at T_C is suppressed and becomes broader when applied magnetic field is $H \geq 20$ kOe and no other anomalies have been observed at magnetic fields for all samples except for x = 0.25, and 0.4. Interestingly, additional phase transitions have been observed below T_C for x = 0.25 at 20 kOe ≤ H ≤ 30 kOe. A strong first order phase transition has been observed for x = 0.4 at 40 ≤ H ≤ 80 kOe, and for H ≥100 kOe the compound exhibits only a ferromagnetic to paramagnetic transition at T_C.

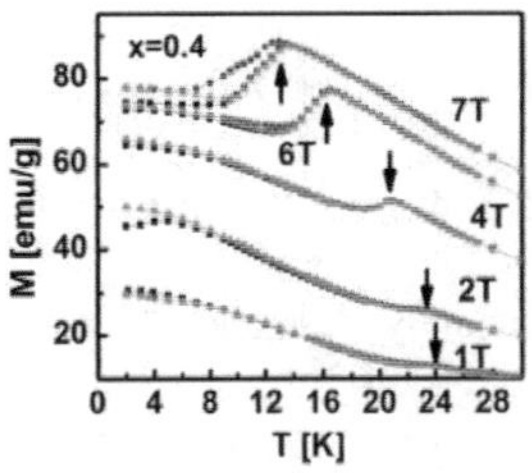

Fig. 1: M(T) curves for $Pr_{0.6}Er_{0.4}Al_2$ measured at different magnetic fields.

Consistent with C_P measurements, the thermal hysteresis between FCC and FCW M(T) observed at certain applied magnetic fields confirms the field induced first order phase transition for x = 0.4. It is most noticeable for H = 70 kOe (Fig. 1) and disappears when H ≥ 100 kOe. In addition, C_P of $Pr_{1-x}Er_xAl_2$ shows Schottky anomaly at T< 2K, resulting in large value of nuclear specific heat C_N that arises due to the splitting of the nuclear hyperfine levels. Increase in Er concentration suppresses the low temperature Schottky anomaly and it becomes negligible for x = 0.95.

CONCLUSIONS

Crystal structure, heat capacity, magnetization measurements revealed lattice instabilities in the mixed light and heavy lanthanide series of $Pr_{1-x}Er_xAl_2$ alloys.

REFERENCES

1. Khan M. *et al.*, J. Mater. Chem. C. **1**, 896 (2017)
2. Pathak A. *et al.*, Phys. Rev. Lett. **110**, 186405 (2013)
3. Pathak A. *et al.*, Phys. Rev. B. **94**, 224406 (2016)

ACKNOWLEDGMENTS

The Ames Laboratory is operated for the U. S. Department of Energy by Iowa State University of Science and Technology under contract No. DE-AC02-07CH11358. This work was supported by the Department of Energy, Office of Basic Energy Sciences, Materials Sciences Division.

Magnetic Properties of the $SmNi_{4-x}Fe_xB$ Solid Solution.

Michael T. Onyszczak[a,b], Tej Lamichhane[a,b], Valentin Taufour[b,c], Andriy Palasyuk[b], Sergey L. Bud'ko[a,b], Paul C. Canfield*[a,b]

[a]Department of Physics and Astronomy, Iowa State University, Ames, IA 50011, U.S.A. *canfield@ameslab.gov
[b]Ames Laboratory, Iowa State University, Ames, IA 50011, U.S.A.
[c]Department of Physics, University of California, Davis, CA 95616, U.S.A.

1. INTRODUCTION

The magnetic properties of $SmCo_5$ have been extensively studied and optimized for uses in permanent magnets,[1] but many of its lesser studied structural derivatives such as $Sm_{n+1}T_{3n+5}B_{2n}$ (T = transition metal, $n = 0, 1, 2, \ldots$) have significant magnetic properties of their own.[2] In the $SmNi_4B$ system, it has been determined that the Ni does not contribute to the anisotropy of the system,[3] but the compound still displays high coercivity of 28 kOe[4] at 5K. However, the $SmNi_4B$ system has low Curie temperature (T_c = 39K) and low magnetic saturation ($0.15\mu_B$ per Sm at 5K). The addition of iron has shown the ability drastically increase both the Curie temperature to 322 K and the saturation moment to 1.35 μ_B per Sm at 5K.[5]

2. EXPERIMENTAL METHODS

Alloys with the nominal compositions of $SmNi_{4-x}Fe_xB$ with x=0, 0.5, 1.0, 1.5, & 2.0 were prepared via arcmelting under an argon atmosphere with excess to compensate for small losses associated with samarium's high vapor pressure. The samples were then sealed in amorphous silica tubes under an argon atmosphere and annealed at 1000°C for 72 hours and then quenched in ice water. Then x-ray, micro-structure, and magnetization measurements were performed.

3. RESULTS AND DISCUSSION

The $SmNi_{4-x}Fe_xB$ system appears to have an end to its solid solution at x=1.5 as seen in Figure 1, but the system retains its magnetic properties beyond its solid solution. The micro-structure of $SmNi_2Fe_2B$ reveals sharp grain boundaries with impurity phases mostly lying on these boundaries. At the end of the solid solution $SmNi_{2.5}Fe_{1.5}B$ shows the improved magnetic properties of a Curie Temperature of 550K, coercivity of 5kOe and saturation magnetization of $2\mu_B$ per Sm at 300K.

4. CONCLUSIONS

The addition of iron to $SmNi_4B$ shows dramatic increases to all magnetic properties, and magnetic properties were able to remain even beyond the limits of the solid solution.

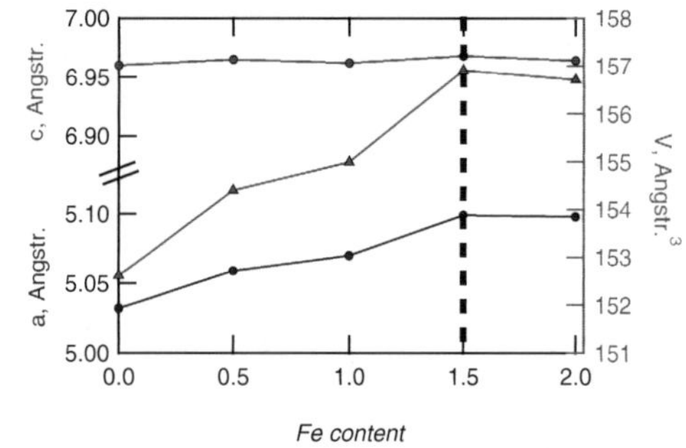

Figure 1: Dependence of Lattice parameters of $SmNi_{4-x}Fe_xB$ on Fe Content with dashed line indicating extent of solid solution

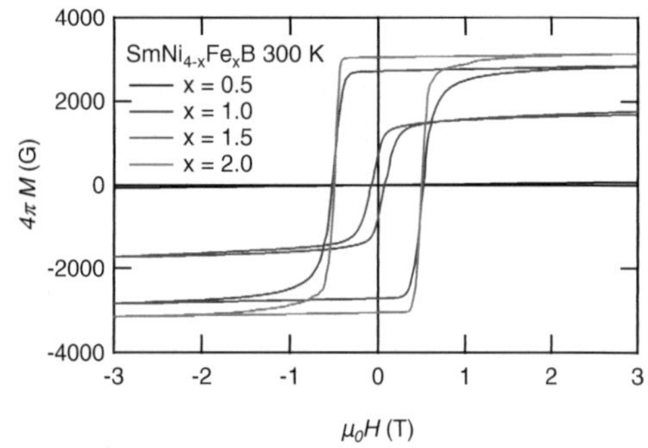

Figure 2: Magnetic saturation and coercivity of $SmNi_{4-x}Fe_xB$ at 300K

5. REFERENCES

1. Buschow, K. H. J. et al., J. Appl. Phys., 40 (1969), 4029
2. Ido H. et al., Physica B, 177 (1992), 265
3. Caraballo R.J. et al., J. Appl. Phys., 116 (2014), 116
4. Mazumdar C. et al., Appl. Phys. Lett., 77 (2000), 895
5. Yao J. et al., J. Magn. Magn. Mater., 419 (2016), 479

6. ACKNOWLEDGEMENTS

This research was supported by the Critical Materials Institute, an Energy Innovation Hub funded by the U.S. Department of Energy, Office of Energy Efficiency and Renewable Energy, Advanced Manufacturing Office.

Magnetic Properties of the $SmNi_{4-x}Fe_xB$ Solid Solution.

Michael T. Onyszczak[a,b], Tej Lamichhane[a,b], Valentin Taufour[b,c], Andriy Palasyuk[b], Sergey L. Bud'ko[a,b], Paul C. Canfield*[a,b]

[a]*Department of Physics and Astronomy, Iowa State University, Ames, IA 50011, U.S.A. *canfield@ameslab.gov*
[b]*Ames Laboratory, Iowa State University, Ames, IA 50011, U.S.A.*
[c]*Department of Physics, University of California, Davis, CA 95616, U.S.A.*

1. INTRODUCTION

The magnetic properties of $SmCo_5$ have been extensively studied and optimized for uses in permanent magnets,[1] but many of its lesser studied structural derivatives such as $Sm_{n+1}T_{3n+5}B_{2n}$ (T = transition metal, $n = 0, 1, 2, \ldots$) have significant magnetic properties of their own.[2] In the $SmNi_4B$ system, it has been determined that the Ni does not contribute to the anisotropy of the system,[3] but the compound still displays high coercivity of 28 kOe[4] at 5K. However, the $SmNi_4B$ system has low Curie temperature (T_c = 39K) and low magnetic saturation ($0.15\mu_B$ per Sm at 5K). The addition of iron has shown the ability drastically increase both the Curie temperature to 322 K and the saturation moment to 1.35 μ_B per Sm at 5K.[5]

2. EXPERIMENTAL METHODS

Alloys with the nominal compositions of $SmNi_{4-x}Fe_xB$ with x=0, 0.5, 1.0, 1.5, & 2.0 were prepared via arcmelting under an argon atmosphere with excess to compensate for small losses associated with samarium's high vapor pressure. The samples were then sealed in amorphous silica tubes under an argon atmosphere and annealed at 1000°C for 72 hours and then quenched in ice water. Then x-ray, micro-structure, and magnetization measurements were performed.

3. RESULTS AND DISCUSSION

The $SmNi_{4-x}Fe_xB$ system appears to have an end to its solid solution at x=1.5 as seen in Figure 1, but the system retains its magnetic properties beyond its solid solution. The microstructure of $SmNi_2Fe_2B$ reveals sharp grain boundaries with impurity phases mostly lying on these boundaries. At the end of the solid solution $SmNi_{2.5}Fe_{1.5}B$ shows the improved magnetic properties of a Curie Temperature of 550K, coercivity of 5kOe and saturation magnetization of $2\mu_B$ per Sm at 300K.

4. CONCLUSIONS

The addition of iron to $SmNi_4B$ shows dramatic increases to all magnetic properties, and magnetic properties were able to remain even beyond the limits of the solid solution.

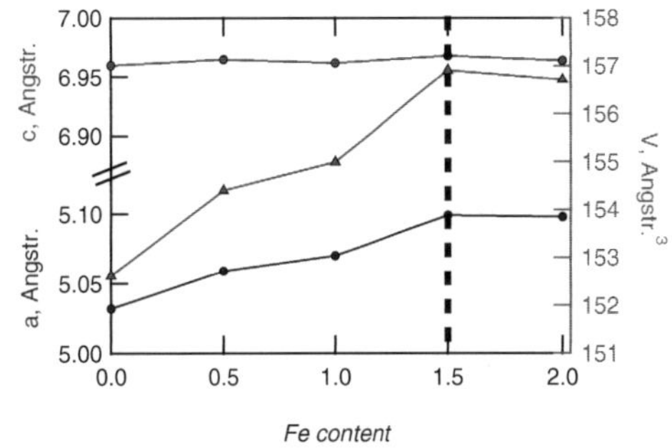

Figure 1: Dependence of Lattice parameters of $SmNi_{4-x}Fe_xB$ on Fe Content with dashed line indicating extent of solid solution

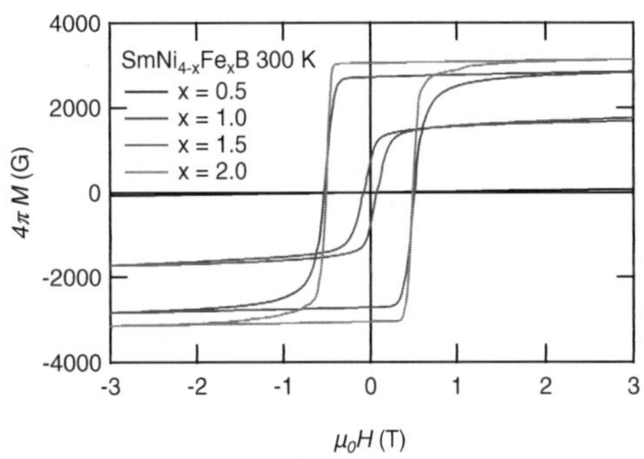

Figure 2: Magnetic saturation and coercivity of $SmNi_{4-x}Fe_xB$ at 300K

5. REFERENCES

1. Buschow, K. H. J. et al., J. Appl. Phys., 40 (1969), 4029
2. Ido H. et al., Physica B, 177 (1992), 265
3. Caraballo R.J. et al., J. Appl. Phys., 116 (2014), 116
4. Mazumdar C. et al., Appl. Phys. Lett., 77 (2000), 895
5. Yao J. et al., J. Magn. Magn. Mater., 419 (2016), 479

6. ACKNOWLEDGEMENTS

This research was supported by the Critical Materials Institute, an Energy Innovation Hub funded by the U.S. Department of Energy, Office of Energy Efficiency and Renewable Energy, Advanced Manufacturing Office.

Magnetic states and exchange interactions in GdNiAl

X. B Liu , D. Paudyal, Y. Mudryk, and V. K. Pecharsky
Ames Lab, DOE, Iowa State University, Ames, IA 50011 USA liux@ameslab.gov

INTRODUCTION

GdNiAl crystallizes in the Fe_2P type hexagonal structure with Gd at 3g, Al at 3f and Ni at 2c and 1b sites. It orders ferromagnetically in the liquid nitrogen temperature range,[1,2] and undergoes an isostructural transition with discontinuous anisotropic lattice parameters change around 220 K in the paramagnetic state upon cooling.[2] The isostructural transition is also observed in other RNiAl compounds, which is ascribed to the existence of a 'forbidden' c/a ratio. By coupling this structural transition with the magnetic ordering, one may be able to realize a large magnetocaloric effect. Here we explore the GdNiAl system from electronic structure, exchange interactions, and Heisenberg model Hamiltonian calculations.

METHODS

The electronic structure calculations have been performed employing the linear muffin-tin orbital method.[3] The generalized gradient approximation of Perdew–Burke–Ernzerhof (PBE) has been used for the exchange-correlation functional. The Gd 4f electrons are treated within the density functional theory by incorporating the onsite electron correlation parameter (DFT+U). The inter-site exchange coupling parameters, J_{ij}, of the Heisenberg model have been calculated using the linear-response method.

RESULTS AND DISCUSSION

Starting with experimental crystallographic data,[2] the high (H) and low (L) temperature structures are optimized. The calculated density of states (DOS) shows that GdNiAl is a typical metal. The majority 4f states are completely occupied by seven electrons and are located around 7 eV below the Fermi level (E_F) while the minority 4f states are empty and are located at 6.5 eV above E_F. The calculated Gd magnetic moments are 7.3 μ_B, and are slightly larger than the Gd 4f value of 7 μ_B due to the small spin polarized Gd 5d moment. Including the small induced antiparallel moments on Ni and Al sites, the calculated total magnetization per formula unit is 7.20 μ_B.

To understand the magnetic states, the intersite exchange coupling parameters have been calculated for GdNiAl with both H and L structures. As shown in Fig. 1, the main contribution to the exchange interactions is from the Gd-Ni_1 (with Ni at the 2c site) atom pair while the Gd-Gd and Gd-Ni_2 (with Ni at the 1b site) exchange interactions are comparable. Further, the nearest Gd-Ni_2 and Gd-Gd exchange is practically insensitive to the unit cell size in GdNiAl (Fig. 1). The Gd-Ni exchange interactions are almost identical while the Gd-Gd exchange interactions are slightly enhanced as the high temperature structure transforms to the low temperature one. The weak dependence of the magnetic interactions upon structural distortion may be responsible for the decoupled magnetic- and structural- transitions in GdNiAl.

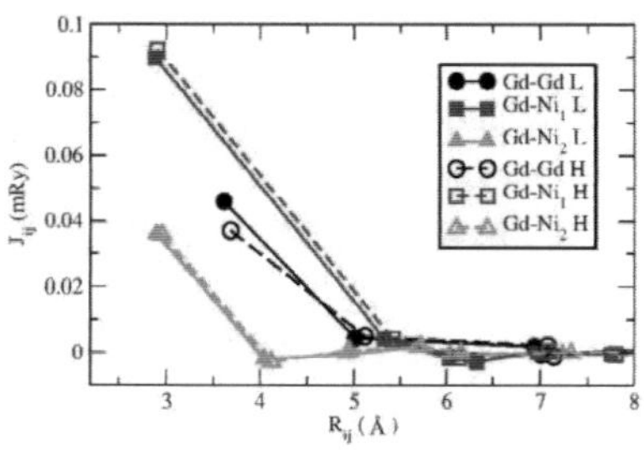

Fig. 1 Gd–Gd, Gd-Ni_1 and Gd-Ni_2 exchange coupling parameters J_{ij} in GdNiAl with low temperature structure (filled symbols) and high temperature structure (open symbols).

The mean field Curie temperature T_C for GdNiAl with L and H phases are 73 K and 76 K, respectively (the experimental value is 60 K). These results also confirm that the magnetic ordering is practically insensitive to the structural distortion.

CONCLUSION

We have calculated and analyzed magnetic exchange interactions in GdNiAl by employing advanced density functional methods coupled with the Heisenberg model. The mean field estimate of Curie temperature closely agrees with the experimental value.

REFERENCES

1. Merlo F. et al., J Alloy Compd. 266:22-25, 1998
2. Jarosz J. *et al.*, J. Magn. Magn. Mater. 208:169-208, 2000
3. O.K. Andersen, Phys. Rev. B 12: 3060-3063, 1975

ACKNOWLEDGMENTS

CaloriCoolTM is supported by the Advanced Manufacturing Office of the Office of Energy Efficiency & Renewable Energy and managed jointly through the Advanced Manufacturing and Building Technologies Offices of the U.S. Department of Energy. Ames Laboratory is operated for the U.S. Department of Energy by Iowa State University of Science and Technology under Contract No. DE-AC02-07CH11358.

Magnetism and Magnetocaloric Effect in $(Gd_{5-x}Sc_x)Si_{1.8}Ge_{2.2}$

Kirk Rudolph[1], Arjun Pathak[2], Yaroslav Mudryk[2], and Vitalij Pecharsky[2,3]

[1]Department of Mechanical Engineering, Iowa State University, USA, krudolph@iastate.edu
[2]Ames Laboratory, U.S. Department of Energy, Iowa State University, USA
[3]Department of Materials Science and Engineering, Iowa State University, USA

INTRODUCTION

Since the discovery of the giant magnetocaloric effect (GMCE), the magnetocaloric effect (MCE) has become an increasing area of interest because of its critical role in magnetic refrigeration, the next generation of refrigeration technologies[1,2]. New magnetic refrigeration technologies are estimated to be 20 to 30 percent more energy efficient, eliminate the use of environmentally harmful coolants[3], and still be smaller and quieter than current gas-compression refrigeration systems. Yet, the lack of suitable materials capable of producing a large temperature change at room temperature in economically feasible magnetic fields, while being able to endure the millions of magnetization/demagnetization cycles during the expected lifespan of a refrigerator, has hindered the development of this new technology. This study reports the results of magnetic, calorimetric, and crystallographic investigations of the Sc-doped giant magnetocaloric compound $Gd_5Si_{1.8}Ge_{2.2}$. In addition we discuss the ways to control the phase transition and magnetocaloric effect.

EXPERIMENTAL METHODS

Stoichiometric amounts of each element were weighted and then arc melted together in an argon atmosphere to produce five $Gd_{5-X}Sc_XSi_{1.8}Ge_{2.2}$ samples where x = 0, 0.1, 0.2, 0.3, and 0.5. Magnetic measurements were performed using a superconducting quantum interference device (SQUID) magnetometer (MPMS XL, Quantum Design, USA) in the temperature range from 2 to 300 K and in magnetic fields from 0 to 7 T. Electrical transport properties were measured by Physical Property Measurement System (PPMS, Quantum Design) in magnetic fields from 0 to 5 T using four-probe technique. The heat capacities of the $Gd_{5-X}Sc_XSi_{1.8}Ge_{2.2}$ samples, where x = 0.1 and x = 0.2, were measured using a semi-adiabatic calorimeter. Finally, temperature-dependent x-ray powder diffraction data was collected for the x = 0.2 sample.

RESULTS AND DISCUSSION

Figure 1A shows the magnetization of $Gd_{5-x}Sc_xSi_{1.8}Ge_{2.2}$ measured as a function of temperature during both heating (solid symbols) and cooling (open symbols) in a 0.1 T magnetic field. The first order ferromagnetic-paramagnetic transition temperature increases for x < 0.2 where it reaches a critical concentration and T_C of

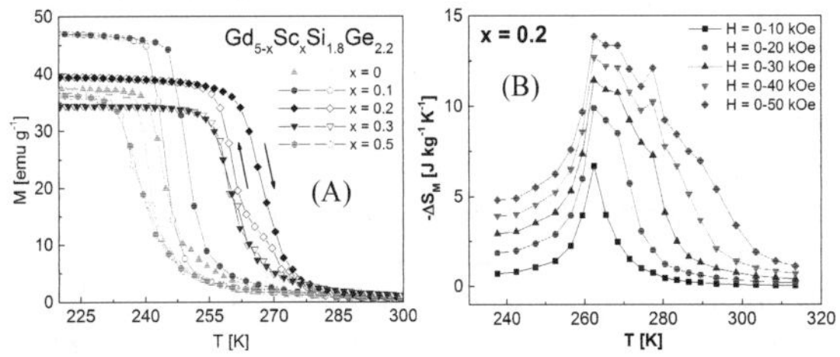

Figure 1. (A) Magnetization of $Gd_{5-x}Sc_xSi_{1.8}Ge_{2.2}$ in 0.1 T magnetic field, (B) MCE in sample with x = 0.2.

266K (heating). Further Sc substitution results in a second-order transition with decreasing T_C. Figure 1B shows magnetic entropy change calculated using the isothermal magnetization data for five different magnetic changes for x = 0.2. There is the progressive decrease in maximum entropy change with Sc composition reflecting the change from first to second order of the transition.

CONCLUSIONS

The substitution of Sc for Gd in the giant magnetocaloric compound $Gd_5Si_{1.8}Ge_{2.2}$ has been studied. Samples of the $Gd_{5-X}Sc_XSi_{1.8}Ge_{2.2}$ series were synthesized and measured for magnetic, electrical, caloric, and crystallographic properties. The addition of the Sc up to x = 0.2 increases the temperature of the first order transition while maintaining relatively large magnetic entropy change. Further addition of Sc changes the first order transition to second order and lowers the temperature of the transition below the initial T_C when x = 0. The second order transition is accompanied by a much smaller magnetic entropy change over a much wider temperature span.

REFERENCES

1. V. K. Pecharsky *et al.*, Phys. Rev. Lett. **78**, 1997
2. G.V. Brown, J. Appl. Phys. **47**, 1976
3. O. Tegus *et al.,* Nature **415**, 2002

ACKNOWLEDGMENTS

The Ames Laboratory is operated for the U. S. Department of Energy by Iowa State University of Science and Technology under contract No. DE-AC02-07CH11358. This work was supported by the Department of Energy, Office of Basic Energy Sciences, Materials Sciences Division as well as the Science Undergraduate Laboratory Internship (SULI) program.

Magnetism in CePtPb and GdPtPb with distorted Kagomé lattice

Soham Manni[1*], Sergey L. Bud'ko[1,2], D. H. Ryan[3], C.D. Boyer[4] and Paul C. Canfield[1,2]

[1*]Department of Physics and Astronomy, Iowa State University, USA, smanni@ameslab.gov
[2] Ames Laboratory, Iowa State University, USA.
[3] Physics Department And Centre for Physics of Materials, McGill University, Canada.
[4] Canadian Neutron Beam Centre, Chalk River Laboratories, Ontario, Canada.

INTRODUCTION

Magnetic frustration in solid state materials often brings intriguing ground states like quantum spin liquids.[1] The research into frustrated magnetism usually circles around magnetic insulators. Our goal is to search for magnetically frustrated ground states in metals. Rare earth intermetallic systems can be a very rich playground for this, with rare earth moments, often visualized as local moments, giving rise to frustrated exchange interaction on a geometrically frustrated Kagomé lattice. In recent years some of the Ce and Yb-based rare earth intermetallics with hexagonal ZrNiAl-type structure (having distorted Kagomé lattice of rare earth ions) have shown very unconventional ground states, often under magnetic field e.g. CeRhSn[2] and YbAgGe[3]. We have studied CePtPb and GdPtPb with the same structure. GdPtPb whose magnetism is not affected by crystal electric field (CEF) splitting, can give us new insight into rare earth magnetism in geometrically frustrated lattice.

EXPERIMENTAL METHODS

We have grown single crystals of LaPtPb, CePtPb and GdPtPb from excess Pb-flux. Magnetic measurements were done using a Quantum Design Magnetic Property Measurement System SQUID magnetometer in between 1.8-300 K temperature and 0 - 55 kOe magnetic field. Electrical transport and thermodynamic measurements were done in Quantum Design Physical Property Measurement System in between 0.05-300 K temperature and 0 - 140 kOe magnetic field.

RESULTS AND DISCUSSION

CePtPb orders aniferromagnetically below 0.9 K. We have studied the change of the magnetic ordering under magnetic field parallel to the Kagomé lattice plane. We have observed partial suppression of magnetic ordering. In GdPtPb, where whole J=7/2 multiplet participates in the magnetic exchange, we find interesting magnetic properties. Analysis of high temperature magnetic susceptibility data reveals that the system is far from frustrated with a frustration parameter of $(\Theta/T_N) = 0.2$-0.3; a value that is much less than 1.0. This unusual value can be understood by mean field theory (MFT) using a two sub-lattice model. Gd S=7/2 local moments order aniferromagnetically below 16 K, possibly having a very unusual magnetic structure for a hexagonal system. This magnetic ordering can be tuned by magnetic field along the plane of the spins. A metamagnetic transition was also observed around 20 kOe. The MFT analysis of low temperature magnetic susceptibility indicates that the GdPtPb is non-collinear antiferromagnet. Unusual magnetic structure is being studied using local probe measurements.

CONCLUSIONS

Our comparative study of single crystals of CePtPb and GdPtPb suggests future research direction for rare earth magnetism on geometrically frustrated lattice in metals. In addition, we have discovered a unique mean field magnet GdPtPb with distorted Kagomé magnetic sub-lattice having most likely a very unconventional magnetic structure.

REFERENCES

1. Balents L., Nature 464:199, 2010.
2. Tokiwa Y. *et. al,*. Sci. Adv. 1: e1500001, 2015.
3. Bud'ko S.L. *et. al.*, PRB, 69: 014415, 2004.

ACKNOWLEDGMENTS

This research is funded by the Gordon and Betty Moore Foundations EPiQS Initiative through Grant GBMF4411 (S.M) and by the U.S. Department of Energy under Contract No. DE-AC02-07CH11358 (S. L. B. and P. C. C.).

Magnetocrystalline Anisotropy Driven Magnetostructural Transformations in RMn_2 (R = Tb, Ho)

Manish K. Kashyap,[1,2*] Timothy A. Hackett,[1] Bruce N. Harmon,[1] Vitalij K. Pecharsky,[1,3] and Durga Paudyal[1]

[1]Ames Laboratory, US Department of Energy, Iowa State University, Ames, IA-50011-3020, U.S.A.
[2]Department of Physics, Kurukshetra University, Kurukshetra-136119 (Haryana), India, *manishdft@gmail.com
[3]Department of Materials Science and Engineering, Iowa State University, Ames, Iowa, 50011-2300, USA

INTRODUCTION

RMn_2 (R = rare earth) compounds mainly form in the cubic Laves phase (C15) structure (Fig. 1) in which rare earth atoms have tetrahedral (Td) site symmetry, i.e., the positions of the carbon atoms in the diamond structure, while the Mn atoms have D_3d local symmetry. The magnetocrystalline anisotropy energy (MAE) in rare earth compounds is fascinating as it affects the shape of hysteresis loop (B-H curve). The MAE is strongly dependent on the direction of magnetization and a concise knowledge of same is helpful in the design of most commercially available magnetic materials. Thus, it is worthwhile to calculate the anisotropy of magnetic materials. Therefore, we aim to investigate electronic and magnetic properties, along with MAE, of ordered RMn_2 (R = Tb, Ho) compounds.

COMPUTATIONAL DETAILS

The calculations of the electronic and magnetic properties of RMn_2 (R = Tb, Ho) compounds were performed using density functional theory (DFT) based full potential linear augmented plane wave (FPLAPW) method[1] as implemented in the WIEN2k package[2]. The added onsite electron correlation in generalized gradient approximation (GGA+U) was used for the 4f-states with inclusion of spin orbit coupling for MAE.

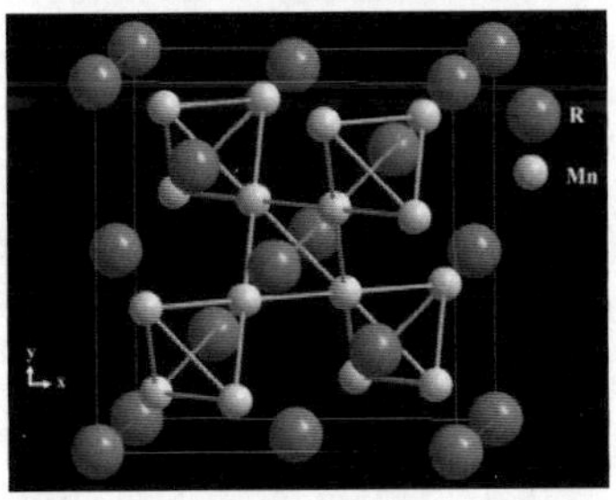

Fig 1: Crystal Structure of RMn_2 (R = Tb, Ho)

RESULTS AND DISCUSSION

The density of states (DOS) (Fig. 2) indicates the instability/stability of cubic laves phase of $TbMn_2$/$HoMn_2$ due the presence of R-4f electrons at the Fermi level. Mn-3d states hybridize with R-4f states to create the magnetic environment and spin polarization in the vicinity of the Fermi level. The spin moments of both compounds indicate the ferrimagnetic ordering. The orbital moment of Tb and Ho in $TbMn_2$ and $HoMn_2$, respectively, closely agree with the value as predicted by Hund's rule (Tb = 3 μ_B and Ho= 6 μ_B).

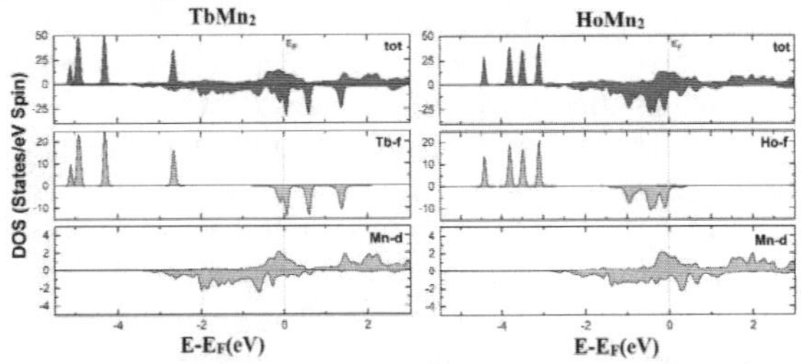

Fig 2: Total and partial DOS of RMn_2 (R = Tb, Ho)

Table1: Calculated Spin and Orbital magnetic moments in μ_B for RMn_2 (R = Tb, Ho)

Case	M_R		M_{Mn}		M_{Int}
	Orbital	Spin	Orbital	Spin	
$TbMn_2$	2.98	6.05	0.02	-2.60	0.66
$HoMn_2$	5.87	3.99	0.02	-1.39	0.44

The MAE for $TbMn_2$/$HoMn_2$ (i.e. 48.99/4.31 meV/f.u.) is found to be highest and uniaxial along the direction of <111>, which indicates that the rhombohedral transformation is the most probable in both compounds.

CONCLUSIONS

Using DFT simulations, it has been shown that $HoMn_2$ is more stable than $TbMn_2$ in Laves phase structure. Indirect 4f-4f exchange interactions and crystal field splitting play a significant role to decide magnetic and structural phases. Our results confirm the magnetocrystalline anisotropy driven rhombohedral ground state in $TbMn_2$ and $HoMn_2$.

REFERENCES

[1]G. K. H. Madsen *et al.*, Phys. Rev. B 64 (2001) 195134
[2]www.wien2k.at

ACKNOWLEDGMENTS

The Ames Laboratory is operated for the U. S. Department of Energy by Iowa State University of Science and Technology under contract No. DE-AC02-07CH11358. This work was supported by the Department of Energy, Office of Basic Energy Sciences, Materials Sciences Division. One of the authors; MKK would like to acknowledge UGC, New Delhi for providing Raman Post-Doctoral Fellowship.

Make Europium Great Again

Thomas Schleid

Institute for Inorganic Chemistry, University of Stuttgart, Pfaffenwaldring 55, 70569 Stuttgart (Germany)
E-Mail: schleid@iac.uni-stuttgart.de

The combination of europium with elements of different electronegativity (*EN* as defined according to *Allred* and *Rochow*[1]) reveals that it is easy to obtain EuF_3 (*EN*(F) = 4.0), Eu_2O_3 (*EN*(O) = 3.5), EuN (*EN*(N) = 3.1), $EuCl_3$ (*EN*(Cl) = 2.9), and $EuBr_3$ (*EN*(Br) = 2.8), all containing Eu^{3+} cations. Beginning with carbon (*EN*(C) = 2.5), the difficulty to yield compounds with trivalent europium increases, so with sulfur (*EN*(S) = 2.5) and selenium (*EN*(Se) = 2.4), Eu_3S_4 and Eu_3Se_4 represent the highest oxidized binary chalcogenides exhibiting a structure with both Eu^{2+} and Eu^{3+} cations simultaneously and at the same crystallographic site. While moving to iodine (*EN*(I) = 2.3), hydrogen (*EN*(H) = 2.2) and tellurium (*EN*(Te) = 2.0), exclusively divalent europium compounds occur, such as EuI_2, EuH_2, and EuTe. All other elements with electronegativities larger than two (phosphorus: *EN*(P) = 2.2), arsenic: *EN*(As) = 2.1, germanium: *EN*(Ge) = 2.0), boron: *EN*(B) = 2.0) are hardly capable to convert europium into simple salt-like binary compounds, which is also true for carbon, where at least EuC_2 according to $\{(Eu^{2+})[|C{\equiv}C|]^{2-}\}$ exists as acetylenide. In these cases, we start to enter the realm of *Zintl* phases with mostly polyanionic arrangements.

Upon reduction with europium or other suitable reducing agents, it is possible to get access to divalent europium compounds even with the highly electronegative elements, e.g. EuF_2, EuO, $EuCl_2$, and $EuBr_2$, but not in the case of EuN. Moreover, tailor-made mixed-valent compounds, like Eu_3O_4 or *Vernier*-type halides ($Eu_{13}F_{32}$, Eu_4Cl_9 and Eu_5Cl_{11}) again with Eu^{2+} and Eu^{3+} cations in similar coordination spheres of the respective anions, can be synthesized. This simple picture changes a bit, when anions of elements with lower electronegativity are put into a covalent binding situation to increase their hardness or the group electronegativity of the complex anion. So thiosilicate(IV) anions stabilize trivalent europium in compounds like $CsEu[SiS_4]$, although S^{2-} anions alone do not. On the other hand, even oxoanions with a very high group electronegativity like $[SiO_4]^{4-}$ allow to reach the divalent oxidation state of europium, as the example of $Eu_2[SiO_4]$ impressively shows.

Extremely interesting features of that kind emerge, if one is able to combine different anions in order to manipulate the actual europium valence state. The admixture of oxide and fluoride anions not surprisingly results in the formation of the oxide fluoride EuOF or *Vernier*-type phases like $Eu_3O_2F_5$ or $Eu_5O_4F_7$, now containing exclusively trivalent Eu^{3+} cations. EuOCl, EuOBr and EuOI, but also Eu_3O_4Br, are similar examples with the heavier halide congeners. Other oxide halides are feasible to stabilize either mixed-valent (Eu_2O_2Br and Eu_2O_2I) or exclusively divalent oxidation-state situations (Eu_4OCl_6, Eu_4OBr_6 and Eu_4OI_6 or Eu_2OI_2). In the latter cases Eu^{2+} behaves like an alkaline-earth metal cation with a size close to Sr^{2+}, but smaller than Ba^{2+}.

Interhalogenide and -chalcogenide compounds are apt to favor either europium(II) fluoride halides ($Eu_7F_{12}Cl_2$ and $EuFX$ with X = Cl, Br and I) or europium(III) oxide chalcogenides (Eu_2O_2Ch, Ch = S – Te). In consideration of hydride anions as halide species softer than fluoride, hydride halides with the compositions $EuHX$ and Eu_2H_3X) occur, displaying Eu^{2+} cations.

Astoundingly enough, the admixture of fluoride and sulfide anions does not result in the formation of EuFS with only Eu^{3+} cations. Here the oxidation leads to the fluoride sulfide $Eu_3F_4S_2$, which represents a mixed-valent situation according to $(EuFS)_2{\cdot}EuF_2$, or to $Eu_{13}F_{30}S$ (= $EuFS{\cdot}(EuF_3)_5{\cdot}(EuF_2)_7$), when the extremely fluoride-rich situation is regarded. But also combinations of sulfide with complex oxoanions can lead to mixed-valent europium compounds, as the examples $Eu_5S[SiO_4]_3$ and $Eu_6S[BO_3]_4$ demonstrate. With slightly different ratios of Eu^{2+} and Eu^{3+} cations, even isostructural halide analoga exist ($Eu_5F[SiO_4]_3$ and $Eu_6Br[BO_3]_4$). On the other hand, in combination with $[SiO_4]^{4-}$ groups even Se^{2-} anions can tolerate Eu^{3+} partners, for instance in $Eu_2Se[SiO_4]$, but successful syntheses of europium(III) oxide fluoride chalcogenides, such as Eu_3OF_5S, Eu_2OF_2Se or $Eu_6O_2F_8Se_3$, with the heavy chalcogenide as minority component, well-known with many other Ln^{3+} cations, still wait to be achieved for the Eu^{3+} case.

The choice of the right partners appears to be very important, since in compounds like $LiEu_2NH_3$ hydride and nitride anions together are able to stabilize mixed-valent europium cations. Especially in terms of nitrides, europium does not behave like a heavy alkaline-earth metal at all, proven by the lack of europium compounds analogous to AE_2NX (AE = Sr and Ba; X^- = e^-, H^-, Cl^- and Br^-), as EuN is almost impossible to reduce. In this case, nitridosilicate anions $[SiN_4]^{8-}$ condensed to a framework help to promote the formation of Eu^{2+} cations with $Eu_2Si_5N_8$ as the most prominent example. On the other hand, extremely stable EuH_2 and EuI_2 are hard to oxidize. Although combinations of more than two anions seem to complicate the situation at first glance, the possible characterization of $Eu_5O_2H_2I_4$ as europium(II) oxide hydride iodide gives hope for more examples.

REFERENCES

1. A. L. Allred, E. G. Rochow: "A Scale of Electronegativity Based on Electrostatic Force", *J. Inorg. Nucl. Chem.* **1958**, *5*, 264–268.

Title: Manipulation of Redox Properties of F-Block Elements Using Redox-Active Ligands

Abstract: Redox-active ligands have been demonstrated to mediate multi-electron chemistry on a variety of transition metals and main group elements, promoting useful bond forming and activation chemistry. Our group has been exploring such frameworks for the elements of the f-block, in hopes to extend this methodology to generate elusive targets. Herein, we present our latest findings on the unique electronic structures of these complexes, as well as their reactivity as entries to unprecedented species bearing uranium-element multiple bonds. These unusual species have interesting chemical properties, and their study gives molecular insight into key intermediates identified in the nuclear fuel cycle. Our complete study combining characterization by spectroscopic, structural, and computational methods will be presented, as well as a discussion of the role of the redox-active ligands in the observed chemistry.

Mechanistic Study of Crystal Growth Pathways of Pr123 Using Biopolymers

Joshua L. Konne[1], Simon R. Hall[2*] and Sean A. Davis[3]
[1]Department of Chemistry, Rivers State University of Science and Technology Nkpolu-Oroworukwo Port Harcourt, Nigeria.
[2,3*]School of Chemistry, University of Bristol, Cantock's Close, BS8 1TS, United Kingdom.
Konne.joshua@ust.edu.ng

INTRODUCTION

Crystal engineering of rare-earth oxides provides emerging opportunities for energy applications such as superconductivity and fuel cell technologies[1]. The substitutions of Y in yttrium barium copper oxides (YBCO) with trivalent rare earths from La-Lu produced similar superconducting phases like YBCO except for Ce, Pr and Tb with more stable tetravalent states[2,3] In Praseodymium barium copper oxide (PrBCO) phases, localization of holes in hybridized Pr-O bonds and long calcination hours in order to eliminate $BaPrO_3$ and $BaCuO_2$ impurities pose some significant challenges to its synthesis and use as superconducting material[4].

However, morphological control via use of biopolymers has been found to be quite promising in time reduction, crystal engineering and elimination of impurities in PrBCO phases[1]. A mechanistic study of the formation, growth, elimination/ exclusion of the tetravalent $BaPrO_3$ impurity phase in PrBCO synthesis using biopolymers such as alginate and chitosan (biopolymers) could provide information for a robust timescale control/redox tuning of PrBCO properties.

EXPERIMENTAL METHODS

1 mL of stoichiometric PrBCO precursor solution was dissolved separately in 10 mL alginate and chitosan solutions stirred to form composite gels which were dried overnight in an oven and calcined at 920 °C for 2hrs at a ramp rate of 5 °Cmin^{-1} in air to form the ceramic phase. Control samples without biopolymer were also prepared using the same protocol. Both samples were subjected to growth Kinetics *via* quench reactions. Samples were characterized using PXRD and SEM.

RESULTS AND DISCUSSION

The XRD of the quench reactions showed that the Pr123 crystals grew through the $BaPrO_3$ intermediate pathway before melting into the ceramic phase at 860 °C as shown in **Figure 1**. However chitosan directed growth went through the $BaCO_3$ route.

Below are some proposed equations for the reactions:-

$$BaCO_3 + PrO_2 + CuO \rightarrow BaPrO_3 + CuO + CO_2\uparrow \text{.......(1)}$$

$$BaPrO_3 + 3CuO + BaO \rightarrow PrBa_2Cu_3O_{7-\delta} \text{...............(2)}$$

for the alginate route while Eq 3 illustrates the chitosan pathway.

$$BaCO_3 + PrO_2 + BaO + 3CuO \rightarrow PrBa_2Cu_3O_{7-\delta} + CO_2\uparrow \text{..(3)}$$

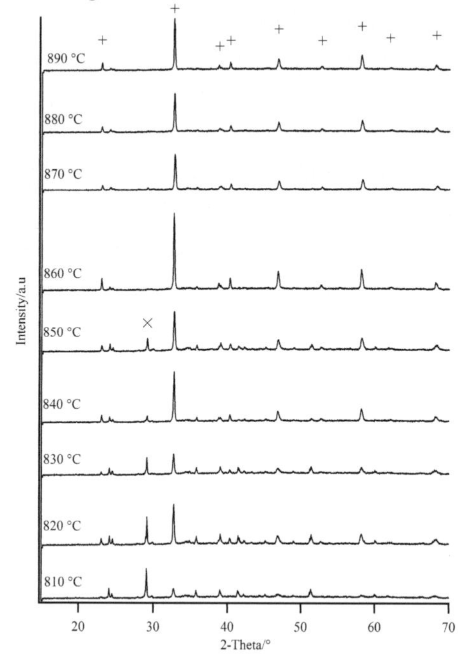

Fig.1 XRD patterns of Pr123 synthesized with alginate quenched at every 10 °C from 810- 890 °C. Pr123 phase is marked (+) and $BaPrO_3$ (X).

CONCLUSIONS

The results of the mechanistic study have shown that the intermediate, $BaPrO_3$ was a key player that linked the transition from the $BaCO_3$ phase to the Pr123 ceramic phase in the alginate-mediated route while interactions of pre-organized copper oxide nanoparticles supported by barium carbonate nanoparticles characterized the chitosan route.

REFERENCES

1. Konne, J.L., Davis, S.A., Glatzel, S. & Hall, S. R. *Chem. Commun. 49*, 5477-5479, 2013.

2. L. Soderholm; K. Zhang; D.G. Hinks; M.A. Beno; J.D. Jorgensen; C.U. Segre; Schuller, I. K. *Nat. Mater. 328*, 604, 1987.

3. E.M. Engler; V.Y. Lee; A.I. Nazzal; R.B. Beyers; G. Lim; P.M. Grant; S.S.P. Parkin; M.L. Ramirez; J.E. Vazquez; Savoy, R. J. *J. Am. Chem. Soc.109*, 2848, 1987.

4. J.-S. Kang; J.W. Allen; Z.-X. Shen; W.P. Ellis; J.J. Yeh; B.W. Lee; Maple, M. B. *JLCM, 148*, 121, 1989.

ACKNOWLEDGMENTS

Authors wish to acknowledge PTDF, Nigeria and RSC, UK for their funding support to JLK and SRH respectively.

Mechanochemical Synthesis: Novel Materials for Magnetic Cooling Applications

Viktor P. Balema,[1] Niraj K. Singh,[2] Shalabh Gupta,[1] Yaroslav Mudryk,[1] Vitalij K. Pecharsky[1,3]

[1] Ames Laboratory of US DOE, Ames, IA 50011, USA, vbalema@ameslab.gov
[2] Aldrich Materials Science, MilliporeSigma, 6000 N. Teutonia Ave., Milwaukee,WI, 53209, USA
[3] Department of Materials Science and Engineering, Iowa State University, Ames, IA 50011, USA

INTRODUCTION

The presentation addresses the mechanochemical approach to solid-state synthesis, which proved to be indispensable for the solvent-free preparation of a great variety of molecular, ionic, and metallic solid materials. The work specifically focuses on the solid-state synthesis of rare earth-based metal organic frameworks (MOFs), which have a potential for becoming the materials of choice for low temperature magnetocaloric systems, which are under development at CaloriCool™ – a newly established research consortium that discovers, develops, and deploys materials for magneto-, electro-, and elastocaloric cooling. CaloriCool™ is a part of the DOE's Energy Materials Network (EMN) designed to accelerate innovation around the clean energy industries by addressing pressing materials challenges ranging from early development to manufacturing

EXPERIMENTAL METHODS

Rare earth based MOFs were synthesized by ball milling of the reactants for different periods of time. The milled samples were analysed by powder X-ray diffraction (PXRD) on a PANalytical X'PERT powder diffractometer using Cu $K_{\alpha 1}$ radiation. The thermos-gravimetric analysis (TGA) and differential scanning calorimetric (DSC) data were collected using METTLER TOLEDO STAR system. The magnetic properties were measured using the superconducting quantum interference device (SQUID, MPMS XL, Quantum Design, USA). The magnetic entropy change was calculated from isothermal magnetization (*M*(H)) curves using Maxwell equations.

RESULTS AND DISCUSSION

In the course of our research, we found out that mechanical processing of benzene 1,3,5-tricarboxylic acid [$C_6H_3(COOH)_3$], with rare earth hydrides n[RH_3, R = Y, Gd][1] or carbonates [$R_2(CO_3)_3 \cdot xH_2O$; R= Gd, Tb and Dy][2], leads to the formation of ordered metal-organic frameworks [$R\{C_6H_3(COOH)_3\}$] that adopt MIL-78 type structure. *M*(*T*) data of the investigated Gd, Dy and Tb-MOFs did not show any apparent onset of long range magnetic ordering down to 2 K. However, the *M*(*H*) data for $Gd\{C_6H_3(COOH)_3\}$ collected at 2K showed deviations from the magnetization behaviour expected for non-interacting Gd^{3+} ions. For the Gd-based MOF structure, the temperature dependence of the isothermal magnetic entropy change (ΔS_M) exhibited a monotonous increase with decreasing temperature and at $T = 3.5$ K it reaches -34.1 J kg^{-1} K^{-1} for a field change (ΔH) of 50 kOe. For the same ΔH the maximum values of ΔS_M for R = Tb and Dy are -5.5 J kg^{-1} K^{-1} and -8.5 J kg^{-1} K^{-1} at 9.5 K and 4.5 K, respectively.[2]

Our further investigations revealed that the mechano-chemical approach can be extended onto other rare earth-based hybrid organic-inorganic materials with substantial magnetocaloric effect at cryogenic temperatures and that water can act as catalyst.

CONCLUSIONS

Mechanochemical approach has been successfully applied to the preparation of MIL-78-type Metal Organic Frameworks starting with rare earth metal carbonates and trimesic acid (TMA) as organic linker. Contrary to the previous reports[3], the presence of a liquid (liquid assisted grinding) is not essential for the process. The investigated Gd-, Tb- and Dy- trimesates show short range magnetic order below 20 K; no long range order is observed down to 2 K. The magneto-caloric effect of the Gd-MOF is comparable to the best known cryogenic refrigerants.

REFERENCES

1. Singh, N. K.; Gupta, S.; Pecharsky, V. K.; Balema, V. P., *J. Alloys Compd.* (2017), 696, 118.
2. Singh, N.K.; Hardi, M.; Balema, V.P. *Chem. Commun* (2013), 49(10), 972.
3. Yuan, W.; O'Connor, J.; James, S. L. *CrystEngComm.* (2010) 12, 3515.

ACKNOWLEDGMENTS

CaloriCool™ is supported by the United States Department of Energy, Office of Energy Efficiency and Renewable Energy though its Advanced Manufacturing and Building Technologies Offices. Ames Laboratory is supported by the Office of Science, Basic Energy Sciences Programs and operated by Iowa State University under contract No. DE-AC02-07CH11358 with the United States Department of Energy.

Metal growth on and under graphene: morphology, growth mode and intercalation

M T Hershberger, D.McDougall, M. Hupalo, X. Liu, C. Z. Wang P. A. Thiel, M. C. Tringides,
Department of Physics and Chemistry Iowa State University Ames Laboratory –USDOE
email: mctringi@iastate.edu

Graphene based electronic and spintronic devices require understanding of the growth of metals on graphene. Several metals (Gd, Dy, Eu, Fe,Pb) deposited on epitaxial graphene were studied with STM, SPA-LEED and DFT. The grown morphology (island density and domain size distributions) was used to extract the metal diffusion and adsorption barriers[1,2]. For practically all metals the grown mode is 3-d as a result of (i) the low ratio of the metal adsorption to metal cohesive energy and (ii) the repulsive interaction of unscreened charges at the metal-graphene interface that favours islands of small "footprint". These experimental results are fully supported with DFT calculations . It is essential to find ways to modify the growth to layer–by–layer for high quality metal contacts and for using graphene as a spin filter. By growing Dy at low temperatures or high flux rates it is found that upward adatom transfer is kinetically suppressed and layer-by-layer is possible[3].

Graphene intercalation provides a novel way to tune its properties. Experiments on graphene partially intercalated with Dy show that nucleation is preferred on the intercalated than on the pristine areas. Difference in doping between the two areas generates an electric field that transforms random to directional diffusion and accounts for the guided nucleation[4]. This can be a general method to control patterning of metallic films on graphene.

Many issues related to the intercalation process itself are poorly understood, i.e. ,the temperature where atoms move below graphene, different intercalation phases, their coverage , the entry points for atoms to move below, the intercalation time etc. SPAL-LEED has been used to study these questions for Dy intercalation in graphene. Spot profiles of several spots (specular, 6sq(3), graphene) are studied as function of temperature and electron energy to deduce the kinetics of intercalation and the layer where atoms reside. This information is essential to tune the emerging properties of intercalated graphene.

References

1.M. Hupalo et al *Advanc. Mater.* 23 2082 (2011) 2.X. Liu, et al. *Progr. Surf. Sci.* 90 397 (2015) 3. D. Mc Dougall et al *Carbon* 108 283 (2016)) 4. X. Liu et al. *Nano Research* 9(5): 1434 (2016)

Mitigating Rare Earths Supply Risks: Insights from Japan's Supply Structure

Marc Schmid[1]

[1]Chair for Economic Policy, Martin-Luther-University Halle-Wittenberg, Germany
marc.schmid@wiwi.uni-halle.de

INTRODUCTION

Supply disruptions of Rare Earths (RE) can not only affect certain industries in consuming countries, but also hamper the development of climate friendly technologies.[1, 2] Japan is the major demander of RE outside of China due to its industry structure with world leading electronic manufacturers and first movers in the field of electric and hybrid vehicles. Mitigation of RE supply disruptions is therefore crucial for the economy of Japan.

Whereas recent contributions to mitigation strategies discuss RE material flows, recycling potentials and the establishment of alternative supply chains outside of China,[3, 4, 5] minor attention has been put on the evaluation of mitigation strategies that aim to diversify a country's supply structure. This contribution provides insights into the reactions of the economy and politics in Japan on the RE crisis, the changes in the RE supply structure of Japan's economy in recent years and examines results and challenges of these changes.

EXPERIMENTAL METHODS

The analysis of Japan's mitigation strategy is based on an in-depth monitoring and evaluation of corporate and governmental publications about mining projects, supply agreements and project investments. Changes in Japan's supply structure are analysed on the basis of trade data of UN comtrade.[6] The base period for the analysis is the year 2011, when prices peaked.

RESULTS AND DISCUSSION

Results show fundamental changes in Japan's rare earths supply structure. Whereas in 2011, more than 80% of the total import value and 70% of Japan's total import amount of rare earth metals and compounds were imported directly from China, these numbers significantly dropped in the following years. In 2015, only 37% of the imports in terms of value and 53% in terms of weight came from China. This development is even more remarkable in comparison to the peak in Chinese share of about 90% in 2008. The supply structure shifted from a high share of imports from China to a relatively well-balanced structure.

Nevertheless, in general, trade statistics lack a differentiation of specific Rare Earth Elements and do not cover all countries. Sustainability of rare earth projects outside of China remains a challenge, as projects are not economically viable as long as prices remain on the current level. Additionally, heavy rare earths are still exclusively produced in China.

CONCLUSIONS

Findings highlight the potentials of supply diversification as a mitigation strategy through Japanese investments in a range of projects. Reactions have led to an overall decreased supply concentration within a relatively short period of time. Japan's supply structure shows changing patterns of dependence on a country level.

Results also demonstrate remaining bottlenecks along the global value chain and question the sustainability of the diversified supply structure of Japan.

REFERENCES

1. Alonso, E. *et al.*, Environ Sci Technol 46:3406-3414, 2012
2. Smith Stegen, K., Energ Policy 79:1-8, 2015
3. Machacek, E., Fold, N., Resour Policy 42:53-64, 2014
4. Sekine, N. *et al.*, J Ind Ecol, doi: 10.1111/jiec.12458, 2016
5. Seo, Y., Morimoto, S., Resour Policy 39:15-20, 2014
6. UN comtrade, http://comtrade.un.org/db/default.aspx, 2017

ACKNOWLEDGMENTS

The author would like to thank the European Union EFRE program for providing financial support to this project.

Mixed-Valent Europium in Two *ortho*-Oxoborates Derivatized with Soft Anions: $Eu_5X[BO_3]_4$ (X = S and Br)

Christian Funk[1], Olaf Reckeweg[1] and Thomas Schleid[1*]

[1] Institute for Inorganic Chemistry, University of Stuttgart, Germany
[1*] Institute for Inorganic Chemistry, University of Stuttgart, Germany, *schleid@iac.uni-stuttgart.de*

INTRODUCTION

Europium can occur in two different oxidation states, namely as Eu^{2+} and Eu^{3+} [1]. One of the best-known divalent europium compounds must be the monoxide EuO [1]. For preparing single crystals mostly flux techniques are used, and a common flux is B_2O_3 [2]. But in the phase system EuO–B_2O_3 there are several phases occurring, such as $Eu_3B_2O_6$, $Eu_2B_2O_5$, EuB_2O_4 and EuB_4O_7 [3]. Divalent europium in oxoborates is well known for its specific luminescence properties, however, as shown for BaB_8O_{13}:Eu^{2+} [3]. Therefore, we tried to synthesize new divalent europium compounds with oxoborate together with other anions.

EXPERIMENTAL METHODS

Synthesis of $Eu_6S[BO_3]_4$

Eu_2O_3, B_2O_3, B and EuS were mixed in an agate mortar filled into niobium crucibles and sealed in secondary fused silica ampoules. After 48 h at 1100 °C their reaction yielded orange needles.

Synthesis of $Eu_6Br[BO_3]_4$

A mixture of Eu_2O_3, B and NaBr as flux was packed into silica-jacketed, arc-welded niobium ampoules and reacted for 72 h at 1100 °C to yield red needles.

RESULTS AND DISCUSSION

$Eu_6S[BO_3]_4$ and $Eu_5Br[BO_3]_4$ crystallize hexagonally in the space group $P6_3mc$ with a = 1058.79(7) and c = 681.96(5) pm (c/a = 0.6441) and a = 1059.06(7), c = 682.37(5) pm (c/a = 0.6443), respectively, for Z = 2 each. There are two independent sites for europium. On one, europium (Eu2) is coordinated octahedrally by six $[BO_3]^{3-}$ anions, four times via one oxygen atom only, but two times via two. So an overall coordination number of eight results. The other europium cation (Eu1) has five $[BO_3]^{3-}$ anions as neighbours and two soft anions (S^{2-} or Br^-). The $[BO_3]^{3-}$ anions coordinate this Eu1 cation as a distorted trigonal bipyramid as shown in Figure 1. Overall it has a coordination number of nine, seven times as oxygen atoms of five $[BO_3]^{3-}$ anions and two soft anions. The latter ones are surrounded octahedrally (CN = 3 + 3) by six europium cations (Eu1) each and reside in europium channels along the c-axis. The europium atoms in $Eu_6X[BO_3]_4$ (X = S and Br) have two different oxidation states. The Eu1 position should bear only Eu^{2+}, while a mixed occupation of Eu^{2+} and Eu^{3+} has to be assumed for Eu2 to secure charge compensation (Table 1).

All elements were confirmed by EDX measurements on a *Cameca SX100* electron-beam microprobe with these compositions for both compounds.

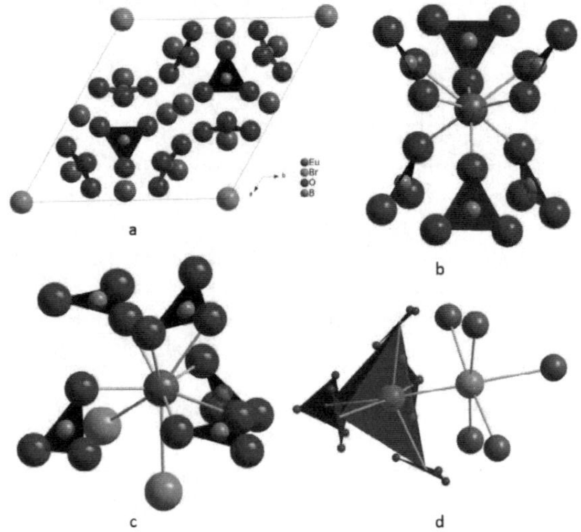

Figure 1. a) Crystal structure of $Eu_6X[BO_3]_4$ (X = S and Br) as viewed along [001], b) Eu2 coordination by six $[BO_3]^{3-}$ units, c) Eu1 coordination by five $[BO_3]^{3-}$ units and two X anions and d) X anion coordinated by six Eu1 cations and a coordination polyhedron of one of these.

Table 1. Selected interatomic distances (d/pm) in $Eu_6S[BO_3]_4$ and $Eu_5Br[BO_3]_4$.

$Eu_5S[BO_3]_4$		$Eu_6Br[BO_3]_4$	
Eu1–O2	256.4 (1×)	Eu1–O2	253.5 (1×)
–O3	256.1 (2×)	–O3	255.4 (2×)
–O1	274.5 (2×)	–O1	271.1 (2×)
–O3'	275.6 (2×)	–O3'	277.2 (2×)
–S	322.2 (1×)	–Br	331.6 (1×)
–S'	363.2 (1×)	–Br'	358.7 (1×)
CN = 8+1		CN = 9	
Eu2–O2	240.4 (2×)	Eu2–O2	241.7 (2×)
–O1	244.9 (1×)	–O1	244.6 (1×)
–O3	245.7 (2×)	–O3	244.4 (2×)
–O1'	254.2 (1×)	–O1'	254.4 (1×)
–O3'	272.0 (2×)	–O3'	271.0 (2×)
CN = 8		CN = 8	

REFERENCES

[1] a) McGuire, T. R.; Shafer, M. W., *J. Appl. Phys.*, *35*, 984–988, 1964. b) Morrish, A. H.; Shafer, M. W. *J., Appl. Phys.*, *36*, 1145–1152, 1965.

[2] Mori, D.; Sonoyama, N.; Yamada, A.; Kanno, R.; Azuma, M.; Takano, M.; Suda, K.; Ishizawa, N., *J. Solid State Chem. 179*, 935–940, 2006.

[3] Machida, K.; Adachi, G; Shiokawa, J., *J. Lumin. 21*, 101–110, 1979.

Modifications of the *f*-element / aminopolycarboxylate complex for efficient differentiation of trivalent actinides from trivalent actinides.

Peter R. Zalupski, Colt Heathman and Travis Grimes

Department of Aqueous Separations and Radiochemistry, Idaho National Laboratory, Idaho Falls, ID 83415, USA

Peter.Zalupski@inl.gov

Efficient separation of trivalent lanthanides (Ln^{3+}) from trivalent actinides (An^{3+}) may be afforded when using an aminopolycarboxylate aqueous complexant such as diethylenetriamine-N,N,N',N",N"-pentaacetic acid (DTPA). In a liquid-liquid distribution system An^{3+}-DTPA complexes remain in aqueous mixture, while the phase transfer of lanthanides is facilitated by a liquid cation exchanger. While the octadentate coordination of *f*-element by DTPA is largely responsible for the accomplished An^{3+}/Ln^{3+} separation it is also liable for the slow liquid-liquid mass transfer. The binding pocket contains seven stable five-membered rings, with three nitrogens and five oxygens coordinating the metal ion. Accordingly, the metal is well shielded from the interactions with the water molecules, resulting in kinetic inertness. Current research efforts in the area of minor actinide recovery are focused on enhancing the phase transfer kinetics of systems containing aminopolycarboxylates. This contribution will relate options for structural modification of aminopolycarboxylates, the corresponding disruption of *f*-element coordination sphere, and its effects on An^{3+}/Ln^{3+} differentiation. Connections between thermodynamic studies on the trivalent *f*-element complexation by aminopolycarboxylates and phase transfer kinetic studies will be extended to seek options for intra-lanthanide differentiation.

Molecular-Scale Insight into the Structure, Stability, and Electrochemistry of Ceria Nanoparticles

Shanna L. Estes[1], Mark R. Antonio[1], Valérie Vallet[2], S. Skanthakumar[1], and L. Soderholm[1*]

[1]Chemical Sciences and Engineering Division, Argonne National Laboratory, USA, *ls@anl.gov
[2]Laboratoire PhLAM, CNRS UMR 8523, CNRS FR 2416, Université de Lille, France

INTRODUCTION

Cerium's unique chemistry has been exploited in numerous applications, ranging from catalysis to oxygen storage. However, structural ambiguities in Ce materials impair efforts to understand these oxidation and catalysis mechanisms, particularly for nanosized ceria (CeO_2), which exhibits increased catalytic efficacy compared with bulk ceria. Guided by the aqueous chemistry of other tetravalent metals (e.g., Th^{IV}, Pu^{IV}, Zr^{IV}), we use a molecular-scale approach to examine the electronic and structural influences driving Ce chemistry. Specifically, we report the isolation, structural characterization, and electrochemical behavior of two polynuclear Ce^{IV} clusters that can be considered as the smallest CeO_2 nanoparticles.

RESULTS AND DISCUSSION

A dinuclear Ce^{IV} cluster (Fig. 1) was crystallized by heating and evaporating a solution of nanosized CeO_2 dissolved in HNO_3. Single-crystal X-ray diffraction identifies three bidentate NO_3^- and three H_2O ligands coordinated to each Ce^{IV}. Additionally, and unlike dinuclear Th^{IV} and Pu^{IV} clusters,[1,2] the Ce^{IV} atoms are bonded via a linear oxo-bridge. Examination of this

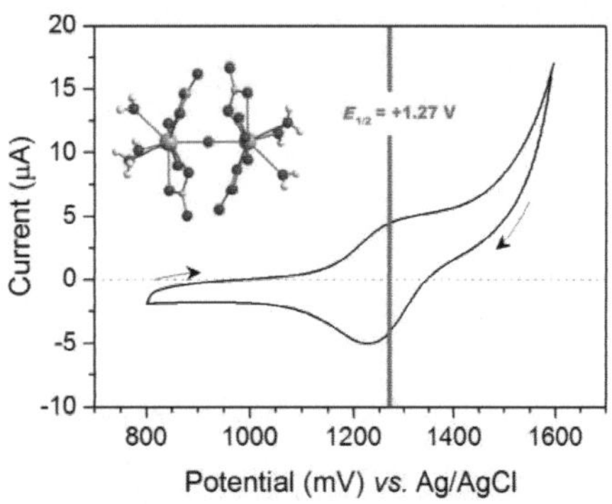

Fig. 1. Representation of the dinuclear Ce^{IV} cluster (Ce, O, N, and H atoms are colored in orange, red, blue, and pink, respectively); cyclic voltammogram for 1 mM of the dinuclear cluster dissolved in 0.1 M HNO_3.

unique bonding environment using density-functional theory is currently underway. Cyclic voltammetry of the dinuclear cluster dissolved in 0.1 M HNO_3 (Fig. 1) confirms that the cluster initially contains 100% Ce^{IV} and suggests that Ce^{IV} is stabilized within the dinuclear cluster relative to the Ce^{IV} monomer (E^0 = +1.52 V vs Ag/AgCl). Under aqueous conditions similar to those used in CeO_2 syntheses, a Ce^{IV} hexanuclear cluster that exhibits an atomic arrangement analogous to the fcc structure observed for bulk CeO_2 was synthesized with varied counterion (NO_3^- and glycine) surface decoration (Fig. 2). Cyclic voltammetry of the hexanuclear cluster dissolved in 0.1 M HNO_3 (Fig. 3) also confirms that the hexanuclear cluster initially contains 100% Ce^{IV} and that Ce^{IV} is stabilized within the hexanuclear cluster relative to both the Ce^{IV} monomer and dinuclear cluster

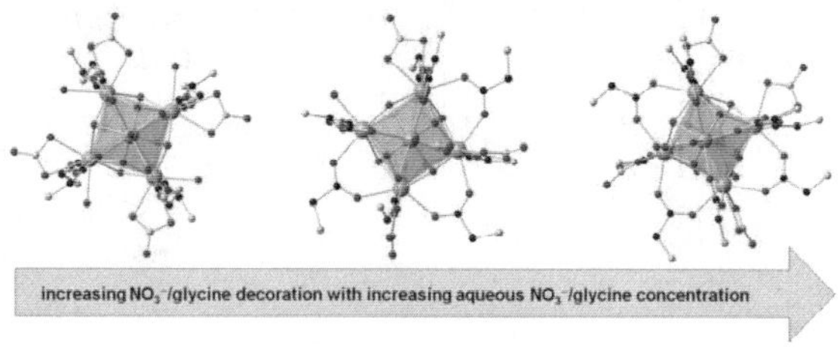

Fig. 2. Representation of varied ligand decoration of the hexanuclear Ce^{IV} cluster (Ce, O, and N atoms are colored in orange, red, and blue, respectively).

(Fig. 1). The electrochemical behavior of these clusters suggests that Ce^{IV} stability increases as cluster nuclearity increases, a trend that is consistent with lower measured electrode potentials for nanosized CeO_2

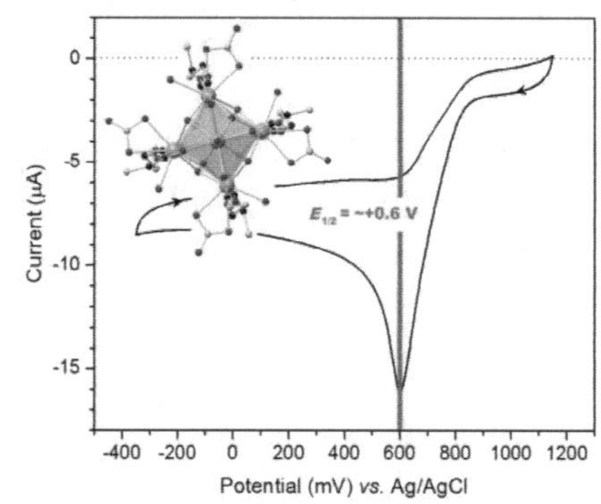

Fig. 3. Representation of the hexanuclear Ce^{IV} cluster (Ce, O, and N atoms are colored in orange, red, and blue, respectively); cyclic voltammogram for 1 mM of the hexanuclear cluster dissolved in 0.1 M HNO_3.

($E_{1/2}$ = ~+0.3 V vs Ag/AgCl).[3] These electrochemical findings combined with the observed stability of the hexanuclear cluster core and the lability of decorating ligands on the hexanuclear cluster surface suggest that variable surface chemistry may explain the enhanced efficacy of nanosized CeO_2.

CONCLUSIONS

The clusters examined herein give new molecular-scale insight into Ce stability and redox chemistry, which limits discussion of possible mechanisms responsible for nanosized CeO_2 efficacy – i.e., neither O-vacancies nor Ce^{III} are observed in the clusters, instead, cluster charge is compensated via varied counterion surface decoration.

REFERENCES

1. Wilson R. E.; Skanthakumar S.; Sigmon G.; Burns P. C.; Soderholm L., Inorg. Chem. 46:2368, 2007
2. Knope K. E.; Skanthakumar S.; Soderholm L., Inorg. Chem. 54:10192, 2015
3. Padmanathan N.; Selladurai S., RSC Adv. 4:6527, 2014

ACKNOWLEDGMENTS

This work is supported at UChicago Argonne, LLC (Argonne National Laboratory), by the U.S. Department of Energy, BES Heavy Elements Program under Contract No. DE-AC02-06CH11357.

Multiple magnetic transitions and magnetic entropy change in pseudo-binary $Er_{0.75}Dy_{0.25}$ polycrystal

R. Nirmala[1*], Ya. Mudryk[2], Durga Paudyal[2], V. K Pecharsky[2,3], K. A. Gschneidner Jr. [2,3]
[1]Department of Physics, Indian Institute of Technology Madras, Chennai 600 036 India
[2]The Ames Laboratory, U. S. Department of Energy, Iowa State University, Ames, Iowa 50011-3020, USA
[3]Department of Materials Science and Engineering, Iowa State University, Ames, Iowa 50011-2300, USA
nirmala@physics.iitm.ac.in

INTRODUCTION

For practical magnetic refrigeration applications, materials must have considerable magnetocaloric effect (MCE) over a wide span of temperature. In general, MCE exhibits a maximum value in the neighbourhood of magnetic ordering temperature (T_C / T_N) of a material and hence composite materials that display multiple magnetic transitions with distributed ordering temperatures are desired. In this regard, intra-lanthanide $R_{1-x}R'_x$ (R, R' = rare earth) alloys are ideal candidates as they undergo series of magnetic transitions. In particular, the end members Er and Dy do undergo a first order transition at low temperatures which could be useful in tapping large MCE[1,2]. Thus the knowledge of MCE in these $Er_{1-x}Dy_x$ alloys is of great interest.

Recently, anomalous first order transitions were reported in rare earth dialuminides, namely $Er_{1-x}Dy_xAl_2$ and $Er_{1-x}Tb_xAl_2$ (x = 0.25), the origins of these are not fully understood yet. It has been suggested that competition between magnetic exchange, crystalline electric fields (CEF), magnetocrystalline anisotropy, magnetoelastic coupling including higher order quadrupolar interactions may be responsible for the occurrence of these intriguing transitions[3,4,5]. In this connection, magnetic properties of $Er_{1-x}Dy_x$ (x =0.25) single crystal have been studied[6]. In this work, we focus on magnetic and magnetocaloric properties of $Er_{1-x}Dy_x$ (x =0.25) polycrystal.

EXPERIMENTAL METHODS

Polycrystalline sample of $Er_{1-x}Dy_x$ (x =0.25) has been prepared by arc melting under inert argon atmosphere. DC and ac magnetization data were measured using a SQUID magnetometer [MPMS XL, Quantum Design, USA]. Heat capacity was measured using a quasi-adiabatic calorimeter between 2 to 300 K in zero field and 0.5 T field.

RESULTS AND DISCUSSION

AC magnetic susceptibility data reveal occurrence of three magnetic transitions at ~27 K, ~43 K and ~106 K respectively [Fig. 1]. These temperatures are close to that observed for single crystal $Er_{0.75}Dy_{0.25}$.[6] These transitions are also marked as peaks in the zero field heat capacity data at 108 K, 41 K and 30 K respectively [Fig. 2].

The key features are (i) the frequency dependent ac susceptibility peak at 27 K and (ii) the disappearance of sharp heat capacity peak at 41 K in 0.5 T field. Isothermal magnetic entropy change vs temperature plot indeed has a larger full-width at half-maximum owing to merger of maxima corresponding to magnetic transitions at 27 K and 43 K [Fig. 3].

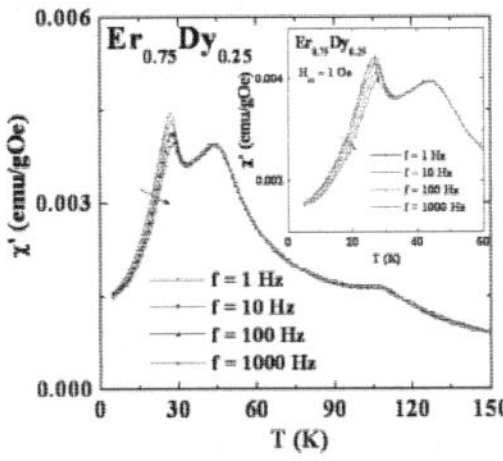

Fig. 1

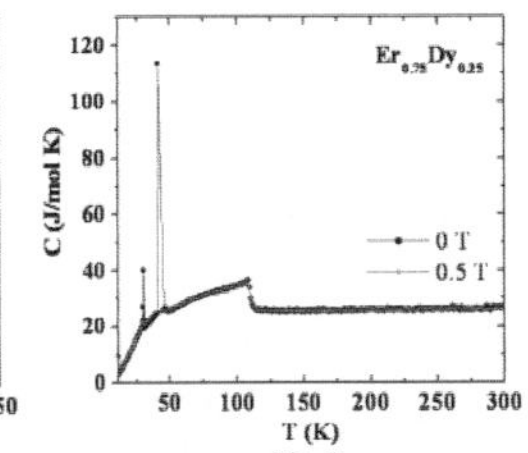

Fig. 2

Fig. 1 AC magnetic susceptibility vs temperature of $Er_{0.75}Dy_{0.25}$ and its frequency dependence [Inset shows magnified low temperature data]. Fig. 2 Heat capacity vs temperature of $Er_{0.75}Dy_{0.25}$ measured in zero applied field and in 0.5 T field.

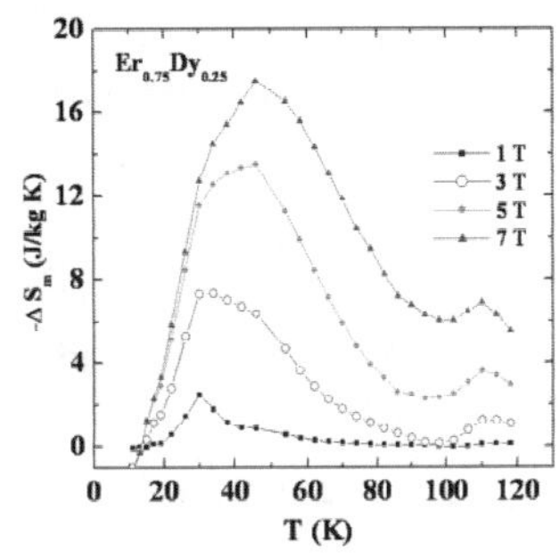

Fig. 3 Isothermal magnetic entropy change vs temperature of $Er_{0.75}Dy_{0.25}$ for various field changes.

SUMMARY

To summarize, the $Er_{0.75}Dy_{0.25}$ pseudo-binary system indeed presents interesting and complex magnetic behaviour in low applied fields within the magnetically ordered state just as intralanthanide dialuminides. This alloy also exhibits substantial MCE over broad temperature range.

REFERENCES

1. Gschneidner Jr. et al, Phys. Rev. Lett. **78**, 4281 (1997) and references therein.
2. Chernysov et al, Phys. Rev. B **71**, 184410 (2005)
3. Mahmud Khan et al, Phys. Rev. B **82**, 064421 (2010).
4. Lima et al, Phys. Rev. B **68**, 134409 (2003).
5. Nirmala et al, Phys. Rev. B **76,** 014407 (2007).
6. Mahmud Khan et al, Phys. Rev. B **84**, 134424 (2011).

ACKNOWLEDGMENTS

The Ames Laboratory is operated for the U. S. Department of Energy by Iowa State University of Science and Technology under contract No. DE-AC02-07CH11358. This work was supported by the Department of Energy, Office of Basic Energy Sciences, Materials Sciences Division.

New Insights in Rare Earth Triflates

Mona Struckmann[1], Jörn Bruns[2], Mathias S. Wickleder[1*]

[1]Institute of Inorganic and Analytical chemistry, Justus-Liebig-University Giessen, Germany,
[2]Institute of General, Inorganic and Theoretical Chemistry, University of Innsbruck, Austria,
mona.struckmann@anorg.chemie.uni-giessen.de

INTRODUCTION

Rare Earth triflates $RE(O_3SCF_3)_3$ are commercially available compounds and widely used as environmentally benign Lewis acids. Nevertheless, their structures remain still elusive. The commercially available compounds are mostly nonahydrates showing the lanthanide cations coordinated homoleptic by water molecules. The triflate anion functions only as a counter anion. Dehydration of these hydrates could change the coordination sphere towards triflate coordination, but this couldn't be proved by structure determinations up to now[1].

Recently, we were able to develop a synthesis towards the exclusive coordination of lanthanide metals by triflate anions.

EXPERIMENTAL METHODS

Lanthanide oxides were loaded into a thick-walled glass ampoule (l = 250 mm, d = 20 mm, thickness of the tube wall = 2 mm). Subsequently, fuming nitric acid, trifluormethanesulfonic acid and trifluormethanesulfonic acid anhydride were added. The ampoule was torch-sealed, placed into a resistance furnace and heated to 120 °C within 24 h. The temperature was maintained for two days and finally reduced to 25 °C during 79 h.

RESULTS AND DISCUSSION

On our way to exclusively triflate coordinated lanthanides we were able to obtain molecules in which the lanthanide ion (Y, Sm-Gd, Ho, Tm-Lu) is coordinated by three water molecules and five triflates in form of a slightly distorted square antiprism.

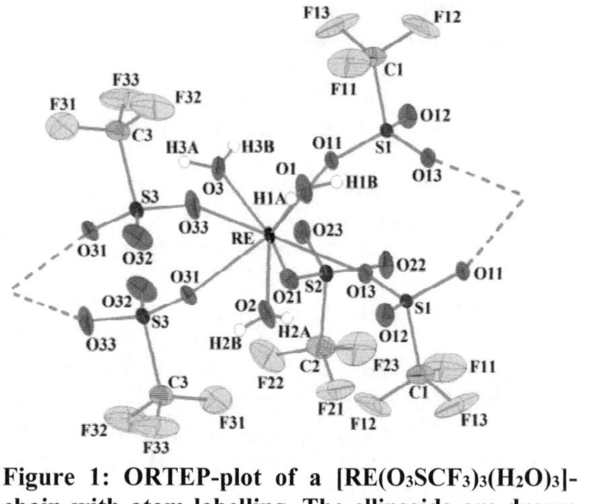

Figure 1: ORTEP-plot of a $[RE(O_3SCF_3)_3(H_2O)_3]$-chain with atom labelling. The ellipsoids are drawn at a 75 % probability level.

Four of the five triflate ligands function as bridging ligands between two lanthanide ions, resulting in infinite chains (figure 1). These chains are separated by the fluoride atoms of the triflate groups.

For the examples of the compounds $(NO)_5[RE(O_3SCF_3)_8]$ (RE = La, Pr, Sm, Eu, Tb, Dy) a homoleptic coordination sphere of triflates can be found. The lanthanide ion is coordinated by eight monodentate triflate ligands leading to a slightly distorted square antiprism. Due to the C_2 symmetry of the $[RE(O_3SCF_3)_8]^{5-}$ complex, according to the RE^{3+} ion placed on the special Wyckoff site 16*e*, only four triflate ligands are unique (figure 2). Charge compensation for the $[RE(O_3SCF_3)_8]^{5-}$ anion is achieved by five NO^+ cations.

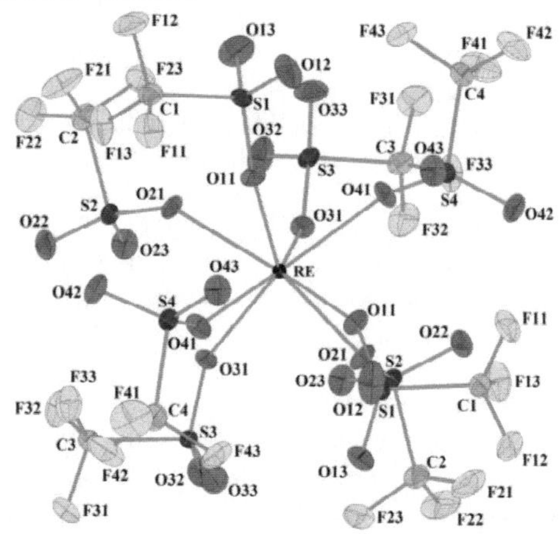

Figure 2: ORTEP-plot of the $[RE(O_3SCF_3)_8]^{5-}$ complex with atom labelling. The ellipsoids are drawn at a 75 % probability level.

CONCLUSIONS

Preparations under harsh conditions lead to new highly triflated lanthanide ions. It could be shown, that mixed coordination spheres consisting of water and triflate ligands could be found for the heavier rare earth ions, whereas the lighter rare earth ions are exclusively coordinated by triflate ions. The preparation of single crystals of both compounds is difficult because the CF_3-groups lead to a strong cleavability of the compounds.

REFERENCES

1. N. Yanagihara, S. Nakamura, M. Nakayama, *Polyhedron* **1998**, *17*, 3625-3631.
2. J. Bruns, S. Krüger, M. Adlung, C. Wickleder, O. Niehaus, R. Pöttgen, T. Klüner, J. Kräuter, M. S. Wickleder, *Chem. Eur. J.* **2015**, *21*, *35*, 12389-12395.

New Reactivity in Actinide Chemistry Facilitated by Supporting Ligand Design

Mary E. Garner, Michael A. Boreen, Bernard F. Parker, Trevor D. Lohrey, Stephan Hohloch, Laurent Maron, and John Arnold*

Department of Chemistry, University of California, Berkeley, California, 94720 *and* Chemical Sciences Division, Lawrence Berkeley National Laboratory, Berkeley, California 94720, USA, arnold@berkeley.edu

INTRODUCTION

The chemistry of early actinide elements continues to be a flourishing field of active research. While a major focus has been nuclear fuel applications, the large size, oxidation state accessibility, and 5*f*-orbital availability in the actinides have distinguished their coordination chemistry from that of other metals in the periodic table.

RESULTS AND DISCUSSION

In this presentation, I will focus on recent results from our laboratory that focus on the synthesis and reactivity of Th and U derivatives incorporating new supporting ligands. Examples include new species featuring An—C bonds, such as the tris(aryl) uranium(III) complexes shown below.

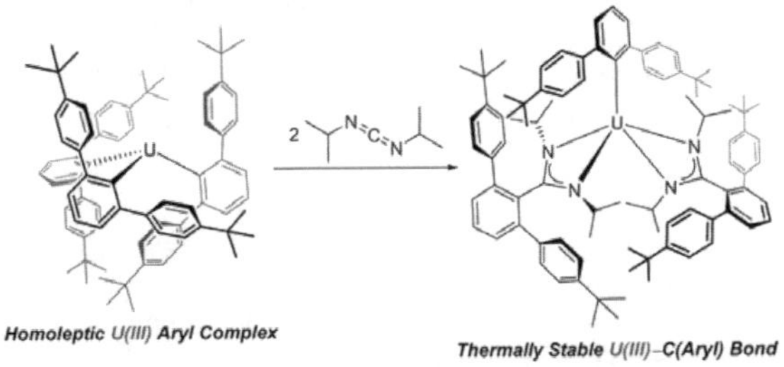

I will also discuss stoichiometric and catalytic reactivity of actinide complexes supported by bis(NHC) ligands, such as the Th derivatives shown below.

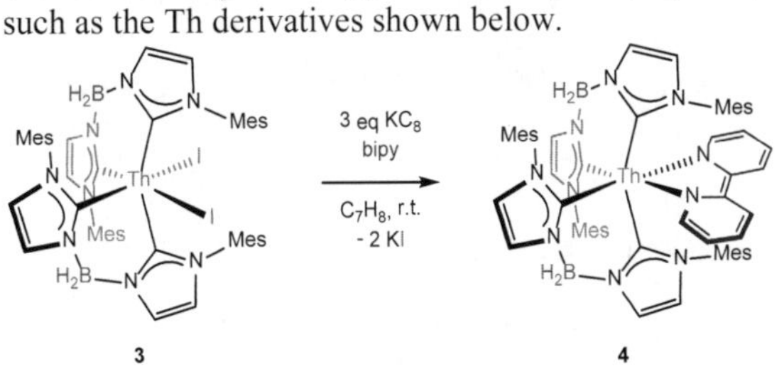

REFERENCES

1. A Homoleptic Uranium (III) Tris (aryl) Complex. MA Boreen, BF Parker, TD Lohrey, J Arnold, Journal of the American Chemical Society, 2016, 138, 15865–15868
2. Carbon–Nitrogen Bond Cleavage by a Thorium-NHC-bpy Complex. ME Garner, S Hohloch, L Maron, J Arnold, Angewandte Chemie International Edition 55 (44), 13789-13792 1 2016
3. A New Supporting Ligand in Actinide Chemistry Leads to Reactive Bis (NHC) borate-Supported Thorium Complexes. ME Garner, S Hohloch, L Maron, J Arnold. Organometallics 35 (17), 2915-2922

ACKNOWLEDGMENTS

We thank the Director, Office of Science, Office of Basic Energy Sciences, Division of Chemical Sciences, Geosciences, and Biosciences Heavy Element Chemistry Program of the U.S. DOE at LBNL under Contract No. DE-AC02-05CH11231 for support of this research.

Non-existent Intermetallics

Gerd H. Meyer[1*,2] and Melissa L. Rhodehouse[1,2]

[1]Department of Chemistry, Iowa State University, Ames, IA, USA, ghmeyer@iastate.edu

[2]Ames Laboratory, USDOE, Ames, IA, USA

With endohedral transition metal atoms, *T*, the rare-earth elements, *R*, form clusters of six to eight *R* atoms. These hetero-bi-metallic clusters, $\{TR_r\}$ (r = CN(*T*) = 6-8), are surrounded by halide ligands, *X*, to build cluster complexes, $\{TR_r\}X_x$. They may be isolated as in the prolific $\{TR_6\}X_{12}$R and $\{TR_6\}X_{10}$ compound types. Edge- and face-sharing of (mainly) octahedral clusters constitute a small number of oligomers, dimers to pentamers. Further connection of clusters via common edges leads to a variety of cluster chains. Coordination numbers (CN) of *T* of seven and eight are also possible: Monocapped prismatic $\{TR_7\}$ and square prismatic as well as antiprismatic $\{TR_8\}$ clusters share common faces to chain structures [1].

Equal or similar coordination environments are also seen in $\{T_tR_r\}$ polar intermetallics, for example CNs of 8 to 10 in $\{Ru_{11}Lu_{20}\}$. Thus, $\{TR_r\}X_x$ cluster complexes may be understood as a symbiotic arrangement of (polar) intermetallic and salt. Consequently, bonding in both polar intermetallics and hetero-metallic cluster complexes is mainly hetero-atomic, *T-R* and *R-X*, as well as *R-X-R*, with minor homo-atomic contributions, *T-T* and *R-R*, and van der Waals *X-X* attractions [2].

There is an unknown, presumably large number of even binary polar intermetallics out there that have not been discovered through phase diagram determinations, often subject to inaccurate phase diagram determinations. One example is $Ru_{11}Lu_{20}$ (35.5 mol-% Ru) which was most certainly falsely addressed as $Ru_{12}Lu_{20}$ = Ru_3Lu_5 (37.5 mol-% Ru) in the phase diagram.

But there are also intermetallics whose synthesis requires more sophisticated routes. We have discovered a number of examples of "non-existent" intermetallics [3] during our efforts to synthesize new rare-earth cluster halides with endohedral transition metal atoms contributing to the electron count. Two examples which we [4] are presently investigating are Co_7Pr_{17} and Pt_3Pr_4 [5]. These do not exist in the presently known phase diagrams and cannot be synthesized by the usual rough methods to synthesize intermetallics, e.g. arc melting.

Flux-growth techniques (salts, ionic liquids, metallic or intermetallic melts, preferably at fairly low temperatures) are more subtle routes to these non-existent intermetallics. These routes are presently explored exhaustively, and a number of new compounds e.g., $Pr_3Pt_4Sn_6$ and $Pr_4Pt_{12}Sn_{25}$ (from tin melts) or known compositions with new structures (Pt_2Pr_3) have been discovered.

REFERENCES

1. Meyer, G. Rare-Earth Metal Cluster Complexes, in: The Rare Earth Elements: Fundamentals and Applications (D. Atwood, Ed.) John Wiley & Sons, Chichester, UK, **2012**, 415-436.
2. Meyer, G. Symbiosis of Intermetallic and Salt: Rare-Earth Metal Cluster Complexes with Endohedral Transition Metal Atoms, in: Handbook on the Physics and Chemistry of Rare Earths (J.-C. G. Bünzli, V. K. Pecharsky, Eds.) **2014**, *45*, 111-178.
3. The expression "non-existent intermetallics" quotes a very stimulating book from 1965 [W. E. Dasent, Non-existent compounds – Compounds of Low Stability, Marcel Dekker, Inc., New York, 1965] which lists (mostly molecular) compounds which had not been synthesized at the time, often subject to low thermo-dynamic stability and lack of synthetic skill.
4. Rhodehouse, M. L., planned PhD thesis; Iowa State University, Ames, IA.
5. Bell, T., PhD Thesis, University of Cologne, Germany, 2014.

ACKNOWLEDGMENTS

This research was supported by the Office of Basic Energy Sciences, Materials Sciences Division, U. S. Department of Energy (DOE). Ames Laboratory is operated for DOE by Iowa State University under contract number DE-AC02-07CH11358.

Non-Innocent Behavior of Cerium Oxide Supports in Catalytic Conversions of Phenolics

Nicholas C. Nelson and Igor I. Slowing*

US DOE Ames Laboratory, Ames, IA, USA
Department of Chemistry, Iowa State University, Ames, IA, USA, islowing@iastate.edu

INTRODUCTION

Phenols are common structural motifs in nature, primary constituents of lignin, and the most abundant products of biomass pyrolysis.[1] Therefore developing catalysts that can convert phenolics from biorenewables into high-value commodity chemicals is relevant in the pursuit of a sustainable economy. One appealing transformation of phenolics is their reduction to produce alicyclic ketones or alcohols, which have several important applications such as synthetic intermediates for high volume polymers. This transformation is typically accomplished using Pd supported on carbon, silica or alumina.[2] While supports are typically chosen to maximize catalyst dispersion and stability, they can also be selected based on their capacity to participate in the reaction (*i.e.* non-innocent supports). Herein we examine the non-innocent behavior of ceria when used as a support for the Pd-catalyzed reduction of phenolics. We also explore two novel approaches for improving the activity of ceria for the same reaction.

EXPERIMENTAL METHODS

High surface area ceria was produced via evaporation induced self-assembly.[3] Modified ceria was prepared by impregnation of the solid with $NaNO_3$ or $(CH_3O)_3PO$ respectively, followed by calcination. The materials were then impregnated with $Pd(CH_3CO_2)_2$ (1 wt% Pd), calcined at 350 °C and reduced under a flow of H_2 at 350 °C. Three different catalytic studies were performed. The first one was performed in batch mode for Pd/CeO_2 using phenol and 1 bar H_2 at r.t. The second study was run in flow using sodium promoted Pd/CeO_2 and isopropanol as transfer hydrogenation reagent. The third study used Pd on phosphated ceria for hydrogenolysis of eugenol and 10 bar H_2 at 100 °C in water. All products were monitored by GC-MS.

RESULTS AND DISCUSSION

Ceria is a reducible support whose fluorite structure facilitates formation of oxygen vacancies. Vacancy formation exposes coordinatively unsaturated sites (*cus*) that display Lewis acidity and can serve as binding and activating sites for oxygen-bearing compounds. The reducibility of ceria is also believed to enable hydrogen spillover from Pd, further enhancing catalytic activity towards reduction of adsorbed species. Significant enhancement of phenol hydrogenation activity was found compared to that of Pd supported on non-reducible oxides such as silica (3x), or on supports capable of spillover but lacking Lewis acidity like carbon (2x), indicating non-innocent character of ceria for the reaction. Enhanced activity in hexane relative to water (2x), the latter capable of competing for *cus* binding sites, provided further evidence of the relevance of oxygen vacancies in the process. Spectroscopic and temperature programmed experiments provided further support to the involvement of *cus* in the reaction.[3] The capacity of ceria to undergo redox cycling also allowed using Pd/CeO_2 as a transfer hydrogenation catalyst replacing H_2 with alcohols as reducing equivalents. Modification of ceria with sodium salts led to formation of subsurface sodium carbonates that increased the number of surface defects (i.e. *cus*) giving even higher catalytic activity (3x that of Pd/CeO_2, Fig. 1a). Additional modification of ceria surface via deposition of phosphoric esters followed by partial hydrolysis led to stable Brønsted acidic sites that acted cooperatively with the Lewis (*cus*) sites to activate complex phenolics simultaneously for C-O bond cleavage and reduction of the aromatic ring (Fig. 1b).

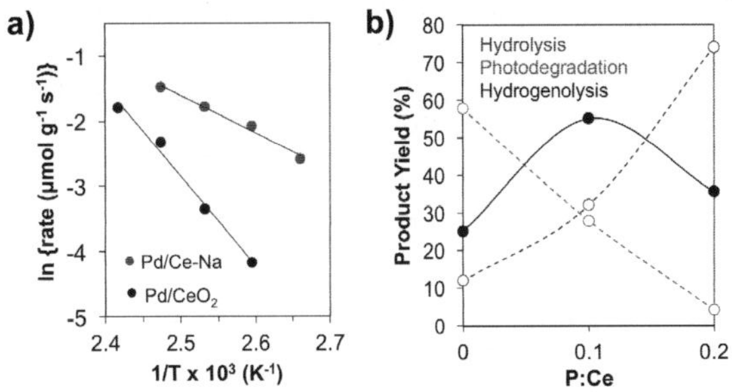

Figure 1. a) Arrhenius plots for the transfer hydrogenation of phenol catalyzed by Pd/CeO_2 with and without Na promotion, b) catalytic activity of Pd/CeO_2 as a function of P-modification.

CONCLUSIONS

Ceria was proven a non-innocent (i.e. active) support for the hydrogenation of phenols, and its activity was improved by incorporation of subsurface sodium carbonates. In addition, hydrogenolytic activity was developed on the material by controlled modification of its surface with phosphates.

REFERENCES

1. Fu, D. *et al.*, Bioresour. Technol. 154:101, 2014
2. Rase, H. Handbook of Commercial Catalysts. CRC press. p. 128, 2000
3. Nelson, N. *et al.*, ACS Catal. 5:2051, 2015
4. Nelson, N. *et al.*, J. Catal. 346:180, 2017
5. Nelson, N. *et al.*, J. Mater. Chem. A 5:*in press*, 2017

ACKNOWLEDGMENTS

This research was supported by the U.S. Department of Energy, through the Ames Laboratory Catalysis Science program (Contract No. DE-AC02-07CH11358).

Non-pincer-type Mononuclear Rare-earth Metal Alkylidene Complexes: Synthesis, Structure and Reactivity

Chen Wang and Yaofeng Chen*

State Key Laboratory of Organometallic Chemistry, Shanghai Institute of Organic Chemistry, Chinese Academy of Sciences, P. R. China, yaofchen@mail.sioc.ac.cn

INTRODUCTION

Mononuclear transition metal alkylidene (or carbene) complexes have attracted intense attention, and a great number of mononuclear transition metal alkylidene complexes have been synthesized in past decades.[1] One exception is those with rare-earth metal ions. Due to HOMO/LUMO orbital energies mismatch between the d^0 rare-earth-metal ions and the alkylidene groups, the formation of mononuclear rare-earth metal alkylidene complexes is unfavorable.[2] The alkylidene groups have a strong tendency to bind more than one rare-earth metal ion or one rare-earth metal ion and two other metal ions.[3] For the purpose of stabilization of mononuclear rare-earth metal alkylidene complexes, bis(iminophosphinoyl) methandiide dianion $[C(PPh_2NR)_2]^{2-}$ and bis(thiophosphinoyl) methandiide dianion $[C(PPh_2S)_2]$ were employed by Cavell, Liddle, Le Floch, Mézailles and Nief.[4] Introducing the pincer-type structure with two hypervalent phosphorous substituents stabilize the mononuclear rare-earth metal alkylidene complexes, but simultaneously reduce their reactivity.[5] How to balance stabilization and reactivity of mononuclear rare-earth metal alkylidene complexes remains a big challenge. Recently, we reported synthesis, structural characterization and reactivity of the first non-pincer-type mononuclear scandium alkylidene complexes.[6] Herein, we report a series of non-pincer-type mononuclear rare-earth alkylidene complexes, including the lanthanum one.

RESULTS AND DISCUSSION

β-diketiminato based tridentate ligand supported rare-earth metal methyl chlorides [LLn(Me)Cl] (L = $[MeC(NDIPP)CHC(Me)(NCH_2CH_2N(Me)_2)]^-$, DIPP = 2,6-($(i\text{Pr})_2C_6H_3$) were treated with $Li[CH(SiMe_3)PPh_2S]$ or $Li[CH(SiPh_3)PPh_2S](THF)$ to afford the non-pincer-type mononuclear rare-earth metal alkylidene complexes $[LLn\{C(SiR_3)PPh_2S\}]$ (**1**: Ln = Y, R = Me; **2**: Ln = Y, R = Ph; **3**: Ln = Lu, R = Me; **4**: Ln = Lu, R = Ph; **5**: Ln = Sm, R = Ph) via salt elimination and alkane elimination. The alkylidene complexes $[LLn\{C(SiR_3)PPh_2S\}]$ (**6**: Ln = Sm, R = Me; **7**: Ln = La, R = Ph) were synthesized from reactions of $[LLn\{C(SiR_3)PPh_2S\}Cl]$ with MeLi. All these complexes were characterized by the single crystal X-ray diffraction, which show the very short Ln-C bonds. Reactivity of these complexes was investigated, and the effects of the metal ion and alkylidene ligand on the reactivity were studied.

CONCLUSIONS

Non-pincer-type mononuclear rare-earth (Y, Lu, Sm, La) alkylidene complexes can be synthesized by using the suitable ancillary ligand and alkylidene ligand. Reactivity of these complexes was influenced by metal ion and alkylidene ligand.

REFERENCES

1. For some representative reviews of the transition metal alkylidene (or carbene) complexes, see: (a) Astruc, D. *New J. Chem.* **2005**, *29*, 42. (b) Elschenbroich, C. *Organometallics*, 3rd ed.; Wiley-VCH, Weinheim, Germany, 2006. (c) Schrock, R. R. *Angew. Chem., Int. Ed.* **2006**, *45*, 3748. (d) Grubbs, R. H. *Angew. Chem., Int. Ed.* **2006**, *45*, 3760.
2. Giesbrecht, G. R.; Gordon, J. C. *Dalton Trans.* **2004**, 2387.
3. For reviews of the rare-earth metal complexes containing bridging alkylidene groups, see: (a) Liu, Z. X.; Chen, Y. F. *Sci. Sin. Chim.* **2011**, *41*, 304. (b) Summerscales, O. T.; Gordon, J. C. *RSC Adv.* **2013**, *3*, 6682. (c) Kratsch, J.; Roesky, P. W. *Angew. Chem., Int. Ed.* **2014**, *53*, 376.
4. For reviews of the pincer-type rare-earth metal carbene complexes, see: (a) Liddle, S. T.; Mills, D. P.; Wooles, A. J. *Chem. Soc. Rev.* **2011**, *40*, 2164. (b) Fustier-Boutignon M.; Mézailles, N. *Top. Organomet. Chem.* **2014**, *47*, 63.
5. Mills, D. P.; Cooper, O. J.; McMaster, J.; Lewis, W.; Liddle, S. T. *Dalton Trans.* **2009**, 4547.
6. Wang, C.; Zhou, J. L.; X, F. Zhao.; Maron, L.; Leng, X. B.; Chen, Y. F. *Chem. Eur. J.* **2016**, *22*, 1258.

ACKNOWLEDGMENTS

The authors would like to thank the National Natural Science Foundation of China (Grant Nos. 21132002, 21325210 and 21421091), the State Key Basic Research & Development Program (Grant No. 2012CB821600), and the Strategic Priority Research Program of the Chinese Academy of Sciences. The authors also would like to thank Prof. Laurent Maron's group for DFT calculations.

Normal-Inverse-Normal Magnetocaloric effect across the first order transition in the electron doped manganite $Ho_{0.15}Ca_{0.85}MnO_3$

R. Nirmala[1*], Moumita Naskar[1], R. Rajivgandhi[1], A. K. Nigam[2], S. Quezado[3] and S. K. Malik[3]
[1]Department of Physics, Indian Institute of Technology Madras, Chennai 600 036 India
[2]Tata Institute of Fundamental Research, Mumbai 400 005, India
[3]Departamento de Física Teórica e Experimental, Universidade Federal do Rio Grande do Norte, Natal,59082-970, Brazil
nirmala@physics.iitm.ac.in

INTRODUCTION

Electron-doped perovskite manganites exhibit interesting physical properties such as colossal magnetoresistance, charge and orbital ordering, just like their hole-doped counterparts. In addition, some of them also show coexisting crystal structural and magnetic phase separation[1]. Neutron diffraction experiments on $R_{0.15}Ca_{0.85}MnO_3$ (R = Y, La and Sm) reveal that there is a cross-over from room temperature paramagnetic phase (orthorhombic) to a low temperature the low temperature antiferromagnetic (mixed monoclinic-orthorhombic) phase. The low temperature magnetic phase consists of a major C-type antiferromagnetic phase in monoclinic structure and a minor G-type antiferromagnetic phase together with a ferromagnetic component in orthorhombic structure[1, 2]. The structural phase separation survives even in 17 nm size nanocrystalline $La_{0.15}Ca_{0.85}MnO_3$ [3].

In the present work, we study the magnetic and magnetocaloric properties of $La_{0.15}Ca_{0.85}MnO_3$, We find a first order magnetic transition followed by region of metamagnetic transition across these the magnetocaloric effect (MCE) changes sign.

EXPERIMENTAL METHODS

Polycrystalline $Ho_{0.15}Ca_{0.85}MnO_3$ oxide (Orthorhombic, *Pnma*) has been prepared by solid state reaction and characterized by powder X-ray diffraction and SEM-EDAX experiments. DC magnetization data were measured using a SQUID magnetometer [MPMS XL, Quantum Design, USA] and vibrating sample magnetometer [PPMS Dynacool, Quantum Design, USA].

RESULTS AND DISCUSSION

DC magnetization data obtained in 100 Oe field reveal an antiferromagnetic transition at 114 K (T_N) [Fig. 1a]. The paramagnetic susceptibility follows Curie-Weiss law. The paramagnetic Curie temperature is +83 K indicating the presence of competing ferromagnetic interactions in the system. The effective paramagnetic moment is 5.4 μ_B/f.u. This transition signified by interesting hysteresis between field-cooled cooling and warming magnetization [Figs. 1a and b].

From the field dependent magnetization, M(H) data obtained in 10 K to 130 K temperature range, isothermal magnetic entropy change(ΔSm) is calculated. Isothermal magnetic entropy change vs temperature plot indicates interesting sign-changing MCE across the paramagnetic to antiferromagnetic and antiferromagnetic to field-induced ferromagnetic transitions [Fig. 2]. Metamagnetic transition occurs at temperatures below 115 K(~T_N). The magnetic moment at 2 K in 140 kOe field is 1.6 μ_B/f.u indicating that it is not fully ferromagnetically ordered state. The virgin curve lies outside the envelope curve in the M(H) data obtained at 2 K. The occurrence of normal-inverse-normal MCE is similar to that observed across the magnetostructural transition in $Ni_{50}Mn_{33.13}In_{13.90}$ single crystal[4].

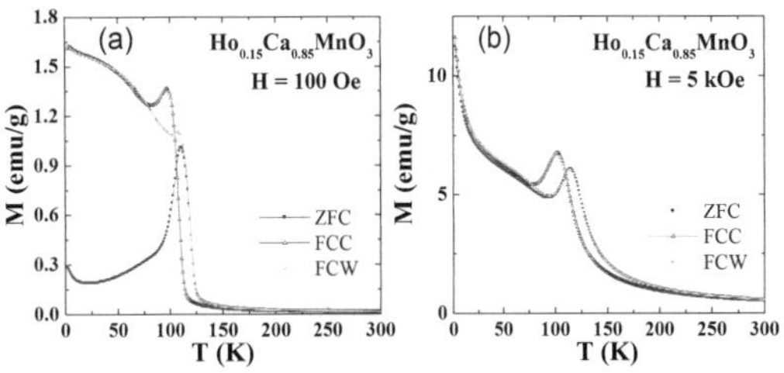

Fig. 1 DC magnetization vs temperature of $Ho_{0.15}Ca_{0.85}MnO_3$ in zero-field-cooled and field-cooled states in (a) 100 Oe and (b) 5 kOe applied fields.

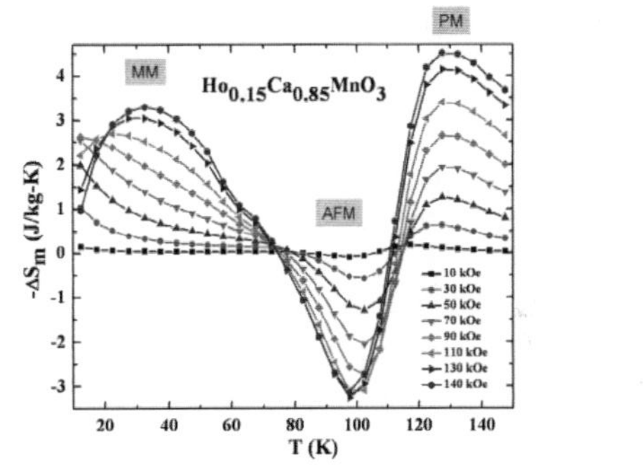

Fig. 2 Isothermal magnetic entropy change vs temperature of $Ho_{0.15}Ca_{0.85}MnO_3$ for various field changes.

SUMMARY

To summarize, the $Er_{0.75}Dy_{0.25}$ pseudo-binary system indeed presents interesting and complex magnetic behaviour in low applied fields within the magnetically ordered state just as intralanthanide dialuminides. This alloy also exhibits substantial MCE over broad temperature range.

REFERENCES

1. Tong. et al, Appl. Phys. Lett. **2008**, 93, 202504.
2. Mahendiran et al, Phys. Rev. B **62,** 11644 (2000).
3. Rini Thomas et al, J. Appl. Phys. **111,** 07D729 (2012).
4. Zhang et al, Phys. Rev B **76**, 132403 (2007).

ACKNOWLEDGMENTS

R. N thanks UGC-DAE-CSR for project support. R. N and M. N thank SAIF, IITMadras for extending SEM-EDAX facility. M. N thanks R. Mondal for help during synthesis. S.K.M. thanks CAPES for the award of a fellowship, MEC and MCTI (Brazil) for financial support during the course of this work.

Novel highly symmetrical class of lanthanide tetrakis-complexes $[Na(H_2O)_4][LnL_4]$ (HL -1,3-bis(1,3-dimethyl-1H-pyrazol-4-yl)-1,3-propandione) – synthesis, crystal structure and unique photophysical properties.

Ilya V. Taydakov[1-3 *], Igor Ch. Avetisov[2], Konstantin A. Lyssenko[3],
[1]S.I. Vavilov Department of Luminescence, P.N. Lebedev Institute of Physics of the Russian Academy of Sciences, Russia
[2]D. Mendeleev University of Chemical Technology of Russia, Russia
[3]A.N. Nesmeyanov Institute of Organoelement Compounds of the Russian Academy of Sciences, Russia
*taidakov@gmail.com

INTRODUCTION

Tetrakis-complexes are the least of all studied among all lanthanide 1.3-diketonates[1]. In the same time peculiarities of they structures make them an attractive objects for investigation of photophysical, crystallographic and magnetic properties. Recently we have developed a general synthetic approach to 1.3-diketones, bearing pyrazole moiety[2]. One of these ligands - 1,3-bis(1,3-dimethyl-1H-pyrazol-4-yl)-1,3-propandione appeared to be unique compound because of its complexation capability. Here we wish to report about new isostructural series of tetrakis-complexes based on this ligand and their unusual properties.

EXPERIMENTAL METHODS

An interaction between the above mentioned ligand, $Ln(NO_3)_3$ (Ln=Pr-Lu) and NaOH in water-ethanol media at reflux temperature gave $[Na(H_2O)_4][LnL_4]$ complexes as solvates with 1 mol of EtOH as single organic-soluble products (Fig. 1).

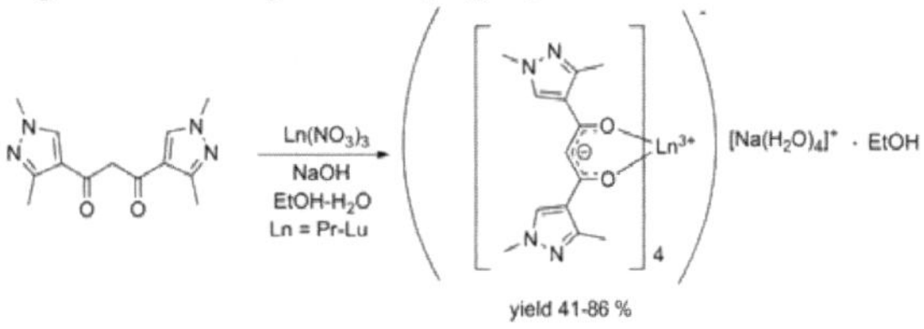

Fig. 1. Synthesis of $[Na(H_2O)_4][LnL_4]$ solvate.

Structures of complexes were determinated by single crystal XRD experiments after recrystallization from CH_2Cl_2.

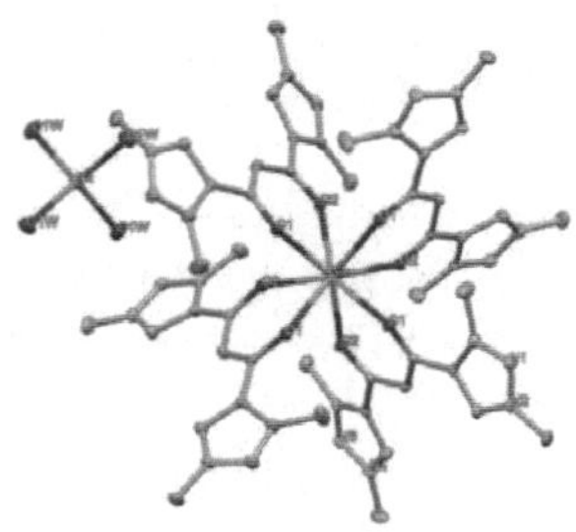

Fig. 2. Molecular structure of $[Na(H_2O)_4][LnL_4]\times0.75$ CH_2Cl_2.

An exchange of solvated molecules was obtained. The structure of Sm complex is given in the Fig. 2.

Integrated photophysical investigation was made for the range of compounds.

RESULTS AND DISCUSSION

The maximum yields were achieved if stoichiometric relation of reagents (4:1:4 mol) was used. In other cases the same complex was isolated along with insoluble hydroxo-salts of lanthanides.

Complexes with $[Li(H_2O)_4]^+$ cathions were obtained if NaOH was substituted by LiOH. Other alkali metals hydroxides gave complicated mixture of products or tetrakis-complexes with completely different structures. The ligand is unique – small variation in its structure (e.g., lack of CH_3-group at C(3) position of pyrazole ring) leads to formation of products of ligand degradation[3].

According to the data of XRD all complexes are isostructural both in powder and single crystal forms. The structures are highly symmetrical (space group *P4/n*, tetragonal cell).

All complexes are strongly luminescent in solid state and in solutions. Due to high symmetry of ligand field emission spectra are unusual. Europium compound spectrum does not show splitting of lines, while uncommonly complicated splitting were detected for Sm, Tb or Dy complexes. The investigation of magnetic properties is in progress.

CONCLUSIONS

The fist series of lanthanides isosructural complexes with 1.3-diketone bearing $[Na(H_2O)_4]^+$ cathion on external coordination sphere were prepared and extensively investigated. Due to high symmetry of ligands environment these compounds are revealed interesting objects for luminescent and magnetochemistry investigations.

REFERENCES.

1. Binnemans K. Handbook on the Physics and Chemistry of Rare Earths. 35:111-272, 2005.
2.Taydakov I. *et al.*, Eur. J. Inorg. Chem. 6: 1074-1082, 2015.
3.Taydakov I. *et al.*, Tet.Lett., 54: 1704-1706, 2013.

ACKNOWLEDGMENTS

This work was partially supported by the Russian Foundation for Basic Research (projects 16-03-00832a, 16-03-00691a) and by the Russian Science Foundation (grant № 14-13-01074).

Nucleation and Growth Kinetics for Near-Surface Intercalation of Rare Earth Islands in Graphite

Yong Han[1,2*], A. Lii-Rosales,[1,3] Y. Zhou,[1] M. Kim,[1] M. C. Tringides,[1,2] C.-Z. Wang,[1,2] P. A. Thiel,[1,3,4] and J. W. Evans[1,2]

[1]Ames Laboratory – U. S. Department of Energy, Ames, Iowa 50011, USA
[2]Department of Physics and Astronomy, Iowa State University, Ames, Iowa 50011, USA
[3]Department of Chemistry, Iowa State University, Ames Iowa 50011, USA
[4]Department of Materials Science and Engineering, Iowa State University, Ames Iowa 50011, USA
*Email: yhan.ameslab@gmail.com

INTRODUCTION

Intercalation compounds have long attracted interest because intercalation is a way to alter and tune the transport and magnetic properties of a bulk compound over a wide range.[1] Compounds formed by elemental metals in bulk graphite are particularly well-known where the intercalants are alkali metals, alkaline earths, and rare earths.[2, 3] However, intercalation at surfaces of bulk layered compounds has received very little attention. The specific possibility of metal intercalation at the surface of graphite presents several attractive possibilities. For example, it could be an opportunity to produce new surface properties in catalysis, magnetism, or friction at the surface of a cheap, robust, and plentiful material. It could be a way to form a protected two-dimensional layer of a reactive substance. In this vein, the rare earths are notably susceptible to oxidation. Motivated by these observations, the current study develops modelling for the nucleation and growth kinetics of near-surface intercalated islands of rare earths on graphite.[4]

THEORY AND SIMULATION METHODS

Density Functional Theory (DFT) using the plane-wave based VASP code is provides insight into key energetic parameters for the system of interest. Continuum deposition-diffusion equation analysis provides insight into the kinetics of the nucleation of intercalated islands. Stochastic lattice-gas (LG) modelling behaviour, which is analysed by kinetic Monte Carlo (KMC) simulation, provides more detailed and quantitative insight into both nucleation kinetics and also the growth of intercalated islands.

RESULTS AND DISCUSSION

Our modelling of the formation of intercalated rare earth islands in the gallery between the top and underlying graphene layer of graphite is based on the following picture. This process involves deposition of atoms onto the surface. Some fraction of these can then enter the gallery through well-separated point-like defects in the top layer, and subsequently diffuse within the subsurface gallery. Our focus is on the nucleation and growth kinetics of a single intercalated island fed by atoms from a nearby isolated point-source defect.

Continuum diffusion equation analysis provides some insight into the nucleation kinetics. In particular, we can determine the characteristic nucleation time and also the nucleation distance from the defect in terms of the key parameters, the rate of intercalation and the diffusion rate for intercalated atoms in the gallery. However, complementary tailored stochastic LG modelling provides a more comprehensive and quantitative characterization of both the nucleation and growth processes. For nucleation, we assess the broad distribution of nucleation times and nucleation positions. The LG modelling also characterizes island growth shapes. Motivation and support for the model is provided by STM observations of intercalated Dy islands on graphite formed at 800-900 K.[5] However, the model has more general applicability for intercalation of other metals.

CONCLUSIONS

A detailed characterization has been provided for the nucleation and growth of intercalated rare earth islands associated with the introduction of intercalated atoms in a novel "point-source" like geometry (corresponding to an isolated defect in the top layer of graphite). Thus, the formulation is quite different from standard treatments of island nucleation and growth on surfaces where atoms are uniformly added to the surface by random deposition.

REFERENCES

1. M. S. Dresselhaus and G. Dresselhaus, Adv. Phys. 51:1-186, 2002.
2. T. Enoki, M. Suzuki and M. Endo, *Graphite Intercalation Compounds and Applications*. Oxford University Press, Oxford, 2003.
3. M. E. Makrini, D. Gérard, P. Lagrange and A. Hérold, Physica B+C 99:481-485, 1980.
4. Y. Han *et al.*, in preparation.
5. Y. Zhou, A. Lii-Rosales, M. Kim, M. Wallingford, M. C. Tringides, C.-Z. Wang, and P. A. Thiel, submitted, 2017.

ACKNOWLEDGMENTS

This work was supported by the U. S. Department of Energy (USDOE), Office of Science, Basic Energy Sciences, Materials Sciences and Engineering Division. Research was performed at the Ames Laboratory, which is operated by Iowa State University under contract # DE-AC02-07CH11358.

On the Magnetic Properties of Ln^{3+} Based Nanoparticles

Frank C.J.M. van Veggel[1]

[1)]Department of Chemistry, University of Victoria
Victoria, BC, Canada, V8W 3V6
fvv@uvic.ca

INTRODUCTION

One of our main research thrusts in the field of colloidal nanomaterials and their applications is based on the trivalent lanthanide ions, which have unique optical and magnetic properties. We cover the range from synthesis and characterisation to (in vivo) applications, with a recent focus on MRI (Magnetic Resonance Imaging) for cancer diagnostics. The colloidal stability is typically obtained by a "soft" layer of organic molecules. To this purpose, we have developed synthesis techniques to make core-shell architectures which allows the magnetic properties to be optimized separately. Additionally, we do surface modifications by organic chemistry to impart compatibility with biological media and to impart binding specificity (e.g. through antibodies).

EXPERIMENTAL METHODS

The nanoparticles are made but was is called an arrested precipitation of $NaLnF_4$ nanoparticles, initially coated with a monolayers of oleates. Surface modification then transfer these colloidally stable nanoparticles to aqueous media, including pH buffers and serum enhanced growth media.

RESULTS AND DISCUSSION

The synthesis and basic characterisation of the nanoparticles will only briefly be discussed, but the core-shell synthesis will be discussed in some detail. Some emphasis will be on the challenges to prove the actual formation of core-shell structures (and the Ln^{3+} dopant distribution). The main focus of the talk is on their surface modification and use as magnetic probes for MRI. Some recent in vivo MRI studies will be included showing their potential as MRI contrast agent at high magnetic fields in brain, breast and prostate cancer.

CONCLUSIONS

Results show that potent MRI contrast agents can be made for T1 and T2 weighed imaging, with specificity obtained by bioconjugation of antibody.

REFERENCES

(1) Frank C.J.M. van Veggel; Upconversion Ln^{3+}-based Nanoparticles for Optical Bio-imaging (chapter 7) in "Luminescence of Lanthanide Ions in Coordination Compounds and Nanomaterials" Ana de Bettencourt Dias (editor), John Wiley & Sons, Ltd, Chichester, West Sussex, UK, 2014 (ISBN: 978-1-119-95083-7).

(2) J. Pichaandi and F.C.J.M. van Veggel; Near-Infrared Emitting Quantum Dots: Recent Progress on Their Synthesis and Characterization *Coord. Chem. Rev.* 2014, **263-264**, 138-150.

(3) F.C.J.M. van Veggel; Near-Infrared Quantum Dots, their delicate synthesis, challenging characterisation, and exciting potential applications *Chem. Mater.* 2014, **26**, 111-122.

(4) B. Blasiak, F.C.J.M. van Veggel, and B. Tomanek; Applications of nanoparticles for MRI cancer diagnosis and therapy *J. Nanomaterials* (review) 2013, **2013**, 148578 (12 pages).

(5) A. Gautam and F.C.J.M. van Veggel; Synthesis of Nanoparticles, their Biocompatibility, and Toxicity Behavior for Biomedical Applications *J. Mater. Chem. B* 2013, **1**, 5186-5200.

(6) N.J.J. Johnson and F.C.J.M. van Veggel; Sodium Lanthanide Fluoride Core-Shell Nanocrystals: A General Perspective on Epitaxial Shell Growth *Nano Research* 2013, **6**, 547-561.

ACKNOWLEDGMENTS

NSERC is gratefully acknowledged for funding. In addition we like to acknowledge our collaborators: Prof. B. Tomanek (Univ. of Alberta) and Dr. B. Blasiak (Univ. of Calgary).

On the Periodicity of Oxidation States in d- and f-Element Compounds

Shu-Xian Hu,[1] Wei Huang,[1] W. H. E. Schwarz,[1] Mingfei Zhou[2*] and Jun Li[1*]

[1]Theoretical Chemistry Center, Department of Chemistry, Tsinghua University, Beijing 100084, China

[2]Department of Chemistry, Fudan University, Shanghai 200433, China

*E-Mail: junli@tsinghua.edu.cn; mfzhou@fudan.edu.cn

INTRODUCTION

The oxidation state is a central concept in chemistry [1-3]. The highest known oxidation state in the Periodic Table is +VIII for neutral compounds, as exemplified in tetra-oxides MO_4 (M = Ru, Os, Ir, Xe) and +IX in IrO_4^+ cation [4]. The chemistry of lanthanides (Ln = La – Lu) is dominated by the low-valent +III or +II oxidation state due to the chemical inertness of the valence 4f electrons. The highest known oxidation state of the whole lanthanide series is +IV for Ce, Pr, Nd, Tb, and Dy [5].

RESULTS AND DISCUSSION

We have recently found that the pentavalent lanthanide oxide complexes of PrO_4 and PrO_2^+ species have the unprecedented Pr(+V) oxidation state, demonstrating that +V oxidation state and pentavalent state are viable for lanthanide elements under suitable coordination environment [6]. This new +V oxidation state and pentavalent chemical-valence state for lanthanides are also found in lanthanide nitride-oxides such as NPrO molecule and others recently [7].

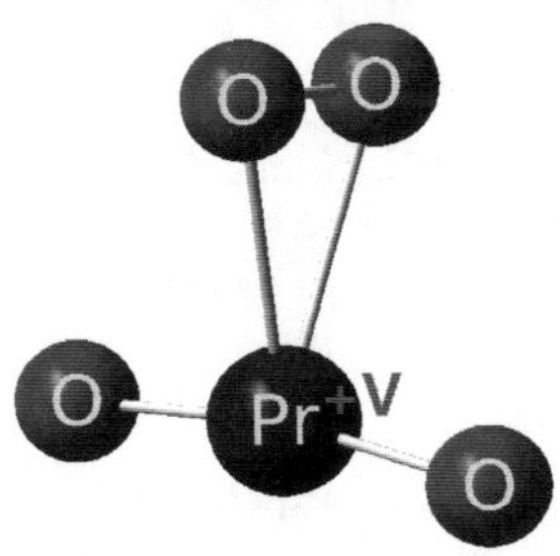

In this talk we will present an overview of the periodic trend of the oxidation states of the d- and f-block elements in the Periodic Table, with particular attention on the +V oxidation state of lanthanides. We will discuss the stability of pentavalent Pr(V) and summarize our understanding of the periodicity of oxidation states and its relationship with the electronic structures. The possibility of finding other high-valent lanthanide complexes will also be discussed.

REFERENCES

[1] P. Karen, P. McArdle, J. Takats, "Towards a Comprehensive Definition of Oxidation State (IUPAC Technical Report)", *Pure Appl. Chem.* 86: 1017-1081, 2014.

[2] H. Xiao, H.-S. Hu, W. H. E. Schwarz, J. Li, "Theoretical Investigations of Geometry, Electronic Structure and Stability of UO_6: Octahedral Uranium Hexoxide and Its Isomers", *J. Phys. Chem. A* 114: 8837–8844, 2010.

[3] W. Huang, W.-H. Xu, J. Su, W. H. E. Schwarz, J. Li, "Oxidation States, Geometries, and Electronic Structures of Plutonium Tetroxide PuO_4 Isomers: Is Octavalent Pu Viable?", *Inorg. Chem.* 52: 14237-14245, 2013.

[4] G.-J. Wang, M.-F. Zhou, J. T. Goettel, G. J. Schrobilgen, J. Su, J. Li, T. Schloeder, S. Riedel, "Identification of an Iridium-Containing Compound with a Formal Oxidation State of IX", *Nature*, 514: 475-477, 2014.

[5] J. Su, S.-X. Hu, W. Huang, M.-F. Zhou, J. Li, "On the Oxidation States of Metal Elements in MO_3^- (M = V, Nb, Ta, Db, Pr, Gd, Pa) Anions", *Sci. China-Chem.* 59: 442-451, 2016.

[6] Q.-N. Zhang, S.-X. Hu, H. Qu, J. Su, G.-J. Wang, J.-B. Lu, M.-H. Chen, M.-F. Zhou, J. Li, "Pentavalent Lanthanide Compounds: Formation and Characterization of Praseodymium(V) Oxides", *Angew. Chem. Int. Ed.* 55: 6896-6900, 2016.

[7] S.-X. Hu, J.-W. Jian, J. Su, X. Wu, J. Li, M.-F. Zhou, "The Pentavalent Lanthanide Nitride-Oxides: NPrO and $NPrO^-$ Complexes with N≡Pr Triple Bonds", *Chem. Sci.* 8: 4035-4043, 2017.

ACKNOWLEDGMENTS

This work is supported by the National Basic Research Program of China (2013CB834603), the National Natural Science Foundation of China (21433005, 21590792, 21571185, 21221062, 21201106 and 21501189) and NASF (U1530401). The calculations were performed using the Supercomputer Center of the Computer Network Information Center, the Chinese Academy of Sciences, and the Computational Chemistry Laboratory of Department of Chemistry, Tsinghua University.

One-step synthesis of nanostructured RE-Nitrides from RE-fluorides *via* gas-solid mechanochemistry

Jacob Rabe[1], Shalabh Gupta[2] and Vitalij K. Pecharsky[1,2]

[1]Department of Materials Science and Engineering, Iowa State University,
Ames, Iowa, 50011-2300, USA, jarabe@iastate.edu, Shalabh@ameslab.gov
[2]Division of Materials Science and Engineering, Ames Laboratory of US DOE, Iowa State University,
Ames, Iowa, 50011-3020, USA

INTRODUCTION

Rare-earth mononitrides (RENs) are ferromagnetic semiconductors that may have potential for applications in spintronic and low temperature magnetic refrigeration [1,2]. Common methods of synthesis involve reaction of RE-metals with gaseous N_2 or NH_3 at high temperatures that present a variety of scale-up issues, including formation of accidental impurities, and provide little control over particle size and microstructure. In the past, REN's have been synthesized via reactive milling of respective elements under N_2 atmosphere at room temperature [3,4]. While the method provides a direct route to synthesize pure RENs, it is essential that the RE metal precursors are protected from air at all times. Here we describe a new synthesis method in which the rare-earth fluoride salts (REF_3) are mechanochemically reduced by lithium metal under N_2 atmosphere yielding REN and LiF in a one-step process.

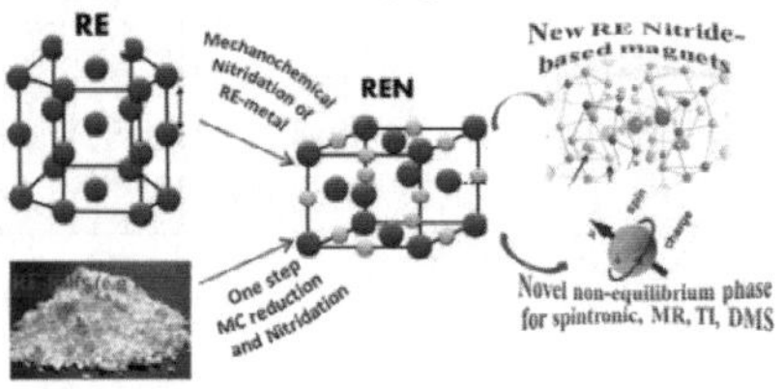

Schematic diagram showing possible mechanochemical routes of nitridation and applications

EXPERIMENTAL METHODS

RENs of Nd, Gd–Er, Lu and a few mixed RE nitrides were prepared either by ball-milling at room temperature or cryomilling REF_3 with Li-metal. In a typical process, REF_3 and Li-metal were taken in 1:3 molar ratio and the mixture was ball milled in a SPEX8000 mill for 60–120 min at room temperature under N_2 gas present at a pressure of slightly over 1 bar. The vials were opened intermittently inside the N_2 glove box to replenish the N_2 atmosphere. Once the reaction was completed, black fine powder was recovered, and further annealed under Ar atmosphere at 500 °C. The powders were then analyzed using XRD for phase analysis and to determine the lattice parameters of RENs.

RENs powders were also produced by the reaction of REF_3 and Li-metal in a cryomill (6870D Freezer/Mill®) operating at 77K. In this case the milling vessel was submersed in a liq. N_2 bath which provides the source of nitrogen. The XRD data were collected on a PANalytical diffractometer with Cu-$K_{\alpha 1}$ radiation and lattice parameters were obtained using the PowderCell 2.4 software.

RESULTS AND DISCUSSION

In all the cases presented here, the XRD analysis of the fine black powders obtained after 120 min of milling at RT shows only cubic REN (NaCl, Fm-3m) and LiF as crystalline products, indicating that the conversion was essentially complete. The refined lattice constant of as-synthesized RENs show a decreasing trend scaling well with the decreasing ionic radii of the rare-earth cation (+3, VI coordinated) across the lanthanide series from La-Lu. Furthermore, the lattice constants for the three mixed RENs prepared in this study also follow the trend precisely indicating that solid-solutions are formed and are homogeneous.

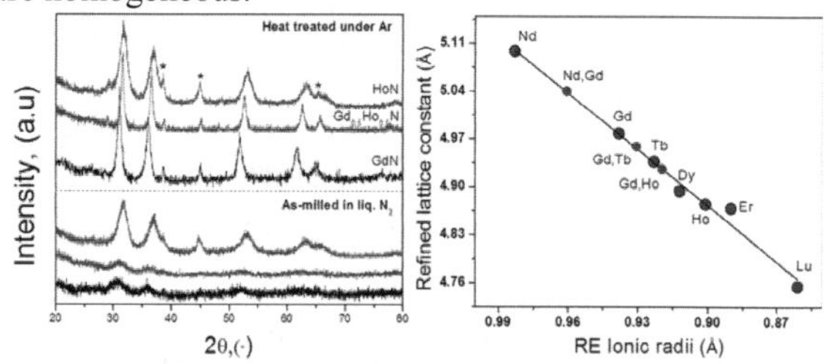

Fig. 1. (Left) X-ray diffraction patterns of representative GdN, $Gd_{0.5}Ho_{0.5}N$ and HoN as obtained after (bottom) cryomilling; and (top) after heat treatment. (*) denotes Bragg peak positions for LiF formed as by-product during the mechanochemical redox reaction. (Right) Refined lattice parameter plotted against ionic radii of RE

CONCLUSIONS

A facile and scalable one-step process is developed and successfully utilized to synthesize rare-earth mononitrides from their respective metal fluorides. Besides being rapid and efficient, the simultaneous mechanochemical reduction and nitridation process eliminates high temperature requirements, bypasses the separate metal reduction step, and avoids contamination from oxide and other impurities.

REFERENCES

1. F. Natali, *et al.* Prog. Mater. Sci. **58**, 1316 (2013)
2. Shinde K.P. *et al.* Dalton Trans., **44**, 20386 (2015).
3. Jaques B. *et al.* J. Alloys and Compd. **619**, (2015), 253.
4. Alanko G. *et al* J. Alloys and Compd. **620**, 413 (2015)

ACKNOWLEDGMENTS

The Ames Laboratory is operated for the U. S. Department of Energy by Iowa State University of Science and Technology under contract No. DE-AC02-07CH11358.

Orbital Mixing and Multiconfigurational Ground States in Lanthanide Organometallics

Corwin H. Booth,[1*] Robert L. Halbach,[2] Grégory Nocton,[1,2,3] Laurent Maron[4] and Richard A. Andersen[1,2]

[1*]Chemical Sciences Division, Lawrence Berkeley National Laboratory, USA, chbooth@lbl.gov
[2]Department of Chemistry, University of California, Berkeley, USA
[3]LCM, CNRS, Ecole Polytechnique, Université Paris-Saclay, France
[4]LPCNO, UMR-5215, Université de Toulouse-CNRS, INSA, France

INTRODUCTION

Although 4f orbitals were widely considered not to be involved in bonding in lanthanide organometallic systems, Fulde and co-workers conjectured in 1989 that orbital mixing could contribute to the bonding in certain Ce and Yb compounds.[1] A theoretical description was then developed[2] using configuration interaction methods that have been supported by, for example, x-ray absorption spectroscopy and magnetic susceptibility measurements.[3-5] In this presentation, work in this area will be reviewed, highlighting the successes in both Ce and Yb organometallics and the special remaining challenges in Ce compounds.

EXPERIMENTAL METHODS

Recent results describing the crystal structures, x-ray absorption near-edge structure (XANES) spectra, and magnetic susceptibility of cerium tetra(tropolonate), cerium tetra(acethylacetonate), and cerium bis(tetraazamacrocycle) cerium will be presented and contrasted to previous results on a large series of bis(cycloepentadienyl) ytterbium compounds and on bis(cyclooctatetraene) cerium, also known as cerocene.

RESULTS AND DISCUSSION

Previous studies on various ytterbium organometallic compounds have shown remarkable agreement between Complete Active Space (CAS) calculations and experiment.[4,5] In particular, Yb L_3-edge XANES measurements of the f-occupancy (n_f) are generally within a few percent of the calculated values. Likewise, the relative energy levels and magnetic state of the ground state as determined by the magnetic susceptibility are also in agreement with calculation.

The situation in Ce organometallics is more ambiguous. The canonical example is cerocene. The consensus currently acknowledges the importance of configurational mixing.[2,3,6] Although some discrepancy exists between calculations regarding n_f,[6] these have been ascribed[7] to an energy-invariant rotation between the choice of single- or multi-electron wavefunctions, leading to either a predominantly Ce(IV) ground state with significant 4f covalency or a predominantly Ce(III) ground state with Ce(IV) admixture.

While the work on cerocene still qualitatively highlights the importance of multiconfigurations, the quantitative agreement with experiment is not as clear as in the work on the Yb organometallics. Expanding this research to the formally Ce(IV) compounds described above, we find that, like cerocene, they all exhibit temperature-independent paramagnetism (TIP). Likewise, CASSCF calculations indicate a singlet ground state, but in contrast to cerocene, n_f is less than 0.2. However, the energy gap to the first excited state, which presumably is responsible for the TIP through a Van Vleck mechanism, increases with decreasing susceptibility, in support of the calculations.

Interestingly, while the calculation and the magnetic results agree well regarding the energy gaps and the ground magnetic state, the Ce L_3-edge XANES indicate an n_f closer to 50%. In addition, the spectra do not change monotonically with either the calculated first excited state gap or the susceptibility. This disagreement is in contrast to the larger range of results on Yb organometallics where all three properties are in agreement.[4,5] This result is, however, in agreement with the description[7] whereby the Ce f-orbital occupancy is not uniquely determined within the CAS formalism.

CONCLUSIONS

Theoretical and experimental results on Ce and Yb organometallic compounds demonstrate the unusual role of f-electrons in bonding. While a wide range of behaviour in the Yb organometallics is modeled well by populating multiple configurations with different f-orbital occupancies, the bonding is less well described in Ce organometallics, although the energetics are relatively accurate. This discrepancy may be due to the greater extent of the f-orbital in Ce compounds.

REFERENCES

1. Neumann C.-S. and Fulde P., Z. Phys. B 74:277, 1989.
2. Dolg M. *et al.*, Chem. Phys. 195:71, 1995
3. Booth C. H. *et al.*, Phys. Rev. Lett. 95:267202, 2005.
4. Booth *et al.*, J. Am. Chem. Soc. 131:6480, 2009.
5. Booth *et al.*, J. Am. Chem. Soc. 132:17537, 2010.
6. Kerridge A. *et al.*, J. Phys. Chem 113:2896, 2009.
7. Moossen O. and Dolg M., Chem. Phys. Lett. 594:47 2014.

ACKNOWLEDGMENTS

Work was supported by the U.S. Department of Energy (DOE) under contract DE-AC02-05CH11231.

Organothorium Pnictogen Chemistry

Justin R. Walensky

[1]Department of Chemistry, University of Missouri, Columbia, MO, USA
walenskyj@missouri.edu

INTRODUCTION. The organometallic chemistry of the actinides has experienced a renaissance in the past ten years but the coordination chemistry of actinides with pnictogens has lagged significantly. Recently, the chemistry of pnictogens has received more attention.[1,2,3] We have started to examine the coordination chemistry and reactivity of organometallic thorium complexes with phosphorus and arsenic ligands.[4]

RESULTS AND DISCUSSION. We started with the synthesis of primary phosphido and arsenido, Figure 1, complexes with thorium, $(C_5Me_5)_2Th[EH(2,4,6\text{-}R_3C_6H_2)]_2$ by either salt metathesis or protonolysis. The reactivity of these complexes has been investigated with ^{t}BuNC, $KN(SiMe_3)_2$, as well as $Ni(CO)_2(PPh_3)_2$.

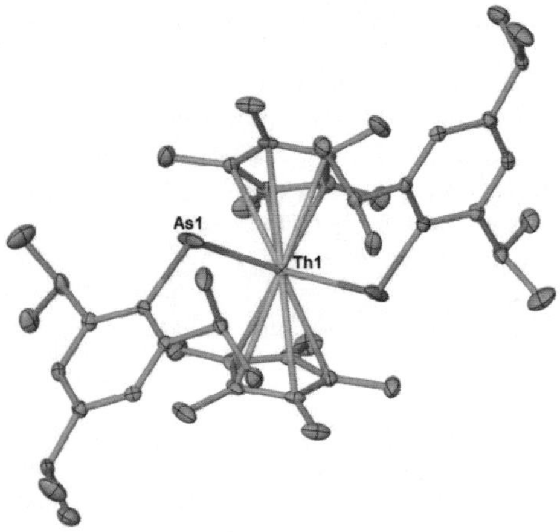

Figure 1. Thermal ellipsoid plot of $(C_5Me_5)_2Th[AsH(2,4,6\text{-}^iPr_3C_6H_2]_2$ shown at the 50% probability level. Hydrogen atoms have been omitted for clarity.

REFERENCES

[1] Gardner, B. M.; Balazs, G.; Scheer, M.; Tuna, F.; McInnes, E. J. L.; McMaster, J.; Lewis, W.; Blake, A. J.; Liddle, S. T. Nat. Chem. 2015, 7, 582-590.

[2] Gardner, B. M.; Balazs, G.; Scheer, M.; Wooles, A. J.; Tuna, F.; McInnes, E. J. L.; McMaster, J.; Lewis, W.; Blake, A. J.; Liddle, S. T. Angew. Chem. Int. Ed. 2015, 54, 15250-15254.

[3] Wildman, E. P.; Balazs, G.; Wooles, A. J.; Scheer, M.; Liddle, S. T. Nature Commun. 2016, 7, 12884.

[4] Behrle, A. C.; Castro, L.; Maron, L.; Walensky, J. R. J. Am. Chem. Soc. 2015, 137, 14846-14849.

ACKNOWLEDGMENTS

J.R.W. gratefully acknowledges support for this work from the U.S. Department of Energy, Office of Science, Early Career Research Program under Award Number DE-SC-0014174.

Outer-sphere Coordination of Counter-ions in Lanthanide(III) Complex with Diglycolamide Ligands

Derek M. Brigham, Bruce A. Moyer, Lætitia H. Delmau, Vyacheslav Bryantsev, Alexander S. Ivanov and Ross J. Ellis

Oak Ridge National Laboratory, USA, brighamdm@ornl.gov

INTRODUCTION

The efficient separation of the rare earth ions has been identified as one of the most important technological challenges of our time.[1] Solvent extraction allows for high throughput of material. However, the low separation factors and high acid-base consumption of cation exchanging ligands employed industrially provide incentive for improvements. Neutral diglycolamide (DGA) ligands have been investigated extensively as extractants for f-elements.[2] Our study investigates the separation of adjacent lanthanides praseodymium and neodymium employing the solvating extractant *N,N,N′,N′*-tetra(*n*-octyl)diglycolamide (TODGA). The structure of the DGA ligands with the lanthanides was investigated with extended X-ray absorption fine structure (EXAFS) studies as well as theoretical studies employing density functional theory (DFT) calculations and classical molecular dynamics simulations.

RESULTS AND DISCUSSION

Liquid-liquid extraction studies of Pr^{3+} and Nd^{3+} out of nitric and hydrochloric acid by TODGA show significantly stronger extraction out of nitrate media. Slope analysis of the extraction systems containing phase modifier indicates a 1:3 metal-to-ligand stochiometry regardless of counterion. To investigate the coordination structure of the extracted species and the effect of anion on this structure, Nd L3-edge EXAFS measurements were employed. The EXAFS modeling suggests the presence of a homoleptic $[Nd(TODGA)_3]^{3+}$ inner-sphere complex. It also indicates a trefoil-shaped ion cluster with outer-sphere coordination of the counter-ions. DFT calculations for the complex of tetraethyl DGA with Pr^{3+} and Nd^{3+} provide bond distances that closely match those found in the EXAFS modeling for the TODGA complex. Molecular dynamics simulations of $Nd(Cl_3)(TODGA)_3$ show the same trefoil-shaped cluster with the chloride ions located in clefts formed between the three coordinating TODGA ligands.

CONCLUSIONS

From the liquid-liquid extraction, X-ray absorption spectroscopy (XAS), and simulation studies we find new structural insight into the outer-sphere ion clusters that form between lanthanide(III) ions and diglycolamide ligands. We found that the structure of the inner-sphere DGA-Ln(III) complexes, the stoichiometry of the extracted species, and selectivity between adjacent lanthanides is independent of anion type and organic solvent environment. Conversely, the strength of the extraction was significantly impacted by anion type and solvent character, which we attributed to outer-sphere effects in stabilizing the ion cluster in the low dielectric-constant solvent.

REFERENCES

1. Sholl, D. S. *et al.*, *Nature* **2016**, 532, 435-437.
2. Ansari, S. *et al.*, *Chem. Rev.* **2012**, 112, 1751-1772.

ACKNOWLEDGMENTS

Research funded by the Critical Materials Institute, an Energy Innovation Hub funded by the U.S. Department of Energy, Office of Energy Efficiency and Renewable Energy, Advanced Manufacturing Office.

PARASHIFT probes: the importance of coordination environment on the NMR shifts of lanthanide complexes.

Kevin Mason[1*], Elizaveta A. Suturina[2], Ilya Kuprov[2], Michele Vonci[3], Nicholas F. Chilton[3], Eric J. L. McInnes[3] and David Parker[1]

[1*]Department of Chemistry, Durham University, South Road, Durham UK, DH1 3LE, kevin.mason@durham.ac.uk
[2]Department of Chemistry, University of Southampton, Highfield, Southampton UK SO 17 1BJ.
[3]School of Chemistry, The University of Manchester, Oxford Road, Manchester, M13 9PL.

INTRODUCTION

Paramagnetic lanthanide complexes can be used as chemical shift agents in magnetic resonance imaging. By shifting a reporter group away from the endogenous water and fat signals we can selectively observe complex resonances with zero background. The benefit of the fast relaxation induced by the paramagnetic lanthanide complexes is a much shorter image/spectral acquisition time. The relative magnitude of the shift and relaxation can be roughly predicted by established methods, namely Bleaney and BRW theories. Recently, there have been reports of complexes that do not follow the spectral behaviour predicted by these theories.[1, 2] A more precise and inclusive theory that correlates magnetic and ligand field anisotropy is needed. Thus, an in-depth understanding of the solution dynamics, molecular structure, electronic structure and magnetic properties of well-defined series of lanthanide complexes is required.

RESULTS AND DISCUSSION

A number of isostructural lanthanide series based on 1,4,7-triazacyclononane (9-N_3) and 1,4,7,10-tetraazacyclododecane (12-N_4) (Fig.1) have been investigated. These systems define an array of coordination environments that allow an assessment of the underlying contributions responsible for deviations in the current theories. Newly attained X-ray crystal structures of the 12-N_4 complexes allow an accurate determination of bond lengths and angles of the coordination environment in solid state that can then be used in conjunction with quantum chemical calculations to rationalise the solution NMR data.[3]

Pure shift ^{1}H NMR and EXSY spectroscopy allowed the dynamic exchange between the (RRR)-Δ-(δδδδ) and (RRR)-Λ-(λλλλ) TSAP diastereoisomers of [Ln.L^8] to be measured. By reviewing Bleaney's theory using an *ab initio* ligand field approach it has been shown that the ligand field in an isostructural 9-N_3 series can vary across the series. This calculation has been backed up with evidence from NMR, EPR, SQUID magnetometry as well as optical techniques.

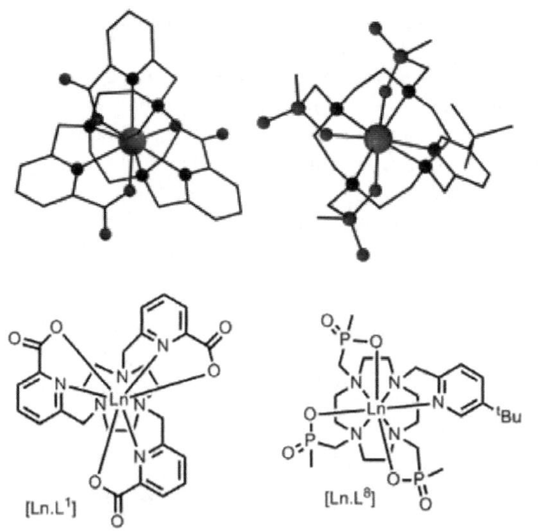

Fig. 1 Molecular structure of (left) 9-N_3-based [Ln.L^1] and (right) 12-N_4-based [Ln.L^8].

CONCLUSIONS

A combination of solid state, solution studies and quantum calculations has been applied to a wide range of lanthanide systems. The results shed light on the structure and solution dynamics of the PARASHIFT NMR contrast agents, an important part of ascertaining why they are so suited for the role. The 9-N_3 complexes which do not conform to Bleaney's theory have been extensively investigated and the subtle variation in electronic structure explains the psuedocontact shift behaviour.

REFERENCES

1. Funk, A. M.; Finney, K.-L. N. A.; Harvey, P.; Kenwright, A. M.; Neil, E. R.; Rogers, N. J.; Senanayake, P. K.; Parker, D. Chemical Science, 6 (3): 1655-1662, 2015.
2. Funk, A. M. ; Harvey, P.; Finney, K.-L. N. A.; Fox, M. A.; Kenwright, A. M.; Rogers, N. J.; Senanayake, P. K.; Parker, D. Phys. Chem. Chem. Phys., 17 (25), 16507-16511, 2015.
3. Mason, K.; Rogers, N.J.; Suturina, E. A.; Aguilar, J. A.; Batsanov, A. S.; Yufit, D. S.; Kuprov, I; Parker, D. Inorg. Chem., Accepted, 2017.

ACKNOWLEDGMENTS

The authors would like to thank ESPRC (Grant no: EP/N006909/1) for providing financial support to this project.

Periodicity and the Chemistry of Protactinium

Richard E. Wilson[1*] and Valerie Vallet[2]

[1]Chemical Sciences and Engineering Division, Argonne National Laboratory, Argonne Illinois USA
[2]University of Lille, CNRS, UMR 8523 –PhLAM- Physique des Lasers Atomes et Molecules, 59000 Lille France
rewilson@anl.gov

INTRODUCTION

Prior to their identification as a series of f-elements, the actinides were thought to be a fourth row of the transition metal series, principally because the oxidation states of the early actinide elements were similar to those encountered in their transition metal homologues. The rich redox chemistry observed in the actinides is in contrast to the predominantly trivalent oxidation states encountered in the lanthanide series. Unlike the lanthanides and transition metals, the chemistry of the early actinide elements is governed by the interplay of both the 5f and 6d electrons. The relative energies of the electrons and orbitals changes as Z is increased causing a periodic stabilization of the 5f orbitals relative to the 6d resulting in different chemistries being observed for these elements, influencing molecular structure and oxidation reduction properties. In moving from Th, to Pa, to U, the chemistry and bonding transitions from that of considerable d-character at Th to predominantly f-character at U. Protactinium is at the chemical intersection of this phenomenon, and studying its chemistry is critical to understanding the roles of both the 5f and 6d electrons in the chemistry of the actinide elements. We are exploiting both experimental and computational methods to evaluate the chemistry of protactinium, its relationship to the transition metal d-chemistries of Nb and Ta, and its relationship to the f-chemistries of the later actinides.

Most recently we have focused on the chemistry of protactinium in alkaline aqueous solution and its comparison to the polyoxometalate chemistry of niobium and tantalum. Both experimental and computational studies have focused on the formation of actinide polyoxometalates and in particular the role of the 5f electrons and orbitals in these systems.

RESULTS AND DISCUSSION

Prior studies using tracer concentrations of Pa(V) have suggested that an anionic complex of Pa exists under alkaline conditions[1]. Such a complex would suggest that Pa is an amphoteric metal, like Nb and Ta. Experiments in our laboratory were unable to reproduce the soluble anionic complex observed previously. Attempts however were made to access this unknown species by the preparation of soluble protactinium peroxo complexes, much like those known for Ta and Nb as shown in Figure 1a. Unstable solutions of Pa in alkaline peroxide were observed but we were unable to crystallize any species from these solutions.

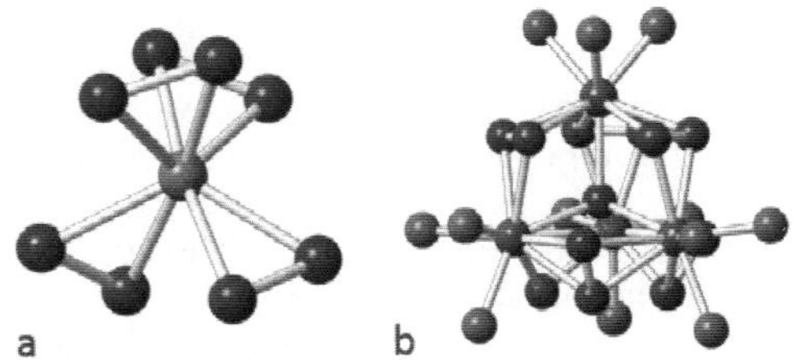

Alternatively, using our previously reported fluoride complexes[2-3] of Pa(V), we were able to synthesize and crystallize a polynuclear Pa complex like that shown in Figure 1b. Quantum chemical computations were conducted to evaluate the stability and electronic structure of this complex and such complexes formed with the transition metals.

Additionally, we explored the stability and electronic structure of a series of actinide polyoxometalates synthesized *in-silico* and compared them to the known transition metal polyoxometalates.

CONCLUSIONS

These results demonstrated a series of clear periodic trends in the electronic structure as the chemistry transitioned between transition metal and actinide chemistries. Importantly significant differences in the role of the 5f and 6d orbitals and electrons were observed between protactinium and uranium.

REFERENCES

1.In *Physico-chimie du protactinium*, Physico-chimie du protactinium, Orsay, France, Bouissieres, M.; Muxart, M., Eds. Centre National de la Recherche Scientifique: Orsay, France, 1965; p 351.

2.De Sio, S. M.; Wilson, R. E., Structural and Spectroscopic Studies of Fluoroprotactinates. *Inorg. Chem.* **2014,** *53* (3), 1750-1755.

3.Wilson, R. E.; De Sio, S.; Vallet, V., Structural and Electronic Properties of Fluoride Complexes of NbV, TaV, and PaV: The Influence of Relativistic Effects on Group V Elements. *Eur. J. Inorg. Chem.* **2016,** *2016* (35), 5467-5476.

ACKNOWLEDGMENTS

This work was performed at Argonne National Laboratory, operated by UChicagoArgonne LLC for the United States Department of Energy, Office of Science, Office of Basic Energy Sciences under contract no DE-AC02-06CH11357.

Phosphinoboranate Reactivity and Coordination Chemistry with Uranium and Lanthanides

Anastasia V. Blake, Zachary Theiler, Scott R. Daly

Department of Chemistry, University of Iowa, USA, scott-daly@uiowa.edu

INTRODUCTION

Actinide and lanthanide borohydride complexes have exceptional chemical properties and fascinating structures.[1] Their inner-coordination spheres are comprised mostly or exclusively of metal-bound hydrogen atoms. Many are highly volatile, which led to their use in thin film deposition processes (CVD) and separations. A historically important example of the latter involved $U(BH_4)_4$, which was considered for gaseous ^{235}U enrichment during the Manhattan project because of its exceptionally high volatility. Ultimately, UF_6 was selected to avoid issues associated with the large neutron cross-section of boron-10.

We recently began exploring the chemistry of *f*-element borohydride complexes with soft-donor chalcogen ligands. A broad objective of our work is to understand how different ancillary ligands affect metal-ligand soft-donor binding, which is implicated in effective lanthanide/actinide separations. Actinide borohydride complexes are somewhat unusual because they are known to form isolable complexes with weakly donating neutral soft donors such as thioethers. We postulated that covalent metal-ligand bonding between borohydrides and *f*-elements, as proposed by Ephritikhine, Gagliardi, and others,[2,3] may promote increased metal-chalcogen binding affinity. Here we describe our preliminary efforts to test this hypothesis.

RESULTS AND DISCUSSION

To investigate the bonding preferences of *f*-elements with borohydrides and soft chalcogen donors, we explored the reactivity of $K[EP^tBu_2BH_3]$ (where E = S, Se, or Te) with uranium and lanthanide halides (Scheme 1). The $K[EP^tBu_2BH_3]$ salts were first reported by Wagner *et al.* in 2006,[4] but their coordination chemistry is relatively unexplored. We postulated that having chalcogen and borohydride groups in the same ligand would provide a means to investigate their relative binding preference for the metal. The ligands have the option of binding the metal via borohydride, chalcogen, or both. For example, it has been shown that $[SP^tBu_2BH_3]^-$ binds to $CpFe(CO)_2$ via the sulphide only.[5]

Scheme 1

In a representative reaction, $UI_3(1,4\text{-dioxane})_{1.5}$ was treated with three equivalents of $K[SeP^tBu_2BH_3]$. The reaction unexpectedly yielded a mixture of products, and crystals grown from the mixture revealed that the ligand disproportionated to form a new diselenophosphinate $U(Se_2P^tBu_2)_4$. Similar reactions with $K[SP^tBu_2BH_3]$ and $K[TeP^tBu_2BH_3]$ also appeared to yield complex mixtures.

Given the propensity of the $[EP^tBu_2BH_3]^-$ ligands to disproportionate, and seeking to know the fate of the borane, we tested the same reaction with the chalcogen removed. Treating $UI_3(1,4\text{-dioxane})_{1.5}$ with three equivalents of $K[P^tBu_2BH_3]$ again resulted in ligand disproportionation to yield dimeric $U_2(H_3BP^tBu_2BH_3)_6$, the first example of an actinide phosphinodiboranate (Fig. 1). We have since prepared this compound by a more direct synthesis using $K[H_3BP^tBu_2BH_3]$.

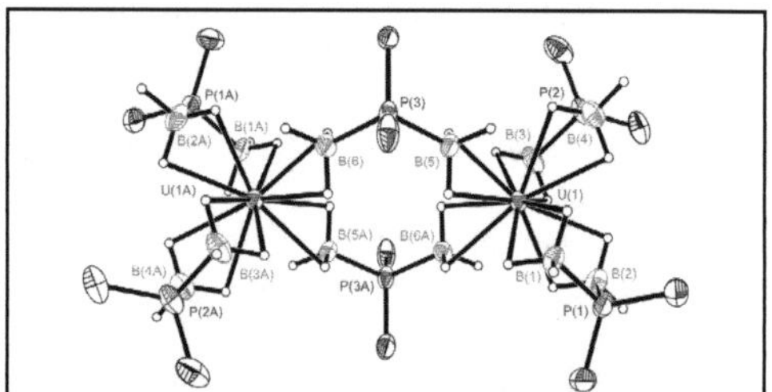

Fig. 1. Structure of $U_2(H_3BP^tBu_2BH_3)_6$. The ^{t}Bu groups were omitted for easier viewing.

CONCLUSIONS

We have prepared several new uranium complexes via disproportionation of the phosphinoboranates $[EP^tBu_2BH_3]^-$ and $[P^tBu_2BH_3]^-$. The first example of an actinide phosphinodiboranate was isolated and authenticated by single-crystal X-ray diffraction studies. Reactions with lanthanide halides are currently in progress and will be described.

REFERENCES

1. Ephritikhine, M. Chem. Rev. 97: 2193-2242, 1997.
2. Ephritikhine, M. *et al.* Inorg. Chem. 48:221-230, 2009.
3. Gagliardi, L. *et al.* J. Phys. Chem. C 116: 23194-23200, 2012.
4. Wagner, M. *et al.*, Eur. J. Inorg. Chem. 5138-5147, 2006.
5. Wagner, M. *et al.*, Eur. J. Inorg. Chem. 1989-2003, 2007.

ACKNOWLEDGMENTS

The authors would like to thank the University of Iowa for providing financial support. AVB would like to thank the Nuclear Regulatory Commission for a graduate fellowship.

Photocatalytic Carbon–Carbon Bond Formation Using a Luminescent Eu(II)-Containing Complex and Visible Light

Michael E. Cross, Matthew D. Bailey, Tyler C. Jenks, and Matthew J. Allen*

Department of Chemistry, Wayne State University, United States of America
* mallen@chem.wayne.edu

INTRODUCTION

Divalent lanthanides are useful reducing agents in organic synthesis. SmI_2 has found routine use in synthetic organic chemistry, owing to its highly reducing ground state;[1] however, divalent europium has seen less use due to its relatively positive ground state potential. In this work, we present the use of a nitrogenous cryptand that chelates $EuCl_2$ in methanol and facilitates the intense absorption of visible light that lead to reductive coupling reactions that do not proceed in the dark. We also present the use of a sacrificial reducing agent to make the system catalytic.

RESULTS AND DISCUSSION

A nitrogenous derivative of 2.2.2-cryptand bathochromically shifts the absorption of $EuCl_2$ from the UV to the visible. This absorption band is accompanied by bright yellow luminescence in solution. Based on luminescence decay measurements this excited state has a lifetime of 16.3 ± 0.5 µs, and a calculated reducing potential of –2.9 V based on the Rehm–Weller formalism.[2]

The coupling reaction of two equivalents of benzyl chloride to form 1,2-diphenylethane was evaluated as a synthetic target, inspired by recent work with tetravalent cerium.[3] The reaction proceeded when the nitrogenous ligand, $EuCl_2$, and blue illumination from light-emitting diodes; however, removal of any one of these does not lead to observable product. To make the system catalytic, zinc was used as a sacrificial reductant. Control reactions indicated that zinc is not able to promote the reductive coupling on its own.

Stern–Volmer analysis was performed using a variety of alkyl and aryl substrates, and indicated that the excited state is quenched when there is an extended pi system on the substrate, but not in the absence of a pi system. These studies also provide evidence that the mechanism is proceeding through an outer-sphere mechanism.

CONCLUSIONS

The europium-containing complex presented promotes reductive coupling reactions in the presence of light and is catalytic in europium. The system is tolerant of water and some air. We suspect that this system will be of interest to synthetic chemists in addition to the community of scientists studying the rare earth elements.

REFERENCES

1. Miller R. S. *et al.*, J. Am. Chem. Soc. 122:7718–7722, 2000
2. Flowers R. A. II, Synlett 10:1427-1439, 2008
3. Yin H. *et al.*, J. Am. Chem. Soc. 137:9234–9237, 2015

ACKNOWLEDGMENTS

The authors thank the National Science Foundation (CHE-1564755) for providing financial support to this project.

Photoluminescence properties and energy level of RE (RE = Pr, Sm and Tb) in $Y_4Si_2O_7N_2$

Guo-Jun Wang[a], Guang-Xing Xiang[a], Zhi-Jun Zhang[a,*], Hubertus T. Hintzen[b] and Jing-Tai Zhao [a,*]

[a]*School of Materials Science and Engineering, Shanghai University, Shanghai, 200072, P. R. China*
[b]*Luminescent Materials Research Group, Delft University of Technology, Mekelweg 15, 2629 JB Delft, the Netherlands.*
Corresponding author: Zhi-Jun Zhang, Jing-Tai Zhao
Email: zhangzhijun@shu.edu.cn, jtzhao@shu.edu.cn

Abstract:

RE^{3+} (RE = Pr, Sm and Tb)-activated $Y_4Si_2O_7N_2$ were synthesized by a solid-state reaction method at high temperature, and their photoluminescence properties were studied. The direct Pr^{3+} $4f^2 \rightarrow 4f^1 5d^1$ excitation at 275 nm leads to strong $4f^1 5d^1 \rightarrow 4f^2$ emission band (300-450 nm) and typical $4f^2 \rightarrow 4f^2$ emission (450-700 nm) lines, respectively. The charge transfer band of Sm^{3+} was observed at a rather lower energy of 4.68 eV compared to oxide, and Sm^{3+}-activated sample shows a bright red emission originating from $^4G_{5/2} \rightarrow {}^6H_J$ (J = 5/2, 7/2 and 9/2) transitions. The direct Tb^{3+} $4f^8 \rightarrow 4f^7 5d^1$ excitation leads to $^5D_3 \rightarrow {}^7F_J$ (J = 6, 5, 4, 3) (blue) and $^5D_4 \rightarrow {}^7F_J$ (J = 6, 5, 4, 3) (green) line emissions in Tb^{3+}-doped sample, and cross-relaxation depended on Tb concentration also has happened. The energy level diagram containing the position of $4f$ and $5d$ energy levels of all Ln^{2+} and Ln^{3+} ions relative to the valence and conduction band of $Y_4Si_2O_7N_2$ has been established and researched, based on the data presented in this work, and further provides a platform for studying the photoluminescence properties as well as the valence stability of the lanthanide ions.

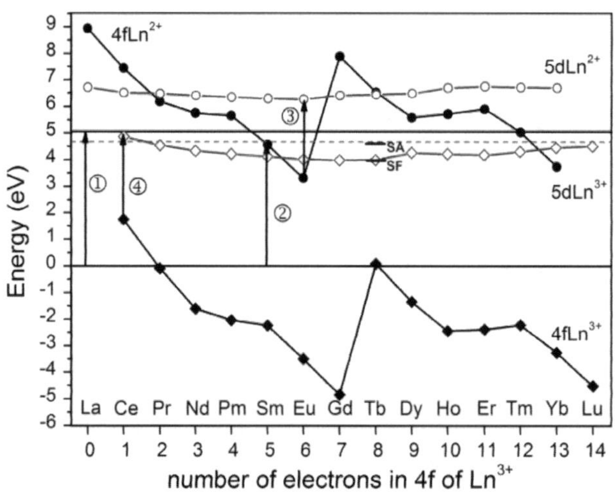

Fig. 4 Energy level scheme showing the $4f$ ground states of the trivalent (filled square) and divalent ions (filled circle) and lowest energy $5d$ states of the trivalent (open square) and divalent ions (open circle) relative to the valence and conduction band of $Y_4Si_2O_7N_2$. The red dashed horizontal line is the optical band gap of $Y_4Si_2O_7N_2$. The arrows (1-4) indicate the experimentally determined transitions used to construct this scheme.

Photophysical Characterization of a Library of Luminescent Lanthanide Complexes.

Daniel Kovacs,[1] Julien Andres[1] and K. Eszter Borbas[1*]

[1*]Department of Chemistry, Ångström Laboratory, Uppsala University, Sweden, eszter.borbas@kemi.uu.se

INTRODUCTION

Luminescent lanthanide (Ln(III)) complexes are extensively used for imaging in complex media.[1,2] As Ln luminescence arises from formally forbidden f-f-transitions, excitation is usually performed through a light-harvesting chromophore ('antenna') placed in close proximity of the metal ion.[3] A full understanding of the processes that take place upon antenna excitation is a prerequisite for the optimization of the luminescent properties of the complexes.[4] As emissions are narrow and metal-specific, components of Ln mixtures are readily identified, which makes such emitters ideal for multiplex detection.[5,6] To increase the scope of multiplex detection, it would be desirable if good emitters based on the lesser-used Lns (i.e. other than Eu and Tb) with new emission wavelengths were identified, as well as ligands/antennae that could sensitize several Lns with comparable efficiencies.

RESULTS AND DISCUSSION

With the dual aim of identifying highly emissive non-Eu or Tb-based complexes and gaining an in-depth understanding of their photophysics, a small library of ligands were synthesized. The ligands were based on either the dota or the do3a chelating framework, and carried coumarin or carbostyril sensitizing antennae. The two different chelate structures provided complexes with either long (do3a) or short (dota) Ln-antenna distances. Antenna substituents were varied systematically. Specifically, groups with varying electron-donating and withdrawing power were introduced. The antennae had a wide range of singlet and triplet energies, as well as redox properties. All ligands were complexed with Eu, Tb and Gd, dota-based ligands were also complexed with Sm and Dy, and, in some cases, with Yb and Nd.

The following parameters were determined for all the complexes: residual antenna fluorescence, antenna singlet and triplet energies, Ln emission quantum yield, Ln emission lifetime and number of Ln-coordinated water molecules. These data enabled us to propose a FRET-type energy transfer pathway in those complexes that had a long antenna-Ln distance. Additionally, for complexes with short antenna-Ln distances, the well-known triple-mediated sensitization was also found; in these complexes, the FRET pathway is likely contributing to the sensitization.

The library screen returned a number of bright Tb and Eu complexes, as well as quite emissive Sm and Dy species. All tested Nd and Yb complexes had sensitized emission in the near infrared. Antenna residual fluorescence quantum yields were significantly lower for Eu-complexes than for their Gd-analogues. This was proposed to be the result of a combination of singlet-mediated sensitization and electron transfer from the excited antenna to the Eu(III). Cyclic voltammetry experiments confirmed the feasibility of the latter process. Furthermore, all other luminescent Ln complexes (Ln = Tb, Dy, Sm, Yb, Nd) also had diminished fluorescence quantum yields, even if the effect was not as striking as for Eu. Electron transfer to the hard-to-reduce Tb(III) is unlikely, so for Tb complexes the antenna fluorescence decrease is presumably due to singlet sensitization. For Lns with intermediate redox potentials (Dy, Nd), electron transfer might be viable, as the ligand can shift the Ln(III)/Ln(II) reduction potential by several 100 mV.[7,8]

CONCLUSIONS

The photophysical characterization of a large group of structurally related Ln complexes enabled the identification of several good emitters, includeing some with underutilized Ln centers. Electron transfer from the excited antenna to the Ln(III) ion was an important quenching process for Eu(III) and Sm(III), and may be relevant for additional Lns. These findings suggest that the antenna redox properties are important parameters for the design of highly luminescent Ln complexes.

REFERENCES

1 Sy, M.; Nonat, A.; Hildebrandt, N.; Charbonniere, L. J. Chem. Commun. 52:5080, 2016.

2 Montgomery, C. P.; Murray, B. S.; New, E. J.; Pal, R.; Parker, D. Acc. Chem. Res. 42:925, 2009.

3 de Bettencourt-Dias, A. In Luminescence of Lanthanide Ions in Coordination Compounds and Nanomaterials; John Wiley & Sons Ltd: 2014, p 1.

4 Bunzli, J.-C. G. Coord. Chem. Rev. 293-294:19, 2015.

5 Song, B.; Vandevyver, C. D. B.; Chauvin, A.-S.; Buenzli, J.-C. G. Org. Biomol. Chem. 6:4125, 2008.

6 Pershagen, E.; Borbas, K. E. Coord. Chem. Rev. 273-274:30, 2014.

7 Ekanger, L. A.; Mills, D. R.; Ali, M. M.; Polin, L. A.; Shen, Y.; Haacke, E. M.; Allen, M. J. Inorg. Chem. 55:9981, 2016.

8 Gál, M.; Kielar, F.; Sokolová, R.; Ramešová, Š.; Kolivoška, V. Eur. J. Inorg. Chem. 3217, 2013.

ACKNOWLEDGMENTS

The authors would like to thank the Swedish Research Council (Grant no: 2013-4655) and Stiftelsen Olle Engkvist Byggmästare for providing financial support to this project.

Probing the bonding in dimers of tetravalent Cerium and actinide homologues

Valérie Vallet*

Laboratoire PhLAM, CNRS UMR 8523, Université de Lille, 59655 Villeneuve d'Ascq Cedex, France,
valerie.vallet@univ-lille1.fr

INTRODUCTION

Cerium can be stabilized in solution in a tetravalent form in the presence of nitrates, predominantly as ceric ammonium nitrate (CAN) assigned as a monomeric hexanitrate cerium structure. However, recent synchrotron X-rays and Raman spectroscopy have challenged this picture, and propose instead an unusual Ce–O–Ce dimeric linkage observed when CAN is dissolved in water.[1]

To rationalize the formation of this dimeric species, we will scrutinize the structure, spectroscopy (Raman, infrared) and the bonding of this peculiar dimeric species, in particular with respect to 4f/5d cerium participation to the Ce-O-Ce bond.

THEORETICAL METHODS

The structures and vibrational spectra of various isomers of the cerium dimer have been density functional theory methods. Natural bond orbital (NBO) and quantum theory of atoms-in-molecules (QTAIM) approaches have been used to analyze the nature of the Ce-O-Ce chemical bond.

RESULTS AND DISCUSSION

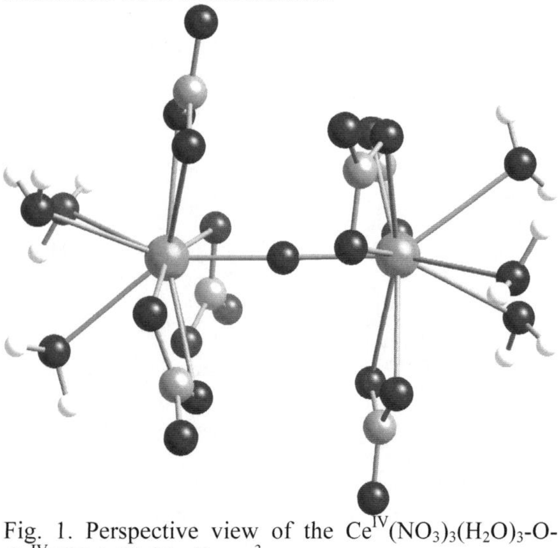

Fig. 1. Perspective view of the $Ce^{IV}(NO_3)_3(H_2O)_3$-O-$Ce^{IV}(NO_3)_3(H_2O)_3$ dimer.[3]

The calculations reveal that in the cerium-nitrate dimer, the Ce-μ_2O bond distance is uniquely short around 2.05 Å, and is best described by strong and covalent Ce-O-Ce two and three center bonds.

These bridging bonds are stronger that those observed in ceria or in the recently synthesized and characterized $[Ce_6(\mu_3\text{-}O)_4(\mu_3\text{-}OH)_4]^{12+}$ hexamers.

To discuss the oxidation state of cerium in these systems, multiconfigurational calculations have also been performed, and confirmed that cerium is tetravalent.

We have also examined the reduced form of this dimer and found that reduction of one or both cerium centers leads to the formation of stable CeIII/CeIV and CeIII/CeIII dimers but with very different Ce-O-Ce bonding characters.

Similar analysis has also been conducted for tetravalent homologues in the actinide series to check whether equivalent An-O-An nitrate dimers might exist.

CONCLUSIONS

Our results shed light on the observed oxo-bridged cerium dimer structure that is found in nitric acid solution, making cerium very stable in its tetravalent state. From a theoretical perspective, tetravalent actinides can also form such dimers.

REFERENCES

References must be numbered. Keep the same style.

1. Demars, T.J. et al., Angew. Chemie. 54:7534-7538, 2-16.
2. Estes, S. L. et al. J. Phys. Chem. C. 128: 5810-5818.
3. Estes, S.L. et al., unpublished.

ACKNOWLEDGMENTS

This work has been conducted in collaboration with Shanna L. Estes, Suntharalingam Skanthakuma, Mark R. Antonio, and L. Soderholm from Argonne National Laboratory.

Production, Chelation, and Biodistribution of Actinium Isotopes: A Handle on the Future of Targeted Radiotherapy

Gauthier J-P. Deblonde[1], Dahlia D. An[1], Andrew L. Lakes[1], Julian A. Rees[1], Stacey S. Gauny[1], Corie Y. Ralston[2], Peter B. Rupert[3], Roland K. Strong[3], and Rebecca J. Abergel[1*]

[1]Chemical Sciences Division, Lawrence Berkeley National Laboratory, USA, rjabergel@lbl.gov
[2]Berkeley Center for Structural Biology, Lawrence Berkeley National Laboratory, USA
[3]Division of Basic Science, Fred Hutchinson Cancer Research Center, USA

INTRODUCTION

Targeted alpha-particle therapy holds tremendous potential as a cancer treatment, since it offers the potential of delivering a highly cytotoxic dose to targeted cells while minimizing damage to the surrounding healthy tissue, due to the short range and high linear energy transfer of alpha particles.[1, 2] A few actinide isotopes have recently emerged as promising short-lived radionuclides that emit multiple α particles in their decay chains, dramatically increasing the potential delivered dose.[3-5] In particular, Ac-225 can act as an *in vivo* alpha-generator radionuclide and is of great interest for new therapeutic applications.[3, 6] To create a targeted alpha therapeutic, one must assemble 3 basic parts: a targeting moiety, a radionuclide chelating ligand, and an appropriate radionuclide such as Ac-225. Though sound in theory, and despite promising therapeutic potential established in pre-clinical and clinical studies, such designs have been slow to emerge. Reasons for this protracted development are many, including limited supply of the radioisotope, insufficient understanding of its biodistribution and biodosimetry, poor retention of alpha-emitting daughter products at the target site, as well as inadequate chelation, one of the major drawbacks.

RESULTS AND DISCUSSION

To seek further development of Ac-225 bioconjugate therapeutics, ongoing efforts aim at addressing all of those limitations. The U.S. Department of Energy's Isotope Program has been exploring new pathways for the production of the radioisotope Ac-225 at accelerator facilities to address a potential increased need for medical applications. However, the presence of co-produced, long-lived Ac-227 ($t_{1/2}$ = 21.8 y) is observed in about 0.15-0.3 activity percent in the accelerator-derived product at the end of target bombardment. Up to 0.5% activity percent values due to Ac-227 could be anticipated in a research/clinical setting. The first goal of our work is to clearly understand the coordination and biological chemistry of actinium isotopes so that adequate ligands can be used for the radiolabeling of antibody conjugates. To that end, we utilized the previously described macro-molecular recognition of f-element siderophore complexes by mammalian proteins to structurally characterize actinium chelates, thereby providing the first Ac-227 crystallographic data set. Further work aims at delineating the biodistribution of Ac-225, its short-lived daughter products, and potential trace contaminants such as Ac-227 that may be co-produced in new larger-scale accelerator-derived processes. The biokinetics of unchelated (in citrate solutions), chelated (with classic macrocyclic, amino-polycarboxylic acids, or new hydroxypyridinone ligands under development), and bio-conjugated Ac-225 and Ac-227 were determined in mice. The distribution profiles highlighted significant differences between isotopes that must be addressed in order to ensure bioconjugated Ac-225 is safe for use in the clinic.

CONCLUSIONS

Ac-225 is certainly one of the most promising isotopes for the development of new targeted alpha-therapy strategies, however, there is still very little information on the chemistry and fate of the radioisotope, once administered. A combination of biochemical and spectroscopic approaches on both *in vitro* and *in vivo* systems was used to study the selective binding and recognition of Ac isotopes, providing a handle on the needs for new chelating options in the design of targeted radio-therapeutic constructs.

REFERENCES

References must be numbered. Keep the same style.
1. Mulford D. A. *et al.*, J. Nucl. Med. 46:199-204, 2005
2. Kim Y. S. *et al.*, Tumor Biol. 33:573-590, 2012
3. McDevitt M. R. *et al.*, Science 294:1537-1540, 2001
4. Aghevlian S. *et al.* Adv. Drug Deliv. Rev. 109:102-118, 2017
5. Read E. D. *et al.*, Targeted Oncol. 10:15-26, 2015
6. Ravandi F. *et al.*, Blood 122:1460-1460, 2013

ACKNOWLEDGMENTS

This work was supported by the U.S. Department of Energy, Office of Science, Office of Basic Energy Sciences, Chemical Sciences, Geosciences, and Biosciences Division and Office of Nuclear Physics, Isotope Program at the Lawrence Berkeley National Laboratory under Contract DE-AC02-05CH11231. The Advanced Light Source (ALS) is supported by the Director, Office of Science, Office of Basic Energy Sciences, of the U.S. Department of Energy under Contract No. DE-AC02-05CH11231.

Pyridine Containing Schiff Base Ligands for Quick Detection of Actinide Contamination

Emily E. Hardy[1*], Kevin M. Wyss[1] and Anne E. V. Gorden[1]

[1*]Department of Chemistry and Biochemistry, Auburn University, USA, eeh0012@auburn.edu

INTRODUCTION

Renewed interest in fundamental f-element coordination has been sparked by continued interest in applications of nuclear fuels for energy.[1] Promising chemistry has been observed using ligand environments with nitrogen and oxygen donors. Examples of this reactivity are found in demonstrating increased actinide covalency and f-element catalysts for otherwise unattainable transformations.[2] The pentadentate coordination environment of a pyridine containing Schiff base ligand scaffold was designed to accommodate the larger atomic radius and preferred coordination geometry of uranium. In the development of improved, rapid detection of the environmental contaminant uranyl (UO_2^{2+}), two new complexes utilizing this scaffold have been synthesized from successive condensation reactions and subsequent metal complexation.

EXPERIMENTAL METHODS

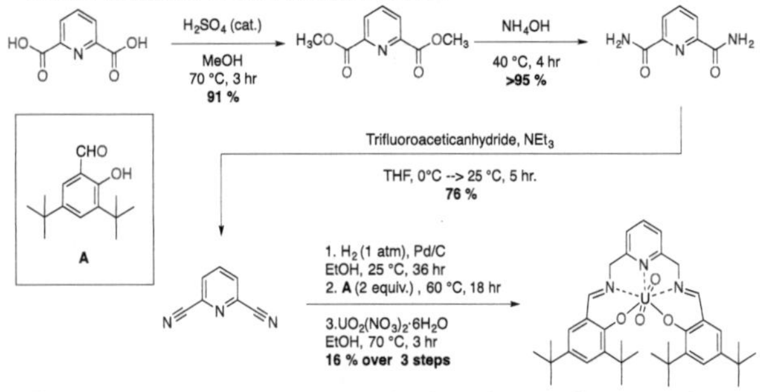

Figure 1: The general synthetic scheme for complex UO_2**1**. Complex UO_2**2** was synthesized by replacing salicylaldehyde **A** with 2-hydroxynapthaldehyde.

RESULTS AND DISCUSSION

Here, two pentadentate Schiff base ligands and the corresponding uranyl UO_2(VI) complexes (UO_2**1**, UO_2**2**) were designed to accommodate the larger atomic radius of uranium -completely occupying its equatorial plane. The expansion and concurrent change in the π-orbital overlap affords differentiation between actinide and transition metal contaminants. One limitation of the tetradentate binding geometry previously studied by our group[3] is the required expansion of the binding pocket to bind uranyl can require longer exposure times. The faster binding kinetics (within seconds for UO_2**1** as compared previous tetradentate systems) can be explained by this reduced need for ligand rearrangement. Solid state structures of UO_2**1**, UO_2**2** are shown in Figure 2.

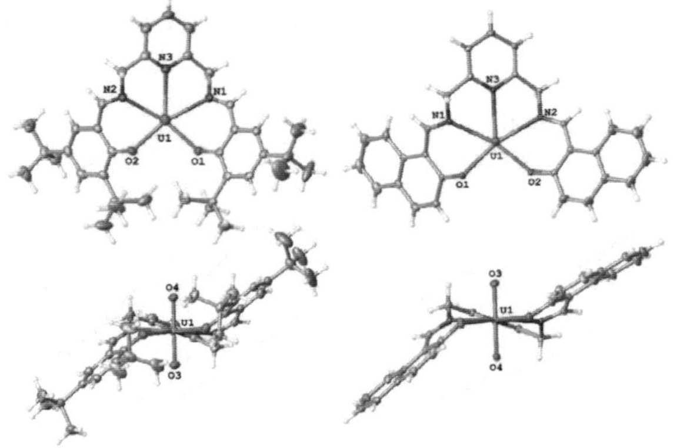

Figure 2: Projections of UO_2**1** and UO_2**2**. U shown in green, carbon in grey, hydrogen in white, oxygen in red, nitrogen in blue. Interstitial solvent removed for clairty.

CONCLUSIONS

Two pentadentate 2,6-bis[1-[(2-hydroxy-3,5-ditert-butylphenyl)imino]ethyl]pyridine] (1) and 2,6-bis[1-[(2-hydroxynapthyl)imino]ethyl]pyridine ligands and corresponding UO_2(VI) complexes (UO_2**1**, UO_2**2**) were designed and synthesized to accommodate the larger atomic radius of uranium and fully occupying its equatorial plane. The fast binding kinetics and simultaneous change in π-overlap affords a chemosensor that can differentiate between actinides and common transition metal contaminants.

REFERENCES

References must be numbered. Keep the same style.

1. A. R. Fox, S. C. Bart, K. Meyer, C. C. Cummins, Nature 2008, 455, 341; N. H. Anderson, S. O. Odoh, U. J. Williams, A. J. Lewis, G. L. Wagner, J. Lezama-Pacheco, S. A. Kozimor, L. Gagliardi, E. J. Schelter, S. C. Bart, J. Am. Chem. Soc. 2015, 137, 4690.
2. P. L. Arnold, J. H. Farnaby, R. C. White, N. Kaltsoyannis, M. G. Gardiner, J. B. Love, Chemical Science 2014, 5, 756; D. P. Halter, F. W. Heinemann, J. Bachmann, K. Meyer, Nature 2016, 530, 317.
3 B. A. Maynard, J. C. Brooks, E. E. Hardy, C. J. Easley, A. E. V. Gorden, Dalton Trans. 2015.

ACKNOWLEDGMENTS

This work was supported in part by the Defence Threat Reduction Agency, Basic Research Award # HDTRA1-11-1-0044 to Auburn University. The authors would also like to thank Dr. J. Gorden for helpful discussion.

Quaternary Rare-Earth Chalcogenides

Abishek K. Iyer[1], Wenlong Yin[1,2], and Arthur Mar[1*]

[1*]Department of Chemistry, University of Alberta, Canada, arthur.mar@ualberta.ca
[2]Institute of Chemical Materials, China Academy of Engineering Physics, People's Republic of China

Quaternary chalcogenides *RE*–*M*–*M'*–*Ch* (*RE* = rare-earth metal; *M*, *M'* = d- or p-block metals or metalloids; *Ch* = S, Se) represent a potentially enormous class of compounds given the numerous permutations of components. They may have interesting applications as optical, magnetic, or thermoelectric materials. Typically, they exhibit small band gaps, and when this feature is combined with the occurrence of noncentrosymmetric structures, they may be good candidates as infrared nonlinear optical materials. Here we present our investigations of several new quaternary chalcogenides, focusing on the very large family of hexagonal $RE_3M_{1-x}M'Ch_7$ phases, as well as the other series $RE_4InSbSe_9$, RE_7MInS_{13} (*M* = Co, Fe), and $BaRE_2In_2Ch_7$.

R_3Au_9Pn (R = Y, Gd, Tb, Dy, Ho; Pn = Sb, Bi): A Link between $Cu_{10}Sn_3$ and $Gd_{14}Ag_{51}$

C. Celania[1,2*], V. Smetana[1,2], A. Provino[1,3,4], P. Manfrinetti[1,3,4], V. Pecharsky[1,2], and A.-V. Mudring[1,2]

[1] The Ames Laboratory, U.S. Department of Energy, Iowa State University, Ames, IA 50011, USA
[2*] Department of Materials Science and Engineering, Iowa State University, Ames, IA 50011, USA ccelania@iastate.edu
[3] Department of Chemistry, University of Genova, Via Dodecaneso 31, 16146 Genova, Italy
[4] Institute SPIN-CNR, Corso Perrone 24, 16152 Genova, Italy

INTRODUCTION

Rare earth-gold intermetallic compounds are receiving special attention due to the formation of unique structure patterns that result in interesting physical properties. Ternary R-Au-M (R = rare earth, M = groups 13-15) are of special interest because, aside from gold's versatility in structure and bonding within intermetallics, the rare earth element is able to endow the material with interesting magnetic properties. The post transition metal allows for final electron tuning conducive to the formation of novel structure types. Amongst them, pnictides (Pn) are the least explored in rare earth-gold intermetallic systems which prompts us to fill this gap of knowledge. Our recent investigations of the R-Au-Pn system led to the discovery of a new series of compounds with general composition R_3Au_9Pn.[1] Their structures represent an ordered variant of the recently reported $R_3Au_7Sn_3$.[2,3]

EXPERIMENTAL METHODS

Samples were synthesized by either arc melting or high temperature self-flux reactions in a furnace (sealed in welded tantalum ampules and jacketed in fused silica) from stochiometric ratios of the respective elements (all >99.9% purity) under inert conditions. Single crystal X-ray structure analysis allowed determination of the structure. Powder X-ray diffraction was used to check for phase purity. DC magnetization measurements were carried out on a Ho_3Au_9Sb sample.

RESULTS AND DISCUSSION

Single X-ray structure analysis of R_3Au_9Pn (R = Y, Gd, Tb, Dy, Ho; Pn = Sb, Bi; [a = 8.079(1) – 8.251(5) Å, c = 8.977(1) – 9.075(1) Å]) revealed that the compounds crystallize as an ordered variant of the $Gd_3Au_7Sn_3$ type structure which, in turn, can be derived from the $Cu_{10}Sn_3$ type. The structures are characterized by a triangular lattice of R atoms between slab-like layers of Au and Pn centered trigonal prisms, which form an anionic network (see Figure 1). The main difference between the recently discovered $R_3Au_7Sn_3$ and the new R_3Au_9Pn is the change in centering of half of the trigonal antiprisms from Sn to Au. This alteration now allows the drawing of a relationship between the $Cu_{10}Sn_3$ and $Gd_{14}Au_{51}$ structures, where the Au@Au_6 trigonal antiprism is also featured. In fact, $R_{14}(Au, M)_{51}$ ternary phases were frequently observed in tested samples due to this close structural and compositional relationship.

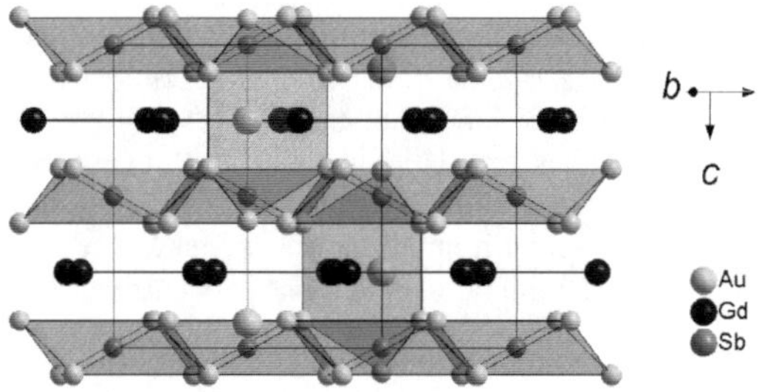

Figure 1. Structure representation of Gd_3Au_9Sb with Sb@Au_6 trigonal antiprisms shaded in red, and Au@Au_6Au_2 bicapped trigonal prisms in green.

Magnetization measurements for Ho_3Au_9Sb demonstrated antiferromagnetic behavior at low temperatures (Néel temperature of 10 K) as well as two metamagnetic transitions toward more ferromagnetic behavior akin to spin canting. Similar transitions were observed for a related $R_3Au_7Sn_3$ representative, $Tb_3Au_7Sn_3$, but with less severe shifts in magnetization, despite the similar triangular RE layers. Density Funtional Theory based electronic structure calculations were used to shed light on the structural and magnetic properties of this materials allowing for a better understanding of the anionic network and the elemental restrictions.

CONCLUSIONS

A new ordered variant of the $Cu_{10}Sn_3$ structure with a general composition R_3Au_9Pn has been observed for R = Y, Gd, Tb, Dy, Ho; Pn = Sb, Bi. Comparison with known variants allowed the formulation of structural relationships to the compositionally close $R_{14}Au_{51}$ structure type. Magnetization data alluded to preferred antiferromagnetic interactions with spin canting resulting in metamagnetic transitions.

REFERENCES

1. Manuscript currently in work for publication
2. Provino, A. *et al.*, J. Mat. Chem. C 2015, 3, 8311-8321.
3. Provino, A. *et al.*, Cryst. Growth Des. 2016, 16, 5657-5668.

ACKNOWLEDGMENTS

Research was supported by the Office of the Basic Energy Sciences, Materials Sciences Division, U. S. Department of Energy (DOE), and the Department of Materials Science and Engineering at Iowa State University (ISU). Ames Laboratory is operated for DOE by ISU under contract No. DE-AC02-07CH11358.

Rare Earth Chemistry in Ionic Liquids and Molten Salts

A.-V. Mudring[1,2]

[1] The Ames Laboratory, U.S. Department of Energy, Iowa State University, Ames, IA 50011, USA
[2] Department of Materials Science and Engineering, Iowa State University, Ames, IA 50011, USA; mudring@iastate.edu

INTRODUCTION

In the quest for safer and greener technologies, ionic liquids (ILs), commonly defined as salts with a melting point below 100°C, offer promising technology solutions because of their unique properties and property combinations which can be realized due to their wide tuneability and modular character. Not only can they be considered as excellent 'green' alternatives to volatile organic solvents (volatile organic compounds, VOCs) because of their generally extremely low vapour pressure. ILs can also offer useful materials properties and unique property combinations such as e.g. wide electrochemical window, excellent thermal stability, miscibility, low combustibility, high catalytic and biologic activity which renders them interesting for a number of special applications. Activities in this field are documented by a large number of scientific publications and patents, the total amount doubling over the last five years. The number of known IL forming cations and anions lead to 10^{19} possible combinations resulting in ILs – each with an individual set of chemical and physico-chemical properties – which underpins their remarkable potential. By combining particular cations and anions, an IL can be designed for a specific application which opens up unique possibilities and enables completely new chemistries that are otherwise not available. This holds specifically true for the chemistry of rare earth elements.

Rare earths belong to the group of critical elements: Rare earth compounds are vital for many every day applications (cell phones, displays, turbines, cars, lighting, health care products etc.). The use of rare earth compounds constitutes an important strategic and dynamic segment of the electronic, chemical and energy industries. However, reduced efforts of the western world to explore and mine rare earths in the past years has led for these important elements to a dependency from external sources – at a time where the demand is growing and it is realized that rare earth offer unique properties for a number of special, advanced applications. Thus, there is a tremendous need for new scientific understanding and expanded research capabilities for rare earth chemistry.

An attractive strategy for improving, enhancing the properties of rare earths (mainly the optical and physical) is the synergetic utilization of ILs. Rare earth containing ILs can serve as soft optical materials with unique optical properties, for example, the IL can be used as a light harvester to enhance the luminescence of the rare earth ions. Due to the high magnetic moment of the rare earth, these ILs represent also new magnetorheological fluids that show a tremendously strong response to applied external magnetic fields which opens up a number of interesting new applications.

RESULTS AND DISCUSSION

A family of bis(trifluoromethanesulfonyl)amide-based ionic liquids of composition $[RE_5(C_2H_5\text{-}C_3H_3N_2\text{-}CH_2COO)_{16}(H_2O)_8](Tf_2N)_{15}$ (RE = Er, Ho, Tm; $C_3H_3N_2$ ≡ imidazolium moiety) featuring the cationic, record quindecim {15+} charged pentanuclear rare earth (RE)-containing cation $[RE_5(C_2H_5\text{-}C_3H_3N_2\text{-}CH_2COO)_{16}(H_2O)_8]^{15+}$ has been synthesized and characterized. In addition, due to the presence of rare earth ions, these ionic liquids show a response to magnetic fields with the highest effective magnetic moment observed so far for an ionic liquid and are a rare examples of ionic liquids showing luminescence in the near-infrared. These ionic liquids also were successfully employed in three-component synthesis of 2-pyrrolo-3'-yloxindole with extremely low (<0.035 mol%) catalysts loading rate.

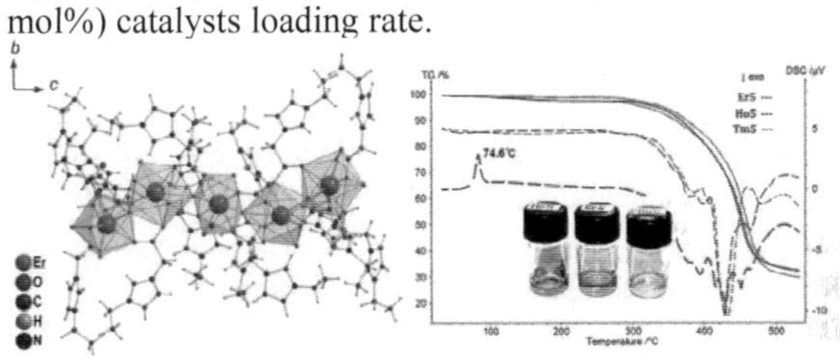

Fig. Structural motive and thermal behaviour.

CONCLUSIONS

With room temperature μ_{eff} (Bohr magnetons) values as high as 21.2 (**1**), 23.3 (**2**) and 16.2 (**3**) the effective moments are the highest ever reported for magnetic ionic liquids (MILs). The ionic liquids do not only respond to a magnetic field, but also show NIR luminescence and extraordinary catalytic activity. They can be used in the three-component synthesis of ethyl 2-methyl-4-(2-oxo-2,3-dihydro-1*H*-3-indolyl)-5-phenyl-1*H*-3-pyrrolecarboxylate at extremely low loadings level (0.029÷0.031 mol%).

REFERENCES

D. Prodius, V. Smetana, S. Steinberg, M. Wilk-Kozubek, Y. Mudryk, V. K. Pecharsky, A.-V. Mudring: Breaking the paradigm: record quindecim charged magnetic ionic liquids, *Materials Horizon* **2017**, *4*, 217-221 and references therein.

ACKNOWLEDGMENTS

Research was supported by the National Science Foundation (NSF CHE-1465071).

Rare Earth Element Distributions and Extraction from Coal Ash

Elliot Roth[1,2*], Megan Macala[1,2], Ronghong Lin[3], Tracy Bank[1,2], Evan Granite[1]

[1]U.S. Department of Energy, National Energy Technology Laboratory, Pittsburgh, Pennsylvania, USA
[2*]AECOM, National Energy Technology Laboratory, Pittsburgh, PA, USA, Elliot.roth@netl.doe.gov
[3]Oak Ridge Institute for Science and Education, Pittsburgh, Pennsylvania, USA

INTRODUCTION

Extracting rare earth elements from coal and coal by-products, such as coal fly ash, has been proposed for many years, but has recently gained significant interest due to price and supply volatility, and expected demand for advanced technologies.[1-4] With the average world coal ash concentration of rare earth elements, including yttrium (REY), being approximately 404 ppm[5] it is obvious that a large amount of coal ash would have to be processed with a high recovery of REY to achieve any appreciable amount of REY. Additionally the individual rare earth elements are not equal in terms of price and criticality,[2] making the distribution, as well as the concentration of the rare earth elements important. In this study the distribution of the rare earth elements is investigated in size separated fractions of a pulverized coal power plant fly ash. The extractions of REY were investigated using hydrochloric acid, versus using a hydrothermal sodium hydroxide pre-treatment followed by hydrochloric acid leaching.

EXPERIMENTAL METHODS

The fly ash sample was size separated using a GilSonic UltraSiever, model GA-8, with ASTM E-11 stainless steel woven wire cloth sieves. A perkin Elmer Nexion 300D ICP-MS with universal cell technology was used for bulk elemental analysis of the fly ash, size separated samples and leaching solutions. The fly ash samples were digested using a lithium metaborate fusion procedure at 1050°C to form a homogeneous glass. The glass was digested in 5% nitric acid and then further diluted in 2% nitric acid before analysis. Prior to fusion the samples were dried for one hour under nitrogen at 107°C and ashed for five hours under air at 550°C using a LECO Thermogravimetric Analyzer (TGA 701). Extraction experiments were performed using 2M HCl solution with a 10:1 liquid to solid ratio by weight. The HCl solution was analysed using ICP-MS to determine element concentrations. Sodium hydroxide pre-treatment using a 5M NaOH solution with a 10:1 liquid to solid ratio, was heated in a hot block to ~93°C for the desired duration. The solid sample was washed three times with water, dried, and then leached with 2M HCl solution.

RESULTS AND DISCUSSION

The sample that was primarily studied was collected by LTI-NETL and is described as UN: 345, a PC power plant fly ash from Ohio.[6] The weight percent recovered for the size separated fly ash and rare earth element (REE) and REY concentrations, shown in Table 1, indicate a slight enrichment in REY as the particle size decreases. REY distributions in the fly ash show a medium rare earth and heavy rare earth enrichment normalized to the upper continental crust.

Table 1: Weight Percent and REY Concentrations for Size Separated Fly Ash

	Wt. % (size fractions)	REE (ppm)	REY (ppm)
Fly Ash		358	431
(+) 100 mesh	1.3%	275	328
100-140 mesh	2.2%	256	300
140-200 mesh	5.5%	304	358
200-230 mesh	3.8%	320	379
230-270 mesh	3.9%	336	398
270-325 mesh	4.9%	336	398
325-400 mesh	5.8%	342	407
400-450 mesh	4.5%	359	426
450-500 mesh	6.4%	356	423
500-635 mesh	10.4%	366	438
minus 635 mesh	51.4%	378	457

Preliminary HCl extractions on the as-received fly ash show very limited extraction of REY (~4%), however using a NaOH pre-treatment for four hours greatly enhanced the recovery of REY to greater than 70%, and with an eight hour pre-treatment ~90% extraction of the REY was achieved. These results could indicate that the glassy fraction of the fly ash is attacked using NaOH solution[7] enhancing REY extraction during subsequent HCl leaching.

CONCLUSIONS

Coal fly ash is a potentially large, already mined source for REY, which can be slightly concentrated in REY by size separation. Extraction of REY using room temperature 2M HCl is very limited. However, using a pre-treatment process such as hydrothermal NaOH solution greatly increases the extraction to ~90%.

REFERENCES

1. Dutta, T. *et al.*, Environ. Res. 150:182-190, 2016
2. Seredin, V. *et al.*, Int. J. Coal Geol. 94:67-93, 2012
3. Hower, J. *et al.*, *Minerals* 6:32, 2016
4. Granite, E. *et al.*, *The Bridge,* 46, 56-57, 2016
5. Ketris, M. *et al.*, Int. J. Coal Geol.78:135-148, 2009
6. https://edx.netl.doe.gov/ree/ accessed 2-24-2017
7. Murayama, N. *et al.*, Int. J. Miner. Process., 64:1-17, 2002

ACKNOWLEDGMENTS

This technical effort was performed in support of the National Energy Technology Laboratory's ongoing research under the RES contract DE-FE0004000. The author would like to thank Dr Bret Howard for technical support and Mary Anne Alvin for program management

Rare Earth Solid Solution of $Yb_{14-x}RE_xZnSb_{11}$: Structure and Properties

Elizabeth L. Kunz Wille[1*], Na Hyun Jo[2], Susan M. Kauzlarich[1] and Paul C. Canfield[2]

[1]Department of Chemistry, University of California Davis, Davis, CA 95616, USA
[2]Ames Laboratory and Department of Physics and Astronomy, Iowa State University, Ames, IA 50011, USA
*presenting author email: ewille@ucdavis.edu

ABSTRACT

Zintl phases have been a subject of much exploration in the past few years due to their unique magnetic properties. Zintl phases are characterized by a complete donation of electrons from cations to anions, and anions form bonds to satisfy their octet, which allows for isoelectronic substitution but also allows for substitutions of small amounts of elements with dissimilar oxidation states. In particular, the magnetic properties of $Yb_{14}MnSb_{11}$ have been investigated as this phase is considered a partially compensated Kondo system and nearly half metallic. The structure can be described according to Zintl electron counting as consisting of $14Yb^{2+}$ cations and $MnSb_4^{9-}$ tetrahedral cluster, Sb_3^{7-} linear anion and 4 isolated Sb^{3-} anions. In this case, only the $MnSb_4^{2-}$ tetrahedron has a magnetic moment. Experimentally, this compound orders ferromagnetically at about 52 K[1,2]. The Yb cation is Yb^{2+} and the tetrahedral cluster can be described as $[d^5 + \text{hole}]$[3,4], providing a moment of about 4 μ_B and the magnetic coupling is between the tetrahedral units. $Yb_{14-x}RE_xMnSb_{11}$ has been investigated and substitutions of RE^{3+} in this p-type material have the effect of lowering the carrier concentration and the rare earth moment interacts with the tetrahedral unit in distinctive ways, depending upon whether it is a light or heavy rare earth 3+ cation. For example, in the light rare earths such as Pr, Nd, and Sm, the saturation moments are consistent with the 4 μ_B observed for $Yb_{14}MnSb_{11}$, however, the Curie temperature is reduced[5-7]. This result suggests that while the rare earth adds an additional electron to the system, the light rare earths do not exhibit local moments in the ordered regime. In contrast, when RE = Gd, the Gd moment orders ferromagnetically with $[MnSb_4]^{9-}$ and the full moment of Mn^{2+} is evident. The different magnetic behavior associated with the RE cation is attributed to subtly different interactions between the 4*f*-orbitals on the RE with the cluster[5]. In an attempt to better understand and elucidate the effect of the RE^{3+} on magnetic properties, the Zn^{2+} analogue is used in place of Mn^{2+}. $Yb_{14}ZnSb_{11}$ is an ideal system to explore because it can be prepared as high quality single crystals via flux reaction. It shows intermediate Yb valence fluctuations with a spin fluctuation temperature of approximately 85 K[8]. The resistivity shows a slope change near the transition temperature, 85 K. The presence of Yb^{2+} and small amounts of Yb^{3+} in $Yb_{14}ZnSb_{11}$ is thought to result from a need for charge balance following the Zintl concept. Single crystals of $Yb_{14-x}RE_xZnSb_{11}$ were produced in a Sn-flux synthesis[9]. Physical properties were measured on single crystals. Upon increase of x for the RE = Y phase, the signature for valence fluctuation is suppressed and is gone by x = 1, while the samples retain their metallic behavior. The synthesis and magnetic and transport properties of the single crystals of $Yb_{14-x}RE_xZnSb_{11}$, RE = Y, Gd, Ce will be presented and discussed in comparison with other rare earth doped analogues such as $Yb_{14-x}RE_xMnSb_{11}$.

REFERENCES

1. Chan, J. Y.; Olmstead, M. M.; Kauzlarich, S. M.; Webb, D. J., Structure and Ferromagnetism of the Rare-Earth Zintl Compounds: $Yb_{14}MnSb_{11}$ and $Yb_{14}MnBi_{11}$. *Chemistry of Materials* **1998,** *10* (11), 3583-3588.
2. Fisher, I. R.; Wiener, T. A.; Bud'ko, S. L.; Canfield, P. C.; Chan, J. Y.; Kauzlarich, S. M., Thermodynamic and transport properties of single-crystal $Yb_{14}MnSb_{11}$. *Physical Review B* **1999,** *59* (21), 13829-13834.
3. Holm, A. P.; Kauzlarich, S. M.; Morton, S. A.; Waddill, G. D.; Pickett, W. E.; Tobin, J. G., XMCD Characterization of the Ferromagnetic State of $Yb_{14}MnSb_{11}$. *Journal of the American Chemical Society* **2002,** *124* (33), 9894-9898.
4. Hu, Y.; Chen, C.-W.; Cao, H.; Makhmudov, F.; Grebenkemper, J. H.; Abdusalyamova, M. N.; Morosan, E.; Kauzlarich, S. M., Tuning Magnetism of $[MnSb_4]^{9-}$ Cluster in $Yb_{14}MnSb_{11}$ through Chemical Substitutions on Yb Sites: Appearance and Disappearance of Spin Reorientation. *Journal of the American Chemical Society* **2016,** *138* (38), 12422-12431.
5. Holm, A. P.; Ozawa, T. C.; Kauzlarich, S. M.; Morton, S. A.; Dan Waddill, G.; Tobin, J. G., X-ray photoelectron spectroscopy studies of $Yb_{14}MnSb_{11}$ and $Yb_{14}ZnSb_{11}$. *Journal of Solid State Chemistry* **2005,** *178* (1), 262-269.
6. Hu, Y.; Bux, S. K.; Grebenkemper, J. H.; Kauzlarich, S. M., The effect of light rare earth element substitution in $Yb_{14}MnSb_{11}$ on thermoelectric properties. *Journal of Materials Chemistry C* **2015,** *3* (40), 10566-10573.
7. Grebenkemper, J. H.; Kauzlarich, S. M., Magnetic and structural effects of partial Ce substitution in $Yb_{14}MnSb_{11}$. *APL Materials* **2015,** *3* (4), 041503.
8. Fisher, I. R.; Bud'ko, S. L.; Song, C.; Canfield, P. C.; Ozawa, T. C.; Kauzlarich, S. M., $Yb_{14}ZnSb_{11}$: Charge Balance in Zintl Compounds as a Route to Intermediate Yb Valence. *Physical Review Letters* **2000,** *85* (5), 1120-1123.
9. Canfield, P. C.; Fisk, Z., Growth of single crystals from metallic fluxes. *Philosophical Magazine Part B* **1992,** *65* (6), 1117-1123.

ACKNOWLEDGMENTS

Financial support for E.K.W. from the Nuclear Energy University Program (NEUP) is gratefully acknowledged. This work was supported by the U.S. Department of Energy, Office of Science, Office of Workforce Development for Teachers and Scientists, Office of Science Graduate Student Research (SCGSR) program. The SCGSR program is administered by Oak Ridge Institute for Science and Education for the DOE under contract DE-SC0014664. Work at the Ames Laboratory was supported by the US Department of Energy, Basic Energy Sciences, Division of Materials Sciences and Engineering under Contract No. DE-AC02-07CH11358. N. H. J. was supported by the Gordon and Betty Moore Foundation EPiQS Initiative (Grant No. GBMF4411).

Rare Earths-Silicide/Silicon Composite Electrodes with Excellent Anode Performance for Lithium-Ion Battery

Hiroki Sakaguchi[1*,2], Hiroyuki Usui[1,2], and Yasuhiro Domi[1,2]

[1*]Department of Chemistry and Biotechnology, Graduate School of Engineering, Tottori University, Japan, sakaguch@chem.tottori-u.ac.jp

[2]Center for Research on Green Sustainable Chemistry, Tottori University, Japan

INTRODUCTION

Silicon (Si) has attracted much attention as anode material due to its huge theoretical capacity, while there are critical disadvantages: a low electronic conductivity and significant volume changes during Li-insertion/extraction. We have revealed that Si's anode performance can be improved by being composite electrodes consisting of elemental Si and transition metal silicides [1-4]. In particular, we found that lanthanum silicide (La–Si) is very effective for improvement in the performance [1,2]. In contrast, the effect of other rare-earth metal silicides on the performance has never explored yet. In this study, we chose samarium (Sm), gadolinium (Gd), and dysprosium (Dy) as rare-earth elements to be rare-earth metal silicides (*Ln*–Si), and investigated the anode performance of *Ln*–Si composite electrodes prepared by a gas-deposition method. In addition to this, the effect of nickel substitution on the electrochemical performance of a lanthanum silicide ($LaSi_2$)/silicon (Si) composite electrode was studied.

RESULTS AND DISCUSSION

Figure 1 shows dependence of discharge capacities for these composite electrodes on charge–discharge cycle numbers. All the *Ln*–Si composite electrodes exhibited improved performances, while the Si electrode showed a rapid capacity decay induced by pulverization of its active material layer. Even after 1000 cycles, the Gd–Si/Si electrode maintained a very high capacity of 840 mA h g^{-1}, which clearly demonstrating its excellent performance [4]. The discharge capacity of Gd–Si/Si is much higher than the graphite's theoretical capacity of 372 mA h g^{-1}. To evaluate the contribution of these silicides to the composite electrodes' capacities, we synthesized silicide electrodes composed of Sm–Si, Gd–Si, or Dy–Si. These silicide electrodes showed very small capacities of around several ten mA h g^{-1} though the capacities stably maintained for a long cycle, as in the case of La–Si electrodes [1,2]. This result suggests that the composite electrodes derive nearly all capacities not from silicides but from elemental Si. Electrical conductivities of *Ln*–Si powders were measured under a uniaxial press using a two probes method. The *Ln*–Si powders showed three orders of magnitude higher conductivity than a Si powder, and that there was no difference in the conductivity was observed among the *Ln*–Si powders. On the other hand, Gd–Si exhibited a more elastic property in comparison with other silicides, indicating

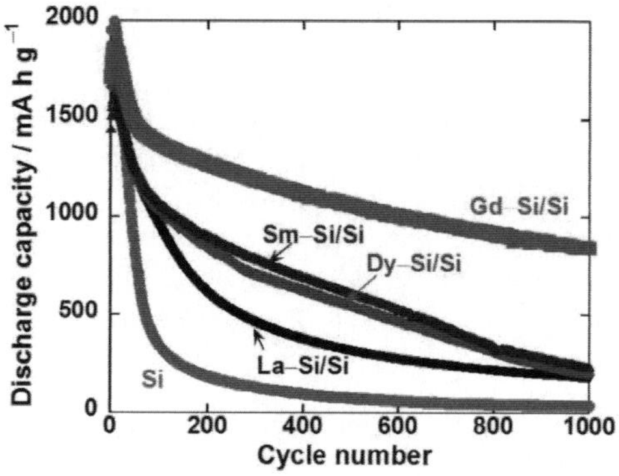

Fig. 1 Cycling performances of *Ln*–Si/Si composite electrodes (*Ln*: La, Sm, Gd, and Dy).

that it released effectively the stress from Si during Li-insertion/extraction and suppressed silicon's pulverization.

The effect of nickel substitution on the performance of a lanthanum silicide ($LaSi_2$)/silicon (Si) composite electrode was studied. The results of X-ray diffraction analysis showed that $LaSi_2$ forms a substitutional solid solution with Ni, and that only the Si site in $LaSi_2$ is substituted by Ni. A $LaNi_{0.10}Si_{1.90}$/Si (70:30 wt.%) composite electrode showed a superior long-term cycle performance compared to electrodes composed of Si alone and $LaSi_2$/Si composite. The $LaNi_{0.10}Si_{1.90}$/Si electrode exhibited an excellent high-rate performance with a reversible capacity of 2240 mA h g(*Si*)$^{-1}$ at a rate of 10 *C*. The results of computational chemistry demonstrated that $LaNi_{0.25}Si_{1.75}$ favors Li migration in the pathway compared to $LaSi_2$.

REFERENCES

[1] Sakaguchi H. and Usui H., *Electrochemistry*, **80**:45-50, 2012.

[2] Usui H., Shimizu M., Sakaguchi H., *J. Power Sources*, **235**:29-35, 2013.

[3] Usui H., Nouno K., Takemoto Y., Sakaguchi H. *et al.*, *J. Power Sources*, **268**:848-852, 2014.

[4] Usui H., Nomura M., Nishino H., Kusatsu M., Murota T., Sakaguchi H., *Mate. Lett.* **130**:61-64, 2014.

[5] Domi Y., Usui H., Takemoto Y., Yamaguchi K., and Sakaguch H., *J. Phys. Chem. C*, **120**:16333-16339, 2016.

ACKNOWLEDGMENTS

This work was partially supported by the Japan Society for the Promotion of Science (JSPS) KAKENHI.

Rare Mixed-valent Rare Earth Metal Complexes

Glen B. Deacon,[1] Peter C. Junk,[2] and Jun Wang[2]

[1]School of Chemistry, Monash University, Victoria, 3800, Australia
[2]College of Science & Engineering, James Cook University, Townsville, Qld, 4811, Australia, peter.junk@jcu.edu.au

INTRODUCTION

Over the path of the advances of organolanthanoid chemistry, the researchers have been so far focusing on the chemistry of complexes in which each metal element has singular oxidation state in individual compounds. The mixed-valent metal complexes, where the compound has two or more oxidation states for each metal element, though established for transition metals, have been extremely sparsely explored for organo-rare earth metal complexes.[1] The mixed-valent lanthanoid metal complexes are expected to vary from relevant singular valent metal complexes in their properties.

RESULTS AND DISCUSSION

In this work, we report the synthesis of bulk mixed valent ytterbium(II/III) *N*, *N'*-diphenylethylenediaminato complex $[Yb_3(PhNC_2H_4NPh)_4)(thf)_2]$ through convenient redox transmetalation/protolysis (RTP) reactions. The relevant RTP reactions with calix[4]pyrroles ($H_4N_4Et_8$) gave readily the divalent metal complexes $[Ln^{II}_2(N_4Et_8)(thf)_4]_2$ (Ln = Yb,[2] Eu). These divalent precursors were oxidised by ethyl iodide forming the mixed valent europium(II/III) complex $[(thf)_2Eu^{II}(N_4Et_8)Eu^{III}(\mu\text{-}I)]_2$ (the structure shown right) in good yield, but the analogous ytterbium species $[(thf)_2Yb^{II}(N_4Et_8)Yb^{III}(\mu\text{-}I)]_2$ in low yield. The mixed valent Sm(II/III) metal complex $[\{(thf)_2Sm^{II}(N_4Et_8)Sm^{III}(thf)\}_2\{\mu\text{-}(OC)_{10}W_2\}]$ was obtained by the oxidation of $[Sm^{II}_2(N_4Et_8)(thf)_4]_2$ with $W(CO)_6$.

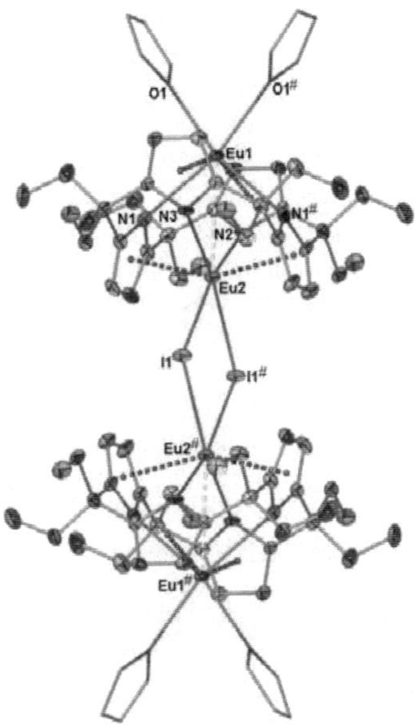

Figure 1. Xray structure of the europium(II/III)complex $[(thf)_2Eu^{II}(N_4Et_8)Eu^{III}(\mu\text{-}I)]_2$

ACKNOWLEDGMENTS

The authors acknowledge the Australian Research Council for support of this work.

REFERENCES

[1] K. Norton, T. J. Emge and J. G. Brennan, Inorg. Chem., 2007, 46, 4060-4066.
[2] G. B. Deacon, M. G. Gardiner, P. C. Junk, Josh P. Townley and J. Wang, Organometallics, 2013, 32, 1370-1378.

Ratiometric nanothermometers based on Eu^{III} and Tb^{III} complexes grafted to pNIPAM

Josiane A. Sobrinho[1], Gilmar A. Brito Júnior[1], Italo O Mazali[1] and Fernando A. Sigoli[1*]

[1] Laboratory of Functional Materials, Institute of Chemistry, University of Campinas, Brazil
*fsigoli@iqm.unicamp.br

INTRODUCTION

Temperature plays an important role in many biological processes and deregulations of this physical property at bio-micro-environment are normally correlated to cellular dysfunctions or pathologies. In this context, the design of temperature probes with high sensitivity, biocompatibility, and accuracy at subcellular level is of great interest for the understanding of cellular events and development of diagnostic and therapeutic techniques for some diseases.[1] For this purpose, fluorescent molecular thermometers have received considerable attention. Since they work at molecular level and fluorescence sensing can provides a sensitive and selective detection method, these thermometers enable the mapping of temperature variations in extremely small spaces such as living cells. Materials based on pNIPAM (poly(N-isopropylacrylamide)), for instance, have been considered as promising candidates for molecular thermometry. It is a water-dispersable and biocompatible polymer that undergoes a phase transition (ca.32 °C) near to the physiological temperature. These temperature-induced structural changes can provide a fluorescence enhancement of fluorophores incorporated into the pNIPAM, even with a small variation in temperature.[2] Among these fluorophores, lanthanide(III) complexes have several advantages over organic molecules, due to their intrinsic spectroscopic properties, such as narrow f-f transitions bands, long emission lifetimes and large pseudo-Stokes shift.[3] This work shows the synthesis and characterization of ratiometric luminescent molecular thermometers based on lanthanide complexes grafted on pNIPAM nanoparticles as potential candidates for thermal sensing.

EXPERIMENTAL METHODS

The 4-(allyloxy)dipicolinic acid (*adp*) ligand was synthesized from the chelidamic acid (*che*) through a nucleophilic substitution reaction with allyl bromide. The obtained molecule was characterized by ^{1}H and ^{13}C NMR, FTIR and UV-Vis. A copolymer containing the thermoresponsive monomer (NIPAM) and the grafted *adp* chelate monomer was synthesized by an emulsion polymerization method. A polymer without *adp* was also prepared for comparison. After purified, an $EuCl_3$ and/or $TbCl_3$ aqueous solution were add to an aliquot of the *adp*-functionalized polymer solution, which was kept under stirring at room temperature for 24 h. The lanthanides-containing polymers were characterized by DLS, UV-Vis and PL spectroscopy.

RESULTS AND DISCUSSION

DLS measurements of the aqueous hydrogel solution shows that polymeric nanoparticles have an average hydrodynamic diameter of 50.2±0.3 nm at 37 °C, with a polydispersity index (PDI) of 0.074, indicating a narrow size distribution and high monodispersitivity. The measured zeta potential is -33.2±0.4 mV, ensuring the stability of the colloidal dispersion. The excitation spectrum of the Eu^{III}-*adp-functionalized* polymer (λ_{em}= 612 nm, $^5D_0 \rightarrow {}^7F_2$) shows an intense band at 275 nm when compared to the non-functionalized polymer. This result indicates an energy transfer from the ligand to Eu^{III} and its coordination by the *adp*-functionalized polymer. PL measurements, as a function of the temperature, shows a reversible process and an emission intensity enhancement of 3.5 times of the Eu^{III} $^5D_0 \rightarrow {}^7F_2$ transition when the temperature is increased from 35 °C to 50 °C. The same behaviour is observed for the Tb^{III}-*adp-functionalized* polymer however with a less pronounced emission intensity increase of the $^5D_4 \rightarrow {}^7F_5$ transition (543 nm) over the same temperature range. As the pNIPAM phase transition results in different responses of the lanthanide ions emission, a Tb^{III}-Eu^{III}-adp-functionalized polymer was prepared and the luminescent intensity ratio (LIR) of the emissions at 612 and 543 nm was obtained, resulting in a ratiometric and reversible system with a maximum relative thermal sensitivity (S_m = 8.4 % °C^{-1}, 40 °C) in the biological temperature range.

CONCLUSIONS

This work illustrates the preparation and characterization of a new ratiometric luminescent nanothermometer based on lanthanide complexes grafted on pNIPAM nanoparticles. This reversible and ratiometric thermometer has one of the highest values of the relative thermal sensitivity (S_m = 8.4 % °C^{-1}, 40 °C) reported in the literature for lanthanide complexes in the biological temperature range. Therefore, the system is a potential candidate for intracellular temperature probing.

REFERENCES

1. Bai T. *et al.*, Small. 12:4590-4610, 2016
2. Chen C. *et al.*, Chem. Commun. 47: 994-996, 2011
3. Heffern M. *et al.*, Chem. Rev. 114:4496-4539, 2014

ACKNOWLEDGMENTS

The authors would like to thank CNPq, CAPES, INOMAT and FAPESP for financial support

Rational Conception of Dysprosium-based Complex featuring Different Single-Molecule Magnets: Magnetic Interaction and Isotopic Effects

Fabrice Pointillart[1], Boris Le Guennic[1], Olivier Cador[1] and Lahcène Ouahab[1]

[1] Institut des Sciences Chimiques de Rennes UMR 6226 CNRS-UR1, Université de Rennes 1, France
fabrice.pointillart@univ-rennes1.fr

INTRODUCTION

Lanthanide-based Single Molecule Magnets (SMMs) are fascinating objects for both fundamental science and potential applications.[1] The most used lanthanide to elaborate such molecular objects is undoubtedly the Dy^{III} ion. The mononuclear Dy-based SMMs present a large magnetic hysteresis loop but closed in zero field due to the fast relaxation of the magnetization by quantum effect while the polynuclear Dy-based SMMs present a narrower hysteresis loop but opened in zero field when the magnetic interaction are significant.

RESULTS AND DISCUSSION

So, it is natural to try to combine both mononuclear and polynuclear (with significant magnetic interaction) SMMs in the same compound in order to obtain as large as possible magnetic hysteresis loop. To do so, we start from two systems synthesised in our group: i) a binuclear Dy^{III}-based SMM ($[Dy(tta)_3(\mathbf{L}^1)]_2$ with $\mathbf{L}^1$ = tetrathiafulvalene-3-pyridine-*N*-oxide and tta^- = 2-thenoyltrifluoroacetonate anion)[2] in which the antiferromagnetic coupling between axial magnetic moment leads to the opening of the magnetic hysteresis loop at zero field which closes in field (double butterfly) and ii) a mononuclear Dy^{III}-based SMM ($[Dy(tta)_3(\mathbf{L}^2)]$ where $\mathbf{L}^2$ = 4,5-bis(propylthio)-tetrathiafulvalene-2-(2-pyridyl)benzimidazole-methyl-2-pyridine)[3] with hysteresis opened in field only (butterfly). The magnetic dilution and isotopic enrichment in nuclear spin free Dysprosium have been studied.[4] Then during this presentation, it will be exposed how the carefully designed TTF-based ligand $\mathbf{L}^3$ ($\mathbf{L}^3$ = 2-{1-Methylpyridine-N-oxide-4,5-[4,5-bis(propylthio)tetrathiafulvalenyl]-1*H*-benzimidazol-2-yl}pyridine) can guaranty the auto-assembly of the two previously presented systems in a unique molecule. Among other properties, the SMM behaviour of the resulting tetranuclear complex $[Dy_4(tta)_{12}(\mathbf{L}^3)_2]$[5] will be presented and explained.

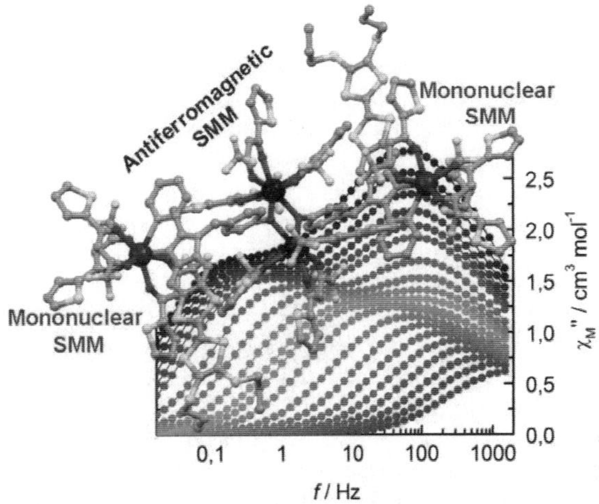

"Figure 1 shows the molecular structure of $[Dy_4(tta)_{12}(\mathbf{L}^3)_2]$ complex and its out-of-phase signal of the magnetic susceptibility"

CONCLUSIONS

Memory effect at zero magnetic field can be observed after vanishing of the quantum effect by isotopic enrichment of nuclear spin free ^{164}Dysprosium as well as by the presence of significant magnetic interaction. Finally rational design of the ligand $\mathbf{L}^3$ guaranties the auto-assembling of mononuclear and dinuclear species in a unique molecule.

REFERENCES

1. Woodruf D. N. *et al.*, Chem. Rev. 113:5510-, 2013
2. Pointillart F. *et al.*, Chem. Eur. J. 117: 10397-10404, 2011
3. da Cunha T. T. *et al.*, J. Am. Chem. Soc. 135: 16332-16335, 2013
4. Pointillart F. *et al.*, Angew. Chem. Int. Ed. 54: 1504-1507, 2015
5. Pointillart F. *et al.,* Chem. Eur. J. 21: 16929-16934, 2015

ACKNOWLEDGMENTS

This work was supported by the CNRS, Rennes Métropole, Université de Rennes 1, Région Bretagne, FEDER and Agence Nationale de la Recherche (N° ANR-13-BS07-0022-01).

Reactions of Dysprosium on the Graphite (0001) Surface

Ann Lii-Rosales[1,2*], Yinghui Zhou[1,5], Minsung Kim[1], Mark Wallingford[1], Michael C. Tringides[1,3], Cai-Zhuang Wang[1,3], and P. A. Thiel[1,2,4]

[1*]Ames Laboratory, US Department of Energy, USA, ylii@iastate.edu
[2]Department of Chemistry, Iowa State University, USA
[3]Department of Physics and Astronomy, Iowa State University, USA
[4]Department of Materials Science and Engineering, Iowa State University, USA
[5]Present address: Department of Physics, Xiamen University, China

INTRODUCTION

Magnetic properties of dysprosium (Dy) make it attractive for electronic and magnetic devices. It is of fundamental importance and interest to understand reactions between Dy and a substrate on the nanoscale. Here, we present various interactions of Dy on two types of graphite surfaces at different temperatures: a pristine surface vs. a surface that is pre-treated with Ar^+ ion bombardment. In *bulk* systems, carbide formation competes with intercalation at elevated temperatures.[1] In the present study, we delineate conditions that induce *surface* Dy carbide formation and its characteristics. Moreover, *surface* intercalation of Dy can occur under specific conditions. Surface intercalated Dy rafts exhibit an unexpected structure that is distinct from its bulk counterpart (graphite intercalation compound, GIC)[1] and also clearly distinguishable from the surface carbide.

EXPERIMENTAL METHODS

All experiments were performed in an ultrahigh vacuum (UHV) chamber. A commercial sample of highly oriented pyrolytic graphite (HOPG) was used as a substrate for Dy deposition. A pristine graphite sample was prepared by tape-cleavage in air followed by transfer into UHV to outgas at 800 K. An *e*-beam heater was used for sample heating. For a bombarded graphite surface, the above preparation was followed by backfilling the chamber with Ar gas, which was ionized by an ion gun for bombardment. The bombarded sample was annealed at 900 K to remove embedded Ar gas. Dy was deposited onto the sample via physical vapor deposition from an *e*-beam evaporator. A scanning tunnelling microscope (STM) was used for imaging at 300 K.

RESULTS AND DISCUSSION

Figure 1 shows diverse morphologies of Dy on graphite surface at elevated temperatures. Upon deposition of Dy on pristine graphite at 650-800 K, Dy forms faceted islands that are 1.5-2.5 nm tall and have triangular or hexagonal footprints (Fig. 1a). After annealing at 1000 K for 10 minutes, formation of Dy carbide is evident by etch pits (an example is marked by white arrow in Fig. 1a') that surround carbide islands. Based on mass balance of C from graphite substrate to Dy islands, we estimate the carbide stoichiometry to be Dy-rich Dy_2C, which differs from a previous report.[2] On an Ar^+ ion bombarded surface at 800 K, however, Dy forms clusters at defects that are often accompanied by islands with a consistent height of 0.61 ± 0.03 nm (circled in Fig. 1b). High resolution STM images reveal that these Dy islands are covered by a graphitic lattice, forming surface intercalated rafts (Fig. 1b'). Structured moiré patterns (Fig. 1b") are observed atop faceted islands. Defects from ion bombardment serve as entry portals for Dy intercalation. DFT calculations on Dy rafts show that instead of a *dilute*, $(\sqrt{3}\times\sqrt{3})R30°$ arrangement of Dy atoms relative to graphite lattice as in stage-1 Dy-GIC, our Dy rafts are *dense* and are composed of three contiguous layers of Dy sandwiched between graphitic layers.

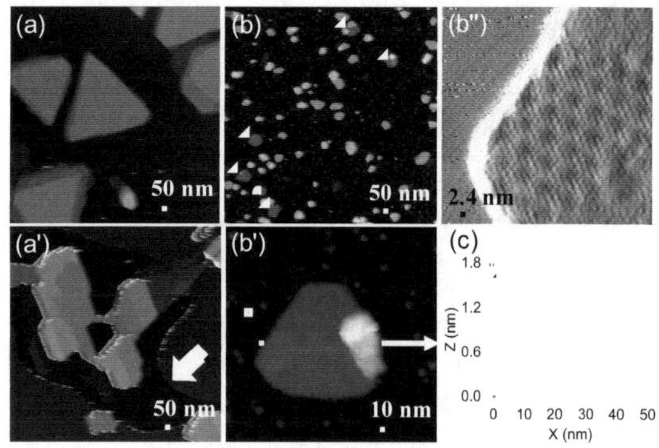

Figure 1. STM images of Dy on graphite. (a) Dy deposited on pristine graphite at 650-800 K. (a') Dy carbide islands that form after annealing at 1000 K. Arrow points to a ring that surrounds a carbide island, indicating graphite erosion. (b) Dy deposited on bombarded graphite at 800 K. Intercalated Dy rafts are circled. (b') Example of a Dy raft with the area in white box enlarged in (b") showing moiré patterns on top of the raft. Line profile in (c) shows the raft height of (b').

CONCLUSIONS

The Dy/graphite system is rich in morphology that spans formation of faceted metallic islands, Dy carbide with etch pits, and intercalated surface Dy rafts. Both experimental and theoretical work support the distinct surface Dy raft from its bulk counterpart. Defects created from ion bombardment serve as entry portals to graphite galleries. Surface intercalated Dy rafts have a good contact with graphite sheets, which is an important and fundamental aspect for future applications.

REFERENCES

1. Hagiwara R. *et al.*, Carbon. 34:1591-1593, 1996
2. Gorovikov S.A. *et al.*, Surf. Sci. 474:98-106, 2001

ACKNOWLEDGMENTS

This work was supported by the Division of Materials Sciences, Basic Energy Sciences, U.S. Department of Energy (DOE). The research was performed at Ames Laboratory, which is operated for the U.S. DOE by Iowa State University under contract No. DE-AC02-07CH11358.

Reactivity Comparison of Complexes of $4f^n5d^1$ and $4f^{n+1}$ Ln^{2+} Ions with Cyclooctatetraene

Megan E. Fieser,[1,2] Chad T. Palumbo,[1] Joseph W. Ziller,[1] William J. Evans[1]

[1]*Department of Chemistry, University of California, Irvine, California 92697, United States*
[2]*Department of Chemistry, University of Minnesota, Minneapolis, Minnesota 55455, United States*; fiese004@umn.edu

INTRODUCTION

Recently, complexes of the entire series of lanthanide metals have been isolated in the +2 oxidation state: [K(2.2.2-cryptand)][Cp'$_3$Ln] (Cp' = $C_5H_4SiMe_3$), **1-Ln**.[1,2] In this ligand field, these Ln^{2+} complexes fall into two categories, with the La, Ce, Pr, Nd, Gd, Tb, Dy, Ho, Er, and Lu complexes being characterized as containing $4f^n5d^1$ ions (in relation to the starting $4f^n$ Ln^{3+} complex), while the Sm, Eu, Tm, and Yb complexes were characterized as traditional $4f^{n+1}$ Ln^{2+} ions.[1-6] Previous studies have shown that complexes of the $4f^n5d^1$ Ln^{2+} ions are able to reduce naphthalene, eq 1,[7] while most traditional $4f^{n+1}$ ions cannot. It was desirable to find a

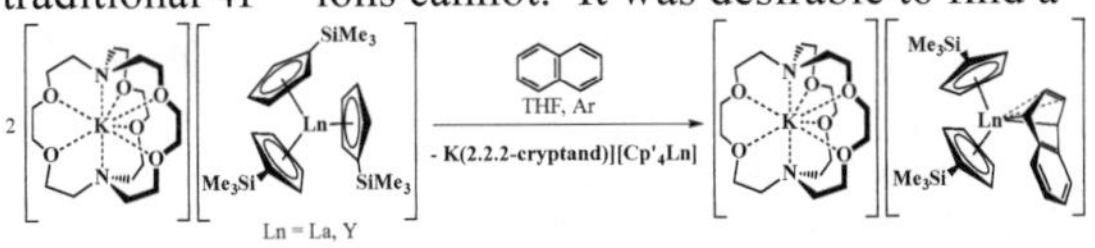

(1)

reagent that Ln^{2+} complexes with both electron configurations could reduce to compare reaction products. 1,3,5,7-Cyclooctatetraene (C_8H_8) was chosen as a substrate since it was successful in differentiating the reactivity of recently isolated U^{2+} and Th^{2+} complexes.[8,9] Therefore, the Ln^{2+} complexes **1-Ln** (Ln = Y, La, Ce, Pr, Nd, Sm, Eu, Dy, Tm, Yb) were treated with C_8H_8. This presentation will compare the reactivity between the $4f^{n+1}$ ions (Sm, Eu, Tm, Yb) and the $4d^1$ (Y) and $4f^n5d^1$ (La, Ce, Pr, Nd, Dy) ions in **1-Ln**.

EXPERIMENTAL METHODS

Reactions with C_8H_8 were done in THF. The **1-Ln** complex was generated quickly in situ before the addition of excess C_8H_8. Reactions were also performed between C_8H_8 and isolated **1-Ln** and reactivity was not found to differ.

RESULTS AND DISCUSSION

Most of the Ln^{2+} complexes studied (**1-Ln**, Ln = Y, La, Ce, Pr, Nd, Sm, Dy, and Tm) reacted readily with C_8H_8. Similar to the reactivity studies with naphthalene, a Ln^{3+} byproduct, [K(2.2.2-cryptand)][Cp'$_4$Ln], was observed for all but La and Sm, although it is the likely byproduct for those metals as well. So far reduced C_8H_8 complexes have been isolated only for the larger metals. Interestingly, the structures of these complexes appear to vary depending on the size of the metal and not the configuration of the Ln^{2+} starting complex. For La and Ce, complexes of the formula [K(2.2.2-cryptand)][Cp'$_2$Ln(C_8H_8)] were isolated, which represent the first examples of a metal complex surrounded by two cyclopentadienyl ligands and a $(C_8H_8)^{2-}$ ligand, eq 2. For slightly smaller metals, Pr, Nd and Sm, the $(C_8H_8)^{2-}$

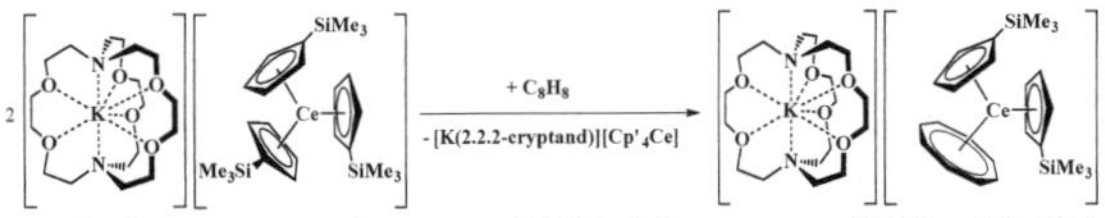

(2)

sandwich complexes, [K(2.2.2-cryptand)][Ln(C_8H_8)$_2$], were isolated, eq 3. **1-Eu** and **1-Yb**, which are the least reducing Ln^{2+} complexes, were not reactive with C_8H_8.

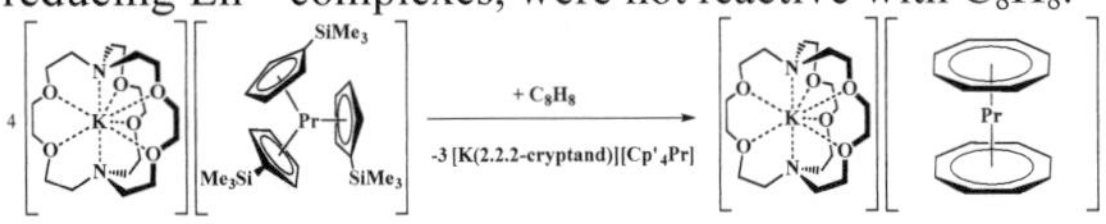

(3)

CONCLUSIONS

Reactivity studies of [K(2.2.2-cryptand)][Cp'$_3$Ln] complexes with C_8H_8 indicate that the products formed are more affected by the metal size rather than by the electron configuration of the divalent ion. Instead, the electron configuration has a large effect on the reducing ability of the Ln^{2+}, indicating the new $4f^n5d^1$ complexes can provide analogous reduction chemistry with a wider range of possible substrates than the traditional $4f^{n+1}$ complexes.

REFERENCES

1) MacDonald, M. R. *et al.*, *J. Am. Chem. Soc.* **2013**, *135*, 9857.
2) Fieser, M. E. *et al.*, *J. Am. Chem. Soc.* **2015**, *137*, 369.
3) Cassani, M. C. *et al.*, *Chem. Commun.* **1997**, 1563.
4) Hitchcock, P. B. *et al.*, *Angew. Chem., Int. Ed.* **2008**, *47*, 1488.
5) MacDonald, M. R. *et al.*, *J. Am. Chem. Soc.* **2011**, *133*, 15914.
6) MacDonald, M. R. *et al.*, *J. Am. Chem. Soc.* **2012**, *134*, 8420.
7) Kotyk, C. M. *et al.*, *Organometallics* **2015**, *34*, 2287.
8) Langeslay, R. R. *et al.*, *Chem. Sci.* **2015**, *6*, 517.
9) Windorff, C. J. *et al.*, *Chem. Eur. J.* **2016**, *22*, 772.

ACKNOWLEDGMENT

The U.S. National Science Foundation is gratefully acknowledged for support of this research (CHE-1565776).

Reactivity of a Uranium(IV) Dialkyl Complex Supported by a Rigid NON-Donor Ligand: Alkyl Exchange and C–H Activation of Pyridines

Nicholas R. Andreychuk and David J. H. Emslie*

Department of Chemistry and Chemical Biology, McMaster University, Canada, andreynr@mcmaster.ca

INTRODUCTION

The burgeoning field of non-carbocyclic organoactinide chemistry is being driven forward through the development of new ligand systems that offer diverse structural and electronic motifs, and as a result, exciting and highly sought-after molecular species have become accessible, and new and unusual chemical transformations continue to be uncovered.[1] In that vein, the Emslie group has developed extremely rigid, tridentate diamido NON- and NSN-donor ligands for the support of thermally robust neutral- and cationic organothorium(IV) species,[2] as well as uranium chloro complexes in various oxidation states,[3] and uranium(IV) alkyl complexes.[4]

RESULTS AND DISCUSSION

Recently, we have explored the reactivity profile of the neutral uranium(IV) dialkyl complex $[(XA_2)U(CH_2SiMe_3)_2]$ (**1**; XA_2 = a rigid xanthene-based NON-donor ligand). Dialkyl complex **1** underwent an unusual alkyl exchange reaction upon treatment with neopentyllithium in benzene; such a conversion likely proceeded *via* anionic trialkyl 'ate' intermediates, and complexes of this type proved accessible in coordinating solvents. Dialkyl complex **1** also reacted readily with pyridine derivatives to yield uranium(IV) monoalkyl species featuring cyclometalated pyridine-based ligands *via* σ-bond metathesis.

This contribution will highlight recent developments in this area, including the synthesis, spectroscopy, and crystallographic characterization of new complexes, and mechanistic studies.

REFERENCES

1. For example, see: King, D. M., Liddle, S. T., *Coord. Chem. Rev.*, **2014**, *266-267*, 2-15.
2. See: Cruz, C. A., Emslie, D. J. H., Harrington, L. E., Britten, J. F., Robertson, C. M., *Organometallics*, **2007**, *26*, 692-701.
3. Vidjayacoumar, B., Ilango, S., Ray, M. J., Chu, T., Kolpin, K. B., Andreychuk, N. R., Cruz, C. A., Emslie, D. J. H., Jenkins, H. A., Britten, J. F., *Dalton Trans.* **2012**, *41*, 8175-8189.
4. Andreychuk, N. R., Ilango, S., Vidjayacoumar, B., Emslie, D. J. H., Jenkins, H. A., *Organometallics,* **2013**, *36*, 1466-1474.

ACKNOWLEDGMENTS

We would like to thank the NSERC of Canada, the Government of Ontario, and McMaster University for providing financial support to this project.

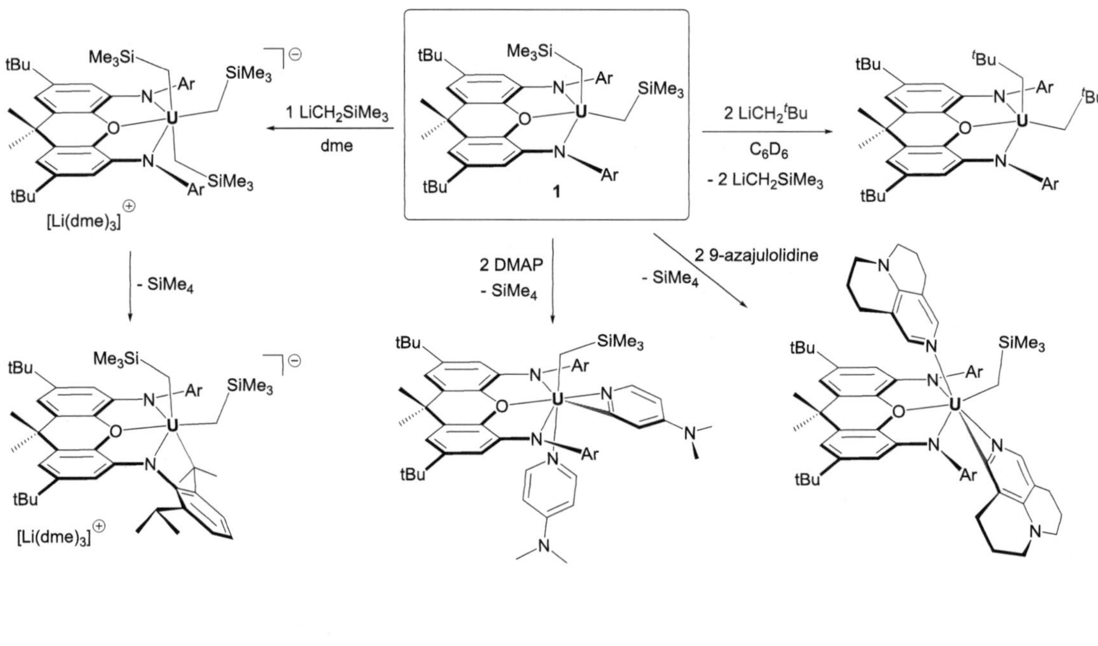

Reactivity of homoleptic trisalkyl lanthanum grafted on mesoporous silica

Smita Patnaik, Zhuoran Wang, Sebastian Manzano, Naresh Eedugurala, Igor I. Slowing, Marek Pruski, Aaron D. Sadow*

Department of Chemistry, Iowa State University, USA
spatnaik@iastate.edu

INTRODUCTION

Development of catalytic systems for achieving efficient, selective, and environmentally benign reactions are always challenging. Organolanthanide complexes have been shown to be effective in hydroelementation (e.g., hydroamination, hydrosilylation, hydroboration) of olefins.[1] Supporting these complexes on surfaces is advantageous for separation and recycling of the catalyst. Several studies on homogeneous rare-earth metal-catalyzed hydroamination of alkenes were reported.[1] However, there are few reports of silica-supported catalysts for the hydroamination of aminoalkenes.[2,3] One part of the challenge is that the lanthanide catalyst appears to deactivate on the surface at elevated temperatures.[4] Our focus is to improve these systems by performing the hydroamination reactions at lower temperature, while comparing the selectivity of amino dialkenes cyclizations by homoleptic rare-earth alkyl $Ln\{C(SiHMe_2)_3\}_3$ in solution and grafted on mesoporous silica nanoparticles. In the grafted complexes, the surface can influence the reactivity of metal center. Therefore, we are also interested in understanding the characteristic properties of the reactive surface species in the hydroamination reaction.

EXPERIMENTAL METHODS

Synthesis of $La\{C(SiHMe_2)_3\}_n$@MSN

$La\{C(SiHMe_2)_3\}_3$ (0.250 g, 0.353 mmol) was stirred with MSN_{550} material (0.390 g) at room temperature in benzene (10 mL) for 20 h to produce $La\{C(SiHMe_2)_3\}_n$@MSN_{550} (0.410 g). This material was characterized by DRIFT, EA, and solid state $^1H/^{13}C$ NMR spectroscopies.

RESULTS AND DISCUSSION

$La\{C(SiHMe_2)_3\}_3$ was grafted on high surface area SBA-15 type mesoporous silica nanoparticles previously heated under vacuum at 550 °C (MSN_{550}) to give $La\{C(SiHMe_2)_3\}_n$@MSN_{550}.

Surface Grafting Reaction

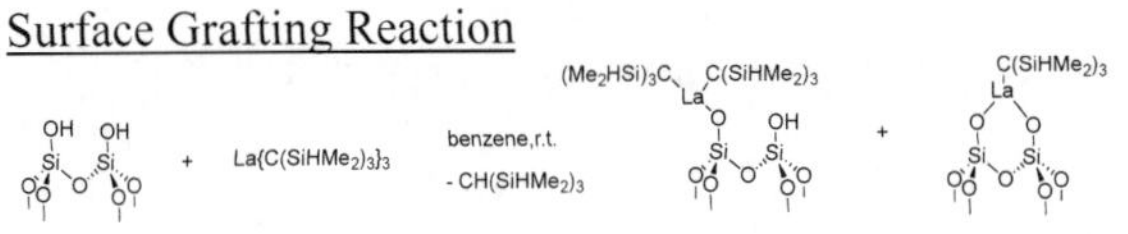

Diffuse reflectance infrared (DRIFT) spectra were collected for MSN_{550} and $La\{C(SiHMe_2)_3\}_n$@ MSN_{550}. MSN_{550} contained an absorption at 3745 cm^{-1} assigned to isolated SiOH, which is not detected in the IR spectrum of $La\{C(SiHMe_2)_3\}_n$@ MSN_{550}, indicating that isolated silanols are consumed in the grafting process. The Si-H region of $La\{C(SiHMe_2)_3\}_n$@MSN_{550} contained a band at 2108 cm^{-1} assigned to the terminal SiH stretching mode and a band at 1858 cm^{-1} assigned to La←H–Si.

Catalytic Hydroamination of Amino dialkene

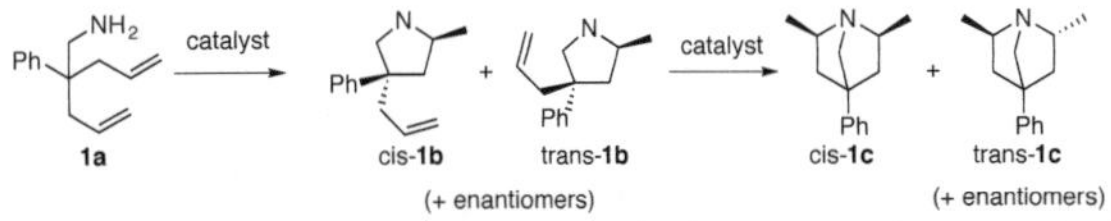

We attempted to compare catalytic conversions under homogeneous and heterogeneous conditions by using $La\{C(SiHMe_2)_3\}_3$ as a fully soluble catalyst precursor and $La\{C(SiHMe_2)_3\}_n$@MSN as a single-site surface-grafted precursor. Reaction of **1a** with the heterogeneous catalyst gives a combination of monocyclic and bicyclic products at r.t. after 24 h.

CONCLUSIONS

Comparisons to hydroaminations catalyzed by $La\{C(SiHMe_2)_3\}_3$ reveal that there are significant effects of the surface on the catalytic site, in contrast to an yttrium silizido-grafted system.[3]

REFERENCES

1. Marks T. J. *et al.*; *J. Am. Chem. Soc.* 118:9295-9306, 1996.
2. Anwander R. *et al.*; *J. Am. Chem. Soc.* 132:16368-16371, 2010.
3. Sadow A. D. *et al.*; *Organometallics* DOI: 10.1021/organomet.6b00956
4. Basset J. M. *et al.*; *J. Am. Chem. Soc.* 119:12408-12409, 1997.

ACKNOWLEDGMENTS

This research was supported by the U.S. Department of Energy, Office of Basic Energy Sciences, Division of Chemical Sciences, Geosciences, and Biosciences. The Ames Laboratory is operated for the U.S. Department of Energy by Iowa State University under Contract No. DE-AC02-07CH113

Recent Advances in the +2 Oxidation State Chemistry of the Rare Earth Metals

William J. Evans

Department of Chemistry, University of California, Irvine, California 92697-2025, United States; wevans@uci.edu

INTRODUCTION

It has recently been discovered that the +2 oxidation state is accessible in crystallographically-characterizable molecular complexes for all the rare earth metals and not just Eu, Yb, Sm, Tm, Dy, and Nd.[1-6] Complexes of the new +2 ions were discovered using the tris(cyclopentadienyl) coordination environments $(Cp'_3)^{3-}$ and $(Cp''_3)^{3-}$, where Cp' = $C_5H_4SiMe_3$ and Cp'' = $C_5H_3(SiMe_3)_2$. In these ligand fields, reduction of $4f^n$ Ln^{3+} lanthanide complexes, Cp'_3Ln and Cp''_3Ln, formed $4f^n5d^1$ Ln^{2+} products, $(Cp'_3Ln)^{1-}$ and $(Cp''_3Ln)^{1-}$ rather than the traditional $4f^{n+1}$ Ln^{2+} ions of Eu, Yb, Sm, Tm, Dy, and Nd.[1-6]

The synthetic studies of the past few years have led to a coherent generalization of ligand effects on the reduction reactions that lead to these +2 complexes. This postulated that in the coordination environment of three Cp' or Cp'' ligands, crystallograpically-characterizable species can be isolated, but that with more strongly donating ligands like C_5Me_4H, C_5Me_5, and $N(SiMe_3)_2$, the analogous +2 species are too reactive to isolate. On the other hand, reductions of LnA_3 complexes (A = anion) of the latter, more strongly-donating ligands are excellent routes to reduced dinitrogen complexes containing $(N{=}N)^{2-}$ ligands, whereas reductions of Cp' and Cp'' ligated complexes do not form $(N{=}N)^{2-}$ products. This presentation will describe how the recent synthetic results have shown that these generalizations are not correct!

EXPERIMENTAL METHODS

Reduction reactions are typically done in THF with potassium graphite, potassium smear, sodium, lithium, or rubidium. The reactions are always done quickly often using a flash reduction column[2-4] and often at temperatures of −35 °C and below.

RESULTS AND DISCUSSION

Although molecular complexes of Sc^{2+} have been in the literature since 1991,[7] no crystallograpically-characterized examples had been reported. The reductive chemistry of $Sc[N(SiMe_3)_2]_3$ has now been explored and has provided crystalline complexes of Sc^{2+}.[8] The isolation of these complexes disproves one of the generalizations above and has led to a search for the factors involved. These investigations have suggested that the long-held generalization that yttrium has chemistry similar to the late lanthanides may not apply to the chemistry of +2 ions. The reductive reactivity of scandium has further broadened the area of reduced dinitrogen complexes of the rare earths. Products with unexpected compositions and structures have been isolated and will be described.

CONCLUSIONS

Our understanding of the chemistry of the rare earth metals and the actinides in the +2 oxidation state continues to evolve as new syntheses are discovered.

REFERENCES

1. Hitchcock, P. B. *et al.*, Angew. Chem., Int. Ed. 47:1488-1491, 2008
2. MacDonald, M. R. *et al.*, J. Am. Chem. Soc. 135:9857-9868, 2013
3. Fieser, M. E. *et al.*, J. Am. Chem. Soc. 137:369-382, 2015
4. Corbey, J. F. *et al.*, Organometallics 34:3909-3921, 2015
5. Kotyk, C. M. *et al.*, Chem. Sci. 6:7267-7273, 2015
6. Evans, W. J., Organometallics 35:3088-3100, 2016
7. Cloke, F. G. N. et al., J. Chem. Soc., Chem. Commun. 1372-1373, 1991
8. Woen, D. H. et al., Angew. Chem. Int. Ed. 56:2050-2053, 2017

ACKNOWLEDGMENT

The U.S. National Science Foundation is gratefully acknowledged for support of this research (CHE-1565776).

Recent Advances in Triamido-Supported Actinide-Ligand Multiple Bonding

Stephen T. Liddle[1]*

[1]School of Chemistry, The University of Manchester, Oxford Road, Manchester, UK
*steve.liddle@manchester.ac.uk

INTRODUCTION

There is currently great interest in the nature and reactivity of molecular actinide-ligand multiple bonds.[1] This is because the nature and extent of 5f/6d orbital participation in actinide-ligand bonding is still a topic of debate. Indeed, the unique formal orbital-hybridisation and crystal field splitting patterns available to actinides coupled to significant spin orbit and relativistic effects promises novel electronic structure, reactivity and magnetism. However, probing and quantifying these characteristics remains a real challenge experimentally and theoretically.

It could be argued that by definition the best chance to study covalency and electronic structure of actinide-ligand bonds is in multiple bond scenarios.[2,3] Thus, we have undertaken a programme of research targeting actinide-ligand multiple bonds where the ligand is a pnictide or chalcogen donor-based system. The field of non-aqueous uranium chemistry is burgeoning, but arguably the most valuable understanding will emerge from comparative studies where iso-structural series of molecules are prepared where only the central actinide metal is varied; only then can variation in covalency and 5f/6d orbital contributions to the chemical bonding be mapped out across the actinide elements.

In order to do this robust supporting ligands are required, and in that regard we have found that certain triamido-ligand combinations are exceptionally effective at stabilising uranium- and thorium-ligand multiple bonds involving main group fragments of interest in their own right.

RESULTS AND DISCUSSION

We have principally focussed on the application of uranium and thorium triamidoamine complexes where the triamidoamine ligand bears three very sterically demanding triisopropylsilyl groups, Scheme 1. This ligand occupies the vast majority of the actinide coordination sphere in these complexes leaving a well defined pocket within which to install multiple bond linkages whilst avoiding other vacant coordination sites that might enable deleterious decomposition pathways.

This contribution will provide an overview of our progress to date covering oxo, nitride, parent-imido, -phosphinidene, -arsinidene, and arsendio complexes of uranium and thorium,[2-10] and their electronic structure, bonding, reactivity, and magnetism will be discussed. If time allows we will describe our more recent and unpublished results in this area.

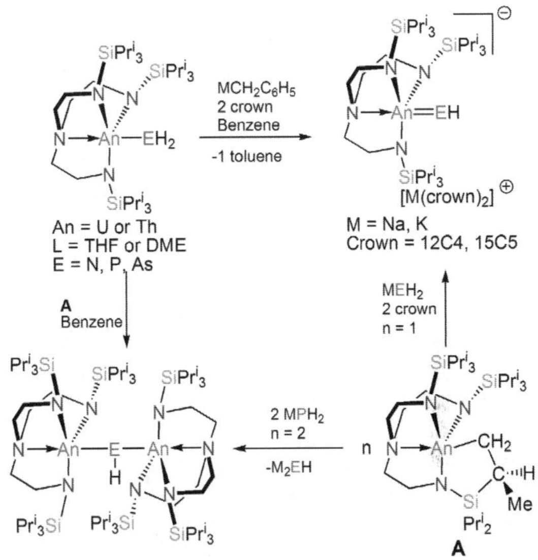

Scheme 1. General overview of actinide-ligand multiple bond linkages that can be prepared in a triamidoamine ligand environment.

CONCLUSIONS

By combining sterically demanding triamido ligand combinations at uranium and thorium novel actinide-ligand multiple bond linkages can be accessed and probed.

REFERENCES

1. Liddle, S. T. Angew. Chem. Int. Ed. 54:8604, 2015.
2. Wildman, E. P. *et al.* Nat. Comm. 8:14769, 2017.
3. King, D. M. *et al.* Nat. Comm. 7:13773, 2016.
4. Wildman, E. P. *et al.* Nat. Comm. 7:12884, 2016.
5. Gardner, B. M. *et al.* Angew. Chem. Int. Ed. 54:15250, 2015
6. Gardner, B. M. *et al.* Nat. Chem. 7:582, 2015.
7. Gardner, B. M. *et al.* Angew. Chem. Int. Ed. 53:4484, 2014.
8. King, D. M. *et al.* J. Am. Chem. Soc. 136:5619, 2014.
9. King, D. M. *et al.* Angew. Chem. Int. Ed. 52:4921, 2013.
10. King, D. M. *et al.* Nat. Chem. 5:482, 2013.
11. King, D. M. *et al.* Science 337:717, 2012.

ACKNOWLEDGMENTS

We gratefully acknowledge continued and generous support from the Royal Society, ERC, EPSRC, and NNL.

Re-establishing an American Rare Earth Industry: The Challenges and Opportunities

Brock O'Kelley

Kroll Institute for Extractive Metallurgy, Colorado School of Mines, U.S.A., okelley@mines.edu

INTRODUCTION

Rare earth elements are relatively unheralded components of our everyday life yet outside a limited number of government agencies, industrial consumers and select universities, few understand the importance of having both resources and the expertise required to extract and separate the chemically similar elements into usable products.

Indeed, RE elements are currently plentiful with very low prices from China. However, as with many elements extracted from the earth, prices and availability have not always been stable and many companies whose products depend on rare earths are never comfortable single sourcing critical components, especially when cyclical availability has been observed. In this session, we will examine:

- A brief history of the RE industry and why expertise is important
- Sources for RE feedstock within the US; ore types/deposits and recycling
- RE processing challenges; separation schemes differ depending on source
- Where do opportunities lie?

Reversible Phase Transformation of γ-Ce_2S_3 by Introducing Oxygen And Use of the Product as a Wave Absorber

Yohei Kubota[1], Shinji Hirai[2*], Eiji Nakamura[3], Kishio Hidaka[4]

[1]Graduate School of Engineering, Muroran Institute of Technology, Japan
[2]Research Center for Environmentally Friendly Materials Engineering, Muroran Institute of Technology, Japan
[3]Santoku Corporation, Japan
[4]Faculty of Engineering, Yamagata University, Japan
[*]hirai@mmm.muroran-it.ac.jp

By introducing oxygen into γ-Ce_2S_3 powder, which is a high-temperature stable phase, it became possible to produce β single phase or β, γ mixed phase Ce_2S_3 based on reversible transformation. Moreover, the wave absorptivity of the resin with β, γ mixed-phase powder dispersed therein proved close to the practical level.

INTRODUCTION

It is known that the phase transformation temperature from tetragonal β-Ce_2S_3 with solid solution ranges from Ce_2S_3 to Ce_3S_4 into the semiconductor Th_3P_4-type cubic γ-Ce_2S_3 is 1423±50 K. However, in the case of β-Ce_2S_3 containing large amounts of oxygen, oxygen acts as a stabilizer of β-Ce_2S_3 and no phase transformation occurs to γ-Ce_2S_3 even if it is heated above the phase transformation temperature. The oxygen-stabilized β-Ce_2S_3 can also be expressed as $Ce_{10}S_{15-X}O_X$ $(0 \leq X \leq 1)$ [1]. Our research group found that the dielectric loss tangent of this β-phase varies with the oxygen content and that a sintered compact using the powder with an oxygen content of 0.95 wt% or more can provide a comparatively high dielectric loss tangent[2]. For example, a sintered compact containing 1.8 wt% oxygen and prepared from CeO_2 powder via CS_2 gas sulfurization (temperature and time of sulfurization: 1473K, 25.2 ks) and pressureless sintering method (temperature and time of sintering: 1473 K, retention time: 1 ks) gave a dielectric loss tangent of 11.4 at a frequency of 1 KHz. By combining a dielectric substance with significant dielectric loss tangent with a polymer, it can be applied to an in-vehicle wave absorber, which is required to be lighter and more compact than a resistance type or magnetic loss type wave absorber.

The authors considered that if the phase transformation from β-Ce_2S_3 to γ-Ce_2S_3 is reversible, β-Ce_2S_3 will be formed by introducing oxygen into γ-Ce_2S_3 at a temperature below the phase transformation temperature. In this study, we aimed to introduce oxygen while heating the γ-Ce_2S_3 powder to the phase transformation temperature, or lower, by using a pulse CVI (Chemical Vapor Infiltration) apparatus to generate β single phase or β, γ mixed phase.

EXPERIMENTAL METHODS

As a starting material, commercially available γ-Ce_2S_3 powder (particle size: 7 μm) was used and oxygen was introduced using a pulse CVI apparatus. A stainless-steel cathode and anode were mounted at both ends of a reaction tube, and a porous alumina basket containing γ-Ce_2S_3 powder was placed in front of the cathode. First, the reaction tube was heated to 1473 K while applying voltage (4.5 kV) between the two electrodes. Next, an operation including, per 1 pulse, evacuation in the reaction tube (10 s), the introduction of Ar-7%H_2 gas containing O_2 (2 s), and subsequent holding (10 s) was repeated between 75 to a maximum of 250 times. The obtained powder was dispersed at a rate of 50 wt% with respect to an epoxy resin and cured by holding it for 43.2 ks while heating at 338 K. For this resin, the wave absorption characteristics from 3 GHz to 19.4 GHz were measured by the coaxial waveguide method.

RESULTS AND DISCUSSION

Due to reversible transformation, the product depends on the number of pulses; a β, γ mixed phase is obtained at 75 to 100 pulses, a β-Ce_2S_3 single phase above 100 pulses up to 150 pulses, and at 250 pulses Ce_2O_2S is formed as an impurity. Since the β, γ mixed phase is produced, even when heating to 1473 K for 3.6 ks without applying voltage, a pulse or voltage application effect was confirmed. It was further found that, whether in β single phase or β, γ mixed phase, the oxygen content of the powder showed a linearly increasing tendency with an increase in the number of pulses. The wave absorptivity of the resin in which β, γ mixed phase Ce_2S_3 was dispersed in epoxy resin was nearly at the practical level, reaching 19 dB in the high-frequency range of 11 GHz to 15 GHz.

CONCLUSIONS

By introducing oxygen into γ-Ce_2S_3 via the pulse CVI method, the production of β single phase or β, γ mixed phase became possible due to reversible transformation. The wave absorptivity of the resin, in which the β, γ mixed phase was dispersed, was 19 dB at 11 GHz to 15 GHz.

REFERENCES

1. Schleid T.*et al.*: J. Less-Common Met., 175:309-319, 1991.
2. Hirai S.*et al.*: Japan Patent No.4551987.

Selective *f*-Element Extraction Agents based on Tripodal Scaffolds

Shannon M. Biros

Department of Chemistry, Grand Valley State University, USA, biross@gvsu.edu

INTRODUCTION

A recent resurgence in research on lanthanide (Ln) and actinide (An) coordination chemistry stems from society's need for alternative energy sources, which may include nuclear power.[1] Specifically, the processing of spent nuclear fuel as well as the remediation of contaminated sites remains an unsolved problem. One strategy for both of these applications involves the separation of the heavy An ions (Np, Pu, Am, Cm) from the lighter An ions and various Ln ions present in waste solutions.

Along these lines, the Ln metals have seen increasing use in applications ranging from optical materials to biosensors due to their unique photophysical properties.[2,3] The need for effective Ln separations is further driven by the growing desire to recycle materials using Ln metals (e.g., hybrid car batteries, magnets, smart phones) combined with an increased focus on the purification of Ln metals from raw sources.

This seminar will present work toward the development of selective extraction agents for Ln and An ions from aqueous media. Current investigations are focused on extractant preparation, characterization of the Ln and An complexes and determination of the extraction properties of these ligands ions out of aqueous media.

EXPERIMENTAL METHODS

Compounds **1** and **2** were prepared according to procedures published by our group.[4,5] Extraction experiments were carried out following established conditions,[6] with an aqueous solution of 10^{-4} M Ln or An nitrate in 1M nitric acid and an organic solution of 10^{-3} M ligand in CH_2Cl_2. Extraction efficiencies were determined by the Arsenazo method[7] or ICP.

RESULTS AND DISCUSSION

Recently, compound **1** (Figure 1) has demonstrated the ability to selectively extract Tb^{3+} out of 1M nitric acid solutions with an extraction efficiency of 23%. This value is remarkable when compared to the single digit extraction efficiencies obtained for the other lanthanide ions (all except Pm^{3+}), Th^{4+} and UO_2^{2+}.[4]

As a way to investigate the influence of ligand size on extraction preferences, we prepared compound **2** (Figure 1) using synthetic chemistry based on well trodden paths.[5] Under the same extraction conditions, this compound demonstrates preferences for the larger lanthanides as well as Th^{4+} and UO_2^{2+}. When the nitric acid concentration is increased to 5M for these experiments, nearly quantitative Th^{4+} and UO_2^{2+} extraction efficiencies are observed.

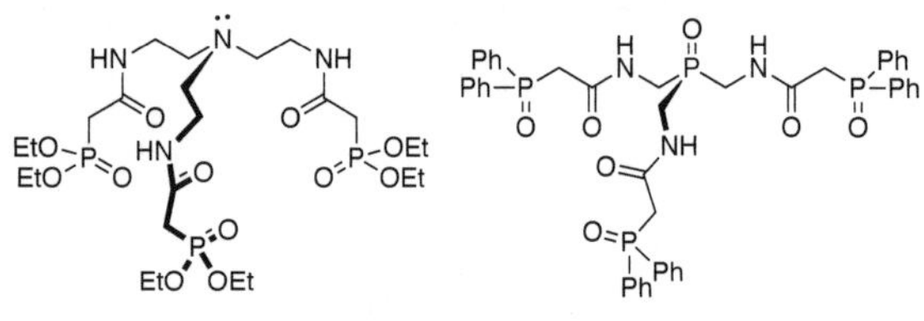

1 **2**

Figure 1. Tripodal CMPO ligands **1** and **2**.

These ligands and their resultant metal complexes have been characterized in solution and the solid state by NMR, IR, MS, and luminescence experiments.

CONCLUSIONS

The efficient extraction and purification of Ln and An ions has implications in the areas of alternative energy, recycling of materials, and isolation of these precious metals from raw sources. We report here a contribution to this field with two new organic ligands that incorporate the well-known CMPO motif. These ligands demonstrate the ability to selectively extract one Ln ion (Tb^{3+}) away from all others, as well as the efficient extraction of the An ions Th^{4+} and UO_2^{2+}.

REFERENCES

1. Roberto, J. and Diaz de la Rubia, T., "*Basic Research Needs for Advanced Nuclear Energy Systems*" Office of Basic Energy Sciences, US DOE, 2006.
2. Atwood, D.A., "*The Rare Earth Elements: Fundamentals and Applications*" 1st ed., 2012.
3. de Bettencourt-Dias, A., "*Luminescence of Lanthanide Ions in Coordination Compounds and Nanomaterials*", 1st ed., 2014.
4. Sartain, H.T. *et al.*, Inorg. Chim. Acta. 426:126-135, 2015.
5. Coburn, K.M. *et al.*, Inorg. Chim. Acta. 449:96-106, 2016.
6. Arnaud-Neu, F. *et al.*, J. Chem. Soc. Perkin Trans. 2 1175-1182, 1996.
7. Ivanov, V.M. *et al.* J. Analytical Chem. 26:519-523, 2001.

ACKNOWLEDGMENTS

We thank Grand Valley State University (Weldon Fund, OURS, CSCE) and The University of Tampa (Dept. of Chemistry, Biochemistry and Physics, SSRF) for financial support of this work. We are grateful to the NSF for student support (REU-1062944) and NMR instrumentation (300 MHz JEOL, CCLI-0087655), as well as Pfizer, Inc. for the generous donation of a Varian Inova 400 FT NMR.

Sensitised Ln (III) Emission and Excited State Dynamics of Cofacial 'Pacman' Porphyrin Terpyridine Complexes

Jane Y. Liew,[1] Matthias Schwalbe,[2] Evan G. Moore[1]

[1] School of Chemistry and Molecular Biosciences, University of Queensland, Brisbane, 4072, Australia.
[2]Institute of Chemistry, Humboldt Universität zu Berlin, Berlin, 12489, Germany.
E-mail: egmoore@uq.edu.au

Luminescence from trivalent lanthanide cations has received significant interest due to their characteristically sharp emission bands and long-lived excited states.[1] However, since direct *f-f* absorptions are Laporte forbidden, a strongly absorbing organic chromophore is often required, acting as an antenna to absorb light and transfer this energy to the 4*f* excited states.

The use of various porphyrin derivatives as sensitisers for Ln(III) emission is well documented,[2,3] in particular since their excited state energy levels are well matched with those of Near-Infra-Red (NIR) emitting cations such as Yb(III) and Nd(III), which have applications in advanced display technologies,[4] telecommunications,[5] and biological imaging.[6]

In this work, we have prepared[7] a cofacial Zn^{II} porphyrin derivative covalently anchored to a terpyridine chelate using a rigid xanthene moiety, [Zn(PXT)], and its interactions with several trivalent Ln(III) cations (Ln = Nd, Yb and Lu) have been examined. Solution stability constants determined via UV-Vis titration demonstrate exclusive 1:1 ML complex formation, which is also evidenced using NMR and MS techniques.

Figure 1. Illustration of 1:1 [Zn(PXT)Ln$(NO_3)_3$] complex formation.

The resulting [Zn(PXT)Ln$(NO_3)_3$] complexes with Ln = Yb and Nd display characteristic metal centered emission in the NIR region, which has been quantified. Moreover, the kinetics of the energy transfer processes from the Zn(II) porphyrin donor to Ln(III) acceptor have also been investigated using ultrafast transient absorption techniques, allowing an insight into the mechanism and efficiency of the antennae effect.

Funding: The Australian Research Council (DP170101895) is gratefully acknowledged.

References: **[1.]** Bünzli, J.-C. G., *Coord. Chem. Rev.* **2015,** *293-294*, 19. **[2.]** Rusakova, N.; Semenishyn, N.; Korovin, Y., *J. Porph. Phthal.* **2010,** *14*, 166. **[3.]** Bulach, V. S., F.; Hosseini, M. W., *Coord. Chem. Rev.* **2012,** *256*, 1468. **[4.]** Pizzoferrato, R.; Lagonigro, L.; Ziller, T.; Di Carlo, A.; Paolesse, R.; Mandoj, F.; Ricci, A.; Lo Sterzo, C. *Chem. Phys.* **2004**, *300*, 217. **[5.]** Bünzli, J.-C. G.; Comby, S.; Chauvin, A.-S.; Vandevyver, C. D. B., *J. Rare Earths* **2007,** *25*, 257. **[6.]** Foley, T. J.; Harrison, B. S.; Knefely, A. S.; Abboud, K. A.; Reynolds, J. R.; Schanze, K. S.; Boncella, J. M. *Inorg. Chem.* **2003**, *42*, 5023. **[7.]** Schwalbe, M.; Wrzolek, P.; Lal, G.; Braun, B;

Eur. J. Inorg. Chem. **2014,** *2014 25*, 4209.

Small Molecule Activation by Multimetallic Uranium Nitrides

Marta Falcone, Lucile Chatelain, Rosario Scopelliti and Marinella Mazzanti*

Institut des Sciences et Ingénierie Chimiques, Ecole Polytechnique Fédérale de Lausanne (EPFL), CH-1015 Lausanne, Switzerland, marinella.mazzanti@epfl.ch

INTRODUCTION

Uranium nitrides are attractive candidates for stoichiometric and catalytic N-transfer reactions, small molecule transformation,[1,2] and for advanced nuclear fuels.[3] In recent years an increasing number of molecular nitride compounds of uranium has been synthetized.[2,4] However, the reactivity of bridging nitrides remains basically unexplored. The N-C bond formation reactions are key steps in the construction of value-added chemical compounds such as amino acids, pharmaceuticals or agrochemicals. Particularly desirable are the synthetic methods using cheap and available C_1 feedstock such as carbon dioxide or carbon monoxide. *Tert*-butoxysiloxides were shown to be very effective ligands in supporting the reactivity of low valent *f* elements with small molecules due to their ability to stabilize heteromultimetallic frameworks.[5-7]

RESULTS AND DISCUSSION

Here we will report the synthesis of multimetallic nitride-bridged diuranium(IV) and diuranium(III) complexes supported by a flexible potassium-siloxide framework. We will also describe the reactivity of the nitride bridged diuranium(IV) complex Cs{(μ-N)[U(OSi(OtBu)$_3$)$_3$]$_2$}, **1**[6] with CO_2, CS_2 and CO, leading to the formation of N-C bonds.[8,9] This reactivity is remarkable in f-element chemistry, affording diverse N-functionalized products (Figure 1), the thiocyanate/trithiocarbonate complex, **2** and the unprecedented dicarbamate product, **3**. The high reactivity of complex **1** is also confirmed by its ability to cleave one of the strongest bonds in nature, the carbon oxygen bond in carbon monoxide, leading to the oxo/cyanide complex, **4**. The product formed has a [CsU$_2$(μ-CN)(μ-O)] core indicating that the three cations cooperate to cleave CO.

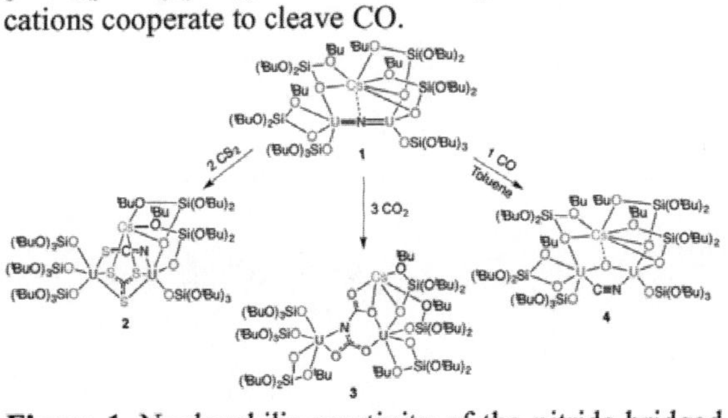

Figure 1. Nucleophilic reactivity of the nitride-bridged di-Uranium(IV) complex with CS_2, CO_2 and CO.

Furthermore, we will show that increasing the electron density on complex **1** by reducing the metal centers leads to a multimetallic M_3UNU (M= Cs^+ or K^+) nitride bridged diuranium(III) complex[10] (Figure 2) which shows extraordinary reactivity towards small molecules. The nature of the alkali cations binding the nitride group is key in the isolation and in the reactivity of these species.

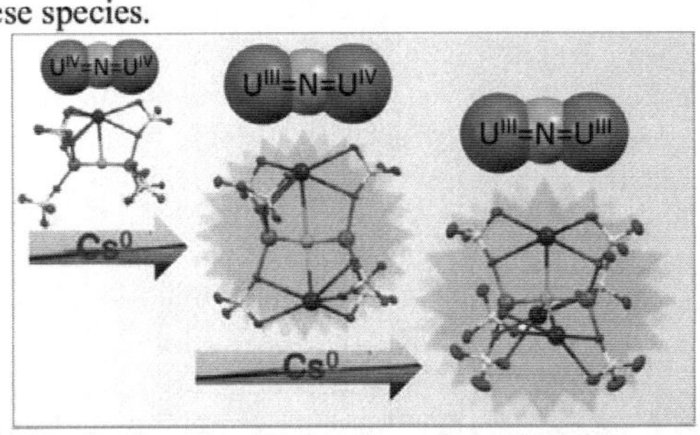

Figure 2. Stepwise Reduction of the U=N=U core to afford a U(III) nitride.

CONCLUSIONS

The results presented open new pathways for the selective synthesis of new organic molecules from metal nitride and the abundant and cheap, CO and CO_2 feedstock and demonstrate that nitride functionalization is readily available to uranium. The presented results will show that the assembly of low-valent metals and alkali ions through bridging nitrides provides an optimized arrangement for the cooperative binding and activation of small molecules. Such novel reactivity of uranium nitrides will certainly inspire new approaches to stoichiometric and catalytic N-C bond and N-H bond formation.

REFERENCES

1. Haber, F. German patent DE 229126, 1909.
2. King, D. M. *et al,* Coord. Chem. Rev. 2014, 266, 2.
3. Silva, G. W. C. *et al,* Inorg. Chem. 2009, 48, 10635.
4. Tsoureas, N. *et al,* Chem. Sci. 2016, *7*, 4624.
5. Mougel, V. *et al,* ACIE 2012, *51*, 12280.
6. Camp, C. *et al,* J. Am. Chem. Soc. 2013, *135*, 12101.
7. Andrez, J. *et al* ACIE 2014, *53*, 10448.
8. Falcone, M., *et al,* ACIE 2016, *55*, 4074.
9. Falcone, M. *et al,* ACIE. 2016, *55*, 12290.
10. Chatelain, L. *et al,* J. Am. Chem. Soc. 2016, *138*, 1784.

ACKNOWLEDGMENTS

"The authors would like to thank the Swiss National Science Foundation and from the Ecole Polytechnique Fédérale de Lausanne (EPFL) for providing financial support to this project".

Small Molecule Activation Induced by Lanthanides

Peter W. Roesky[1]

[1]Institut für Anorganische Chemie, Karlsruher Institut für Technologie (KIT), Germany, roesky@kit.edu

INTRODUCTION

The activation of small molecules has become an emerging research topic over the last years. This research topic ranges from pure academic challenges to industrial applications. The typical examples of small molecules investigated over a long period are O_2 and N_2, which can be used for the formation of C-O and C-N bonds. O_2 is also widely used as oxidation reagent. Due to the discussion of fossil feedstock, the activation of methane and CO_2 faced an increasing research activity in recent years.[1]

Our research interest is focused on the activation of small molecules by lanthanide compounds. We are especially interested in the activation of small molecules, which are important but not in the broad research focus. For some time, we are investigating the reactions of rare earth compounds with SO_2. These results are summarized in the present contribution.

RESULTS AND DISCUSSION

First, an overview on the reaction of SO_2 with di- and trivalent lanthanide compounds is given.

The reaction of $[(\eta^5\text{-}C_5Me_5)_2Ln(THF)_2]$ (Ln = Sm, Eu, Yb) with SO_2 resulted in two major reaction pathways: a) the reductive coupling of two SO_2 molecules to form the dithionite anion $S_2O_4^{2-}$ and b) the nucleophilic attack of one samarocene C_5Me_5 ligand on the sulfur atom of SO_2. As products the dithionite complex $[\{(\eta^5\text{-}C_5Me_5)_2Sm\}_2(S_2O_4)]$ and the sulfinate complex $[\{(\eta^5\text{-}C_5Me_5)Sm\}_2(C_5Me_5SO_2)_4]$ were isolated (Scheme 1).[2] The results obtained can be transferred to other elements such as uranium and zinc.[3-4]

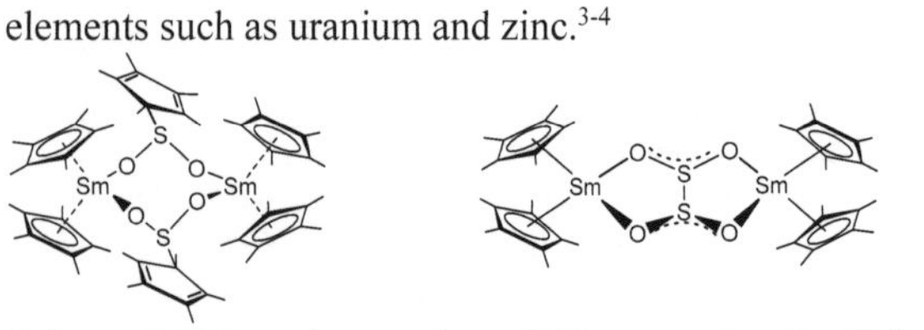

Scheme 1 Selected examples of SO_2 activation by $[(\eta^5\text{-}C_5Me_5)_2Ln(THF)_2]$.[2]

.

In trivalent lanthanide chemistry, we isolated the dinuclear SO_2-bridged complexes, namely $[(\{(^{tBu,tBu}ArO)_3tacn\}Ln)_2(\mu,\kappa O,\kappa O'\text{-}SO_2)]$ (Ln = Sm, Eu; $(^{tBu,tBu}ArOH)_3tacn$ = 1,4,7-tris(3,5-di-*tert*-butyl-2-hydroxybenzyl)-1,4,7-triazacyclononane)), which are the first molecular SO_2 complexes of the lanthanides. In these compounds, the SO_2 ligand is bound in a $\mu,\kappa O,\kappa O'$ binding mode.[5]

Recently, we reported a systematic study on the reactivity of samarocene oxide $[Cp^*_2Sm\text{-}O\text{-}SmCp^*_2]$ with various organic and inorganic anhydrides. Most of these compounds were small molecules such as COS, CS_2, SO_2, SO_3, Re_2O_7 (Scheme 2).[6] In each of these reactions, $[Cp^*_2Sm\text{-}O\text{-}SmCp^*_2]$ acted as mild oxide base, and an insertion reaction into the small molecules is observed. This reactivity was used to prepare under anhydrous conditions some unprecedented organo f-element complexes.[7]

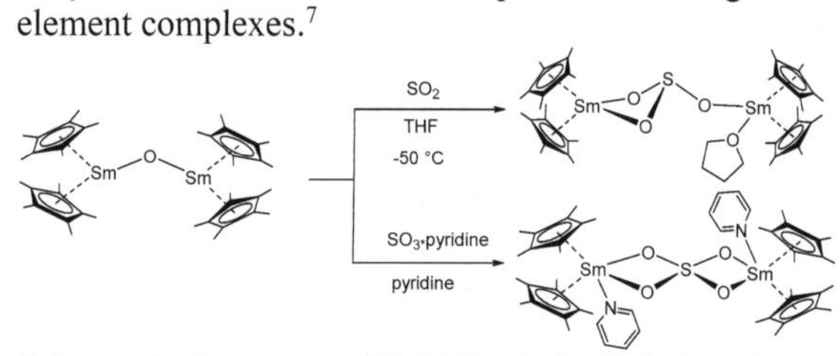

*Scheme 2. Reactions, [Cp*2Sm-O-SmCp*2] with SO_2 and SO_3.*[6]

CONCLUSIONS

During our studies, we showed the activation of SO_2 and other small molecules can be systemized. General reaction pathways, which can be also transferred to other metals, were established. We also learned that surprisingly few organo f-element complexes with classical inorganic anions are known.

REFERENCES

(1) Milani, B.; Licini, G.; Clot, E.; Albrecht, M. *Dalton Trans.* **2016**, *45*, 14419.

(2) Klementyeva, S. V.; Gamer, M. T.; Schmidt, A.-C.; Meyer, K.; Konchenko, S. N.; Roesky, P. W. *Chem. - Eur. J.* **2014**, *20*, 13497.

(3) Schmidt, A.-C.; Heinemann, F. W.; Kefalidis, C. E.; Maron, L.; Roesky, P. W.; Meyer, K. *Chem. - Eur. J.* **2014**, *20*, 13501.

(4) Kelly, R. P.; Kazeminejad, N.; Lamsfus, C. A.; Maron, L.; Roesky, P. W. *Chem. Commun.* **2016**, *52*, 13090.

(5) Benndorf, P.; Schmitt, S.; Koeppe, R.; Ona-Burgos, P.; Scheurer, A.; Meyer, K.; Roesky, P. W. *Angew. Chem., Int. Ed.* **2012**, *51*, 5006.

(6) Schoo, C.; Klementyeva, S. V.; Gamer, M. T.; Konchenko, S. N.; Roesky, P. W. *Chem. Commun.* **2016**, *52*, 6654.

(7) Schoo, C.; Klementyeva, S. V.; Gamer, M. T.; Konchenko, S. N.; Roesky, P. W. *Chem. Commun.* **2016**, *52*, 6654.

ACKNOWLEDGMENTS

The original work was mainly funded by the Deutsche Forschungsgemeinschaft (DFG)

Soft X-ray Spectromicroscopy Studies of f–electron Materials

David K. Shuh[1*]

[1]Heavy Elements Chemistry Group, Chemical Sciences Division,
Lawrence Berkeley National Laboratory, USA *DKShuh@lbl.gov

INTRODUCTION

Soft x-ray synchrotron radiation methodologies are being developed and employed at the Advanced Light Source (ALS) of Lawrence Berkeley National Laboratory (LBNL) to elucidate the electronic structure of f-electron materials. Results from investigations, in most cases coupled to theory, have begun to provide greatly improved fundamental scientific knowledge that provides a deeper, overall understanding of actinide materials and this information can be used as a basis for the enhanced design of special-purpose f-electron materials.[1-3] Importantly, the regular use of the Molecular Environmental Science (MES) scanning transmission x-ray microscope (STXM) as a high value characterization tool for fundamental soft x-ray investigations of f-electron materials has led to further applications in f-electron environmental science, nanoscience, and forensics.[4-6]

EXPERIMENTAL METHODS

The experimental developments at the ALS for f-electron materials have centered on the use of the MES STXM at Beamline 11.0.2 for near-edge X-ray absorption spectroscopy (XAS), and on x-ray emission spectroscopy (XES) at several beamlines, focusing primarily on light atom constituents (C, N, O, F) for ligand K-edge XAS via STXM, and on metal-ion XAS plus light-atom signals from XAS/XES.[7-8] Spectromicroscopy capabilities of the STXM provide a unique and safe means to determine the speciation in a diverse range of f-electron materials, particularly those with nanoparticle characteristics that require spatial resolution approaching the true nanoscale.

RESULTS AND DISCUSSION

Results from several recent STXM spectromicroscopy investigations across a representative range of f-electron materials systems will be discussed. These focus of these studies span fundamental f-electron science from coordination chemistry, environmental science, and nanoscience to forensics. As an example, Fig. 1 shows STXM elemental maps obtained of several uranium oxide particles from a larger specimen distribution.

Complementing STXM spectromicroscopy are results from soft x-ray XAS/XES investigations that have largely centered on the actinide dioxides. These studies have investigated the nature of the electronic structure by probing both low-lying actinide and oxygen edges.

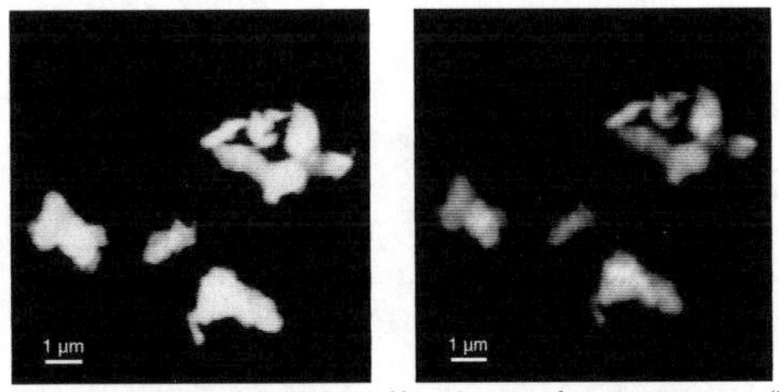

Figure 1. Uranium (yellow) and oxygen (blue) elemental maps obtained from a set of uranium oxide particles with the STXM [from Ref. 6].

CONCLUSIONS

Soft x-ray STXM spectromicroscopy methods give information on electronic structure (chemical bonding and speciation) at a critical length scale from small particles, thin layers, and specially-prepared structures relevant to a range of f-electron materials. Soft x-ray XES yields similar information without high spatial resolution. Theory and simulation of soft x-ray spectra provide essential information and serves to more rapidly connect soft x-ray approaches to new scientific issues.

REFERENCES

1. Bradley, J., *et al.*, J. Amer. Chem. Soc. 132:13914-13921, 2010.
2. Minasian, S. G. *et al.*, J. Amer. Chem. Soc. 135:1864-1871, 2013.
3. Minasian, S. G. *et al.*, Chem. Sci. 5:351-359, 2014.
4. Maloubier, M. *et al.*, Environ. Sci. Technol. 50:10730-10738, 2016.
5. Plakhova, T.V. *et. al.*, J. Phys. Chem. C 120:22615-22626, 2016.
6. Pacold, J. I. *et al.*, to be submitted, 2017.
7. Bluhm, H. *et al.*, Electron Spectros. Rel. Phenom. 150:86-104, 2006.
8. Butorin, S. *et al.*, Anal. Chem. 88:4169-4173, 2016.

ACKNOWLEDGMENTS

This research was supported by the Director, Office of Science (OS), Office of Basic Energy Sciences (BES), Division of Chemical Sciences, Geosciences, and Biosciences (CSGB) Heavy Elements Chemistry of the U.S. Department of Energy (DOE) under Contract Number DE-AC02-05CH11231 at LBNL. Beamline 11.0.2 was supported by the aforementioned Division of CSGB Condensed Phase and Interfacial Molecular Sciences of the U.S. DOE under Contract Number DE-AC02-05CH11231 at LBNL. The ALS is supported by the Director, OS, BES, of the U.S. DOE under Contract No. DE-AC02-05CH11231 at LBNL.

Solution Growth of New Rare-earth Poor Ce/Co/Mg Permanent Magnets.

Tej N. Lamichhane[a,b], Valentin Taufour[b,c], Andriy Palasyuk[b], Qisheng Lin[b], Sergey L. Bud'ko[a,b], Paul C. Canfield[a,b]

[a]*Department of Physics and Astronomy, Iowa State University, Ames, IA 50011, U.S.A.*
[b]*Ames Laboratory, Iowa State University, Ames, IA 50011, U.S.A.*
[c]*Department of Physics, University of California, Davis, CA 95616, U.S.A.*

1. INTRODUCTION

Due to the revolution in Information Technology, demand of portable digital devices has increased extensively. The efficiency and miniaturization of those devices greatly depends up on the efficiency of the installed permanent magnets. Current rare-earth based commercial magnets are considered "critical" because of the key constituent of them i.e. Sm, Nd, and Dy are rare and pose a supply security risk. Finding Ce based, transition metal rich, ferromagnetic compounds is the one of the possible solutions to the current criticality issue of rare-earth based permanent magnets. Ce based ferromagnetic compounds are interesting from the theoretical and experimental perspective because the Ce ion exhibits local moment bearing Ce^{3+} ion, non-magnetic Ce^{4+} ion, mixed valencies and itinerant magnetism. Due to these diverse magnetic properties of Ce based compounds, there is a great potentiality of finding the new Ce based ferromagnetic compounds and possibility of improving the magnetic properties of the existing ferromagnetic materials. Here we will summarize some of our explorations on effect of substituting elemental Mg in various Ce-Co compounds.

2. EXPERIMENTAL METHODS

Mg is an element with high reactivity and vapour pressure. Due to this high vapour pressure, Mg can not be easily arc-melted with other metals. Homogeneous melts of various, Ce/Co/Mg compositions were formed at 1200° C using 3-capped Ta tubes[1] sealed in SiO_2 tube under partial pressure of Ar and centrifuged to decant the melt in the catch side of the tube and air quenched to room temperature. Then the centrifuged homogeneous ingot was annealed at 900° C for 24 h. The ingot were characterized via X-ray diffraction (XRD), Energy Dispersive Spectroscopy (EDS), Wave length Dispersive spectroscopy (WDS) and X-ray Photo-emission Spectroscopy (XPS). In addition, single crystalline samples were grown out of a ternary melt[1]. The magnetic and the anisotropic properties of these materials were studied using Quantum Design Vibrating Sample magnetometer and Magnetic Property Measurement System.

3. RESULTS AND DISCUSSION

Introduction of Mg into Ce-Co binary phases in many instances results in appreance of solid solutions and ternary phases with Ce/Mg atomic mixtures instead of Ce/Co ones, especially in binary Laves phase $CeCo_2$ (structure type $MgCu_2$, *Fd-3m*), or those containing structural fragments of the Laves phase, e.g., $CeCo_3$ (*R-3m*), Ce_2Co_7 (*P*6_3*/mmc*), Ce_5Co_{19} (*R-3m*). Mg also changes the magnetic nature of Ce-Co binary phases. For example, in some cases small doping of the Mg causes the development of the pronounced ferromagnetism in other non-magnetic phases at room temperature. Mg might facilitates the high temperature performance of these magnetic materials. The detailed investigations on role of Mg on the development and variation of magnetic properties of various Ce/Co/Mg phases are underway and will be reviewed.

4. CONCLUSIONS

The addition of Mg into Ce-Co phases preferentially substitutes Ce atoms in Laves-type structural fragments and often gives rise or improves ferromagnetic nature of the parent binary phases.

5. REFERENCES

1. P. C. Canfield and I. R. Fisher, Journal of Crystal Growth 225, 155 (2001)

6. ACKNOWLEDGEMENTS

We would like to thank Dr. T. Kong for useful discussions. Dr. Warren Straszheim is acknowledged for doing SEM and WDS on various samples. This research was supported by the Critical Materials Institute, an Energy Innovation Hub funded by the U.S. Department of Energy, Office of Energy Efficiency and Renewable Energy, Advanced Manufacturing Office.This work was also supported by the office of Basic Energy Sciences, Materials Sciences Division, U.S. DOE. This work was performed at the Ames Laboratory, operated for DOE by Iowa State University under Contract No. DE-AC02-07CH11358.

Solution Structure Trumps Energy – Determining Factors for the Luminescent Properties of Multimetallic Kinetically Inert Lanthanide(III) Complexes

Thomas Just Sørensen[1*]

[1]Nano-Science Center & Department of Chemistry, University of Copenhagen, Denmark TJS@chem.ku.dk

INTRODUCTION

The optical properties of lanthanide ions are important in applications as diverse as medical imaging and data transport. The optical properties in solids containing these ions have seen intense scrutiny, while studying dynamic systems in solution have been hampered by the inherent lability of lanthanide(III) complexes. Working around this issue, we have studied larger architectures of kinetically inert lanthanide(III) complexes.[1-6] It is possible to investigate the effect of solution structure on the properties of these systems, as the constitution is not dynamic. Further, the effects from conformational dynamics at the lanthanide centre is limited in the octadentate binding pockets, which removes the complications that arise from the varied photophysics lanthanide centres exhibit in different coordination environments and symmetries.

RESULTS AND DISCUSSION

In this talk, our results are summarised and the photophysical properties of system similar to those shown in Figure 1 will be rationalised. Furthermore, the implications for the design of responsive or functional luminescent lanthanide(III) complexes will be detailed. We have shown that the solution structure determines whether oxygen responsive probes have a modest or a respectable action,[6] and whether a potassium selective molecule will actually bind potassium or not.[7]

Our investigation of these structure-property relationships rely on the synthesis of large lanthanide containing architectures. These complicated molecules can be synthesised using kinetically inert complexes. The kinetic—and thermodynamic—stability of these is of such a magnitude that heterometallic complexes can readily be made by using lanthanide(III) complexes as molecular building blocks. This approach has the merit of allowing highly defined systems to be prepared, avoiding the problem of similarity between lanthanide(III) ions that effectively dooms any approach to specific (as opposed to merely selective) complexation under thermodynamic control. Kinetically inert lanthanide(III) complexes based on the 1,4,7,10-tetraazacyclododecane-1,4,7-triacetic acid (DO3A) motif can then be used in copper catalyzed azide alkyne click, diazotization, peptide coupling, and Ugi reactions to form multimetallic and heterometallic lanthanide(III) containing architectures.

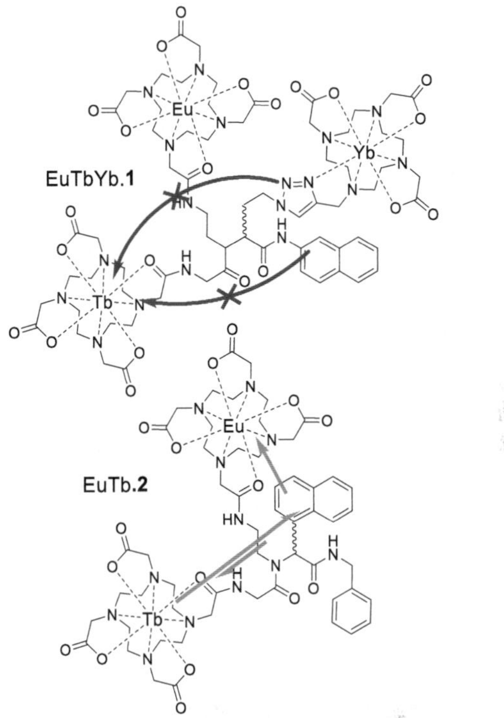

Figure 1. Observed excited state energy transfer pathways in two multimetallic kinetically inert lanthanide complexes.

1 Sørensen, T. J. *et al.* Chem Commun 49, 783-785, 2013.
2 Tropiano, M. *et al.* Chem Eur J 19, 16566-16571, 2013.
3 Hueting, R., Tropiano, M. & Faulkner, S. RSC Adv. 4, 2014.
4 Tropiano, M. *et al.* J Luminesc 167, 296-304, 2015.
5 Watkis, A., Hueting, R., Sorensen, T. J., Tropiano, M. & Faulkner, S. Chem Commun 51, 15633-15636, 2015.
6 Sørensen, T. J., Kenwright, A. M. & Faulkner, S. Chem Sci 6, 2054-2059, 2015.
7 Junker, A. K. R., Tropiano, M., Faulkner, S. & Sørensen, T. J. Inorg Chem 55, 12299-12308, 2016.

"The author would like to thank the Villum Foundation and the Carlsberg Foundation".

Solution thermodynamics of hydropyridinonate 4*f* and 5*f* complexes.

Gauthier J-P. Deblonde[1], Manuel Sturzbecher-Hoehne[1], Dahlia D. An[1], Wibe A. de Jong[2], Peter B. Rupert[3], Roland K. Strong[3], and Rebecca J. Abergel[1]

[1]Chemical Sciences Division, Lawrence Berkeley National Laboratory, Berkeley, California 94720, USA, GDeblonde@lbl.gov
[2]Computational Research Division, Lawrence Berkeley National Laboratory, Berkeley, CA 94720, USA
[3]Division of Basic Science, Fred Hutchinson Cancer Research Center, Seattle, WA 98109, USA

INTRODUCTION

Hydropyridinonates are a class of ligands inspired from endogenous siderophores that play a major role in iron biology. A synthetic octadentate analog, called 3,4,3-LI(1,2-HOPO), has caught attention for its efficacy at chelating bivalent UO_2^{2+}, trivalent lanthanides (Ln^{3+}), Am^{3+}, Cm^{3+} and tetravalent Pu^{4+} ions under physiological conditions and for its low toxicity compared to classical ligands such as ethylenediaminetetraacetic acid (EDTA) or diethylenetriaminepentaacetic acid (DTPA).[1] The 3,4,3-LI(1,2-HOPO) ligand is also undergoing clinical evaluation for its use as a medical countermeasure against internal contamination with radionuclides. Although studied for many years, several thermodynamic constants still need to be measured for this flagship chelator. The present work addresses this issue by evaluating the interactions between the 3,4,3-LI(1,2-HOPO) ligand and a series of metal ions, including Sc, Y, Cf and Bk.

RESULTS AND DISCUSSION

Taking the Ce^{4+}/Ce^{3+} system as a surrogate system for Bk^{4+}/Bk^{3+}, we were able to predict the stabilization of Bk^{4+} by 3,4,3-LI(1,2-HOPO) in aqueous solutions. The ligand-induced and spontaneous stabilization of Bk(IV) was then confirmed experimentally, using ^{249}Bk, by UV-vis spectrophotometry, mass spectrometry, high-performance liquid chromatography, and protein recognition.[2] The 3,4,3-LI(1,2-HOPO) system represents the first case of long term stabilization of Bk(IV) in aqueous solutions under mild conditions. Based on the stability constants experimentally determined for other tetravalent ions (Zr^{4+}, Hf^{4+}, Ce^{4+}, Th^{4+}, and Pu^{4+}), the [Bk^{IV}3,4,3-LI(1,2-HOPO)] complex is expected to have a log β_{110} value higher than 40 which is, by far, the most stable Bk complex ever reported. By comparison the DTPA and dipicolinic acid (DPA) complexes of Bk(III) exhibit log β_{110} and log β_{130} values of 22.8 and 23.1, respectively.[3] None of the classical ligands, (EDTA, DTPA, DPA) are strong enough to stabilize Bk in its tetravalent oxidation state.

The 1,2-HOPO complex of Bk(IV) is also the first chelate amenable to induce luminescence of this heavy actinide in solution. Upon excitation of the ligand and intramolecular sensitization (so-called "antenna effect"), the Bk(IV) metal center emits at 612 nm (Figure 1).

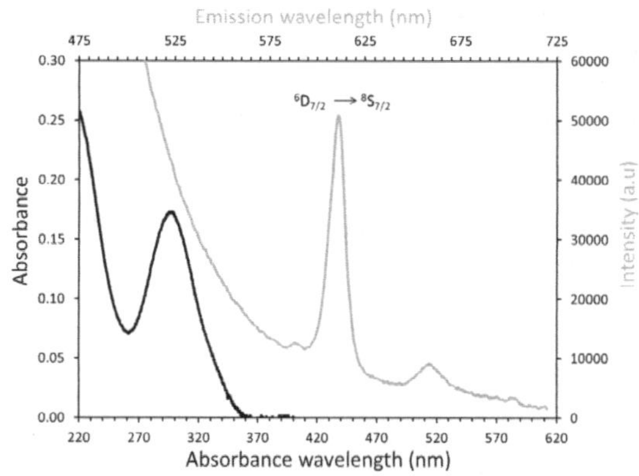

Figure 1: UV-vis absorbance spectrum (blue curve) and Steady-state emission spectrum of a [$^{249}Bk^{IV}$3,4,3-LI(1,2-HOPO)] sample upon excitation at 320 nm in 0.1 M CHES buffer, pH 8.4, 25°C.

CONCLUSIONS

The chemical behavior and spectroscopic properties of Bk with 3,4,3-LI(1,2-HOPO) contrast sharply with those of neighboring actinides (Am, Cm, and Cf) and of all rare earth elements whose only stable oxidation state is +III (except for Ce). This specificity will be discussed in the frame of separation and purification processes for actinides and rare earth elements.

REFERENCES

1 G. J.-P. Deblonde, M. Sturzbecher-Hoehne and R. J. Abergel, *Inorg. Chem.*, 2013, **52**, 8805–8811.
2 G. J.-P. Deblonde, R. J. Abergel, M. Sturzbecher-Hoehne, P. B. Rupert, D. D. An, M.-C. Illy, C. Y. Ralston, J. Brabec, de J. Wibe A. and R. K. Strong, *Nat. Chem.*, 2017.
3 M. A. Silver, S. K. Cary, J. A. Johnson, R. E. Baumbach, A. A. Arico, M. Luckey, M. Urban, J. C. Wang, M. J. Polinski, A. Chemey and others, *Science*, 2016, **353**, aaf3762.

ACKNOWLEDGMENTS

This work was supported by the U.S. Department of Energy, Office of Science, Office of Basic Energy Sciences, Chemical Sciences, Geosciences, and Biosciences Division at the Lawrence Berkeley National Laboratory under Contract DE-AC02-05CH11231

Spectroscopic and Electrochemical Investigations of Heterometallic Lanthanide-Transition Metal Complexes

Trevor Latendresse, Courtney Dickie, Corey Burns, and Michael Nippe*

Department of Chemistry, Texas A&M University, United States, nippe@chem.tamu.edu

Abstract.

Some of the hardest bulk magnets are heterometallic alloys comprised of lanthanides and transition metals, such as $SmCo_5$. Our research is taking inspiration from this fascinating material and is aimed at translating important electronic and geometric aspects from the bulk material into isolated, well-defined, and tunable molecular species with the ultimate goal of developing new routes to hard molecular magnets. Special attention is given to the judicious choice of ligands to allow for the stabilization of heterometallic lanthanide-transition metal complexes and consequences of ligand-based mixed-valence on magnetization dynamics of these systems. Our most recent results related to synthetic routes towards heterometallic lanthanide transition metal complexes will be presented along with their static and dynamic magnetic properties as well as corresponding redox properties.

Structural and Electronic Effects of Ga Doping into $GdCo_2$: An Electronic Structure Study of $GdCo_{2-x}Ga_x$

Asa Toombs,[1] Fang Yuan,[2] Yurij Mozharivskyj,[2] Gordon J. Miller[1]

[1]Ames Laboratory, Iowa State University, USA
[2]Chemistry, McMaster University, Canada, toombsa@iastate.edu

INTRODUCTION

$RECo_2$ (RE= rare earth) crystallize in the cubic $MgCu_2$-type,[1] and are ferrimagnetic but exhibit different magnetic properties depending on the RE.[2] $RECo_2$ (RE=Dy, Ho, and Er) exhibit a first-order magnetic transition and the giant magnetocaloric effect, whereas other $RECo_2$ compounds exhibit a second-order magnetic transition and a conventional magnetocaloric effect.[3] $GdCo_2$ has the highest T_C (400K) of the $RECo_2$ compounds with the T_C dropping rapidly in moving away from Gd in either direction.[2] Gallium was chosen to dope into $GdCo_2$ with the intent of adjusting the T_C to room temperature because such materials are important for commercially viable magnetic refrigerants. $Gd(Co_{1-x}Ga_x)_2$ with x=0, 1/6, 1/3, 1/2, 2/3, 5/6, and 1 were prepared by arc melting and annealing. The resulting structures, determined by X-ray diffraction, were also studied using electronic structure calculations to shed light on any driving forces behind the observed structural changes with varying Ga content. The effects on the Gd and Co magnetic moments, coordination around the Co/Ga sites, and valence electron count will be analyzed.

EXPERIMENTAL METHODS

Electronic structure calculations were performed with the Vienna Ab initio Simulation Package (VASP) using the local spin-density approximation +U (LSDA+U) and the Tight-Binding Linear Muffin-Tin Orbital Atomic Sphere Approximation (TB-LMTO-ASA) using LSDA. Magnetic moments, total energies, and structure optimizations were calculated using VASP, whereas electronic density of states (DOS) and crystal orbital Hamilton population (COHP) curves were calculated with TB-LMTO-ASA.

RESULTS AND DISCUSSION

$Gd(Co_{1-x}Ga_x)_2$ show distinct structural variations along the sequence from $GdCo_2$ to $GdGa_2$, with five structures ranging from cubic → hexagonal → orthorhombic (Pnma) → orthorhombic (Imma) → hexagonal. The Co/Ga environments shift from condensed clusters (Co CN=6) → 3D network (Co/Ga CN=4) → 2D network (Ga CN=3). This work is the first comprehensive computational study on varying the Ga content in this system. Calculated total energies of $Gd(Co_{1-x}Ga_x)_2$ (x=0, 1/2, 1), calculated to assess stability and estimate heats of formation, show that the binary and ternary compounds are favoured over the individual elements and GdCoGa is favoured over the binary compounds. Structure optimizations using VASP also give details on the compound stabilities along with comparing experimental results with energetically competitive structures.

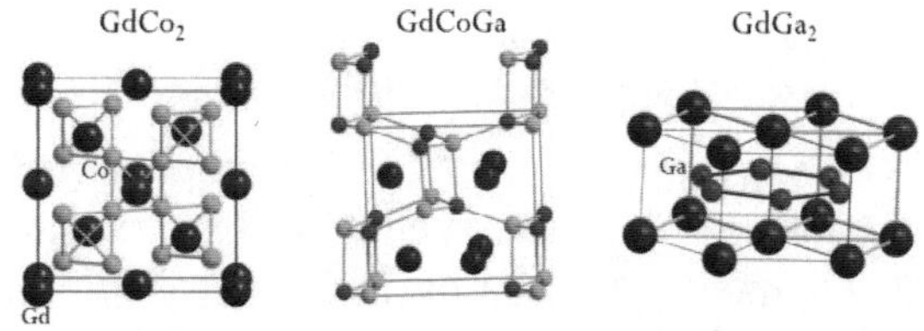

Electronic structure calculations on the stoichiometric compounds and ordered model structures in conjunction with the rigid band approximation give useful insights into the structural variations as a function of Ga content, which affects the valence electron count (VEC) as in GdM_2 (M=Co, Ni, Cu, Zn, Ga). Bonding analyses from COHP curves of $Gd(Co_{1-x}Ga_x)_2$ (x=0, 1/2, 1) show, in general, that Co-Ga interactions are maximized whereas Co-Co and Ga-Ga contacts are minimized for ternary compounds, which are further verified by studies of different ordered superstructures.

CONCLUSIONS

VEC plays a significant role in determining the structures of $Gd(Co_{1-x}Ga_x)_2$. As VEC increases by substituting Ga for Co, the structures obtained resemble the trend of structures in the series of binary GdM2 (M=Co, Ni, Cu, Zn, Ga). Throughout this series, Gd always has CN 12, but the CN of Co/Ga steadily decreases as VEC increases. The rigid band model has limitations when comparing large differences in VEC and compounds with different structures. To gain a better understanding of the mixed occupancy patterns in the ternary $Gd(Co_{1-x}Ga_x)_2$ compounds, super-cells and various colouring models were examined.

REFERENCES

1. Harris I. *et al.*, J. Less Common Met.9:270-280, 1965
2. Khmelevskyi L. *et al.*, J. Phys. Condens. Matter. 12:9453-9464, 2000
3. Duc N. H. *et al.*, Phys. B. 319:1-8, 2002

ACKNOWLEDGMENTS

This research was supported by the Office of the Basic Energy Sciences, Materials Sciences Division, U.S. Department of Energy (DOE). The Ames Laboratory is operated by Iowa State University under Contract No. DE-AC02-0

Structural and Photophysical Properties of Eu(III) Coordination Polymer with an Axial Chiral Ligand

Naoyuki Koiso[1], Yuichi Kitagawa[2], Takayuki Nakanishi[2], Koji Fushimi[2], and Yasuchika Hasegawa[2*]

[1]Advanced Materials Research Laboratory, TOSOH Corporation, Japan
[2*]Faculty of Engineering, Hokkaido University, Japan, naoyuki-koiso-zy@tosoh.co.jp

INTRODUCTION

Lanthanide complexes have been regarded as attractive luminescent materials because of their characteristic narrow emission bands and long emission lifetimes.[1] Particularly, lanthanide coordination polymers with luminescent properties have been widely studied for applications as highly thermo stable materials.[2] Recently, we reported strong luminescent Eu(III) coordination polymers composed of low-vibrational frequency hexafluoroacetylacetonato (hfa) and phosphine oxide ligands.[3] In this study, we prepared a novel Eu(III) coordination polymer with an axial chiral ligand that exhibits high thermal stability and strong luminescence in the solid state and chiroptical properties in solution. We found that there was an effective transformation from a polymer to a monomer structure in solution.

RESULTS AND DISCUSSION

A novel Eu(III) chiral coordination polymer $[Eu(hfa)_3((R)\text{-bidp})]_n$ ((R)-bidp: (R)-1,1'-binaphthyl-2,2'-bis(diphenylphosphinate)), was synthesized by the reaction of $Eu(hfa)_3(H_2O)_2$ and the axial chiral phosphinate ligand (R)-BIDP. Single-crystal X-ray analysis revealed a characteristic helical polymer structure of $[Eu(hfa)_3((R)\text{-bidp})]_n$ with multiple CH/F and CH/π intramolecular hydrogen bonds. Two intrachain CH/F interactions between monomer units were also found. These intrachain hydrogen bonds assist the formation of the helical structure in crystals. The coordination geometry of $[Eu(hfa)_3((R)\text{-bidp})]_n$ was estimated to be a square antiprism structure (8-SAP) by the calculation of continuous symmetry shape measures factor. The coordination polymer shows high thermal stability (decomposition temperature = 320°C). The tight packing structure in $[Eu(hfa)_3((R)\text{-bidp})]_n$ originated from multiple hydrogen bonds increases the thermal stability.

The emission bands of $[Eu(hfa)_3((R)\text{-bidp})]_n$ are observed at around 578, 592, 611, 651 and 699 nm excited at 380nm and are attributed to the f–f transitions of Eu(III). In the solid state, strong luminescence properties (4f–4f emission quantum yields = 76%) due to its tight packing and asymmetric structure (8-SAP) was observed. We found that the emission spectral shape in solution is quite different from that in the solid state. The spectral shapes are based on their coordination geometry. This result suggests structural transformation of the coordination geometry in solution. We consider that $[Eu(hfa)_3((R)\text{-bidp})]_n$ transformed polymer to a monomer structure in solution (Figure 1). The chiroptical properties of the monomer form in liquid media were evaluated by circular dichroism (CD) and circularly polarized luminescence spectra (CPL).

CONCLUSIONS

We investigated a Eu(III) chiral coordination polymer, $[Eu(hfa)_3((R)\text{-bidp})]_n$, with a switchable conformation system. Such characteristic transformation ability of coordination polymer provides new insights into the coordination chemistry and materials science.

REFERENCES

1. Eliseeva S. *et al.*, Chem. Soc. Rev. 39:189-227, 2010
2. Hasegawa Y. *et al.*, RSC Adv. 5:338-353, 2015
3. Miyata K. *et al.*, ChemPlusChem 77: 277-280, 2012

ACKNOWLEDGMENTS

We are particularly grateful for experimental assistance by Professor H. Ito and Assistant Professor T. Seki of Hokkaido University. We appreciate the support of JASCO Corporation for the CPL measurement.

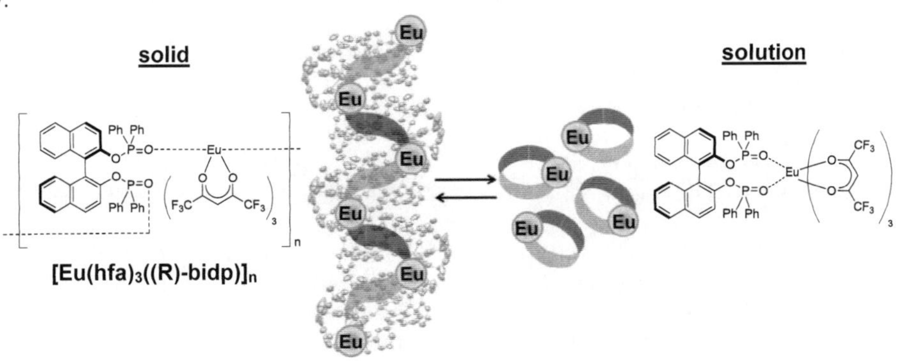

Figure 1. Transformation of the Eu(III) chiral coordination polymer.

Structural Variations and Magnetism in the Transition-Metal-Rich Intermetallics Eu($T1$,$T2$)$_5$In (T = Cu, Ag, Au)

V. Smetana,[1] Y. Mudryk,[1] V.K. Pecharsky,[1,2] and A.-V. Mudring,[1,2]

[1]Ames Laboratory, U.S. Department of Energy, Iowa State University, Ames, Iowa 50011, USA,

[2]Department of Materials Science and Engineering, Iowa State University, Ames, Iowa 50011, USA, smetana@ameslab.gov

INTRODUCTION

Identification of the compositional and structural preferences in polar intermetallics opens wide possibilities for the controlled modification of their physical properties. From this point of view compounds of active metals, gold and post transition elements are very promising as they combine strong electronic correlations with relativistic effects and allow decent compositional modifications through the doping of a third element. These systems demonstrate a wide variability of compounds with complex crystal structures, valuable physical properties and unique bonding characteristics;[1] however, many of them are both poorly investigated and understood. In this study, we focus on a gold rich phase $EuAu_5In$ and two isocompositional specimens with Ag and Cu, investigate the possibility of their mutual substitution and properties modification.

EXPERIMENTAL METHODS

All materials have been prepared from Eu pieces (99.9%), Cu, Ag and Au chunks (99.99%, all Ames Laboratory) and In ingots (99.999%, Alfa Aesar) in arc-welded Ta ampules enclosed in evacuated silica jackets. The samples were heated at 900 °C for 3 h, cooled to 500 °C at a rate of 10 °C/h, then annealed there for 7 days, and cooled by switching off the furnace. Magnetic properties were measured on polycrystalline powders using Quantum Design SQUID MPMS magnetometer.

RESULTS AND DISCUSSION

The series of intermetallic compounds EuT_5In with various combinations of the group 11 elements have been obtained and structurally characterized. The compounds adopt three different crystallographic types[2] with the (average) transition metal radius being the main driving force for the structural preference. $EuCu_5In$ is rhombohedral ($hR42$; R-$3m$, a = 5.0933(7), c = 30.557(6) Å, Sr_5Al_9), $EuAg_5In$ orthorhombic (oP28; $Pnma$, a = 9.121(2), b = 5.645(1), c = 11.437(3) Å, $CeCu_6$), and $EuAu_5In$ tetragonal (tI14; $I4/mmm$, a = 7.1740(3), c = 5.4425(3) Å, $YbMo_2Al_4$). Though no intercalation of other transition metals into the crystal structure of $EuCu_5In$ has been successful, the doping of a fourth element into the other structures led to different outcomes. A minor amount of Cu doped to the $EuAu_5In$ structure results into the crystal structure of $EuAg_5In$ ($CeCu_6$ type), while the latter remained unchanged within a wide compositional range with both Cu and Au doping. On the other hand, $EuAu_5In$ can accommodate a decent amount of Ag (at least 50%) in its crystal structure with no significant transformations. DFT based structural optimizations showed strong site preference for the In positions in all structures.

The exchange of these three nonmagnetic elements was also found to have a significant influence on the type of magnetic ordering. The measurements of magnetic properties of the $EuT_{5-x}Cu_xIn$ alloys (x = 0, 0.25, 1, 2, 2.5, and 5) shows that substitution of Cu by Au leads to the evolution of magnetic structure. The quaternary alloys in this system show multi-step dependence of magnetization as a function of applied magnetic field at 2 K: the quick initial rise of magnetization to $M_1 \sim 2$ to 3 μ_B/f.u. is followed by a second step around 3 T; at higher magnetic fields the magnetization saturates near $M_2 \sim 7$ μ_B/f.u., which indicates divalent state of Eu in this system. The value of initial magnetization increase, M_1, the critical field at which the second step occurs, H_{cr}, and overall shape of the M(H) curve is clearly dependent on the Cu/Au ratio.

CONCLUSIONS

The extensive solid solutions based on EuT_5In (T = Cu, Ag, Au) have been prepared and characterized. The combination of X-ray analysis, theoretical studies and magnetization measurements revealed strong structural preference based on the atomic size, positional irreplaceability for certain element types and dependence of physical properties within a given type on the $T2/T1$ ratio.

REFERENCES

1. Corbett, J. D., Inorg. Chem. 49:13-28, 2010
2. Bigun, I. *et al.*, Chem. Mater. *ASAP*

ACKNOWLEDGMENTS

The research was supported by the Office of the Basic Energy Sciences, Materials Sciences Division, U. S. Department of Energy (DOE), and the Department of Materials Science and Engineering at Iowa State University. Ames Laboratory is operated for DOE by Iowa State University under contract No. DE-AC02-07CH11358.

Study on the Crystal Face Design and Luminescence Properties of Monodispersed YAG Microcrystal

Tao Xu [1], Zhijun Zhang[1,*], JingtaiZhao[1,*]

[1] School of Materials Science and Engineering, Shanghai University, China
zhangzhijun@shu.edu.cn, jtzhao@shu.edu.cn

INTRODUCTION

Rare earth doped $Y_3Al_5O_{12}$ (YAG) exhibits a broad application prospect in solid-state lighting, displays, lasers and etc. Previous studies have been done on solid-phase [1], sol-gel[2], hydrothermal[3] and solvothermal[4] methods for the synthesis of YAG, but the preparation of YAG with better monodispersity is still the focus of our research.

EXPERIMENTAL METHODS

The monodispersed YAG microcrystals were prepared by solvothermal method at high temperature and high pressure, and the crystal faces of YAG microcrystals were designed and adjusted by surfactants. At the same time, the effects of Y / Al ratio and surfactants on the phase, morphology, crystallinity as well as luminescence properties of Ce doped YAG microcrystals were studied systematically.

RESULTS AND DISCUSSION

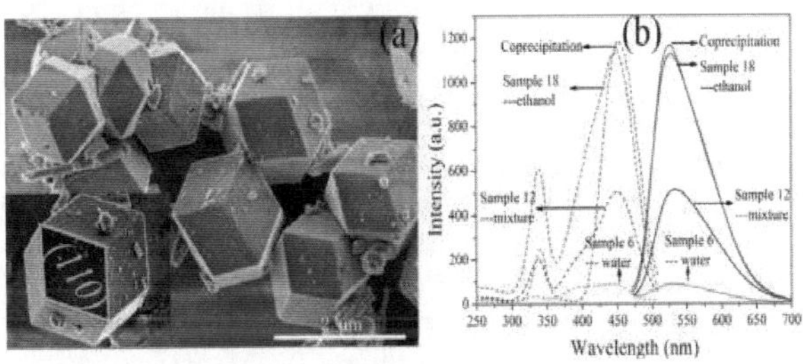

Figure 1. (a) The SEM image of the sample (Y / Al = 3: 5, 270 ° C × 48 h, Ce 2%); (b) The emission spectra of the sample obtained using different solvents.

Figure 1 (a) shows the SEM image of the sample (Y / Al = 3: 5, 270 ° C × 48 h, Ce 2%). The YAG microcrystals with good morphology and very regular dodecahedron were prepared. The emission spectra of the sample obtained using different solvents are shown in Figure 1 (b). Under the excitation of 450 nm, there is a broadband emission peak at 470-700 nm, originating from the $5d \rightarrow 4f$ transition of Ce^{3+}.

CONCLUSIONS

Pure YAG phase can be obtained when Y / Al = 3: 5, and with the addition of the surfactant (EDTA and Na_2SO_4), the size of YAG is larger and the dispersion is better compared with the samples without surfactant. Furthermore, the YAG microcrystal face arrangement is more complete and structured. The YAG:Ce microcrystals with good morphology exhibit good luminescence properties.

REFERENCES

1. Caroline Marlot, Elodie Barraud, *et al.* J Solid State Chem.191: 114 - 120, 2012
2. Hua Gong, Ding-Yuan Tang *et al.* J Cryst. Growth, 362: 52 - 57, 2013
3. Sahraneshin A., Takami S., Minami K., *et al.* Cryst. Growth Ch., 58 (1): 43-50, 2012
4. Li X., Liu H., Wang J. Y., et al. J. Am. Ceram. Soc., 87 (12): 2288-2290,2004

ACKNOWLEDGMENTS

The authors would like to thank Shanghai University 085 project and National Natural Science Foundation of China for providing financial support to this project.

Supercritical Hydrothermal Synthesis and Structural Chemistry of Uranium Silicates and Germanates

Kwang-Hwa Lii

Department of Chemistry, National Central University, Taiwan, R.O.C. liikh@cc.ncu.edu.tw

INTRODUCTION

Most uranium minerals can be classified as oxidized species in which U is fully oxidized to U^{6+}, and reduced species, in which U occurs primarily as U^{4+}. Uranyl silicates are an important group of U(VI) minerals in the altered zones of many uranium deposits.[1] They also form when spent nuclear fuel interacts with water containing silicon under oxidizing conditions. One naturally occurring U(IV) silicate exists, namely coffinite ($USiO_4$), which is an important ore mineral for uranium. Numerous synthetic U(VI) silicates and germanates containing alkali metals and organic amines as countercations have also been reported.[2] In contrast to the U(VI) compounds, the chemistry of materials containing U(V) is considerably less developed owing to the tendency of U^{5+} to either oxidize to U^{6+} or disproportionate to U^{4+} and U^{6+}. In this study uranium silicates and germanates with the oxidation states of uranium from +4 to +6 and mixed-valence compounds with all possible combinations of the oxidation states of uranium have been synthesized under supercritical hydrothermal conditions.

EXPERIMENTAL METHODS

Supercritical hydrothermal syntheses were performed in gold ampules contained in a Leco Tem-Pres autoclave where pressure was provided by water. The apparatus is an externally heated cold-seal pressure vessel for use to 10000 bar and 750 °C. The degree of filling of the autoclave by water at room temperature was about 55%. Solid reactants and aqueous solutions were sealed in gold ampules (inside diameter = 0.48 cm) which were heated in the pressure vessel at about 550 °C for several days. The pressure was estimated to be 1200 bar at the reaction temperature according to the pressure–temperature diagram for pure water. The autoclave was then cooled to 350 at 5 °C/h and quenched in the air at room temperature by removing the autoclave from the furnace. The solid products were separated and structurally characterized by powder and single-crystal X-ray diffraction. U 4f XPS and U L_3-edge XANES spectra were recorded to study the valence states of the uranium atoms in these compounds.

RESULTS AND DISCUSSION

We have synthesized and structurally characterized the first U(V) silicate, $K(UO)Si_2O_6$, and germanate, $Cs_3UGe_7O_{18}$, by a supercritical hydrothermal method.[3a,3b] Following the synthesis of the U(V) compound, a number of mixed-valence uranium silicates and germanates have been synthesized, for example, a mixed-valence U(IV,V) silicate, $Cs_2K(UO)_2Si_4O_{12}$,[3c] U(IV,VI) germanate, $Cs_8U(UO_2)_3(Ge_3O_9)_3 \cdot 3H_2O$,[3d] U(V,VI) germanates, $A_3(U_2O_4)Ge_2O_7$ (A = Rb, Cs),[3e] a salt-inclusion U(V,VI) silicate, $[Na_9F_2][(U^{V}O_2)(U^{VI}O_2)_2(Si_2O_7)_2]$,[3f] and a U(IV,V,VI) silicate, $Na_7U^{IV}O_2(U^{V}O)_2(U^{V/VI}O_2)_2Si_4O_{16}$,[3g] in which three oxidation states of uranium co-exist in one compound. In addition, tetravalent-uranium compounds, $Cs_2USi_6O_{15}$ and $Cs_4UGe_8O_{20}$,[3h,3i] were also synthesized. All members in the family of uranium silicates and germanates with the oxidation states of uranium from +4 to +6 have been observed.

Most of the uranium silicates and germanates exhibit 3D framework structures composed of U−O polyhedra and SiO_4 tetrahedra. Uranium silicates and germanates with 2D layer structures are much less common. Recently, we reported a new pentanary uranyl silicate, $K_2Ca_4[(UO_2)(Si_2O_7)_2]$,[4] which is the first example of uranium silicate with 1D chain structure.

REFERENCES

1. (a) Burns, P. C. *Rev. Mineral.* 1999, *38*, 23-90. (b) Finch, R. J.; Murakami, T. *Rev. Mineral.* 1999, *38*, 24-179, and references therein.
2. (a) Wang, X.; Huang, J.; Jacobson, A. J. *J. Am. Chem. Soc.* 2002, *124*, 15190-15191. (b) Lin, C.-H.; Chiang, R.-K.; Lii, K.-H. *J. Am. Chem. Soc.* 2009, *131*, 2068-2069 and references therein.
3. (a) Chen, C.-S.; Lee, S.-F.; Lii, K.-H. *J. Am. Chem. Soc.* 2005, *127*, 12208-12209. (b) Nguyen, Q. B.; Chen, C.-L.; Chiang, Y.-W.; Lii, K.-H. *Inorg. Chem.* 2012, *51*, 3879-3882. (c) Lee, C.-S.; Wang, S.-L.; Lii, K.-H. *J. Am. Chem. Soc.* 2009, *131*, 15116-15117. (d) Nguyen, Q. B.; Liu, H.-K.; Chang, W.-J.; Lii, K.-H. *Inorg. Chem.* 2011, *50*, 4241-4243. (e) Lin, C.-H.; Lii, K.-H. *Angew. Chem. Int. Ed.* 2008, *47*, 8711-8713. (f) Chang, Y.-C.; Chang, W.-J.; Boudin, S.; Lii, K.-H. *Inorg. Chem.* 2013, *52*, 7230-7235. (g) Lee, C.-S.; Lin, C.-H.; Wang, S.-L.; Lii, K.-H. *Angew. Chem. Int. Ed.* 2010, *49*, 4254-4256. (h) Liu, H.-K.; Lii, K.-H. *Inorg. Chem.* 2011, *50*, 5870-5872. (i) Nguyen, Q. B.; Lii, K.-H. *Inorg. Chem.* 2011, *50*, 9936-9938.
4. Liu, C.-L.; Liu, H.-K.; Chang, W.-J.; Lii, K.-H. *Inorg. Chem.* 2015, *54*, 8165-8167.

ACKNOWLEDGMENTS

The author would like to thank the Ministry of Science and Technology of Taiwan for financial support to this research.

Syntheses and Crystal Structures of Lanthanide Dithiooxamide Complexes

Amanda Bohanon[1] and Eric M. Villa[1*]

[1]Department of Chemistry, Creighton University, 2500 California Plaza, USA, EricVilla@Creighton.edu

INTRODUCTION

The lanthanides (rare earth) metals have many unique characteristics leading them to common uses in LEDs for smart phones and large videoscreens, magnets and magnetic resonance imaging (MRI), lasers, street signs and lighting, as well as anti-counterfeiting techniques and security inks.[1-4] These lanthanide metal ions can bond with organic molecules to form extended networks. Dithiooxamide ($C_2S_2(NH_2)_2$) is a small, organic molecule that is known to react with transition metals to form coordination polymers with the potential for high porosity, catalytic properties and relatively low cost.[5-6] For these reasons, combining lanthanides and dithiooxamide is a smart choice for creating compounds with unique properties. One useful application would be in the formation of cationic frameworks, which have the potential to sequester potentially toxic anions. Herein we explore the reactions between lanthanide cations and the neutral ligand dithiooxamide.

EXPERIMENTAL METHODS

Compounds 1-3 presented here were formed under solvothermal conditions with ethanol.

RESULTS AND DISCUSSION

Compounds 1 & 2 are isotypic to a known praseodymium oxalate phases;[7-8] however, in all cases, using dithiooxamide as a starting material yielded a dark red crystalline product. This is not characteristic of the previously isolated praseodymium oxalates, which are light green. This dark red color is similar to that of the dithiooxamide starting material and to other transition metal oxamide ($C_2(NH_2)_2O_2$) containing compounds.[5-6] Currently, we are working to determine the percentage and location nitrogen in the ligand. The one thing that is certain is that at these synthetic conditions, the sulfur as preferentially hydrolyzed and a substation has occurred (Note: $S_{8(s)}$ is observed in the product mixture). **Compound 3** is a new neutral network: $[Pr_2(C_2X_4)_6(H_2O)_4]$.

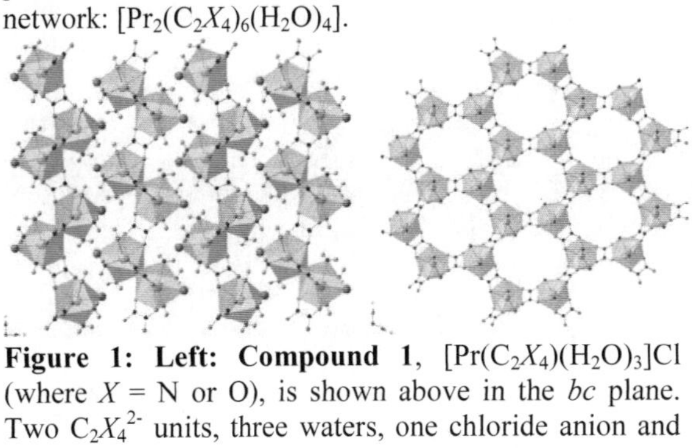

Figure 1: Left: Compound 1, $[Pr(C_2X_4)(H_2O)_3]Cl$ (where X = N or O), is shown above in the *bc* plane. Two $C_2X_4^{2-}$ units, three waters, one chloride anion and one additional praseodymium coordinate each praseodymium polyhedra (shown in yellow). **Right: Compound 2**, $Pr(C_2X_4)_3(H_2O)_3 \cdot 2\,H_2O$, is shown in the *ac* plane. Here, three $C_2X_4^{2-}$ units and three waters coordinate each praseodymium polyhedra, which then forms a pseudo-hexagonal arrangement as viewed down the *b*-axis. Here, the praseodymium are shown in yellow polyhedra, the carbon as black spheres, nitrogen as light blue spheres, oxygen as red spheres, hydrogen as light pink spheres and chloride anions as green spheres (crystal waters omitted for clarity).

While the compounds isolated thus far have a certain degree of hydrolyzation, we are currently working to prevent this reaction. By doing so, we are confident we can isolate useful lanthanide dithiooxamide compounds.

CONCLUSIONS

Lanthanide dithiooxamide complexes represent a potentially profitable area of inquiry due their ability to form cationic networks at a low cost. Currently, we have shown that species similar or isotypic to known oxalate phases are relatively easy to form. We look forward to the interesting materials that form when hydrolysis is prevented.

REFERENCES

1. Roy, S.; Chakraborty, A.; Maji, T. K. *Coord. Chem. Rev.* 45: 139-164, 2014.
2. Eliseeva, S. V.; Bünzli, J. G. *Chem. Soc. Rev.* 39: 189-227, 2010.
3. Hebbink, G. A.; Stouwdam, J.; Reinhoudt, D.; Veggel, F. V. *Adv. Mater.* 14: 1147-1150, 2002.
4. Daiguebonne, C.; Calvez, G.; Bernot, K.; Guillou, O. *Acc. Chem. Res.* 49:844-856, 2016.
5. Son, S.; Ueda, S.; Kanamaru, F.; Koizumi, M. *J. Phy. Chem.* 80: 1780-1782, 1976.
6. Ye, Q. Y.; Nakano, Y.; Frauenhoff, G. R.; Whitcomb, D. R.; Takusagawa, F.; Busch, D. H. *Inorg. Chem.* 30: 1503-1510, 1991.
7. Meng, Y.; Chen, Y.C.; Zhang, Z.M.; Lin, Z. J., Tong, M.L. *Inorg. Chem.* 53: 9052-9057, 2014.
8. Beagley, B.; Pritchard, R. G. *Acta Cryst.*, C44: 174-175, 1988.

ACKNOWLEDGMENTS

Funding for this work was graciously provided by a Clare Boothe Luce Scholarship and the Creighton Center for Undergraduate Research and Scholarship. The authors would like to thank Creighton University and the Creighton University Chemistry Department for continued support of undergraduate research.

Syntheses and Crystal Structures of Lanthanide Periodates

Colin Reedy[1] and Eric M. Villa[1*]

[1]Department of Chemistry, Creighton University, 2500 California Plaza, USA, EricVilla@Creighton.edu

INTRODUCTION

Lanthanides are found in all types of technology. Their unique magnetic and luminescent characteristics make them useful in devices ranging from television to bioanalytical systems.[1-4] When combined with periodate, there is the potential to make interesting crystal structures. Thus far, only one lanthanide periodate structure type is known.[5] Moreover, *in-situ* reduction reactions that yield iodate have the potential to yield new, interesting structure types. Lanthanide iodates have been extensively explored for their unique properties, primarily second harmonic generation.[4,6-7] While these are relatively well documented, the properties of periodates are not due to there being are so few examples of these materials. Herein, we explore the hydrothermal reactions between lanthanide cations and the periodate anion.

EXPERIMENTAL METHODS

Compounds 1 & 2 presented here were formed under mild hydrothermal conditions.

RESULTS AND DISCUSSION

Compounds 1 & 2 are new neodymioum iodate species, which have been formed *via in-situ* reduction of the periodate to iodate. While not the intended complex, this is not the first time this occurrence has been observed.[8]

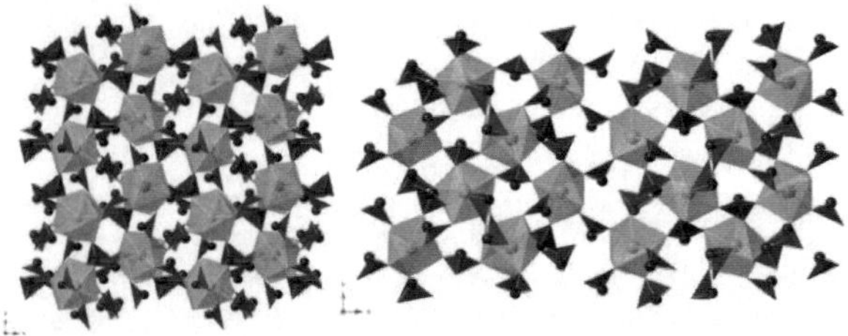

Figure 1: Left: Compound 1, $Nd_2(H_2O)_2(IO_3)_6$, is shown in the *ab* plane. The two crystallographically unique Nd^{3+} centers are both nine-coordinate with each bonding to eight iodate ligands and one water ligand; however, one is an approximate monocapped square antiprism where the other is approximately a muffin geometry. **Right: Compound 2**, $Nd_2(H_2O)_2(IO_3)_6$, is shown in the *bc* plane. The difference is in the coordination of the Nd^{3+} - one is an eight coordinate square antiprism connected by seven iodate and one water; whereas the other is a nine coordinate monocapped square antiprism bonded to eight iodates and one water. In both, the neodymium are shown in orange polyhedra, the iodate as purple trigonal pyramids and the oxygen and hydrogen atoms have been omitted for clarity.

While both compounds isolated are new, they have both been reduced to iodate. Additionally, both exist in centrosymmetric space groups ($P2_1/c$ and $P2_1/n$, respectively). Current explorations are being conducted with sacrificial oxidants to keep the periodate from being reduced. In any case, these new compounds show that *in-situ* reduction has the capabilities to produce new compounds.

CONCLUSIONS

Crystalline materials of both lanthanide periodates and iodates have the potential to yield useful materials. We are currently working to prevent the *in-situ* reduction to iodate, but will continue to logically control and exploit this phenomenon to make new iodate compounds as well and are excited to see the new compounds that form.

REFERENCES

1. Chakhmouradian, A. R., Wall, F. *Elements* 8: 333-40, 2012.
2. Roy, S.; Chakraborty, A.; Maji, T. K. *Coord. Chem. Rev.* 45: 139-164, 2014.
3. Eliseeva, S. V.; Bünzli, J. G. *Chem. Soc. Rev.* 39: 189-227, 2010.
4. Chuanfu, S. Yang, B.; Mao J.G., *J. Sci. China: Chem.* 54: 911-922, 2011.
5. Hector, A. L.; Levason, W.; Webster, M. *Inorg. Chim. Acta.* 298: 43-49, 2000.
6. Assefa, Z.; Ling, J.; Haire, R. G. Albrecht-Schmitt, T.E.; Skyora, R. E. *J. Solid State Chem.* 179: 3653-3663, 2006.
7. Hu, T.; Qin, L.; Kong, F.; Zhou, Y. Mao, J. G. *Inorg. Chem.* 48: 2193-2199, 2009.
8. Hector, A. L.; Henderson, S. J.; Levason, W.; Webster, M. *Z. Anorg. Allg. Chem.* 628: 198-202, 2002.

ACKNOWLEDGMENTS

Funding for this work was graciously provided by the Creighton Center for Undergraduate Research and Scholarship. The authors would like to thank Creighton University and the Creighton University Chemistry Department for continued support of undergraduate research.

Syntheses and Photophysical Properties of Novel Coordination Polymers Composed of Luminescent Europium and Metal Joint Blocks

Masanori Yamamoto[1], Takayuki Nakanishi[2], Yuichi Kitagawa[2], Tomohiro Seki[2], Hajime Ito[2], Koji Fushimi[2] and Yasuchika Hasegawa[2*]

[1]Graduate School of Chemical Sciences and Engineering, Hokkaido University, Japan
[2*]Faculty of Engineering, Hokkaido University, Japan, hasegaway@eng.hokudai.ac.jp

INTRODUCTION

Lanthanide (Ln^{3+}) complexes with *4f-4f* transitions are promising materials for application such as light emitters and molecular sensors [1-2]. In particular, three-dimensional Ln^{3+} coordination polymers (3DLnCPs) composed of Ln^{3+} complexes and joint ligands are expected to open up a new field of photo-functional materials. In the previous study, 3DLnCPs with dicarboxylic acid or bipyridine as joint ligands have been reported [3].

We here focused on transition metal complexes with various coordination structures. Combination of Eu^{3+} complex (luminescent block) and transition metal complex (joint block) promotes conformation of remarkable three-dimensional structures of 3DLnCPs (Figure 1). We selected several metal complexes as joint blocks. In this presentation, preparations and photophysical properties of 3DLnCPs with transition metal complexes as joint ligands are demonstrated.

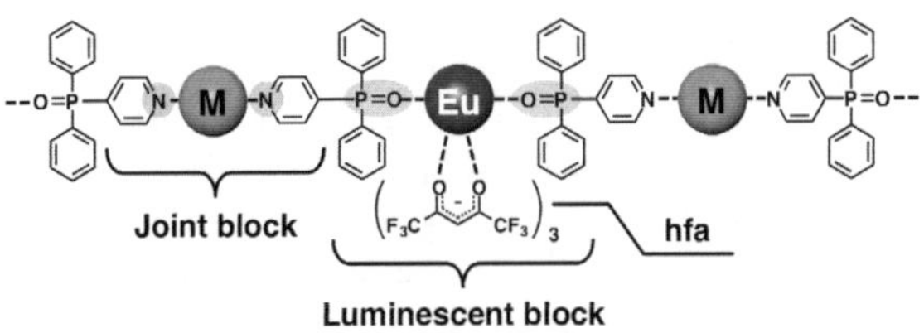

Figure 1. Structure image of Eu^{3+} coordination polymer cross-linked with metal complexes

EXPERIMENTAL METHODS

Phosphine oxide containing pyridyl groups was prepared by the coupling reaction of bromo-pyridine with diphenylphosphine [4]. Transition metal complexes were synthesized by the chelation of the phosphine oxide with metal chloride (palladium chloride, zinc chloride, or aluminum chloride). The 3DLnCPs are obtained by the complexation of transition metal complexes and $Eu(hfa)_3(H_2O)_2$ (hfa: hexafluoroacetylacetone) in organic solvent.

RESULTS AND DISCUSSION

Structure of Eu-Zn is estimated using DFT calculations (B3LYP/6-31G(d)). We consider that Eu-Zn have formed one-dimensional polymer structures, because precursor Zn^{2+} complex is a tetrahedral structure formed by two chloride ions and two phosphine oxide ligands.

The optical properties of these Eu^{3+} coordination polymers in solid state were measured using Fluorolog-3 spectrophotometer, Nd:YAG laser and integrating sphere. The emission spectra of Eu-Zn and precursor $Eu(hfa)_3(H_2O)_2$ are shown characteristic emission bands at 570 - 720 nm. The time-resolved emission profile of Eu-Zn shows single-exponential decay with > 0.80 ms lifetime. The emission quantum yield of the Eu-Zn is larger than those of precursor $Eu(hfa)_3(H_2O)_2$. We consider that the polymer structure of the Eu-Zn leads to large radiative rate constant (k_r) and small non-radiative rate constant (k_{nr}). In the result, the Eu-Zn shows a higher emission quantum yield than precursor $Eu(hfa)_3(H_2O)_2$. In this presentation, we will also report the photophysical properties of Eu-Pd and Eu-Al.

CONCLUSIONS

We have successfully synthesized a novel structural design of Eu^{3+} coordination polymers Eu-Zn, which were composed of luminescent Eu^{3+} and Zn^{2+} complexes. The polymeric structure of Eu-Zn promotes enhancement of luminescent properties related to large k_r and small k_{nr}. The emission quantum yield of the Eu-Zn coordination polymer is larger than that of $Eu(hfa)_3(H_2O)_2$.

Luminescent Ln^{3+} coordination polymers cross-linked with metal complexes are expected to open up new coordination polymers in the frontier field of materials science.

REFERENCES

1. Hasegawa Y. *et al.*, Angew. Chem. Int. Ed. 52:6413-6416, 2013
2. Wang S. *et al.*, Chem. Commun. 48:12059-12061, 2012
3. Du S. *et al.*, J. Mater. Chem. 22:21210-21217, 2012
4. Yamamoto M. *et al.*, Mater Lett. 167:183-187, 2016

ACKNOWLEDGMENTS

This work was supported by a Grant-in-Aid for Scientific Research on Innovative Area of "New Polymeric Materials Based on Element-Blocks" from the Ministry of Education, Culture, Sports, Science and Technology (MEXT), Japan (grant number 2401) and the MEXT through Program for Leading Graduate Schools (Hokkaido University "Ambitious Leader's Program").

Synthesis and Characterization of Homoleptic Bis(dimethylsilyl)benzyl Lanthanide Compounds

Kasuni C. Boteju, Arkady Ellern, Aaron D. Sadow*

Department of Chemistry, Iowa State University, USA
kasuni@iastate.edu

INTRODUCTION

Homoleptic organolanthanide compounds are valuable for synthesis of new catalysts and new materials,[1] yet solvent- or donor-group free, salt-free, and thermally robust organolanthanide compounds are not readily accessed for the larger metal centers.[2] Incorporation of β-SiH in rare earth amides has been used as a strategy of stabilizing lanthanide amides.[3] Thus, a new alkyl ligand was synthesized that contains β-SiH and benzyl functionalities, which can form labile secondary interactions with the rare earth center.

EXPERIMENTAL METHODS

Three equiv. of $KC(SiHMe_2)_2Ph$ (**2**) or **2·TMEDA** and LaI_3THF_4, CeI_3THF_4, PrI_3THF_3, or NdI_3THF_3 provide $Ln\{C(SiHMe_2)_2Ph\}_3$ (Ln = La (**3**), Ce (**4**), Pr (**5**), Nd (**6**). NMR, IR and X-ray diffraction techniques were performed to characterize the rare earth alkyl complexes.

RESULTS AND DISCUSSION

Reductive coupling of $HCPhBr_2$ and $ClSiHMe_2$ provides $HC(SiHMe_2)_2Ph$ (**1**) on preparative scale.

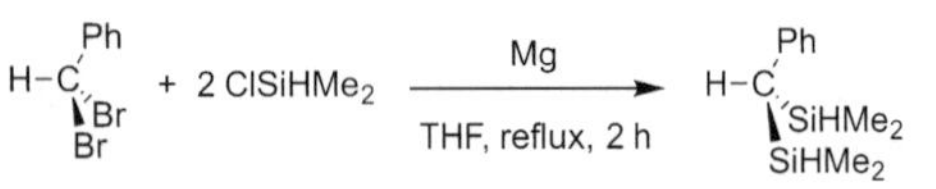

(1)

1, 70.9%

Deprotonation of **1** with KBn at –78 °C provided the major product as $KC(SiHMe_2)_2Ph$ (**2**) assigned to a doublet at 0.47 ppm (Si*Me*) in the ^{1}H NMR spectrum. However, the pure product of **2** was obtained as a TMEDA adduct (**2.TMEDA**) in very low yield from crystallization in pentane at –30 °C. Attempts to deprotonate **1** with KO*t*Bu at –78 °C afforded a mixture of $KC(SiHMe_2)_2Ph$ (**2**) and $KCH(SiHMe_2)Ph$ suggesting a possibility of Si–C cleavage. The reaction between $(Me_2HSi)_3CPh$ and KO*t*Bu provided the desired **2** via Si–C cleavage in excellent yield.

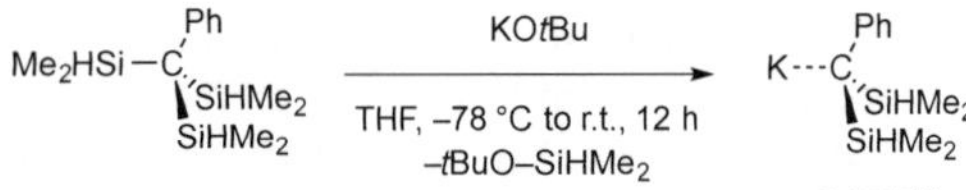

(2)

2, 79.7%

The reaction of **2** or **2.TMEDA** with lanthanide iodides provided new homoleptic organolanthanide complexes in high yields.

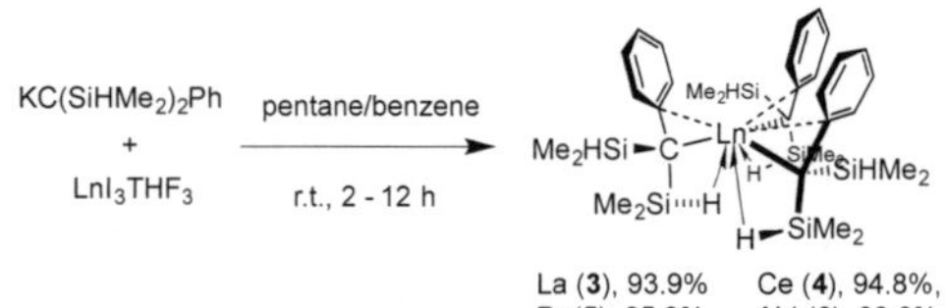

(3)

La (**3**), 93.9% Ce (**4**), 94.8%, Pr (**5**), 85.3% Nd (**6**), 88.6%

The series of compounds yielded pale yellow, orange, yellow and green crystals from pentane at –30 °C. X-ray diffraction studies of all four compounds show both agostic-like interactions (Ln←H–Si) and benzylic-type π interactions between the metal center and alkyl ligand. La compound crystallizes with one molecule per unit cell and shows one agostic-like interaction per ligand, while other three complexes crystallize with two independent molecules per unit cell. One molecule shows one agostic-like interaction per ligand in similar to La complex, while the other molecule shows only two Ln←H–Si interactions. Two SiH peaks are obtained for the La compound at low temperature (–73 °C in toluene d_8): 4.67 ppm ($^1J_{SiH}$ ~180 Hz) and 3.82 ppm ($^1J_{SiH}$ ~120 Hz) for classical and non-classical interactions respectively.

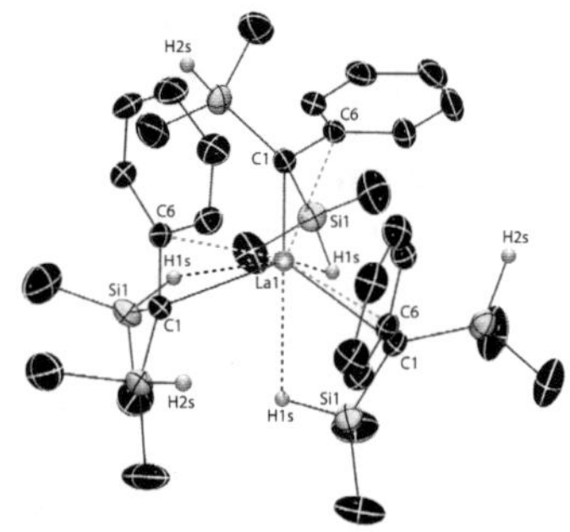

CONCLUSIONS

The reaction involving nucleophilic attack on a Si center to cleave the C–Si bond introduced an alternative route to the desired bis(dimethylsilyl)benzyl ligand. β-SiH and aryl interactions are likely important for the facile isolation of the series of new homoleptic organolanthanides.

REFERENCES

1. Marks T. J. *et al.*, Organometallics. 22:4630-4632, 2003
2. Schumann H. *et al.*, Organometallics. 3:69-74, 1984
3. Anwander R. *et al.*, Organometallics. 16:1813-1815, 1997

ACKNOWLEDGMENTS

This research was supported by the U.S. Department of Energy, Office of Basic Energy Sciences, Division of Chemical Sciences, Geosciences, and Biosciences. The Ames Laboratory is operated for the U.S. Department of Energy by Iowa State University under Contract No. DE-AC02-07CH113

Synthesis and Structure of New Structure Type $Ln_{52}Fe_2Sn_{49}$ (Ln = Gd, Tb, Dy)

Alexander T. Brown, Keaton G. Barron, Randall P. Kirby, Julia Y Chan
Department of Chemistry & Biochemistry, The University of Texas at Dallas, U.S.A.,
alexanderbrown2@utdallas.edu

Abstract
Single crystals of the new structure type $Ln_{52}Fe_2Sn_{49}$ (Ln = Gd, Tb, Dy) are grown from a self-flux method using the Ln-Fe binary eutectic point. The compounds are grown using a reaction ratio of 12 Ln : 3 Fe : 11 Sn and annealed at 1260 °C. The crystal structures as determined at 100 K were solved by the intrinsic phasing method. The $Dy_{51.7}Fe_2Sn_{48.6}$ crystal structure is modeled in space group *Fmmm* with lattice parameters a = 16.2992(6) Å, b = 16.9088(6) Å, c = 39.0656(14) Å, V = 10766.5(7) Å^3 with Z = 4. $Ln_{52}Fe_2Sn_{49}$ has a large cell volume, structural complexity, and consists of ten Ln, two Fe, and fourteen Sn unique crystallographic sites. In ambient conditions, $Ln_{52}Fe_2Sn_{49}$ is silver with a metallic luster for the first 12 h, then a passivation layer forms on the crystal surface. Finally, the electrical properties are presented.

Synthesis and Understanding on Lanthanide Single-Molecule Magnets

Bingwu Wang[1*] and Song Gao[1]

[1]1Beijing National Laboratory for Molecular Science, State Key Laboratory of Rare Earth Materials Chemistry and Applications, College of Chemistry and Molecular Engineering, Peking University, Beijing 100871, P. R. China.
wangbw@pku.edu.cn

Single-molecule magnets (SMMs) exhibit the slow relaxation of magnetization, magnetic hysteresis under blocking temperature for the molecular origin. For the potential application in the high-density information storage, quantum computing and molecular spintronics, SMMs attract the interest of the scientists in inorganic chemistry, coordination chemistry, organometallic chemistry, magnetic physics and materials science [1].
The lanthanide based compound is one of the most important type of SMMs. For the strong magnetic anisotropy of lanthanide ions, chemists have synthesized a number of lanthanide SMMs with various composition and geometrical structures. The high relaxation barrier and blocking temperature are two key properties in the chemical design of SMMs. Herein, we introduce our synthesis and theoretical understanding on some lanthanide SMMs. Various coordination atoms, coordination bond character and local symmetry of lanthanide ions are used in the designation of lanthanide SMMs. The multiple ab initio calculation methods, CASSCF, CASPT2 and MRCI are used to compute the electronic structures of ground and low lying excited states to understand the magneto-structural correlation of lanthanide molecules. [2, 3, 4]

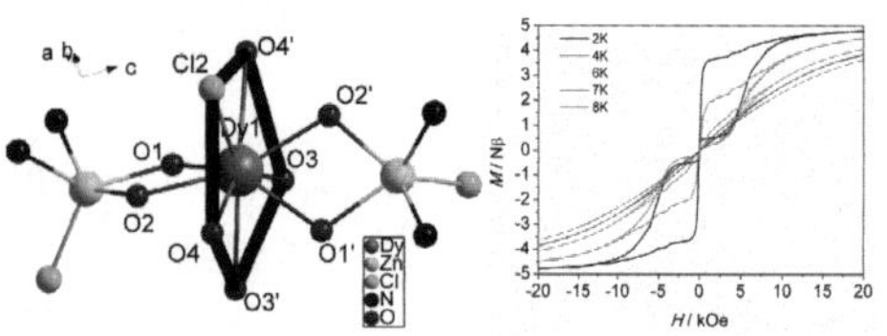

Figure 1. the core structure and hysteresis of $DyZn_2$ SMMs, the outer backbone of ligands and counter anions are omitted for clarity.

REFERENCES

1. Sessoli R., Gatteschi D., Caneschi A. and Novak M. A.*Nature*, **1993** 365, 141

2 Sun, W. B.; Yan, P. F.; Jiang, S. D.; Wang, B. W.; Zhang, Y. Q.; Li, H. F.; Chen, P.; Wang, Z. M.; Gao, S., *Chem. Sci.,* **2016,** *7* , 684.

3 Xiong, J.; Ding, H. Y.; Meng, Y. S.; Gao, C.; Zhang, X. J.; Meng, Z. S.; Zhang, Y. Q.; Shi, W.; Wang, B. W.; Gao, S., *Chem. Sci.* 2017, 8, 1288.

4 Meng, Y. S.; Jiang, S. D.; Wang, B. W.; Gao, S., *Acc. Chem. Res.* **2016**, 49, 2381

ACKNOWLEDGMENTS

"The authors would like to thank the NSFC and the National Basic Research Program of China. for providing financial support to this project".

Synthesis of anisotropic hybrid nanoparticles containing f-block elements.

Bastian Trepka[1*] and Sebastian Polarz[2]

[1*]Chemistry Department, University of Konstanz, Germany, bastian.trepka@uni-konstanz.de
[2]Chemistry Department, University of Konstanz, Germany

INTRODUCTION

Inorganic-organic hybrids are fascinating since they combine the properties of both individual phases in one material. The organic phase allows chemical modification due to the high diversity of functional groups whereas via the inorganic phase optical, electronic or magnetic properties can be introduced to the material. Very often also synergistic effects are observed. One example are upconversion phenomena that appear when fluorophores are combined with absorbing organics. New or enhanced effects like changing the electronic properties can evolve, as the hybrids are reaching a confined state. One versatile approach to synthesize nanostructured hybrid materials is the nonaqueous sol-gel route based on benzyl alcohol.[1] Many metal precursors are established so far, but are mainly containing d-block metals. Rare earth metals are known for strong and outstanding magnetic, electronic and optical properties. Therefore, in our work we are exploring new precursors containing f-block elements that can be converted to nanostructured hybrid materials.

EXPERIMENTAL METHODS

The particles synthesized in our work were produced by a non-aqueous sol-gel route as introduced by Niederberger *et al.*[1]

RESULTS AND DISCUSSION

As an example, a single hybrid particle resulting from the as mentioned synthesis is shown in Fig.1. The particle has a lamellar structuration. High contrast lamellae are of inorganic and low contrast lamellae of organic nature. It is well known that elimination of organic molecules during the synthesis leads to the formation of metal-oxygen-metal bonds.[1-2]
However, in the case of using rare earth elements as metals, the formation of pure metal oxide particles is unfavourable and instead coordination polymers are formed.[2]

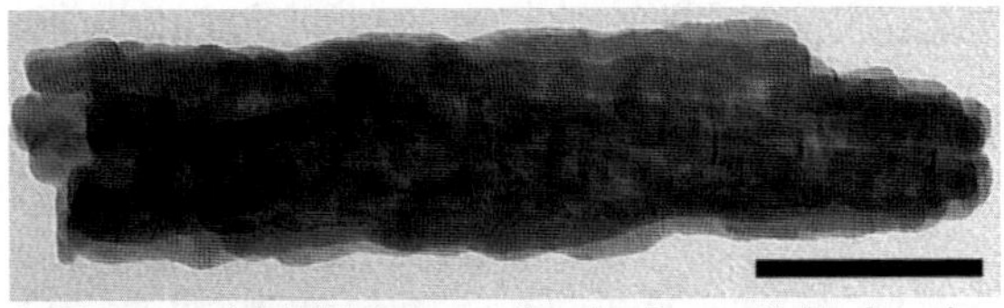

1. Figure: a) TEM image of a single particle showing a lamellar structuration (scale bar 100 nm).

CONCLUSIONS

Using rare earth elements in the benzyl alcohol route leads to the formation of well-defined nanostructures. The particles show a lamellar structuration due to the generation of coordination polymers. This confirms the successful generation of a hybrid material. The influence of rare earth ions on the structure as well as their properties is currently under investigation.

REFERENCES

1. Niederberger M. *et al.*, Prog. Solid State Chem. 33(2-4): 59-70, 2005
2. Karmaoui M. *et al.*, Chem. Mat. 18(18): 4493-4499

ACKNOWLEDGMENTS

Financial support by the DFG (SFB1214) is gratefully acknowledged.

Synthesis of U(VI)bis(imido) Complexes from UCl_4: Investigations of Oxidation Processes in Uranium Chemistry

James M. Boncella[1*] Neil C. Tomson[1], Nickolas H. Anderson[1], Aaron M. Tondreau[1] and Brian L. Scott[2]

[1*]Chemistry Division, Los Alamos National Laboratory, USA, boncella@lanl.gov
[2]Materials Physics and Applications Division, Los Alamos National Laboratory, USA

INTRODUCTION

For a number of years, we have been exploring the chemistry of uranium(IV-VI) mono- and bis(imido) imido complexes. These studies of the synthesis, spectroscopy and fundamental transformations of complexes that have uranium-nitrogen multiple bonds have been useful in developing an understanding of the role of M-L covalency in actinide chemistry.[1-3] In this talk the conversion of UCl_4 to high oxidation state U imido complexes will be discussed.

RESULTS AND DISCUSSION

Recently, we have discovered a simple and convenient method for the preparation of U(VI)bis(imido) complexes starting from UCl_4. This method involves the treatment of UCl_4 with metal amide sources to generate U(IV) tetrakis(amido), bis(amido)/imido or bis(imido) intermediates. The speciation of these intermediates depends upon the nature of the amide ligand, and the solvent or Lewis base that is present in the original reaction mixture. Various U(IV) amido or imido/amido species can be isolated from these solutions depending upon conditions.

Treatment of the U(IV)/amide mixtures with an oxidant (*e.g.* I_2 or PhSSPh) results in conversion of these intermediates to the desired U(VI)bis(imido)$(X)_2$ (X= I, SPh) complexes. When these reactions are monitored by NMR spectroscopy, a new paramagnetic species is formed upon oxidation which subsequently disappears as the final diamagnetic U(VI) bis(imido) complexes are formed. This novel method for synthesizing U(VI)bis(imido) complexes is more convenient than previously published methods and facilitates further elaboration of these compounds, which will be discussed.

CONCLUSIONS

It is likely, that the species that have been identified when UCl_4 reacts with amide sources are in equilibrium with one another. Addition of the oxidant to these solutions intercepts the U(IV) bis(imido) intermediate and generates a U(V) intermediate that is further oxidized to the U(VI) product. This suggests that the oxidation of the U(IV) species to the U(VI) products proceeds via single electron transfer steps rather than by a concerted 2 e^- transfer process. Overall, these observations further reinforce the notion that U(VI)bis(imido) species represent thermodynamic sinks in these systems. It is the stability of the U(VI)=N multiple bond that provides the driving force necessary for conversion to the high oxidation state.

REFERENCES

1. R. E. Jilek, N. C. Tomson, R. L. Shook, B. L. Scott, J. M. Boncella *Inorg. Chem.* **2014**, *53,* 9818-26.
2. R. E. Jilek, L. P. Spencer, R. L. Lewis, B. L. Scott, T. W. Hayton, J. M. Boncella *J. Am. Chem. Soc.* **2012**, *134*, 9876-78.
3. L. P. Spencer, P. Yang, S, G. Minasian, R. E. Jilek, E. R. Batista, K. S. Boland, J. M. Boncella, S. D. Conradson, D. L. Clarke, T. W. Hayton, S. A. Kozimor, D. K. Shuh, M. P. Wilkerson *J. Am. Chem. Soc.* **2013**, *135*, 2279 - 90.

ACKNOWLEDGMENTS

The authors would like to thank the US DOE Division of the Office of Basic Energy Sciences, Heavy Element Chemistry program; contract DE-AC52-06NA25396 for support of this research.

Synthesis, Structure, Bonding and Properties of $Pr_5Co_2Ge_3$ and $Pr_7Co_2Ge_4$

Kaiser Aguirre,[1] Scott Saunders,[2] Yong Liu,[1] Valentin Taufour,[1,2,3] Paul Canfield,[1,2] Gordon J. Miller,[1,4] Qisheng Lin[1,4*]

[1]Division of Materials Sciences and Engineering, Ames Laboratory, U.S. Department of Energy
[2]Department of Physics, Iowa State University, U.S.A.
3. Department of Physics, University of California Davis, U.S.A.
[4]Department of Chemistry, Iowa State University, U.S.A. *qslin@ameslab.gov

INTRODUCTION

Rare earth intermetallics represent one important class of advanced functional materials germane to magnetic properties.[1] Among all rare-earth (Sc, Y, La-Lu)-transition metal (Ti–Ni)-germanium intermetallic compounds, which include ~1000 entries in the Pearson Database, only ~30 entries are found to be rare-earth rich (> 50 at.%). For the Pr-Co-Ge system, Pr_3CoGe_2 is the only compound found on the rare-earth corner of the ternary phase diagram.[2] Herein, we focus on the syntheses, crystal structures, and bonding of two unprecedented rare-earth rich Pr-Co-Ge phases. Both structures feature unprecedented all-metal planar heteroatomic clusters with geometries resembling organic molecules.

EXPERIMENTAL METHODS

$Pr_5Co_2Ge_3$ crystals were grown using self-flux method.[3] $Pr_7Co_2Ge_4$ crystals were obtained from mixture of stoichiometric elements through arc melting followed by annealing treatments. Both crystals were structurally characterized by single crystal X-ray diffraction. Susceptibility, resistivity, and specific heat were evaluated by PPMS and thermal stability by DSC.

RESULTS AND DISCUSSION

The structure of $Pr_5Co_2Ge_3$ (*C2/m*) features two unprecedented species for complex polar intermetallics: the ethylene-like Co_2Ge_4 clusters and polyacene-like $(Co_2Ge_2)_n$ chains, whereas $Pr_7Co_2Ge_4$ ($P2_1/c$) contains 1,2,4,5-tetramethylclohexane-like Co_4Ge_6 clusters (Figure 1). Within these clusters, Co-Co and Co-Ge interactions exhibit strong covalency according to COHP and ELF analyses. However, polar-covalent bonding between rare earth and Co or Ge atoms is also significant in both structures. $Pr_5Co_2Ge_3$ and $Pr_7Co_2Ge_4$ exhibit different magnetic orderings related to the different arrangements of Co-Ge clusters.

The observation of molecular clusters in polar intermetallics is remarkable because to date, such units have only been observed in valence-type Zintl phases,[4] whereas complex two-dimensional layers or three-dimensional frameworks are encountered in intermetallic structures containing transition metals. The discovery of these Pr-Co-Ge phases encourages us to examine the phase spaces of these compounds and to explore the occurrence of other such clusters.

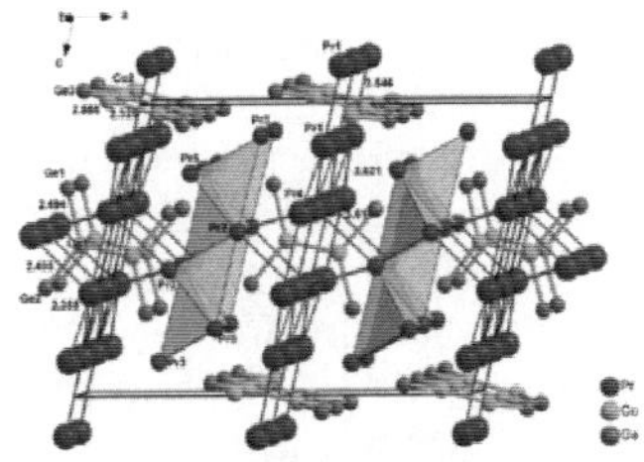
(a)

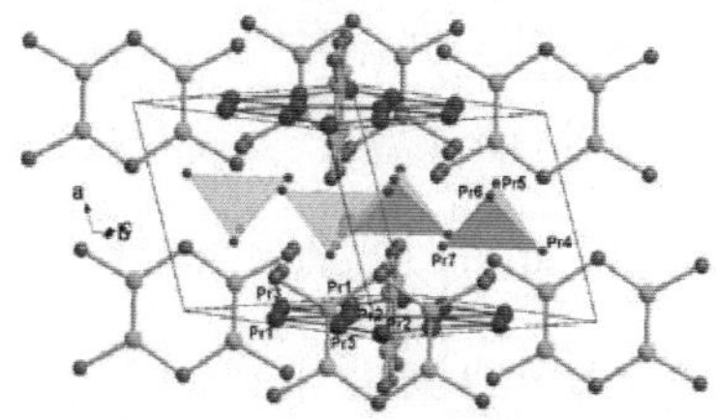
(b)

Fig.1 Structures of (a) $Pr_5Co_2Ge_3$ and (b) $Pr_7Co_2Ge_4$.

CONCLUSIONS

Two unprecedented Pr rich Pr-Co-Ge phases were discovered and their structures determined by single crystal X-ray diffraction. Both metal-rich structures feature molecular clusters, which are often seen in valence compounds.

REFERENCES

1.A. Szytula, J. Leciejewicz, *Handbook of Crystal Structures and Magnetic Properties of Rare Earth Intermetallics*, CRC Press, Boca Raton, Fl, 1994.
2. *Pearson's Handbook of crystallographic data for intermetallic phases*. Metals Park, Oh 44073: American Society for Metals; 1985.
3. P. C. Canfield, I. R. Fisher, *J. Cryst. Growth*, 225 (2001) 155-161.
4. Kauzlarich, S. M., *Chemistry, Structure, and Bonding of Zintl Phases and Ions*. VCH: New York, 1996.

ACKNOWLEDGMENTS

This work was supported by the U.S. Department of Energy (DOE), Office of Science, Basic Energy Sciences, Materials Science and Engineering Division. The research was performed at Ames Laboratory, which is operated for the U.S. DOE by Iowa State University under contract # DE-AC02-07CH11358. K.A. is grateful to the DOE for the assistantship and opportunity to participate in the U.S. Department of Energy Office Sciences, Science Undergraduate Laboratory Internship (SULI) program.

Systematic investigation of the crystalline, magnetic and calorimetric properties of $Gd_{1.02}Sc_{0.98}Ge_{1-x}Sb_x$ compounds

Francois Guillou, Arjun Pathak, Yaroslav Mudryk, and Vitalij K. Pecharsky

The Ames Laboratory, U.S. Department of Energy, Iowa State University, Ames, Iowa 50011, USA,
fguillou@ameslab.gov

INTRODUCTION

The ternary RTX materials, where R = rare earth, T = transition metal, X = p-block element, are among the most extensively investigated rare-earth intermetallics. Of particular interest is the GdScGe material crystallizing in the CeScSi-type structure, as it is ferromagnetic with a strikingly high Curie temperature (T_C~ 350 K).[1] On the other hand, GdScSb crystallizing in the closely related CeFeSi-type orders antiferomagnetically at T_N~ 55 K.[2] Theoretical reports suggested that the different magnetic orders in GdScGe and GdScSb are intimately linked to the crystal structure type.[3] In this work, we performed a systematic experimental study of the crystal structure and physical properties as a function of the concentration in $Gd_{1.02}Sc_{0.98}Ge_{1-x}Sb_x$ intermetallics.

EXPERIMENTAL METHODS

Polycrystalline $Gd_{1.02}Sc_{0.98}Ge_{1-x}Sb_x$ samples were prepared by arc-melting, followed by high temperature annealing. Powder x-ray diffraction patterns were measured with a PANAnalytical diffractometer using monochromatized Cu $K_{\alpha 1}$ radiation. DC and AC magnetic properties were recorded using Quantum Design MPMS (7 T) and PPMS (14 T). Heat capacity measurements were performed in a home-built semi-adiabatic heat-pulse calorimeter with magnetic field capabilities.

RESULTS AND DISCUSSION

Twelve $Gd_{1.02}Sc_{0.98}Ge_{1-x}Sb_x$ chemical compositions were prepared to cover the entire range from $x = 0$ to 1. Except for the end member $x = 1$, powder XRD showed that the samples are free of impurity phases. The materials with $x < 0.7$ are found to crystallize in the CeScSi-type structure in accordance with the results of the literature for GdScGe, whereas the compositions with $x \geq 0.9$ adopt the CeFeSi-type as for GdScSb. In the intermediate range $0.7 \leq x \leq 0.8$, the two crystalline types are found to coexist.

Figure 1 presents the magnetization of 3 selected compositions. As illustrated by the *M*(*T*) curve for $Gd_{1.02}Sc_{0.98}Ge_{0.9}Sb_{0.1}$, a ferromagnetic transition is observed at low Sb content. The Curie temperature is found to decrease with the increase in Sb, while the saturation magnetization does not evolve significantly. At high antimony content, a T_N close to that of GdScSb is observed.

More unexpected is the magnetism of the composition just below the region where both crystal structure types coexist. Neither a magneto-structural transition nor the coexistence of an antiferromagnetic and a ferromagnetic phases are observed. For x= 0.7, one observes an antiferromagnetic-like behaviour at low magnetic field. The application of a large magnetic field points to a field-induced transition toward a ferromagnetic state.

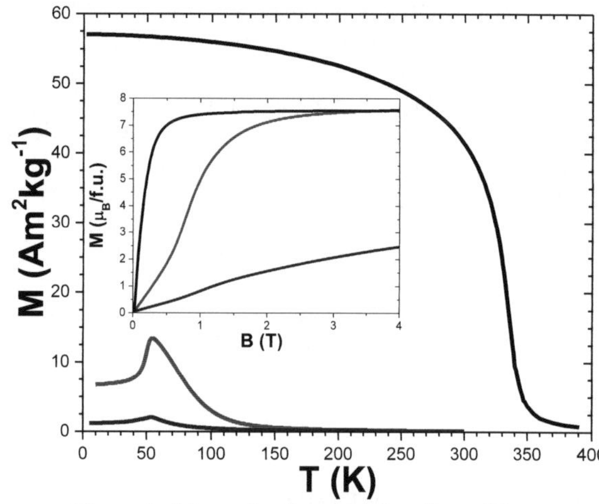

Figure 1: Magnetization as a function of the temperature (FC) in B= 0.1 T for $Gd_{1.02}Sc_{0.98}Ge_{1-x}Sb_x$ with x= **0.1**, 0.7, **0.9**. In inset, M(B) curves at 5 K.

Heat capacity measurements were performed for several x values. In the case of $x = 0.7$, no pronounced anomaly is observed neither in low nor in finite magnetic field. A possible glassy ground state is addressed by AC magnetic measurements.

CONCLUSIONS

The thorough investigation of the structural and physical properties in the $Gd_{1.02}Sc_{0.98}Ge_{1-x}Sb_x$ series reveals that Sb substitution leads to a gradual conversion from CeScSi to CeFeSi-type structure. Compared to the parent GdScGe, the Curie temperatures progressively decrease with x. Just before the boundary between the two crystalline types, a long range ferromagnetic order is no longer observed, which is at odds with the anticipated behaviour from earlier theoretical reports.

REFERENCES

1. Manfrinetti P. *et al.*, J. Alloys Compd. 334: 9, 2002.
2. Provino A. *et al.*, J. Alloys Compd. 587: 783, 2014.
3. Ritter C. *et al.*, J. Phys.: Condens. Matter 26: 366001, 2014

ACKNOWLEDGMENTS

The Ames Laboratory is operated for the U. S. Department of Energy by Iowa State University under Contract No. DE-AC02-07CH11358. This work was supported by the Department of Energy, office of Basic Energy Sciences, Materials Sciences Division.

Taking Advantage of the Chemistry of Heterocycles in Actinide Selective Coordination of U(VI) (UO_2^{2+}) and Th(IV): Crystal packing, Reactivity, and π–π interactions

Anne E. V. Gorden[1*], Emily E. Hardy[1], Kevin M. Wyss[1], Madeline A. Eddy[1], and Rebecca J. Keller[1]

[1*]Department of Chemistry and Biochemistry, Auburn University, Auburn, Alabama USA, anne.gorden@auburn.edu

INTRODUCTION

Increasing the use of nuclear power is one idea to address the demands of our highly energy driven society while still reducing global dependence on fossil fuels.[1] The potential for detrimental environmental effects from contamination due to a spill or radiological event limits public acceptance of the use of 5f elements, the actinides, as nuclear fuels. Current methods of characterization of trace materials require purification of samples prior to measurement. New materials are required that can coordinate, sense, and isolate actinides. In an effort to develop inexpensive ligands for colorimetric sensors, we have incorporated imine aza-donors into an ethylene backbone for rapid detection of uranyl. Here, we report imine aza-donors incorporated into an ethylene backbone for coordination of uranium as uranyl (U(VI) UO_2^{2+}), thorium (Th(IV)), and lanthanides to further decrease detection limits and reduce signal to noise in rapid detection methods, and we characterize these systems, their unique packing, and pi-pi interactions.

EXPERIMENTAL METHODS

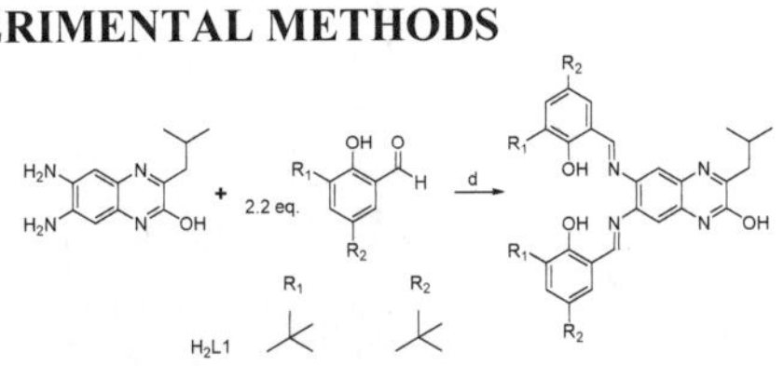

Figure 1: Synthesis of ***salqu*** ligand (**1**).

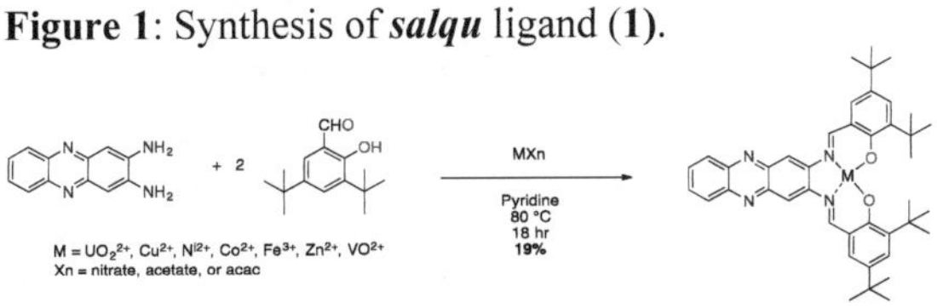

Figure 2: Synthesis of the ***salphenzine*** ligand (**2**).

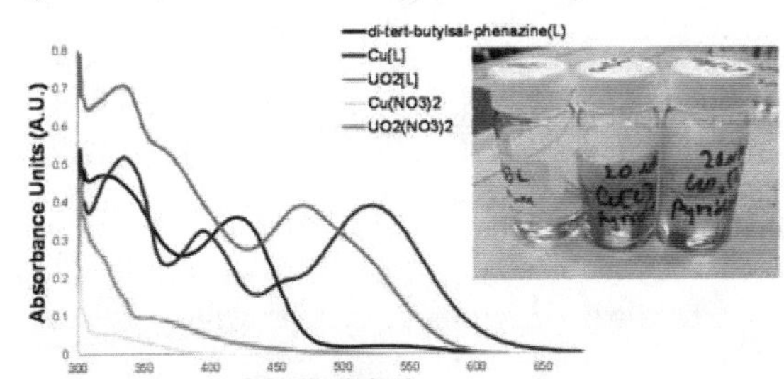

Figure 2: UV-Vis spectra demonstrating the colorimetric response of **2** with the addition of Cu^{2+} or UO_2^{2+} salts. (Inset: 20 μmol solution of ligand vs. with Cu^{2+} or UO_2^{2+} salts)

RESULTS AND DISCUSSION

Two tetradentate Schiff base ligands and the corresponding UO_2(VI) complexes (UO_2**1**, UO_2**2**) were characterized. The change in π-orbital overlap results in a dramatic color change - making actinide and transition metal complexes readily distinguishable by spectroscopy.

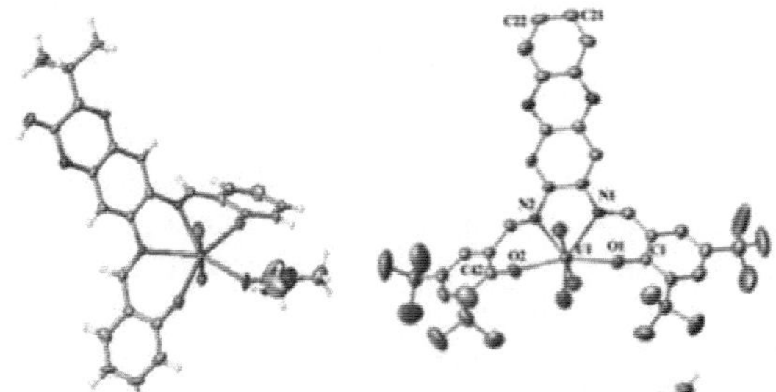

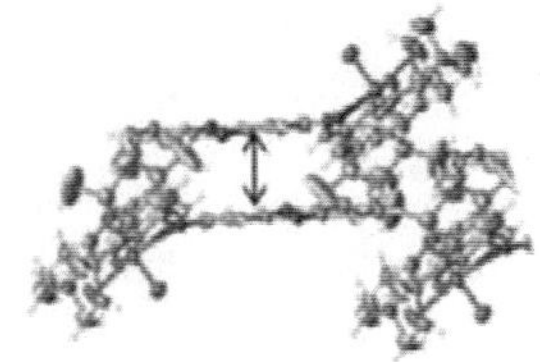

Figure 2: Projections of UO_2**1 (left)** and UO_2**2 (right)**, with the long range ordering caused by pi-pi stacking of the ligand shown. U shown in green, carbon grey, hydrogen in white, oxygen red, and nitrogen blue. Coordinating and interstitial solvents have been removed for clarity.

CONCLUSIONS

The uranyl complex for (**2**) has been found to have a hypsochromic shift in the UV-Vis spectrum while other potentially competing metals have no shift, or a bathochromic shift, as do cobalt and copper. This result suggests the change in pi-overlap of these ligands or electrochemical redox response of similar systems can be useful in a chemosensor.

REFERENCES

1. Gorden, A. E. V.; DeVore, M. A. II; Maynard, B. A. *Inorg. Chem.*, **2013**, 52, 3445. Tutson, C.D.; Gorden, Anne. E. V. *Coord. Chem. Rev.* **2017**, 27.
2. DeVore, M. A., II; Kerns, S. A.; Gorden, Anne E.V. *Eur. J. Inorg. Chem.*, **2015**, 5708.
3. Maynard, B. A.; Brooks, J. C.; Hardy, E. E.; Easley, C. J.; Gorden, A. E. V. *Dalton Trans.* **2015**, 4428.
4. Hardy, E. E.; Eddy, M.A.; Maynard, B.A.; Gorden, A E.V. *Dalton Trans.* **2016**, 14243.

ACKNOWLEDGMENTS

This work was supported in part by the Defense Threat Reduction Agency, Basic Research Award # HDTRA1-11-1-0044 to Auburn University. The authors would like to thank Dr. J. Gorden for helpful discussions.

Technospheric Mining of Rare Earth Elements From Landfilled Stocks of Industrial Process Residues: Phosphogypsum And Red Mud

Oliver Ko[1], Sable Reid[1], Gisele Azimi[1,2*]

[1]Chemical Engineering and Applied Department, University of Toronto, Canada
[2*]Materials Science and Engineering Department, University of Toronto, Canada g.azimi@utoronto.ca

INTRODUCTION

Rare earth elements (REEs) have unique physicochemical properties that make them an indispensible component of many technologically advanced products. Based on the historic data, the demand for some REEs is increasing at a rate that the supply will fall behind in the near future. Although the global reserve for REEs is considerable, most of them are located in China.[1] Because of the recent export quotas; the global supply chain has become vulnerable. To satisfy their demand, many courtiers have initiated the technospheric mining of secondary REE-containing resources. These activities include: 1) extractions from stocks of landfilled industrial process residues, and 2) urban mining of end-of-life products, such as e-waste.[2]

The current study puts the emphasis on developing novel processes to recover REEs from two major industrial process residues: phosphogypsum (PG) and red mud.[3,4] Although the concentration of REEs in these sources is small, the production volume is extensive. Thus, the total amount of REEs that can be recovered from these sources is considerable.

PG is the by-product of phosphoric acid production through sulphuric acid digestion of phosphate rock. In the current study, we obtained the PG sample from Agrium Inc., a fertilizer company in Canada. Red mud is the solid residue generated in the Bayer process for alumina production. In this study, we obtained the red mud sample from Rio Tinto located in Canada.

EXPERIMENTAL METHODS (if relevant)

First, a thorough characterization of PG and red mud samples was performed using various techniques including aqua regia digestion, ICP-OES, SEM-EDS, XRD, and XRF to determine the composition, morphology, and crystal structure. Then, systematic leaching experiments were performed and the effect of various leachants and operating conditions in terms of temperature, concentration, S/L, and residence time was investigated. Reproducibility tests (four separate experiments) showed that the experimentally measured data are accurate to within ± 5%.

RESULTS AND DISCUSSION

Figure 1 presents the SEM-EDS elemental mapping of a red mud particle. Based on aqua regia digestion results, the red mud sample contains 0.03-0.05 wt% REEs, including cerium, lanthanum, neodymium, yttrium, and scandium.

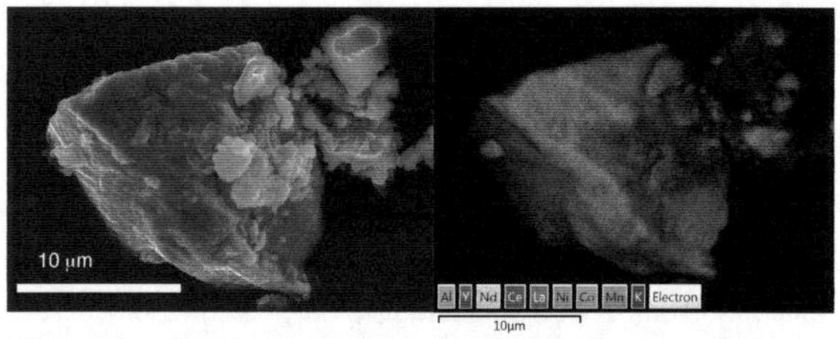

Figure 1 – SEM-EDS mapping of a red mud particle.

Figure 2 present the leaching efficiency of various REEs for three different leaching agents (H_2SO_4, HCl, HNO_3) at various operating conditions: temperature between 25 and 90 °C, concentration between 0.5 and 3 M, and 20 min residence time.

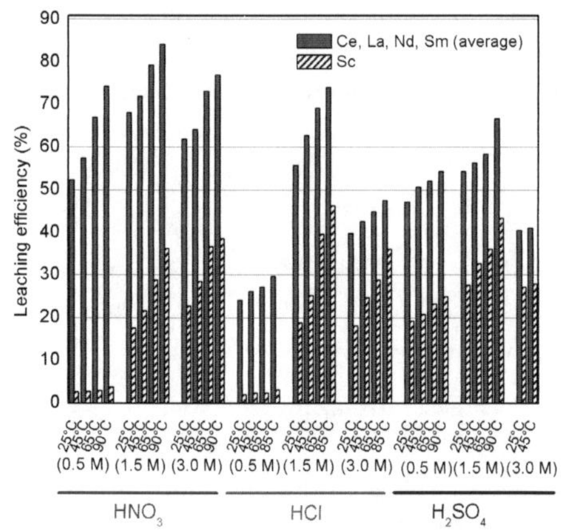

Figure 2 – Leaching efficiency of REEs from red mud.

CONCLUSIONS

Red mud and PG can be used as secondary resource for REEs. Based on the leaching results, HCl is the best leachant, resulting in highest Sc and Nd extraction.

REFERENCES

1. Chen, Z. J. Rare Earth 29: 1-6, 2011.
2. Binnemans, K. *et al.,* J. Clean. Prod. 99: 17-38, 2015.
3. Walawalkar, M. *et al.,* Hydrometallurgy, 166: 195-204, 2016.
4. Walawalkar, M. *et al.,* Ind. Eng. Chem. Res. 55: 12309-12316, 2016.

ACKNOWLEDGMENTS

The authors would like to thank Natural Sciences and Engineering Research Council of Canada (NSERC), Agrium Inc., and Rio Tinto for providing financial support to this project (Grant no: 501018).

The citric acid leaching of rare earths from eudialyte

H. Lim[1*] , D. Ibana[2], J. Eksteen[1]

(1. Western Australia School of Mines, Curtin University, 7 Conlon Street, Waterford,6152, Australia; 2. Western Australia School of Mines, Locked Bag 17, Kalgoorlie, 6433, Australia)

Abstract

Eudialyte is a minor mineral of zirconium but currently gaining significance as an alternative source of rare earths. The conventional method of rare earths extraction from eudialyte is by the use of sulfuric acid, either by direct leaching or via sulfation-baking followed by water leaching. Either way requires acid dosages of 100-750 kg/T. Such a large requirement for acid adds substantial cost to chemicals for leaching and neutralisation of effluent discharges.

In this study, citric acid, an environmentally benign chemical, was utilized to recover the rare earths yttrium, lanthanum and cerium from a eudialyte ore. A 2^{6-2} fractional factorial design of experiment, which denotes six factors at two experimental settings, was used in the conduct of the leaching studies. Analysis of leaching results by ANOVA method indicated that pH and leaching temperature were the significant factors for high REE lixiviation. The optimum conditions, devised as a result of this investigation were a pH of 2-3, leaching at 60°C for 8 h using 0.8 M citric acid on dilution of 100 $ml.g^{-1}$ and grind size of -106+75 μm. Under these conditions, average leaching recovery for the three rare earth elements was 82%. Examination of leach residue by the use of scanning electron microscopy revealed that precipitation of dense silica layers on eudialyte surfaces limited the complete decomposition of eudialyte. The co-leaching of zirconium was low at 39% as brought about by hydrolysis reactions.

The findings of this investigation showed that citric acid can be used as an alternative leaching agent for REE extraction from eudialyte, offering an environmentally advantageous approach compared to the conventional sulfuric acid treatment route.

* Corresponding author: hazel.lim@postgrad.curtin.edu.au, + 61 4 08089086

The crystal structure and magnetism of $R_{10}Pd_{21}$ compounds (R = Pr, Nd, Sm, Gd-Lu, Y)

Pietro Manfrinetti[1], **Alessia Provino**[1], **Nediadath S. Sangeetha**[2], **Sudesh K. Dhar**[2]

[1*]Department of Chemistry, University of Genova, Italy, chimfis@chimica.unige.it
[2]DCMPMS, TIFR, Homi Bhabha Road, Mumbai, India

Dedicated to the memory of Prof. Karl Gschneidner Jr. and his enthusiasm every time he became aware of a new family of rare earth compounds.

INTRODUCTION

The investigation of intermetallic compounds, especially those formed by rare earth (R) elements with special emphasis on the identification of new phases for their potential use in technological applications, covers an important area of scientific research. While for most of the R intermetallics reported in literature the crystal structure (as well as the physical properties) have been determined[1], there are some phases for which an unambiguous characterization is still lacking, even though they apparently show simple stoichiometries. Examples are in the R-Pd binary systems[4], where the evidence of formation of two compounds with apparent stoichiometries "R_2Pd_3" and "RPd_2" are reported for the majority of R elements though neither the crystal structure nor the exact composition of both these phases was ever determined. The "R_2Pd_3" phases were investigated in our previous work[2] and identified to be R_3Pd_5 (orthorhombic Pu_3Pd_5-type). In the present work we have now investigated the formation, crystal structure and physical properties of the "RPd_2" phases.

RESULTS AND DISCUSSION

The exact stoichiometry of the "RPd_2" phases is found to be $R_{10}Pd_{21}$. The crystal structure has been determined by X-ray powder diffraction and refined by Rietveld method. The $R_{10}Pd_{21}$ compounds crystallize in the monoclinic $Sm_{10}Pd_{21}$-type (*mS*124-*C*2/*m*, Z = 4)[3]; Rietveld refinement profiles are shown in **Figure 1**. Their formation has been confirmed for Y, Pr, Nd, Sm, Gd to Lu, including Yb. Sc does not form the 10:21 phase. Lattice parameters and unit cell volume strictly follow the lanthanide contraction (**Figure 2**). These phases form by a peritectic reaction; the formation temperature, T_f, of some of them has also been measured and compared with the values reported for the "RPd_2" phases for which literature data were available. The increase of both the volume contraction and T_f on going from $Pr_{10}Pd_{21}$ to $Lu_{10}Pd_{21}$ suggests an increase of the thermodynamic stability along the series from the lighter Pr to the heavier Lu. Physical properties (magnetic susceptibility, magnetization, heat capacity and magnetoresistance) have been measured for some of the members of the series. They order antiferromagnetically at low temperature; the highest T_N is 10 K for $Tb_{10}Pd_{21}$.

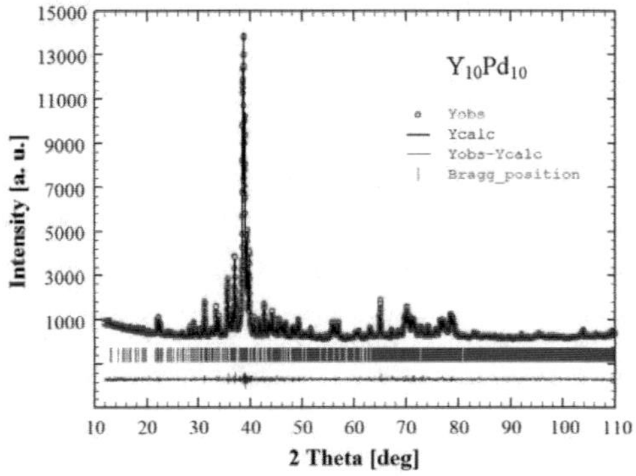

Figure 1. Rietveld refinement plot of $Y_{10}Pd_{21}$.

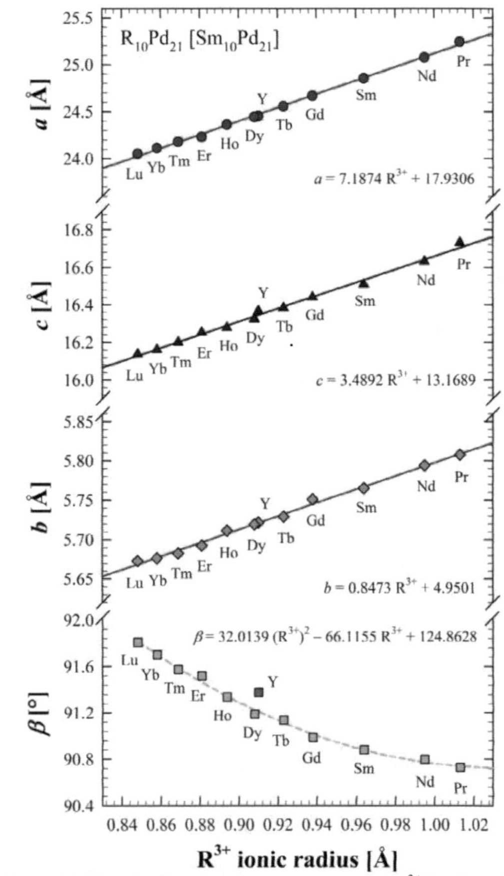

Figure 2. Trend of lattice parameters *vs*. the R^{3+} ionic radius for the $R_{10}Pd_{21}$ compounds.

REFERENCES

1. P. Villars, K. Cenzual, Pearson's Crystal Structure Database for Inorganic Compounds, 2015/16, ASM International, Materials Park, OH, USA.
2. A. Provino *et al.*, Crystal Growth & Design 16, 6001-6015, 2016.
3. M. L. Fornasini *et al.*, Acta Crystallogr. B35, 1950-1953, 1979.

The influence of multi-configurational electronic states on the properties and reactivity of organlanthanides

Violaine Goudy,[1] Arnaud Jaoul,[1] Marie Cordier,[1] Carine Clavaguéra,[2] and Grégory Nocton[*1]

[1*]LCM, CNRS, Ecole polytechnique, Université Paris-Saclay, Route de Saclay, 91128 Palaiseau Cedex, France, greg.nocton@polytechnique.edu

[2] Laboratoire de Chimie Physique, UMR 8000 CNRS/Univ. Paris-Sud, Univ. Paris-Saclay, 91405 Orsay, France

INTRODUCTION

Complexes containing redox non-innocent ligands have been well developed in the last decade with transition metal ions and have led to very important chemical transformations at lowest environmental and economic costs.[1] Nonetheless examples with f-element are very rare and the field is almost empty with lanthanides.[2,3] This is unfortunate since divalent lanthanides are excellent sources of single electron and would provide a good control of the ligand reduction because of strong electron correlation in these systems.[4,5] Thus, we have embarked in developing this field with organolanthanides. The synthesis of original complexes containing lanthanides, redox non-innocent ligands and transitions metals will be presented: the first providing reversible electron(s) source(s) (*remote control*), the second acting as electron(s) reservoir and controlling the electron correlation strength ("*dimmer switch*"), and the last being the site of the selective reaction.

RESULTS AND DISCUSSION

The bipyrimidine and the taphen bismethyl palladium complexes, (bipym)$PdMe_2$, **1**, and (taphen)$PdMe_2$, **2**, were synthetized from (tmeda)$PdMe_2$ in thf (Scheme 1). These two palladium complexes were combined with an organolanthanide fragment, $Cp*_2Yb(OEt_2)$, in toluene to yield the original heterobimetallic complexes $Cp*_2Yb(bipym)PdMe_2$, **3**, and $Cp*_2Yb(taphen)PdMe_2$, **4**, respectively. Both complexes are dark toluene solutions from which dark brown crystals of **3** and **4**, suitable for X-ray diffraction can be obtained. These crystals structures will be presented and discussed.

Scheme 1.

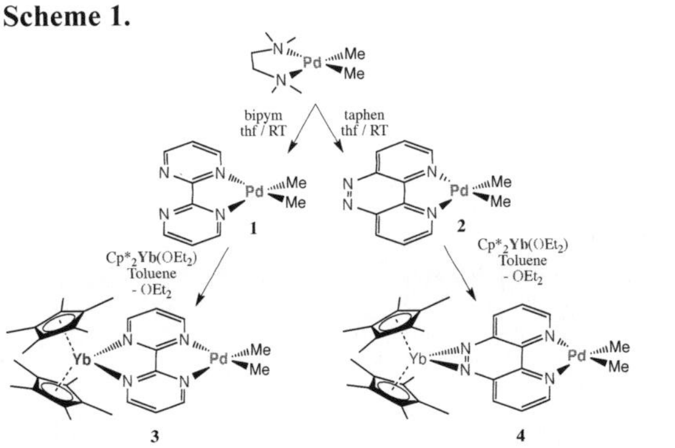

Moreover, we will discuss the electronic structure studies including the magnetic studies of the two later heterometallic complexes (**3** and **4**) as well as their reactivity with methyl iodide. Both complexes yield the Pd(IV) species but these have different stability. The relation with between electronic structure and reactivity will be discussed as well.

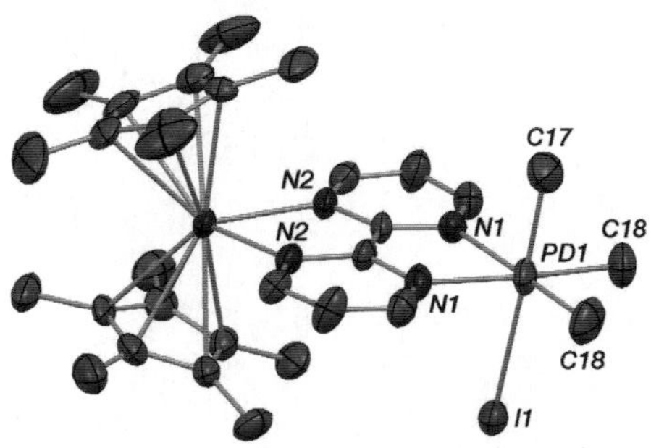

Figure 1. ORTEP of $Cp*_2Yb(bipym)PdIMe_3$. Thermal ellipsoids are set at 50% probability level, hydrogen atoms have been remove for clarity. Ytterbium is in green.

CONCLUSIONS

In conclusion, we will present the syntheses of original heterobimetallic complexes that combine a divalent organolanthanide fragment, $Cp*_2Yb$, with different N-heteroaromatic ligands (bipym and taphen) and a reactive palladium bisalkyl fragment $PdMe_2$. The modification of the bridging ligand induces different electronic structures. These different electronic structures lead to different stability for the Pd^{IV} complexes obtained from oxidative addition of the heterometallic complexes.

REFERENCES

[1] Chirik, P. J. *et al.* Wieghardt, K. Science, 327:794, 2010

[2] Nocton, G *et al.* J. Am. Chem. Soc., 136:8626, 2014.

[3] Anderson, N. H *et al.* Nature Chem., 6:919, 2014.

[4] Nocton, G. *et al.* Organometallics, 32:1150, 2013.

[5] Jacquot, L. *et al.* Organometallics, 33:4100, 2014.

ACKNOWLEDGMENTS

This work was funded by the French National Research Agency (Grant # ANR-15-CE29-0019) and by ERC (Grant # 716314). CNRS and Ecole polytechnique are thanks for funding as well for PhD fellowships.

The Legacy of Karl Gschneidner in Rare Earths and Critical Materials

Alexander H. King

Critical Materials Institute, The Ames Laboratory, U.S.A., alexking@ameslab.gov

ABSTRACT

Scientific awareness of the rare earths emerged with the discovery of a black mineral, originally named ytterbite, in 1787. It was not until the end of the 19th century, however, that the rare earth elements began to be chemically isolated and systematically studied, and well into the 20th century before they were all available for study or use. As with many new materials, the rare earths found a variety of uses, and up to the middle of the 20th century they appeared in products such as gas mantles, carbon arc lamps, spark generators, and pigments for ceramic glazes and glass. Karl Gschneidner began his professional career following the award of his Ph.D. in 1957, and advanced technical uses of the rare earths started in the 1960's, which saw the development of new fluid cracking catalysts and efficient red light emitting phosphors. Computer memories, permanent magnets, rechargeable batteries, lasers, magnetostrictive actuators, high-temperature superconductors, magnetocaloric materials, and many others emerged, as the understanding of the rare earths grew. The modern uses of the rare earths are intimately linked to the scientific contributions made by Karl Gschneidner over his long and distinguished career. We will trace some of those linkages in this presentation, and also touch upon the Karl's influence on the "critical material" status of the rare earths.

The Nanoscale Effect in the Upconversion Dynamics of β-$NaYF_4$:Yb,Er

Mary T. Berry, P. Stanley May, Md Yeathad Hossan

Department of Chemistry, University of South Dakota, USA, mberry@usd.edu

INTRODUCTION

At moderate excitation power densities, nanocrystals of β-$NaYF_4$:Yb^{3+},Er^{3+} are found to exhibit lower NIR-to-visible upconversion (UC) internal quantum efficiencies (IQE) compared to high-quality bulk materials, and exhibit more rapid UC dynamics, typical of quenching, when excited with a pulsed source near 980 nm.[1,2] The addition of a protective shell coating is known to increase the IQE of the nanocrystals and retard the UC dynamics. We have created parallel models for UC in bulk materials, and core and core-shell nanomaterials, and compared the results with experimental measurement in order to assess the contributions from the quenching of individual Yb^{3+} and Er^{3+} excited electronic states to the overall quenching of the UC luminescence.

EXPERIMENTAL METHODS

Core β-$NaYF_4$:17%Yb^{3+},3%Er^{3+} UC nanocrystals were coated with an $NaYF_4$ inert-shell to yield core-shell nanocrystals. The shell addition was confirmed by TEM. Optical characterization included integrated and time-resolved emission spectroscopy, with pump power/energy densities on the order of 10-10^2 W/cm^2 and 10-10^2 mJ/cm^2. Modeling of the UC dynamics in the nanomaterials was based on a previous model developed for the bulk material.[3]

RESULTS AND DISCUSSION

The 24 x 56 nm core nanoparticles grew an inert shell of $NAYF_4$ yielding a 24 x 63 nm core-shell particle. With a shell thickness of 5 nm on the six orthoaxial faces and 3.5 nm on the two axial faces, the core-shell particles were significantly brighter than the core particles, but still exhibited quenching relative to the bulk. The models show that the quenching of the UC emission and the rapid decay dynamics following pulsed excitation can be attributed almost entirely to quenching of Yb^{3+} ($^2F_{5/2}$) and Er^{3+} ($^4I_{11/2}$) at surface sites. Total coverage, with shell thicknesses greater than ~5 nm, appears to block the surface quenching and restore bulk UC behavior to the nanomaterials.

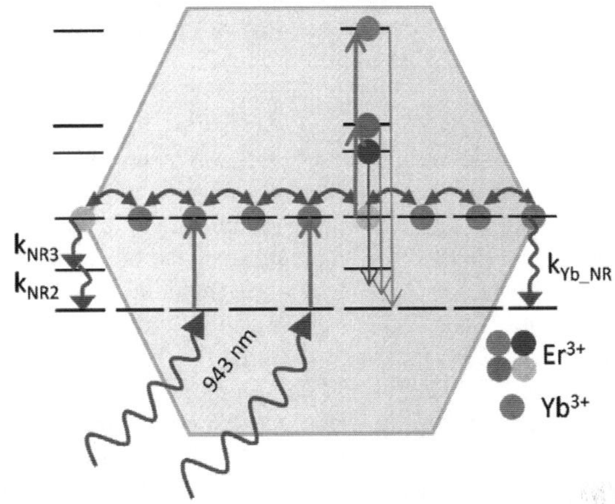

CONCLUSIONS

In β-$NaYF_4$:17%Yb^{3+},3%Er^{3+}, energy migration at the 1 μm level and surface quenching of the Yb^{3+} ($^2F_{5/2}$) and Er^{3+} ($^4I_{11/2}$) reservoir states is the dominant cause of reduced IQE in nanomaterials. This effect can be entirely removed by addition of an inert shell.

RERERENCES

(1) Boyer, J.-C.; van Veggel, F. C. J. M. *Nanoscale* **2010**, *2*, 1417.

(2) Fischer, S.; Bronstein, N. D.; Swabeck, J. K.; Chan, E. M.; Alivisatos, A. P. *Nano Letters* **2016**, 16, 7241.

(3) Anderson, R. B.; Smith, S. J.; May, P. S.; Berry, M. T. *The Journal of Physical Chemistry Letters* **2014**, *5*, 36.

ACKNOWLEDGEMENTS

The authors acknowledge financial support from NSF-CRIF, NSF-PFI, SDBOR-SPACT, and NASA EPSCoR.

The roles of 4f and 5f orbitals in bonding: a magnetochemical, crystal field, density functional theory, and multi-reference wavefunction study

Wayne W. Lukens,[1*] Manfred Speldrich,[2] Ping Yang,[3] Thomas Duignan,[4] Jochen Autschbach,[4] and Paul Kögerler[2,5]

[1*]Chemical Sciences Division, Lawrence Berkeley National Laboratory, Berkeley, CA, USA, wwlukens@lbl.gov.
[2]Institute of Inorganic Chemistry, RWTH Aachen University, Aachen, Germany.
[3]Theoretical Division, Los Alamos National Laboratory, Los Alamos, NM, USA.
[4]Department of Chemistry, University at Buffalo, State University of New York, Buffalo, NY.
[5]Jülich-Aachen Research Alliance (JARA-FIT) and Peter Grünberg Institute (PGI-6), Jülich, Germany.

INTRODUCTION

The electronic structures of actinide and lanthanide complexes have been the subject of considerable recent attention due largely to improvements in computational techniques that allow accurate calculations and to advanced spectroscopic techniques, including ligand K-edge X-ray absorption spectroscopy, which allows covalency to be measured directly.1[1-5] The electronic structures of Cp_3Nd ($4f^3$) complexes have been extensively studied by Amberger and co-workers using UV-Vis-NearIR spectroscopy in conjunction with crystal field modeling to determine the splitting of the 4f-orbitals due to interaction with the ligands.[6,7] While spectroscopic studies of Cp_3U electronic structure are few, this system has been studied computationally.[5,8,9]

EXPERIMENTAL METHODS

Compounds were prepared using standard Schlenk and glovebox techniques. Crystal field modelling was performed using the package CONDON. DFT calculations were performed using PBE0 exchange-correlation functional implemented in the Amsterdam Density Functional (ADF 2014.07) program. Relativistic *ab-initio* wavefunction calculations were carried out with a developer's version of the Molcas program based on release 8.1

RESULTS AND DISCUSSION

Analogous Nd(III) and U(III) tris-cyclopentadienyl complexes and their isocyanide adducts were characterized by EPR spectroscopy and magnetic susceptibility. The crystal field (CF) model was parameterized by fitting EPR and susceptibility data, yielding an accurate description of f orbital splitting. The isocyanide derivatives were also studied using DFT, resulting in f-orbital splitting that is consistent with CF model, and by multi-reference wavefunction calculations that support the electronic structure analysis derived from the crystal-field calculations.

The 5f orbitals, but not the 4f orbitals, are significantly involved in bonding to the isocyanide ligands. The main interaction between isocyanide ligand and the metal center is a σ-bond, with additional 5f to π* donation for the uranium complexes. While interaction with the isocyanide π*-orbitals lowers the energies of the $5f_{xz^2}$ and $5f_{yz^2}$ orbitals, spin-orbit coupling greatly decreases the population of $5f_{xz^2}$ and $5f_{yz^2}$ in the ground state.

CONCLUSIONS

Using the magnetic data allows the f-orbital splitting of Cp''_3U and its base adducts to be evaluated experimentally for the first time. Both experiment and theory clearly show that the 4f-orbitals do not participate significantly in bonding except for the $4f_{y(3x^2-y^2)}$ orbital. On the other hand, crystal field theory, DFT, and *ab-initio* CAS wavefunction calculations show that the 5f orbitals appear to participate in bonding to the isocyanide ligands as well as the Cp ligands.

REFERENCES

1. Minasian, S.G *et al* JACS, 134:5586-5597, 2012
2. Loble, M.W *et al* JACS, 137:2506-2523, 2015
3. Prodan, I.D. *et al* Phys Rev B, 76:4, 2007
4. Neidig M.L. *et al* Coord Chem Rev. 257:394-406, 2013
5. Kaltsoyannis, N. Inorg. Chem., 52:3407-3413, 2013
6. Apolstolidis, C. Z. *et al* Anorg Allg. Chem. 636:201-208, 2010
7. Amberger, H.D. *et al*, JOMC, 689:3143-3157, 2011
8. Strittmatter, R.J. *et al* JACS, 113:552-559, 1991
9.Gendron F. *et al* Inorg. Chem. 53-13174-13187, 2014

ACKNOWLEDGMENTS

The authors thank Prof. Richard A. Andersen for guidance on synthesis and Norman M. Edelstein and Nicola Magnani for helpful discussions on crystal field modeling. DFT calculations were performed at the William R. Wiley Environmental Molecular Sciences Laboratory, a national scientific user facility sponsored by the U.S. DOE BER and located at Pacific Northwest National Laboratory. Portions of this work (WL and PY) were supported by the U.S. DOE, Office of Science, Basic Energy Sciences (BES), Heavy Element Chemistry (HEC) program and were performed at Lawrence Berkeley National Laboratory under contract No. DE-AC02-05CH11231 and at Los Alamos National Laboratory under contract No. DE-AC52-06NA25396. Portions of this work were supported by the ACalNet program (DAAD (German Academic Exchange Service)) and by ERC StG 308051 – MOLSPINTRON. The CAS calculations were carried out by TD and JA at the University at Buffalo Center for Computational Research. JA and TD acknowledge support from the U.S. Department of Energy, BES HEC program, under grant DE-SC0001136 (formerly DE-FG02-09ER16066), and Dr. Frédéric Gendron for help with the CAS computations.

The α-manganese structure in RE_5Mg_{24} type compounds

Horst Borrmann

Max Planck-Institute for Chemical Physics of Solids, Dresden, Germany, borrmann@cpfs.mpg.de

The crystal structure of the α-modification of manganese is beyond doubt the most peculiar one among all metallic elements. Although described in great detail already 90 years ago it is still not well understood.[1] Due to quite large differences among the four distinct crystallographic atom sites α-manganese is sometimes considered a prototypical intermetallic compound. An obvious binary derivative is obtained when both sites with higher coordination number are occupied by one atom type but the remaining two positions by another element. The resulting arrangement was established as Ti_5Re_{24} type structure. Although quite a number of representatives have been identified, only K_5Pb_{24} has been studied by single crystal X-ray methods, verifying unambiguously a well ordered atomic distribution among the sites.[2] To date this still seems to be the only representative formed by main group elements only. Particularly interesting are compounds with general composition RE_5Mg_{24} which have been described for most heavier rare earth elements including yttrium as expected.[3] However, a very careful study at the Ames Laboratory clearly indicated besides Y_5Mg_{24} a phase with mixed occupancy at the site with highest symmetry and thus a limiting composition Y_4Mg_{25}.[4]

We have started an effort to prepare these interesting compounds in well defined ways in order to gain access to high quality structural investigations. First results will be presented and hopefully help to pave the way towards a more detailed understanding of the bonding characteristics and will allow for the analysis of physical properties as well.

REFERENCES

1. Bradley A.J. and Thewlis J., Proc. Royal Soc. London A *115,* 456-471, 1927

2. Fässler T.F., Kronseder C. and Wörle M., Z. anorg. allg. Chem. *625,* 15-23, 1999

3. Kripyakevich A.J., Evdokimenko V.I. and Gladyshevskii E.I., Sov. Phys.-Crystallogr. *9,* 330-331, 1964

4. Smith J.F., Bailey D.M., Novotny D.B. and Davison J.E., Acta Met. *13,* 889-895, 1965

Theoretical Investigation of Crystal Symmetry and Site Dependent Permanent Magnetic Properties of Sm_2Co_{17}

Rajiv Kumar Chouhan[1*] and Durga Paudyal[1]

[1*]Ames Laboratory, US Department of Energy, Ames, IA, USA, rajivchouhan@ameslab.gov

INTRODUCTION

Rare earths (RE) based permanent magnets are important for technological applications due to their high magnetic moment and magnetocrystalline anisotropy (MAE).[1] MAE and magnetic moments are the intrinsic parameters that define the magnetic hardness of the system with respect to microscopic characteristics. The localized 4f-electrons in rare earth elements are a key factor deciding high anisotropy. 4f states provide most of the magnetic anisotropy and 3d states (3d elements) provide most of the magnetic moments in the rare earth transition metal permanent magnets. The site occupancy and crystal environment decide the anisotropy in the system.

METHODS

In these calculations, we employed density functional theory (DFT) implemented within the full-potential linearized augmented plane wave (FP-LAPW)[2] method. The onsite electron correlation (GGA+U with Hubbard U= 6.7 eV and J = 0.7 eV) are implemented in the calculation to take care of occupied and unoccupied 4f states and spin orbit coupling for MAE.

RESULTS AND DISCUSSION

In this work, we performed ab-initio calculations employing in the rhombohedral and hexagonal structures of Sm_2Co_{17}. Calculated results show that the hexagonal form of Sm_2Co_{17} reduces the criticality by 50%. Here, we substituted one Sm site (out of two non-equivalent sites) which is not playing a role in MAE with non-critical element (like La or Ce). Our finding suggested that in hexagonal structure, the RE atoms maintain the atomic picture (uniaxial nature) because of the unaffected local 4f-orbital within the crystal lattice (see figure 1). Calculations show the decrease of magnetic moment by ~1.5 μ_B while going from rhombohedral to hexagonal structure without compromising the magnetic anisotropy of the system.

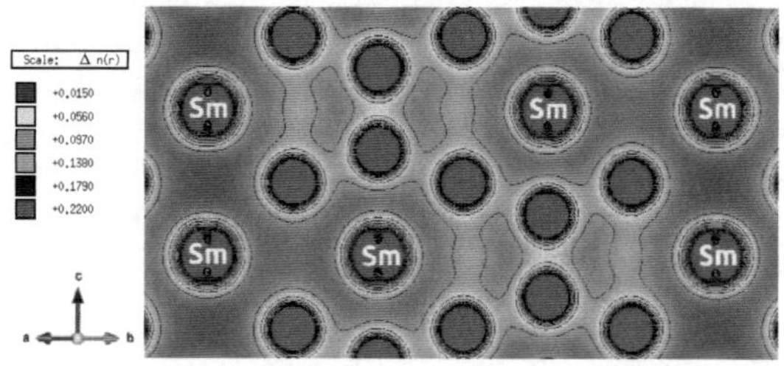

Figure 1: Charge density plot for Sm_2Co_{17} (hexagonal structure) which shows nice hybridization of transition metals (Co) within the layers making the Sm atoms' local 4f-orbitals mostly unaffected.

CONCLUSIONS

Theoretical results predict that the structural tuning of Sm_2Co_{17} along with the site substitution can enhance the magnetic properties of the system. Once the predicted compound is verified experimentally, it can lead to a revolutionary change in the application of permanent magnets with the presence of less critical elements.

REFERENCES

References must be numbered. Keep the same style.
1. McCallum R. *et al.*, Annu. Rev. Mater. Res. 44, 451 2014.
2. Blaha P. *et al.*, WIEN2k, ISBN 3-9501031-1-2 2001.

ACKNOWLEDGMENTS

This work is supported by the Critical Materials Institute, an Energy Innovation Hub funded by the U.S. Department of Energy, Office of Energy Efficiency and Renewable Energy, Advanced Manufacturing Office. The Ames Laboratory is operated for the U. S. Department of Energy by Iowa State University of Science and Technology under contract No. DE-AC02-07CH11358.

Theoretical Studies of Actinide-Ligand Bonding Interactions

Jing Su, Stosh Kozimor, Ping Yang, and Enrique R. Batista
Los Alamos National Laboratory, Los Alamos, NM 87545 USA, erb@lanl.gov

INTRODUCTION

The nature of the chemical bonds is the cornerstone of chemistry. Covalent and ionic bonds control the chemical and physical properties of compounds and materials. Therefore, developing a fundamental understanding can lead to the development of improved chemical processes, materials, and reactions. In the actinides, the participation of the different electronic shells in bonding has been a subject of much debate. This debate has fundamental scientific reasons and practical consequences, such as in chemical separations processes for nuclear materials.

Although computational studies could be carried out to calculate orbital mixings, direct experimental validation of this purely quantum mechanical effect was hard to attain. The developments of experimental techniques such as ligand k-edge X-ray absorption spectroscopy have led to a clean hand shake between theoretical calculations and measurements. Whereas this technique had been pioneered for lighter elements, extension to actinide complexes eventually changed the way people understand these systems. In this talk I will recap this exciting development and present studies of actinide-ligand bonds, e.g. Am-Cl bond in $AmCl_6^{3-}$,[1] from the computational side and its mapping to the experimental measurements. I will also discuss how these techniques can be instrumental, and have been applied, to the development of liquid-liquid extractants for spent nuclear fuel reprocessing.

COMPUTATIONAL METHODS

Density functional theory (DFT) calculations were preformed on the ground-state electronic structure $AmCl_6^{3-}$ using the hybrid functional B3LYP implemented in ADF 2014. 11 code. The Cl K-edge XANES spectra of $AmCl_6^{3-}$ were simulated as the Kohn-Sham orbital energy differences from spin-orbit coupled DFT/B3LYP calculations based on transition dipole method. The all-electron Slater basis sets with the quality of triple-ζ plus two polarization functions (TZ2P) were used.

RESULTS AND DISCUSSION

The experimental and theoretically simulated Cl K-edge XANES spectra of $AmCl_6^{3-}$ were displayed in Figure 1. Orbitals that resulted from Am 5*f*- and Cl 3*p*-orbital mixing were measured to contain 0.54(5)% Cl 3*p*-character, and the amount of mixing between Am 6*d*- and Cl 3*p*-orbitals was measured at 10.3(9)% Cl 3*p*-character per bond. Both experiment and theory also agreed that the oscillator strength for the *d*-transitions dwarfed the intensities of transitions at low energy associated with the *f*-orbitals.

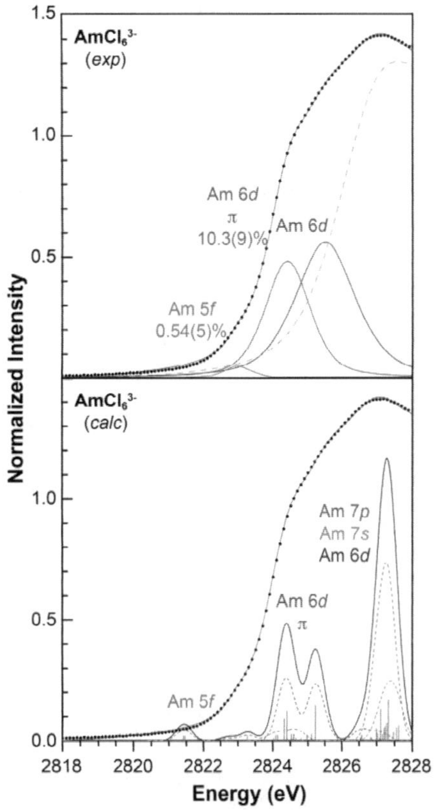

Figure 1 The experimental (•) and DFT calculated (black trace) Cl K-edge XAS spectrum of $AmCl_6^{3-}$.

CONCLUSIONS

Both 5*f*- and 6*d*-orbital mixing with the Cl-3*p* orbitals were observed. However, contributions from the 6*d*-orbitals were more substantial. Our results herein represent a leap forward for characterizing covalency in transplutonium metal-ligand bonding.

REFERENCES

1. Cross J. N. *et al.*, Covalency in Americium(III) Hexachloride. J. Am. Chem. Soc. 2017, Submitted.

ACKNOWLEDGMENTS

Work was supported by the Heavy Element Chemistry program at LANL, sponsored by the Office of Basic Energy Sciences, U.S. Department of Energy and J. Su was partially by postdoctoral Fellowship from the Glenn T. Seaborg institute and the Director's Postdoctoral Fellowship.

Thermo-acoustical properties of Eu(III) and Eu(III)–Lysozyme(HEW) in ternary system at atmospheric pressure

Jignasu P. Mehta,* Kahan I Pandya and Dinesh R Godhani

Analytical Chemistry Division, Department of Chemistry, DST-FIST sponsored Department, Maharaja Krishnakumarsinhji Bhavnagar University, Bhavnagar-364002

E-mail: jpm1569@yahoo.co.in Phone: +912782439852 (O); +919498706147 (M)

INTRODUCTION

Thermodynamic and transport properties of electrolytes and non-electrolytes in aqueous, non-aqueous and mixed solvents are of interest in various technologies. These properties provide an excellent tool to detect solute–solvent and solute–solute interactions[1-4]. The ultrasonic speed may be considered as a thermodynamic property if a negligible amount of ultrasonic absorption of the acoustic waves of low frequency and of low amplitude is observed[5]. Because of their accuracy, ultrasound velocity measurements are useful when dealing with the problems of structure and molecular interactions in liquids[6]. In continuation of our work, herein we report the molecular interactions of Europium and Europium– Hen Egg's White (HEW) Lysozyme under specified experimental conditions. The present study deals with thermodynamic and acoustical properties of Europium alone and Europium–Lysozyme(HEW) in ternary system at T = (313.15, 323.15) K and atmospheric pressure. The results are interpreted in terms of molecular interaction occurring in the solutions.

EXPERIMENTAL METHODS

Ultrasonic sound velocity measurements in DMF and solutions of Eu(III) and Eu(III) –Hen Egg's White (HEW) Lysozyme in DMF were made at T = (313.15, 323.15) K, by using M–81 multi frequency ultrasonic interferometer (2 MHz) (Mittal Enterprise, New Delhi). Single Capillary Pycnometer made of borosil glass having a bulb capacity of 10 cm^3 was used to determine density. The Ubbelohde viscometer with 25 cm^3 capacity was used for the viscosity measurement.

RESULT AND DISCUSSION

A novel approach has been adopted to confirm the coordination of multi-donor polypeptide with lanthanide ions. Ultrasonic sound velocity is significantly useful to determine solute-solute and solute-solvent in the binary or ternary system. We have used ternary system comprises of acetated buffer, TEAP, and DMF as a solvent, while Eu(III) ion and Eu(III)–Lysozyme(HEW). Various acoustical parameters, Gibbs energy (Changed from free energy) of activation, enthalpy of activation and entropy of activation were evaluated at two different temperatures *viz.* T = 313.15.15 and 323.15.15 K at atmospheric pressure. A representative data of ultrasonic sound velocity is also calculated.

Conclusion

Moreover, a comparison between experimental results for the pure system and solutions of Europium and Europium- Lysozyme(HEW) studies has been carried out to analyze carefully how the structural modifications affect the values of Gibbs energy of activation (ΔG), enthalpy of activation (ΔH) and entropy of activation (ΔS) (Δ is converted into normal fonts) in DMF. The positive values of ΔH and ΔS (Δ is converted into normal fonts) in ternary system also reveal that the transition state is associated with bond formation between Eu(III) and Lysozyme(HEW) molecules.

REFERENCES

1. Kim S. *J. Chem. Thermodyn.* 3: 411-417, 1971
2. Syal K. *et al., Acoustica* 81:276-278, 1995
3. Dhondge S.S. *et al., J. Chem. Thermodyn.* 1:1-15, 2008
4. Geetha D. *et al., J. Mole. Liq.* 136: 50-53, 2007
5. Freeman P.I. *et al., Polymer* 1: 20-26, 1960
6. Abe A. *J. Am. Chem. Soc.* 87(9): 1833-1838, 1965

ACKNOWLEDGMENTS

Authors are thankful to the Department of Science and Technology, Government of India **(SR/FTP/CS-12/2005)** for providing financial support during the course of work.

Towards a Better Understanding of the Complex Relationship Between Ln-based MRI Probes and Water

Mauro Botta

Department of Science & Technological Innovation/University of Eastern Piedmont "Amedeo Avogadro", Alessandria, Italy, mauro.botta@uniupo.it

INTRODUCTION

Effective MRI contrast agents on the market or in clinical or pre-clinical trials focus primarily upon $T_{1,2}$ changes. The effectiveness of the contrast depends on many factors, among which the multiple modes of interaction of water molecules with the metal-based probes are particularly relevant. The water exchange dynamics is one of the key parameter to consider in the design of any new MRI probe.[1] In this contribution, I will show and discuss new data on: a) the different rate of water exchange, k_{ex}, for different stereoisomers of GdHPDO3A and GdHPDO3MA (Figure1); b) the modulation of the prototropic exchange and its impact on relaxivity of GdHPDO3A derivatives; c) the rate of exchange of "axial" and "equatorial" water molecules in macrocyclic complexes; d) the unusual change of q and τ_M across the series of $[Ln(AAZTA)]^-$ chelates.

EXPERIMENTAL METHODS

Proton $1/T_1$ NMRD profiles were measured on a fast field-cycling Stelar SMARTracer Relaxometer (Mede (PV), Italy) at magnetic field strengths from 0.00024 to 0.25 T. Additional data points in the range 15-70 MHz were obtained on a Bruker WP80 NMR electromagnet adapted to variable-field measurements. ^{17}O NMR measurements were recorded on a Bruker Avance III spectrometer (11.7 T) equipped with a 5 mm probe and standard temperature control unit.

RESULTS AND DISCUSSION

The kinetic parameters of the water exchange and the value of the other relevant molecular parameters were evaluated by fitting the frequency (^1H) and temperature (^{17}O) dependency of relaxation and shift data. The main results obtained are the following: (a) three acetic pendant arms of HPDO3A were substituted by three methyl acetic arms leaving unaltered the hydroxypropyl pendant arm (Figure 1). This chemical modification has a marked influence on the solution behavior of LnHPDO3MA complexes. Analysis of the ^1H NMR spectra and the global fit of the relaxometric data indicate that the population of the TSAP isomer (fast exchanging) increases and the water exchange rate of the SAP isomer (slow exchanging) accelerates. As a result, the relaxivity of GdHPDO3MA is increased; (b) a proper choice of the electronic properties of the substituent on the hydroxypropyl arm modulates the pK of the coordinating -OH group and changes the prototropic exchange rate. This process provides a significant contribution to relaxivity; (c) two ligands were prepared that possess an identical number and type of donor atoms, while providing Ln(III) complexes containing a coordinated H_2O occupying either an equatorial or a capping position. The k_{ex} of the bound H_2O in the Gd^{3+} complexes were assessed by using ^{17}O NMR R_2 and chemical shifts and ^1H r_1 measurements.[2] The k_{ex} values of the water molecule in the axial and equatorial position differ by three orders of magnitude; (d) the measure and analysis of ^1H R_2 and paramagnetic shift as a function of T for the Ln(III) chelates of AAZTA and derivatives show a two orders of magnitude decrease of the k_{ex} constant on going from Gd^{3+} to Yb^{3+} complexes, accompanied by a decrease of the hydration state from 2 to 1.[3] This is against expectations: a contraction of the ionic radius and a greater steric hindrance should favor an increase of k_{ex}.

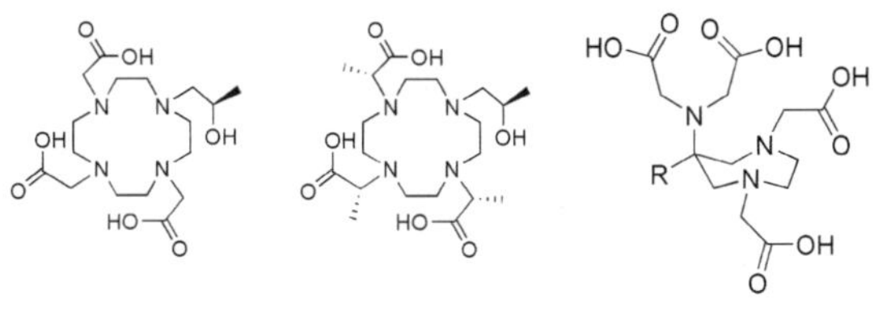

Figure 1 Structure of selected ligands

CONCLUSIONS

The many ways in which water molecules can interact with the paramagnetic lanthanide(III) ions and determine the image contrast is one the issue that is worth further studies. Our recent results have shown that not only we can control the rate of coordinated water exchange but also that it depends on its position in the coordination cage. The value of k_{ex} can vary so much along the Ln series that is possible to use different complexes of the same ligand as T_1 and CEST probes. In addition, the exploitation of the prototropic exchange can provide an additional contribution to relaxivity.

REFERENCES

1. Aime S. *et al.*, Acc. Chem. Res. 32: 941-949, 1999
2. Rodríguez-Rodríguez A. *et al.*, Chem. Eur. J. 23: 1110 – 1117, 2017
3. Karimi S. *et al.*, Inorg. Chem. 55: 6300 – 6307, 2016

ACKNOWLEDGMENTS

This work was supported by funding from local government (Regione Piemonte, PIIMDMT and Nano-IGT projects) and MIUR (PRIN 2012).

Trending Now – Rare Earth Separations

Mark P. Jensen[1*]

[1*]Department of Chemistry and Nuclear Science and Engineering Program, Colorado School of Mines, USA, mjensen@mines.edu

INTRODUCTION

Efficient chemical separations of rare earth elements has always been a challenge because the chemistry of the rare earth ions is dominated by localized electrons, the trivalent oxidation state, and consistent small changes in ionic radii. However, the systematically varying chemical properties of the rare earths and the great variety of physical properties across the range of rare earth elements also make them ideal probes of separations chemistry. This is fortunate as demand for rare earths and are driving new approaches to rare earth processing, purification, and recycle. Recent developments in the separation of rare earth elements will be reviewed and a new optical spectroscopic technique that allows real-time, direct measurements of the absorption spectra of rapidly mixed separation systems containing rare earth elements will be described.

RESULTS AND DISCUSSION

This new spectroscopic approach allows the simultaneous examination of the aqueous and organic phases as well as the aqueous-organic interface throughout the extraction process. This enables simultaneous determination of the distribution ratio as well as the chemical speciation throughout the entire extraction process. This development allows direct kinetic measurements on particular species in rare earth separation systems.

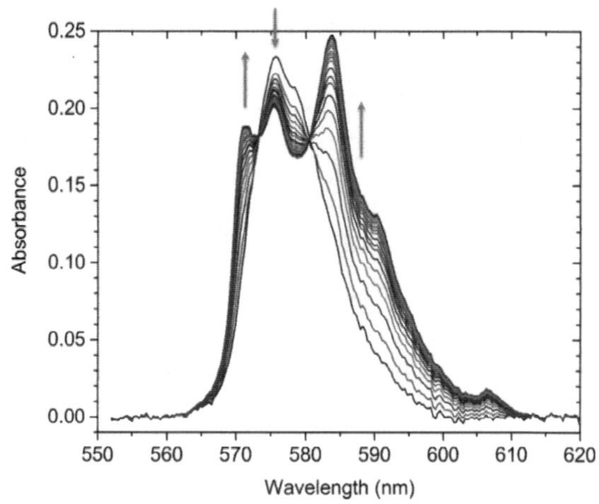

Figure 1. Changes in the optical spectrum the neodymium complexes as neodymium is extracted into 0.2 M bis(2-ethylhexyl)phosphoric acid in *n*-dodecane.

ACKNOWLEDGMENTS

The authors would like to thanks the Department of Energy, Office of Nuclear Energy for supporting this project under award 16-10679.

U(IV) Chloride Complexes Isolated from Acidic Aqueous Media: UCl_6^{2-} and an Unprecedented $U(H_2O)_4Cl_4$ Structural Unit

Jennifer N. Wacker[1], Monica Vasiliu[2], Jeffery A. Bertke[1], David A. Dixon[2], and Karah E. Knope[1*]

[1]Department of Chemistry, Georgetown University, Washington, DC, USA
[2]Department of Chemistry, The University of Alabama, Tuscaloosa, AL, USA
[*]kek44@georgetown.edu

INTRODUCTION

An understanding of actinide-ligand interactions and the identity of the complexes that exist in solution or the solid state is fundamentally important to predicting the overall chemical behaviour of the actinides under scenarios relevant to environmental contamination, separations processes, and nuclear waste management.[1] With respect to the latter, actinide-halide complexes have received considerable attention due to their importance in technological processes, specifically under high halide concentrations. Such species have also been examined as a means of understanding covalency in the f-block and as precursors for non-aqueous transuranic chemistry.[2-4] In this work, we examined the speciation of uranium halide complexes formed in the presence and absence of various hydrogen-bond donors as a means of assessing the impact of secondary coordination sphere interactions on speciation. Two U(IV) complexes including UCl_6^{2-} and $U(H_2O)_4Cl_4$ were isolated from acidic aqueous solutions and are reported.

EXPERIMENTAL METHODS

An aqueous solution of U(IV) in HCl was prepared by dissolving UCl_4 into 1 M HCl. Various organic H-bond donors were added. Evaporation of the solutions led to the isolation of two U(IV) compounds, **U-1** and **U-2** that are built from UCl_6^{2-} and $U(H_2O)_4Cl_4$ structural units, respectively.

Single crystal X-ray diffraction data were collected on a Bruker Quest D8 diffractometer equipped with an IµS X-ray source (Mo Kα radiation; $\lambda = 0.71073$ Å) and a CMOS detector. Raman spectra were recorded for single crystals using a Horiba LabRAM HR Evolution Raman Spectrometer with an excitation line of 532 nm over $\Delta\nu$ 100 – 4000 cm^{-1} using circularly polarized radiation. Infrared spectra were obtained using a Perkin Elmer Spectrum 2 FTIR Spectrometer; powdered samples were collected over $\Delta\nu$ 400 – 4000 cm^{-1}. Computational studies involved utilizing density functional theory benchmarked by coupled cluster CCSD(T) calculations.

RESULTS AND DISCUSSION

Two compounds were isolated and characterized in the presence and absence of H-bond donors. The anionic UCl_6^{2-} structural unit was isolated in the absence of a carboxylic acid, yet in the presence of a carboxylic acid, charge neutral $U(H_2O)_4Cl_4$ structural units were observed. Raman/IR spectroscopies were used to assess solid-state vibrational signatures. These were then compared to their calculated electronic structures. The energetics of the reaction of $U(H_2O)_4Cl_4$ to form UCl_6^{2-} were predicted to be exothermic in aqueous solution with a $\Delta G(aq) = -10$ kcal/mol at the B3LYP level.

Despite the prevalence of An(IV)-chlorides built from $AnCl_6^{2-}$, few solid state examples of An-aquo-halide compounds such as that reported here have been identified in the solid-state. It is worth highlighting that Kiplinger *et al.* isolated a related $Th(H_2O)_4Cl_4$ species,[5] though the $U(H_2O)_4Cl_4$ reported here adopts a different structural arrangement and has not been reported for U. In solution, related species have been observed using EXAFS in aqueous U(IV) chloride solutions where three U(IV)-aquo-chloro species were reported,[6] but the $U(H_2O)_4Cl_4$ complex isolated in this work has not been proposed previously.

CONCLUSIONS

To the best of our knowledge, **U-2** has yet to be reported in descriptions of U(IV) aqueous chemistry. Extensive hydrogen bonding and π -π stacking interactions present in **U-1** and **U-2** suggest that supramolecular or outer coordination sphere interactions may facilitate the crystallization of these species, particularly the U(IV)-aquo-chloro complex observed in **U-2**, and allow for subsequent observation in the solid state, despite the computationally calculated entropic favourability of the UCl_6^{2-} unit.

REFERENCES

1. Maher, K. *et al.,* Inorg. Chem. 52:3510-3532, 2013.
2. Minasian, S.G. *et al.,* Inorg. Chem. 51:5728-5736, 2012.
3. Reilly, S.D. *et al.,* Dalton Trans. 51:9165-9167, 2012.
4. Neidig, M.L. *et al.,* Coord. Chem. Rev. 257:394-406, 2013.
5. Cantat, T. *et al.,* Chem. Commun. 46:919-921, 2010.
6. Henning, C., *et al.* Inorg. Chem. 44:6655-6661, 2005.

ACKNOWLEDGMENTS

The authors would like to thank the Clare Boothe Luce Foundation, the U.S. Department of Energy, Office of Basic Energy Sciences, Heavy Element Chemistry Program (subcontract from Argonne National Laboratory), the Robert Ramsay Fund at the University of Alabama, and the National Science Foundation under grants NSF CHE-1429079 and NSF CHE-1337975.

Under Pressure: New Representatives of the Lanthanoid(III) Chloride Oxidotungstates(VI) $Ln_3Cl_3[WO_6]$ with the Smaller Ln^{3+} Cations (Ln = Dy − Er)

Ingo Hartenbach[1*] and Katharina V. Dorn[1]

[1]Institute for Inorganic Chemistry, University of Stuttgart, Pfaffenwaldring 55, 70569 Stuttgart, Germany
E-Mail: hartenbach@iac.uni-stuttgart.de

INTRODUCTION

Representatives of the rare-earth metal(III) chloride oxidotungstates(VI) are already known with the larger RE^{3+} cations La − Nd and Sm − Tb.[1-5] These compounds, if bearing luminescence-silent cations such as La^{3+} and Gd^{3+}, have proven to be suitable host materials for luminescence applications.[1,5] The representatives with the smaller rare-earth metals (*RE* = Y, Dy − Lu) were impossible to synthesize by solid-state reactions, due to the large coordination number of ten for the trivalent cations. Performing the reactions in a piston-cylinder press, however, yielded the desired results, which are presented here.

EXPERIMENTAL METHODS

Lanthanoid sesquioxides (Ln_2O_3), trichlorides ($LnCl_3$), and tungsten trioxide (WO_3) were mixed together according to equation (1):

(1) $RE_2O_3 + RECl_3 + WO_3 \rightarrow RE_3Cl_3[WO_6]$.

This mixture was inserted into gold ampoules, which were heated to 700 − 800 °C and set under a pressure of 15 − 20 kbar in a piston-cylinder press for four days. Small, transparent and lath-shaped single crystals of the desired products, showing the color of the respective RE^{3+} cations, were yielded among by-products which were identified as lanthanoid(III) hydroxide oxidotungstates $RE(OH)[WO_4]$ (see abstract of Dorn, Schleid, and Hartenbach). Crystals of the target compounds were characterized by means of single-crystal X-ray diffraction as well as single-crystal Raman spectroscopy.

RESULTS AND DISCUSSION

The title compounds crystallize hexagonally in space group $P6_3/m$ with two formula units per unit cell (a = 901 − 906, c = 520 − 523 pm). As mentioned in the introduction, the crystallographically unique RE^{3+} cations are surrounded by six O^{2-} and four Cl^- anions (*C.N.* = 10) in the shape of distorted tetracapped trigonal prisms (Fig. 1). The W^{6+} cations reside in the centers of six oxide anions building up quite unusual isolated trigonal prismatic $[WO_6]^{6-}$ units. The aforementioned building blocks are arranged to form empty channels along [001], lined by Cl^- anions, in the hexagonal unit cell (Fig. 2). By replacing the hexavalent W^{6+} with pentavalent Ta^{5+} cations, a possibility emerges to place small monovalent cations, such as Li^+, in those channels along the c axis, resulting in materials with the potential for ionic conduction.

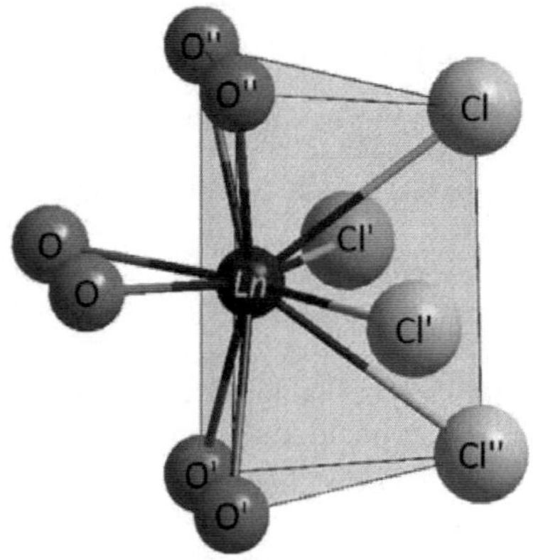

Fig. 1: Anionic coordination sphere around the crystallographically unique Ln^{3+} cations in the crystal structure of the $Ln_3Cl_3[WO_6]$ representatives (Ln = Dy − Er)

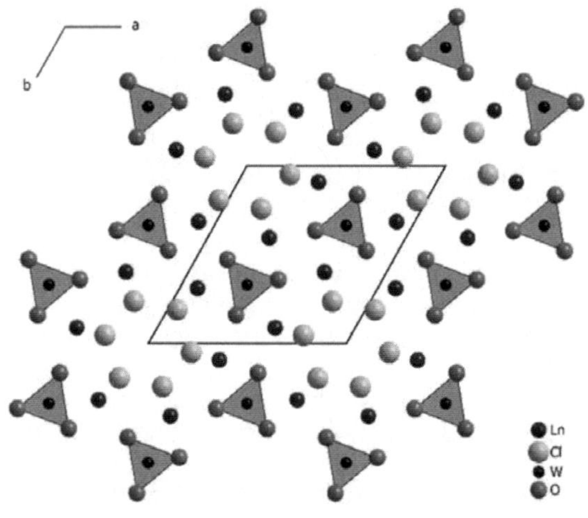

Fig. 2: View at the hexagonal crystal structure of the $Ln_3Cl_3[WO_6]$ representatives (Ln = Dy − Er) along [001].

REFERENCES

1. L. H. Brixner, H.-Y. Chen, C. M. Foris, *J. Solid State Chem. 44*, 99, **1982**.
2. J. B. Parise, L. H. Brixner, E. Prince, *Acta Crystallogr. C39*, 1326, **1983**.
3. T. M. Polyanskaya, S. V. Borisov, N. V. Belov, *Dokl. Akad. Nauk. SSSR 187*, 1043, **1969**.
4. I. Hartenbach, F. Ledderboge, Th. Schleid, *Z. Kristallogr. Suppl. 29*, 5, **2009**.
5. K. V. Dorn, Th. Schleid, B. Blaschkowski, K. Förg, H. A. Höppe, I. Hartenbach *Eur. J. Inorg. Chem. submitted for publication*, **2017**.

Under Pressure: Synthesis and Crystal Structure of the Rare-Earth Metal(III) Hydroxide Oxidotungstates(VI) $RE(OH)[WO_4]$ (RE = Dy and Ho)

Katharina V. Dorn[1], Thomas Schleid[1] and Ingo Hartenbach[1*]

[1] Institute for Inorganic Chemistry, University of Stuttgart, Germany; * hartenbach@iac.uni-stuttgart.de

INTRODUCTION

Single crystals of the title compounds $Dy(OH)[WO_4]$ and $Ho(OH)[WO_4]$ occurred during a new approach to extend the series of rare-earth metal(III) chloride oxidotungstates(VI) with the formula $RE_3Cl_3[WO_6]$ (RE = La–Nd, Sm–Tb)[1,2] towards the smaller rare-earth metals (RE = Dy–Lu) by using a piston-cylinder apparatus[3]. The crystal structure of the first member of these rare-earth metal(III) hydroxide oxidotungstates(VI), $Nd(OH)[WO_4]$, was already published in 1969 by *Klevtsova* and *Borisov*,[4] however, no more further results were announced in the meantime.

EXPERIMENTAL METHODS

Various experiments were performed in gold ampoules filled under ambient conditions into a piston-cylinder apparatus[3] at 700–800 °C and a pressure of 15–20 kbar, which lead to unexpected single crystals of rare-earth metal(III) hydroxide oxidotungstates(VI) with the formula $RE(OH)[WO_4]$ (RE = Dy and Ho).

RESULTS AND DISCUSSION

$Dy(OH)[WO_4]$ and $Ho(OH)[WO_4]$ crystallize monoclinically in space group $P2_1/c$ (no. 14) with four formula units per unit cell (a = 523–525, b = 1237–1241, c = 673–675, $\beta \approx 114.7°$). The crystallographically unique RE^{3+} cations are surrounded by six oxide and two hydroxide anions forming distorted square $[RE(OH)_2O_6]^{11-}$ antiprisms, whereas the W^{6+} cations are located in centers of oxygen tetrahedra according to $[WO_4]^{2-}$ (Fig. 1). The $[REO_6(OH)_2]^{11-}$ polyhedra form layers with the formula $^2_\infty\{[RE(OH)^e_{2/2}O^e_{4/2}O^t_{2/1}]^{6-}\}$ by edge-condensation orthogonal to [010]. These layers form the three-dimensional structure with the W^{6+} cations interconnecting (Fig. 2).

CONCLUSIONS

The title compounds $RE(OH)[WO_4]$ (RE = Dy and Ho) could lead on a pathway towards the synthesis of the so far unknown fluoride derivatives $REF[WO_4]$, which should be isotypic to the rare-earth metal(III) hydroxide oxidotungstates(VI) with RE = Nd[4], Dy, and Ho as well as to the already known rare-earth metal(III) fluoride oxidomolybdates(VI) $REF[MoO_4]$ (RE = Y, Sm–Lu)[5–6] of the smaller rare-earth metals. The aforementioned fluoride derivatives should provide an interesting field of research for luminescence applications due to the fact that they can host luminescent cations like Eu^{3+} or Tb^{3+}.

ACKNOWLEDGMENTS

The authors thank *Prof. Dr. Hans-Joachim Massonne* (Institute for Mineralogy and Crystal Chemistry, University of Stuttgart, Germany) for the opportunity to perform experiments with his piston-cylinder apparatus.

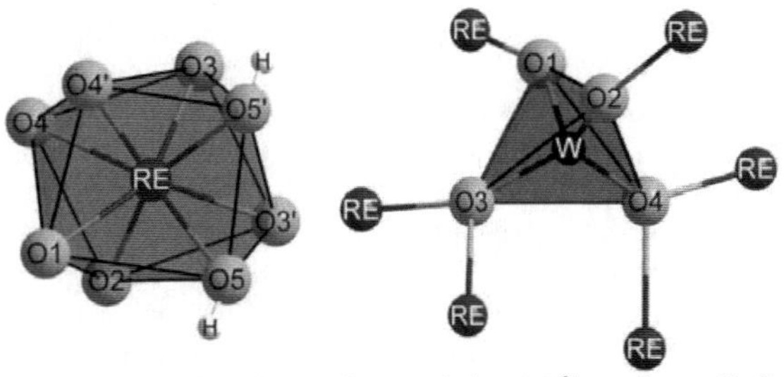

Fig. 1: Coordination sphere of the RE^{3+} cations (*left*) and the tetrahedral $[WO_4]^{2-}$ anions (*right*) in the crystal structure of the $RE(OH)[WO_4]$ representatives (RE = Dy and Ho).

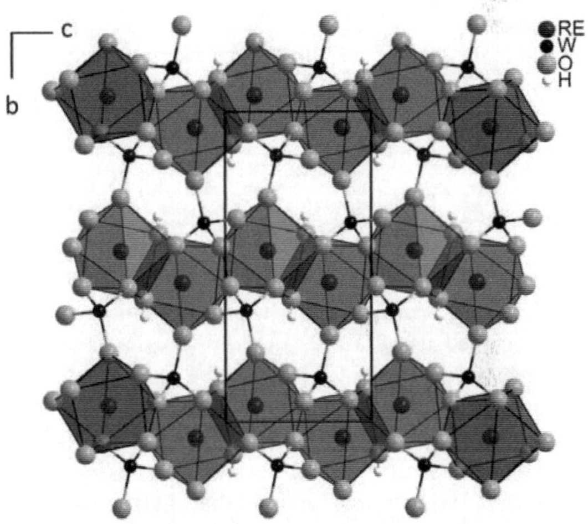

Fig. 2: View at the layers orthogonal to [010] formed by edge-fused $[REO_6(OH)_2]^{11-}$ polyhedra in the crystal structure of the $RE(OH)[WO_4]$ representatives (RE = Dy and Ho).

REFERENCES

1. Dorn K. V., Schleid Th., Hartenbach I., *Z. Anorg. Allg. Chem.* **642**, 1045-1045, 2016.
2. Dorn K. V., Schleid Th., Blaschkowski B., Förg K., Höppe H. A., Hartenbach I., *Eur. J. Inorg. Chem.* 2017, submitted for publication.
3. Massonne H.-J., Schreyer W., *Neues Jb. Miner. Abh.* **153**, 177-215, 1986.
4. Klevtsova R. F., Borisov S. V., *Kristallografiya* **14**, 904-907, 1969.
5. Schleid Th., Strobel S., Dorhout P. K., Nockemann P., Binnemans K., Hartenbach I., *Inorg. Chem.* **47**, 3728-3735, 2009.
6. Hartenbach I., Strobel S., Dorhout P. K., Schleid Th., *J. Solid State Chem.* **181**, 2828-2836, 2008.

Understanding of Magnetic Properties and Magnetocrystalline Anisotropy of Heavy Lanthanides

Adam J. Neuhouser,[1*] T. A. Hackett,[1] and D. Paudyal[1]

[1*]Ames Laboratory, Iowa State University, Ames, IA 50011, USA, ajn@ameslab.gov

INTRODUCTION

The development of high performing lanthanide-containing magnets is of great interest for commercial and industrial applications due to the limited amount of these precious elements.[1] Lanthanide-containing magnets are relevant to technologies in optics, rechargeable hydride batteries, electronics, medical scanners, and a myriad of other technologies. The spin-orbit coupling (SOC) of the 4*f* electrons give lanthanide metals significant magnetic anisotropic properties.[2] Here, we illustrate the electronic structure and report the magnetocrystalline anisotropy energies (MAE) of elements europium through thulium, and emphasize their role in defining the element's magnetic properties. We present the orbital, spin, and total magnetic moments across each heavy lanthanides with respect to the number of 4*f* electrons. The predicted magnetic moments closely agree with the experimental values.

THEORETICAL METHODS

We have employed local spin density approximation including onsite electron correlation and spin orbit coupling (LSDA+U+SOC) to calculate magnetic moments and magnetocrystalline anisotropy properties of the heavy rare-earth metals[3]. The k-space integrations were performed using a minimum 16×16×16 Brillouin zone mesh which was sufficient for the convergence of total energies (10^{-6} Ryd.), charges, and magnetic moments. Electron density plots and fermi surface diagrams of each element were created using XCrySDen. When constructing the Fermi surface diagrams, we increased the mesh to 33×33×8 to obtain higher resolution structures. MAE was calculated as the total energy difference with magnetic moment aligned along the planar and c-axis directions.

RESULTS AND DISCUSSION

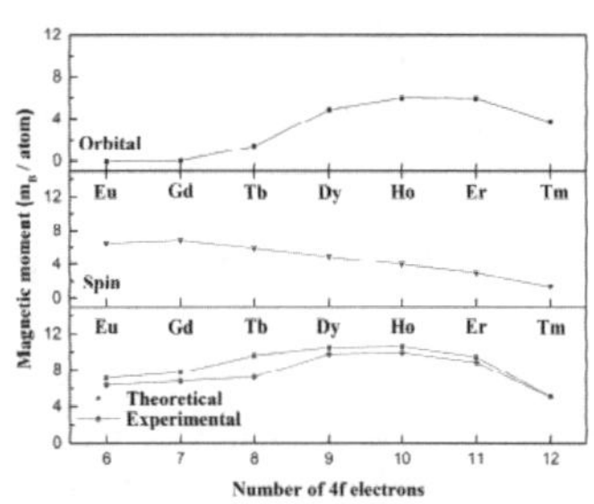

FIG. 1. Orbital, spin, and total magnetic moments of heavy lanthanides.

Figure 1 shows calculated orbital, spin, and total magnetic moments for heavy lanthanides. Both 4*f* orbital and 4*f* spin moments follow the trends of Hund's rule. The electron density plot shows aspherical charges indicating anisotropic characteristics of the heavy lanthanides. The density of states in most of the lanthanides, with exception of Gd^{3+} and Eu^{2+}, indicate crystal field split 4*f* states that give rise to magnetic anisotropy. The plotted Fermi surfaces also indicate how the electronic structure interplays with anisotropy.

CONCLUSIONS

The anisotropic and magnetic properties of heavy lanthanides were calculated and analysed using the all-electron augmented plane wave method. The total magnetic moment of each heavy lanthanides closely match with experimental values from the literature[4]. Employed advanced density functional theory calculations found the charge densities, onsite electron correlation, and spin orbit coupling to be highly dependent on the 4*f* states. The electron density and Fermi surface illustrations support the high magnetocrystalline anisotropy characteristics of heavy lanthanides.

REFERENCES

1. Eliseeva S. *et al.*, New Journal of Chemistry. 35:1165–1176, 2011
2. Skomski R. *et al.*, Journal of Rare Earths. 27(4):675, 2009
3. Blaha P. *et al.*, WIEN2k Manual. 59:399, 1990
4. Coqblin B., The Electronic Structure of Rare-Earth Metals and Alloys. (7), 1977.

ACKNOWLEDGMENTS

The Ames Laboratory is operated for the U. S. Department of Energy by Iowa State University of Science and Technology under contract No. DE-AC02-07CH11358. This work was supported by the Department of Energy, Office of Basic Energy Sciences, Materials Sciences Division and the Science Undergraduate Laboratory Internship (SULI) program. Adam Neuhouser is grateful to the DOE for the assistantship and opportunity to participate in the SULI program. A. N. would like to strongly thank Ed Moxley for all of the technical support provided and Manish K. Kashyap for WIEN2k guidance.

Unravelling the excited state dynamics and luminescnce of the aqueous cerium(III) ion by combined experimental and theoretical approaches

Florent Réal[1], Valérie Vallet, Rafał Janicki[2] and Patric Lindqvist-Reis[3]

[1*]Univ. Lille, CNRS, UMR 8523 - PhLAM - Physique des Lasers Atomes et Molécules, F-59000 Lille, France, email@presenting:florent.real@univ-lille1.fr
[2]University of Wrocław, Faculty of Chemistry, F. Joliot-Curie 14, 50-383 Wrocław
[3]Institut für Nukleare Entsorgung, Karlsruhe Institute of Technology, P.O. Box 3640, 76021 Karlsruhe, Germany

INTRODUCTION

Among the heavy ions, the trivalent cerium ion, considered as a homologue of trivalent actinide ions, has favorable spectroscopic properties (high sensitivity and quantum yield). In dilute aqueous solutions, the five *f-d* transitions of the Ce^{3+}(aq) absorption spectrum[1] were well-attributed (200-260 nm) leading to the conclusion that the mean hydration number is nine in standard conditions; however, the presence of a weak sixth band (295 nm) has been attributed to the admixture of transitions from the aqueous $[Ce(H_2O)_8]^{3+}$ ion species,[1–3] present in minor quantities. More striking feature concerns the origin of the large red-shift in the luminescence spectrum in solution (357 nm) but not observed in crystal (312-330 nm). The hypothesis is that a water molecule is photo-dissociated in solution upon excitation so that emission takes place exclusively from an octahydrated species. [2]

In this work, by combining UV-Vis spectroscopy and relativistic multi-reference calculations, we will explore the excited states dynamics of the aqueous Cerium (III) and shade light on the interplay between the local arrangement, the strength of the metal-ligand interactions in solution and the resulting absorption and emission spectra, properties that can be addressed precisely by EXAFS techniques.

RESULTS AND DISCUSSION

Theoretical $4f^1$-$5d^1$ absorbance spectra of $[Ce(H_2O)_9]^{3+}$, $[Ce(H_2O)_7Cl_2]^+$, $[Ce(H_2O)_8]^{3+}$, and $[Ce(H_2O)_6Cl_2]^+$ cations in the model compounds were calculated using state-of-the-art calculations, including static and dynamic electron correlations effects, scalar and non-scalar relativistic effects and long-range solvent effects, to compute the ground and the excited states.

On several doped-Cerium(III) crystals and Cerium(III) in aqueous solution, UV-VIS spectra were recorded. The excellent agreement, between the QM model and the corresponding experimental spectra (cf Figure 1.a) allow us to explore the dynamic of the excited states.

The QM calculations reveal the existence of a specific 5d state that leads to the departure of one water from the first coordination shell to the second one (cf Figure 1.b). The energy gained is at the origin of the strong Stokes-shift recorded in bulk solution, and that cannot be observed in a crystal environment.

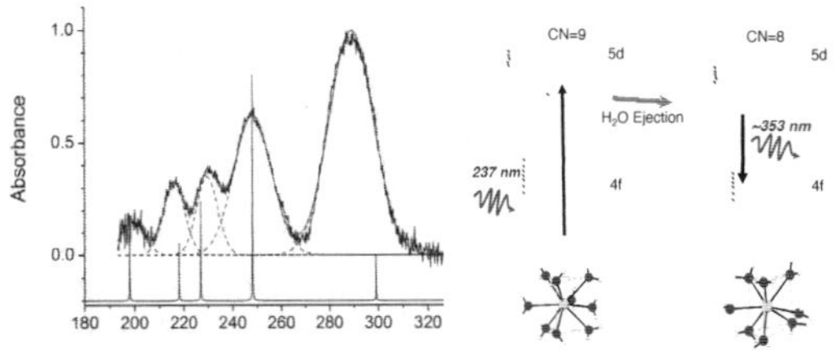

"Figure 1.a shows absorption spectra of Ce3+ in $[Ce(H_2O)_7Cl]Cl_2 \cdot 15$-crown-5$\cdot H_2O$. The spectra are curve fitted with Gaussian component bands, representing the five $4f^1$-$5d^1$ transitions. Also shown are the corresponding theoretical spectra (narrow peaks). Figure 1.b shows the mechanism that corresponds to the ejection of one water of the first-coordination sphere induced by UV irradiation."

CONCLUSIONS

In the present work, we demonstrate that there is a correlation between structural and spectroscopic properties of different cerium(III) hydrates, such as $[Ce(H_2O)_n]^{3+}$ (n=8,9), $[Ce(H_2O)_7Cl_m]^{3-m+}$ (m=1,2) and $[Ce(H_2O)_6Cl_2]^+$. The experimental results were supported by highly accurate quantum chemical electronic structure calculations on representative cerium-hydrated cluster models, extending thus the work of Kotzian and Rösch.[4] This complementary approach enabled us to explore the potential energy landscape of the photo-dissociation reaction of a water molecule within the $[Ce(H_2O)_9]^{3+}*$ cation in the excited electronic states, at the origin of the strong Stokes shift observed in bulk solutions.

REFERENCES

References must be numbered. Keep the same style.
1. Okada, K *et al*, H. *J. Chem. Phys.*, 75:1577, 1981
2. Okada, K. *et al*, *Mol. Phys.* 54:1293–1306, 1985
3. Kaizu, Y *et al. J. Am. Chem. Soc.*, 107:2622–2626, 1985
4. Kotzian, M.; Rösch, N. *J. Phys. Chem.* 96:7288–7293, 1991.

Uranium Complexes with Neutral Soft-Donor Chalcogen and Borohydride Ligands

Anastasia Blake[1], Zachary Theiler[1], and Scott Daly[1]
[1]Department of Chemistry, University of Iowa, USA, anastasia-blake@uiowa.edu

INTRODUCTION

Extraction of actinides (An) from spent reactor fuel and subsequent conversion into shorter lived isotopes (transmutation) can reduce the radiotoxic lifetime of nuclear waste.[1] Methods exist for extracting uranium, neptunium, and plutonium, but separating the remaining long-lived minor actinides (americium and curium) from lanthanide (Ln) fission products remains a challenge. Lanthanides and minor actinides are difficult to separate because they have similar chemical properties, ionic radii, and common oxidation states (+3). However, this separation is necessary if advanced nuclear fuel cycles such as partitioning and transmutation are to be realized. Lanthanides – which are in higher abundance relative to the minor actinides in spent fuel – interfere with the transmutation process due to their higher neutron cross-sections.[1]

In this poster, I will discuss the synthesis and characterization of uranium compounds with ligands containing neutral soft chalcogen donors and borohydrides. These compounds have the potential to be analysed by X-ray absorption spectroscopy (XAS, previously performed by our group) and density-functional theory (DFT) to quantify variations in metal-ligand bond covalency. Comparing metal-ligand covalency in these and similar lanthanide compounds may provide insight into how bonding differences are implicated in effective Ln/An separations.

RESULTS AND DISCUSSION

We prepared a homoleptic U(IV) compound (**1**) containing bis(3,4-dimethyl-2-hydroxybenzyl)(2-(methylthio)ethyl)amine ligands (Fig. 1). The ligand has two vital features, the first of which are hard, anionic oxygen atoms that ensure coordination to uranium. The second important feature is that the neutral thioethers groups are located on labile arms, so they may "choose" to coordinate to the metal. Notably, the neutral sulfur atoms in **1** bind to uranium; structurally similar selenoether ligands are now being investigated.

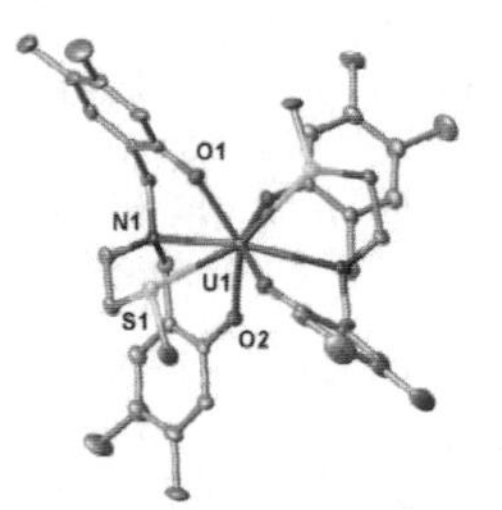

Fig. 1. Structure of **1** from single-crystal XRD.

Actinides can participate in covalent bonding with soft donors,[2] and we postulate that ligands such as the one shown in Fig. 1 can promote effective Ln/An separations. If the pendent soft donor groups prefer to covalently bind to actinides over lanthanides, we expect that the resulting compounds will have different solubilities so that high separation factors can be achieved in liquid-liquid extractions.

Because borohydrides, as well as chalcogens, are thought to form covalent bonds with actinides, we attempted to investigate the binding preference of uranium when both are present in the same ligand (Fig. 2). However, the ligands unexpectedly disproportionated upon reaction with uranium.

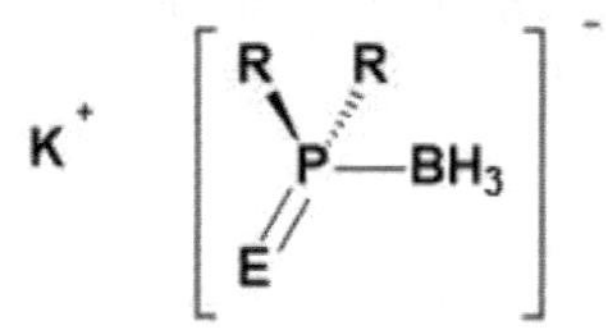

Fig. 2. General structure of the borohydride ligands.[3] R= Ph, ^{t}Bu; E= S, Se, Te

Reactions with $Se{=}P^tBu_2BH_3$, for example, resulted in the formation of a new diselenophosphinate complex. When the same reaction was performed with the di-*tert*-butyl(phosphino)borane ligand with no chalcogen, a uranium phosphinodiboranate complex was formed. Both compounds were characterized by XRD. These results may indicate that uranium prefers to bind to the diselenophosphinate over the borohydride.

CONCLUSIONS

We have prepared and characterized several new uranium complexes with soft-donor chalcogen and borohydride ligands. The compounds shown here, as well as others, will be compared to lanthanide compounds containing the same ligands to evaluate binding differences between 4f and 5f elements. These compounds will also be analysed using XAS to determine if differences in metal-ligand covalency are implicated in preferential soft-donor ligand binding. Overall, we intend to use these preliminary uranium studies as a roadmap to explore the synthesis of transuranic compounds more relevant to challenges in Ln/An separations.

REFERENCES

1. Nash, K. L., Solvent Extr. Ion Exc. 729, **1993**.
2. Minasian, S.G. *et.al.*, J. Am. Chem. Soc. 5586, **2012**.
3. Wagner, M. *et. al.*, Eur. J. Inorg. Chem. 5138, **2006**.

ACKNOWLEDGMENTS

The authors would like to thank the Nuclear Regulatory Commission and the University of Iowa for providing financial support to this project. We also thank Dale Swenson for collecting the XRD data and Justine Olson for providing some of the starting materials.

Uranyl Reduction Facilitated by a Redox-Active, Donor-Expanded Dipyrrin

Nicola L. Bell[1], Polly L. Arnold[1*] and Jason B. Love[1*]

[1]EaStChem School of Chemistry, University of Edinburgh, Scotland, UK
Nicola.Bell@ed.ac.uk

INTRODUCTION

Redox-active ligands can provide alternative redox pathways in metal complex chemistry that are not available using "innocent" ancillary ligands.[1] While these features have been studied extensively in TM complexes, the use of redox-active ligands in actinide chemistry is less developed.[2] This is surprising, as significant advances have been made in the reduction chemistry of the uranyl dication, a reaction important to uranium remediation, *e.g.* forming stable, oxo-functionalised uranyl(V) compounds[3] and reduction of uranyl(VI) β-ketoiminate complexes to U(IV) facilitated by oxo-coordination of boron Lewis acids.[4]

We recently reported the straightforward synthesis of the mono-anionic, tetradentate dipyrrin ligand, L^-. Here, we show how it allows access to new, ligand-mediated, uranyl reduction chemistry: we report a uranyl(VI) complex of L^- and its contrasting inner- and outer-sphere redox chemistry, both of which routes involve the redox-active dipyrrin.

RESULTS AND DISCUSSION

$UO_2Cl(L)$ (**1**) is prepared by two routes: Method A involves reaction of HL with $UO_2\{N(SiMe_3)_2\}_2(THF)_2$ to generate $UO_2\{N(SiMe_3)_2\}(L)$ followed by protonolysis with pyridinium chloride; Method B accesses **1** by reaction between HL and 0.5 equivalents each of $UO_2\{N(SiMe_3)_2\}_2(THF)_2$ and $UO_2Cl_2(THF)_2$.

Acyclic voltammetric study of **1** shows three reversible reductions at −0.97, −1.18 and −2.02 V vs. Fc/Fc^+ assigned as one ligand and two metal-based reductions. DFT calculations show the HOMO of **1** to be ligand-based leading us to assign the first reduction wave as a ligand-based reduction to the radical anion $[UO_2Cl(L)]^{\bullet-}$. The subsequent reductions are assigned to sequential reduction of U(VI) to U(IV). Accordingly, chemical reduction of **1** with one equivalent of the outer-sphere reductant $CoCp_2$ yields NMR-silent crystals of **2** (Scheme 1). The solid-state structure contains short U=O bond lengths (Avg. 1.776 Å) indicative of the U(VI) oxidation state and the EPR spectrum at 223K shows one sharp resonance with g_{iso} = 1.9893 which is assigned to $L^\bullet$.

In contrast, reduction of **1** with 0.5 equivalents of the inner-sphere reductant $[Cp_2Ti^{III}Cl]_2$ yielded a mixture of products comprising a 1:1:1 ratio of **1**:**3**:**4** demonstrating that the U^V complex (**3**) disproportionates to U^{VI} and U^{IV}. Treatment of **1** with two equivalents of Ti(III) cleanly yields **4** (Scheme 1).

The direct reduction of **1** by Ti(III) was not expected due to the high reduction potential of the latter (> −1.27 V) compared with the experimental reduction potential of **1** (−2.02 V). DFT calculations provided a plausible mechanism for the formation of **4** from **1**. As before, initial reduction at the ligand yields $[UO_2Cl(L)]^{\bullet-}$ However, coordination of the highly oxophilic Ti metal to a uranyl oxo induces a positive shift of the reduction potential of the uranium such that ligand to metal charge transfer occurs, yielding **3**. Reduction of this U(V) intermediate to **4** was computed to be exergonic by 12.5 kcal mol^{-1}.

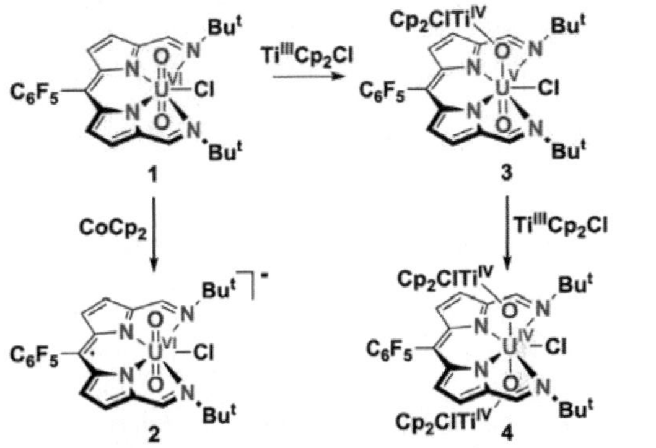

Scheme 1: Reduction of uranyl dipyrrin complexes by inner- and outer-sphere mechanisms.

CONCLUSIONS

The expanded Schiff-base dipyrrin is a non-innocent partner in uranyl reduction. Reduction of **1** by the outer-sphere reductant $CoCp_2$ yields a U(VI) ligand radical anion **2** whilst reduction with an inner-sphere Ti(III) reagent yields the doubly titanated U(IV) product **4**. The redox non-innocent ligand is key to both reduction pathways, with DFT and spectroscopic evidence showing that the first step in both reactions involves a ligand based reduction. Isolation of the U(IV) complex **4** from reaction with the relatively mild reductant Ti(III) further reinforces the understanding that oxo-coordination of the uranyl to the Lewis-acidic Ti(IV) metallocene renders the complex more susceptible to reduction and fits well with previous results on the use of other Lewis acids to mediate $[UO_2]^{2+}$ reduction chemistry.

ACKNOWLEDGEMENTS

We thank the University of Edinburgh and the EPSRC for financial support. Thanks also to Mr James Pankhurst, Dr Stephen Sproules, Dr Louise Natrajan, Prof. Laurent Maron, Dr Markus Zegke, Mr Carlos Lamfsus and Ms. Lucy Platts for their scientific contribution to this work.

REFERENCES

1. Broere, D. L. J.; Plessius, R. et al. *Chem. Soc. Rev.* **2015,** *44*; Luca, O. R.; Crabtree, R. H. *Chem. Soc. Rev.* **2013,** *42*; 2. Anderson, N. H.; Odoh, S. O. et al. *Nat. Chem.* **2014,** *6*; Kraft, S. J.; Fanwick, P. E. et al. *Inorg. Chem.* **2010,** *49*; 3. Arnold, P. L.; Love, J. B. et al. *Coord. Chem. Rev.* **2009,** *253*; Arnold, P. L.; Patel, D. et al. *Nature* **2008,** *451*; 4. Schnaars, D. D.; Wu, G. et al. *J. Am. Chem. Soc.* **2009,** *131*.

Visible and NIR-Emitting Lanthanide Doped Bismuth-Organic Materials

R. Lee Ayscue, Rami Batrice, and Karah E. Knope[1*]

[1]Department of Chemistry, Georgetown University, USA
*kek44@georgetown.edu

INTRODUCTION

Photoluminescent (PL) materials have received significant interest in the past decade due to their breadth of applications in photonic, optical, sensing, and high-tech devices. [1-2] Yet the development of PL compounds that address our energy, lighting, and security needs, while also reducing our dependence on rare earth resources, remains an on-going challenge. [3] To address this challenge while also further developing the chemistry of bismuth, which itself affords promising materials attributes including low cost and low toxicity, our group has begun an investigation into the development of bismuth-based compounds into which lanthanide metal ions may be incorporated with the aim of capitalizing on potential property improvements in bismuth host/lanthanide doped materials. Presented here will be an overview of the synthesis and characterization of a bismuth-organic material (BiOM) that exhibits promising PL behavior and structural features that have not been observed in other main group, d-, or f-block systems. Upon incorporation of lanthanide (Ln) ions via doping, this material exhibits intrinsic visible or NIR metal-based luminescence. Structure-property relationships in this and other recently isolated bismuth-organic complexes, as well as the luminescent behavior of the lanthanide doped phases, will be discussed.

EXPERIMENTAL METHODS

A bismuth-organic material (**BiOM-1**) and its lanthanide-doped analogs were prepared hydrothermally. Single crystals of **BiOM-1** and the Ln doped analogs (**Ln@BiOM-1,** Ln = Nd, Sm, Eu, Tb, Dy, Er, Yb) were isolated from the bulk and single crystal X-ray diffraction studies were performed on a Bruker Quest D8 diffractometer equipped with an IμS X-ray source. Powder X-ray diffraction data were collected on bulk reaction products using Cu-Kα radiation to confirm that the single crystals were representative of the bulk and elemental analyses via combustion confirmed phase purity of a mechanically separated phase. Emission and excitation data were acquired on a Horiba PTI QM-400 system (Horiba PTI) on solid samples at room temperature. Luminescent spectra were additionally collected from single crystals using a Raman microscope equipped with a 355 nm excitation source to confirm the monophasic character of the bulk sample. Metal ion ratios in the lanthanide doped Bi phases were determined by inductively coupled plasma-mass spectrometry.

RESULTS AND DISCUSSION

A bismuth-organic material has been isolated and structurally characterized. The homometallic phase exhibits weak broad-band luminescence. Lanthanide doping into these materials has been achieved by replacing the bismuth precursor with ten mole percent lanthanide (Ln = Nd, Sm, Eu, Tb, Dy, Er, Yb) and the luminescent behaviour of the resulting doped phases has been examined. As shown in Figure 1, the lanthanide doped phases (Ln = Sm, Eu, Tb, Dy) exhibit intrinsic metal-based luminescence in the visible region. Metal based emission is likewise observed for those ions emitting in the NIR.

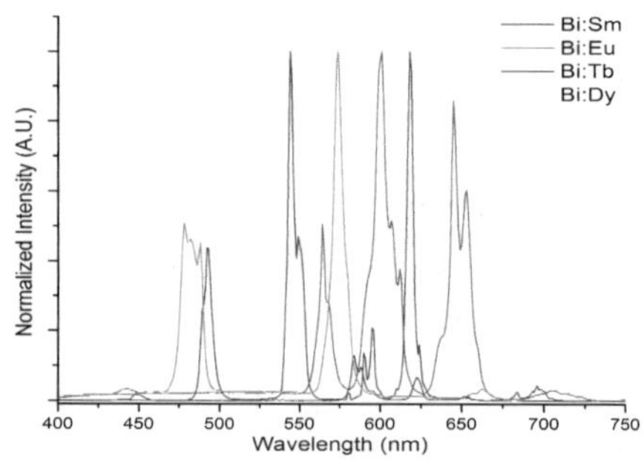

Figure 1. Emission spectra for **Ln@BiOM-1**, where Ln = Sm, Eu, Tb, Dy.

CONCLUSIONS

A novel bismuth-organic material and its lanthanide doped analogs have been prepared and structurally characterized. The luminescent behavior of the doped phases has been examined and visible and NIR-emission is observed in the lanthanide doped phases. These results suggest that bismuth-organic materials may provide suitable hosts for intrinsically luminescent metal ions.

REFERENCES

1. Heine and Müller-Buschbaum, Chem. Soc. Rev., 2013. **42**(24): p. 9232-9242.
2. Bünzli, Coord. Chem. Rev., 2015. **293–294**: p. 19-47.
3. Jaffe *et al. Energy Critical Elements: Securing Materials for Emerging Technologies*. 2011, American Physical Society Panel on Public Affairs & the Materials Research Society.

ACKNOWLEDGMENTS

The authors would like to thank the Clare Boothe Luce Foundation (KEK) and the DoD SMART program (RLA) for funding. We also thank the National Science Foundation under grants NSF CHE-1429079 and NSF CHE-1337975 for their support

Zintl Phases with Rare-Earth Elements

Svilen Bobev[1*]

[1]Department of Chemistry and Biochemistry, University of Delaware, Newark, Delaware 19716, USA
bobev@udel.edu

INTRODUCTION

In the usual connotation of the term, Zintl phases[1] are the compounds of the alkali or alkaline-earth metals and the early post-transition elements. The electropositive metals are the "cations" and they donate their valence electrons to the electronegative elements from groups 13, 14 and 15, which in turn, form covalent bonds to form the anionic building blocks of the resultant structure. The electron transfer is typically considered to be "complete", and all constituent atoms should attain closed-shell configurations.

Many known solid-state compounds with a general formula $AE_{14}MPn_{11}$ (AE = alkaline-earth metal, Eu, and Yb; M = Group 13 elements Al–In, and Pn = Group 15 elements P–Bi) are Zintl phases. Using the archetype $Ca_{14}AlSb_{11}$ as an example, this relatively complicated formula (and bonding arrangement), achieves perfect charge-balance per the Zintl concept as follows: 14 Ca^{2+} cations, 4 isolated Sb^{3-} ions, one $[AlSb_4]^{9-}$ tetrahedron, and a $[Sb_3]^{7-}$ linear anion. Surprisingly, the trivalent Al in this structure can be replaced by Mn, which is nominally stabilized as Mn(II), i.e., a $3d^5$ ion. In fact, there are a number of manganese-bearing analogues already known, namely $AE_{14}MnPn_{11}$ (AE = Ca, Sr, Ba, Yb, and Eu).[2] While initially the latter compounds were believed to be isotypic and *isoelectronic* to $Ca_{14}AlSb_{11}$, subsequent studies showed them to exhibit metal-like electrical conductivities. These findings are in contrast with the expected semiconducting behavior predicted by the Zintl concept, raising doubts over the applicability of the simple rules for electron counting in the cited examples. Apparently, $AE_{14}MnPn_{11}$ are *not* valence precise compounds, and detailed theoretical work, coupled with X-ray magnetic circular dichroism measurements, X-ray photoelectron spectroscopy, Hall, and optical studies on $Yb_{14}MnSb_{11}$, confirmed the presence of Mn^{2+} and an electron hole on the antimony atoms in $[MnSb_4]^{9-}$ (charge averaged to −2.75 per Sb).

RESULTS AND DISCUSSION

Our prior work on related compounds has shown that charge balancing issues also exist for $AE_{14}MPn_{11}$ (M = Zn and Cd), where the d^{10} metal atoms are clearly divalent.[3] Nature solves the electron deficiency in this instance by at least two different mechanisms—employing mixed-valence of the cations, as demonstrated on the example of $Yb_{14}ZnSb_{11}$, or, as in $Sr_{14}Cd_{1+x}Sb_{11}$ and $Eu_{14}Cd_{1+x}Sb_{11}$, which show tendency to incorporate excess Cd at vacant interstitial sites in order to augment their valence electron concentrations. Neither approach, however, brings perfect charge balance (as in the $AE_{14}MPn_{11}$ phases with M = Al, Ga, In, which completely satisfy the Zintl formalism), and these compounds are heavily degenerate *p*-type materials or even poor metals.

All of the above is an indication that the structure in question is very flexible and can be modulated in a variety of ways, via substitutions of both cations and anions. For example, the seemingly trivial substitution of Al^{3+} at the Mn site in $Yb_{14}Mn_{1-x}Al_xSb_{11}$, or via substitution of trivalent RE^{3+} cations (RE = rare-earth metals) in $Yb_{14-x}RE_xMnSb_{11}$, both cases of electron doping, leads to a marked decrease in the hole carrier concentration. Theoretically, replacement of one AE^{2+} by one RE^{3+} is expected to fill the single hole in the $[MnSb_4]^{9-}$ tetrahedral unit, thereby making these compounds intrinsic semiconductors. However, in all reports on solid solutions $Yb_{14-x}RE_xMnSb_{11}$ so far, the incorporation of trivalent rare-earth metal has been limited to ca. 40-80% of the ideal value.

This study is focused on the syntheses, structures, and physical properties of rare-earth metal substituted series $Ca_{14-x}RE_xMnSb_{11}$ (RE = La–Nd, Sm, Gd–Dy; $x \approx 1$). We demonstrate that the exchange of trivalent RE^{3+} for Ca^{2+} is facile, allowing for compositions with x approaching or even exceeding 1 to be obtained. The effects of the extra electron on the structure and properties of the electron deficient $Ca_{14}MnSb_{11}$ host are discussed. Bulk magnetization measurements reveal the onsets of spontaneous ordering at relatively high temperatures, despite the high dilution of unpaired *d*- and/or *f*-electrons, which may suggest clustering and the formation of microscopic magnetic domains.

CONCLUSIONS

The synthesized polycrystalline materials exhibit semiconducting/insulating behavior, indicating the successful electron doping (Ca^{2+} replaced by RE^{3+}) in *p*-type $Ca_{14}MnSb_{11}$. The activation energies were found to be dependent on both the content and the nature and substituting rare-earth metal. These results indicate that the properties can be continuously varied, hence, these materials could be of relevance to the development of new thermoelectrics.

REFERENCES

1. Zintl E. Angew. Chem. 52:1–6, 1939.
2. *Chemistry, Structure and Bonding in Zintl Phases and Ions;* Kauzlarich, S. M. (Ed.), VCH: New York, 1998, and the reference therein.
3. Makongo J. P. A *et al.*, J. Mater. Chem. C 3:10388–10400, 2015.

ACKNOWLEDGMENTS

Financial support for this research came from the US Department of Energy, Basic Energy Sciences through grant DE-SC0008885 and is gratefully acknowledged.

AUTHOR INDEX

AUTHOR INDEX

AUTHOR INDEX

AUTHOR INDEX

AUTHOR INDEX

AUTHOR INDEX